Global environmental risk

TO BE
DISPOSED
BY
AUTHORITY

Note from the editors

The wellspring for this volume is an international workshop, "Understanding Global Environmental Change: The Contributions of Risk Analysis and Management," convened at Clark University's Center for Technology, Environment, and Development (CENTED) 11–13 October 1989. Under the collective sponsorship of the university's Earth Transformed (ET) Program, the Human Dimensions of Global Change Programme (HDGCP), the Society for Risk Analysis (SRA), and the United Nations University (UNU), the workshop brought together some two dozen prominent scholars from diverse disciplines and a dozen countries.

Workshop participants grappled with teasing out the potential lessons that the risk community, long accustomed as it was to dealing with complex problems beset by large uncertainties, might bring to studies of the human dimensions of global environmental change. Rather remarkably, they managed to prune an overwhelming list of worthy topics to focus on several key initiatives that required immediate attention and to which they agreed to commit their personal efforts. The workshop generated a series of recommendations for future research, which UNU subsequently decided to support. This book is the culmination of that work.

Global environmental risk

Edited by Jeanne X. Kasperson and Roger Kasperson

United Nations University Press

TOKYO · NEW YORK · PARIS

Earthscan Publications Ltd
London · Sterling, VA

First published in 2001 by the United Nations University Press and Earthscan
Publications Ltd

ISBN 92-808-1027-8 (UNUP paperback; not for sale in Europe and the
Commonwealth, excluding Canada)
1 85383 801 2 (Earthscan paperback; not for sale in the USA and Canada)
1 85383 800 4 (Earthscan hardback)

The views expressed in this publication are those of the authors and do not
necessarily reflect the views of the United Nations University.

United Nations University Press
The United Nations University, 53-70, Jingumae 5-chome,
Shibuya-ku, Tokyo, 150-8925, Japan
Tel: +81-3-3499-2811   Fax: +81-3-3406-7345
E-mail: mbox@hq.unu.edu
http://www.unu.edu

United Nations University Office in North America
2 United Nations Plaza, Room DC2-1462-70, New York, NY 10017, USA
Tel: +1-212-963-6387   Fax: +1-212-371-9454
E-mail: unuona@igc.apc.org

United Nations University Press is the publishing division of the United Nations
University.

Earthscan Publications Ltd
120 Pentonville Road, London, N1 9JN, UK
Tel: +44 (0)20 7278 0433   Fax: +44 (0)20 7278 1142
http://www.earthscan.co.uk

Cover design by Joyce C. Weston
Printed in the United Kingdom

Library of Congress Cataloging-in-Publication Data

Global environmental risk / Edited by Jeanne X. Kasperson and Roger
Kasperson.
    p.   cm.
Includes bibliographical references and index.
ISBN 92-808-1027-8
1. Environmental risk assessment. 2. Global environmental change. I.
Kasperson, Jeanne X. II. Kasperson, Roger E.
GE145.G57   2001
333.7′14—dc21   2001000693

For Demetri and Kyra

Contents

        Timothy O'Riordan

13   Sea-level rise and the Sea of Japan   397
        Saburo Ikeda and Masaaki Kataoka

Part Four   GLOBAL ENVIRONMENTAL FUTURES   423

        Editors' Introduction   425

14   Risk and imagining alternative futures   429
        Timothy O'Riordan and Peter Timmerman

15   Exploring a sustainable future for Canada   451
        John B. Robinson

16   Social visions of future sustainable societies   467
        Patricia Benjamin, Jeanne X. Kasperson, Roger E.
        Kasperson, Jacque L. Emel, and Dianne E. Rocheleau

        References   506

        Contributors   563

        Index   566

Figures and tables

Figures

Tables

Acknowledgements

The editors owe a great deal to participants at an international workshop, "Understanding Global Environmental Change: The Contributions of Risk Analysis and Risk Management," convened at Clark University in 1989. This sterling assemblage of colleagues – Harold Brookfield, William Clark, Rob Coppock, Exequiel Ezcurra, Silvio O. Funtowicz, Gordon T. Goodman, Saburo Ikeda, N. S. Jodha, Robert W. Kates, Ashok Khosla, Diana Liverman, Giandomenico Majone, Walther Manshard, James K. Mitchell, Elmer Offenbacher, Timothy O'Riordan, Cheri Ragaz, Steve Rayner, Kenneth Richards, Galina Sdasyuk, Kirk R. Smith, Michael Tiller, Peter Timmerman, and B. L. Turner, II – first stimulated the undertaking of this volume. Many of them ended up as authors or coauthors of individual chapters – but even those who did not influenced the final product immensely.

At a crucial juncture, Roland Fuchs, then Vice-Rector of the United Nations University (UNU), encouraged the preparation of *Global Environmental Risk* and facilitated that institution's provision of needed support. Juha Uitto, Senior Programme Officer at UNU, carried through on this original commitment. More recently, Dr Manfred Boemeke, Director of UNU Press, and Managing Editor Janet Boileau have guided the volume through the final stages of production. We also salute Heather Russell, whose superb copy-editing is evident in every chapter.

At our home institution, many people have contributed generously with their time and talents. Anne Gibson, director of Clark University's cartographic laboratory created the handsome maps and other graphics displayed throughout this volume. Miriam Berberian, Betty Jean Perkins, and Octavia Taylor, have put up with our absences "to work on the book" and dispensed generous doses of patience, support, and humour at every stage. And our indispensable accomplice, Lu Ann Pacenka, has come through yet again and transformed another of our tangled webs into a real book.

In this book we have sought to wed two important fields – risk analysis and the human dimensions of global environmental change. The relatively mature field of risk analysis, owing to considerable experience in assessing and managing complex problems with large uncertainties, is well positioned to contribute to the analysis and solution of global environmental problems. We encourage integrated, holistic approaches to addressing the myriad threats to the global environment. In so doing, we invite others to join us in taking up the work of two visionary pioneers, who have boldly charted potential routes for confronting global environmental risks – Gordon T. Goodman and Robert W. Kates.

Jeanne X. Kasperson
Roger E. Kasperson
Worcester, Massachusetts, USA
Stockholm, Sweden

1

Introduction: Global environmental risk and society

Roger E. Kasperson, Jeanne X. Kasperson, and Kirstin Dow, with contributions from Exequiel Ezcurra, Diana M. Liverman, James K. Mitchell, Samuel J. Ratick, Timothy O'Riordan, and Peter Timmerman

Global environmental risk is about threat; it is also about opportunity. On the one hand, stratospheric ozone depletion, potential global warming, and the continuing loss of bio- and cultural diversity signal the human capacity to transform – perhaps irretrievably – the planet on which humans depend. On the other hand, the growing worldwide recognition of this destructive power, and the initiatives undertaken since the Earth Summit in 1992 and the Kyoto Convention of 1997, carry the promise that the use and occupation of the earth may shift at the millennium to more sustainable trajectories. Global environmental risk also involves the recognition that the biosphere is both fragile and resilient: fragile, in that what supports life of present and future peoples and what these people value in their interactions with nature is often easily transformed – or even lost; resilient, in that the planet has an enormous capability to absorb abuse and to recover and that human societies are often enormously adaptive. Addressing global environmental risk, then, is about finding new pathways for averting ongoing destructive human activities and for creating a life-sustaining planet.

This book examines ways of identifying, conceptualizing, organizing, and addressing global environmental risk. Environmental risk analysis brings powerful concepts, methodologies, and experience to these tasks. At the same time, global environmental risks pose new challenges and fundamental problems. Clearly, new theories and analytic structures, as well as new syntheses across diverse ways of conceptualizing and re-

1

sponding to these threats, are in order. How society responds may well reveal as much about human capabilities, fragilities, and values as about ecosystems and biogeochemical processes. Human response to global risk means understanding the processes of change, recognizing and learning from the lessons of history, criticizing and restructuring the economic and political order, creating new institutions, and fashioning a wider and more enduring sense of environmental citizenship and reciprocity with nature.

We begin this book by taking stock of the distinctive challenges posed by global environmental risks, the ability of knowledge systems to identify and characterize such threats, and the capability of societies to address the management challenges. In particular, we specify distinctive properties of these risks and the challenges posed for societal response systems. To highlight connections among often-divided elements of societal response (as well as human driving forces and environmental changes), to identify realms of interdisciplinary interest, and to situate a number of key issues that confront timely and effective political responses to these risks, we undertake an overview of the issues facing societal response to global environmental change.

The nature of global environmental risks

To explore the issues posed by global environmental risks, clarity is needed as to the meaning of "global environmental change" and "global environmental risk." Much of the current debate about risks to the planet and to the global community focuses on greenhouse warming as the archetypical threat to the global environmental system.[1] Previously, we have defined the type of global environmental change represented by climate change as *systemic*, in that environmental change at any locale can affect characteristics of the environment anywhere else, or even the global system itself (Turner et al. 1990b). Accordingly, initial attention in the International Geosphere–Biosphere Programme (IGBP), the 1992 Earth Summit, and the Kyoto Convention of 1997 focused on atmospheric, marine, and biological systems in which human-induced perturbations have the potential to alter the conditions of the earth system. Climate change, stratospheric ozone depletion, and biodiversity in particular have commanded much of the attention of the world's scientific and policy communities through the 1990s; indeed, many members of the public and policy makers equate such risks with global environmental change.

But a second type of risk, which we have termed *cumulative environmental change*, may well eclipse systemic changes in both long-term and (certainly) short-term consequences. This type of global environmental

Table 1.1 **Types of global environmental change**

Type	Characteristic	Example
Systemic	Direct impact on globally functioning system	(a) Industrial and land-use emissions of greenhouse gases (b) Industrial and consumer emissions of ozone-depleting gases (c) Land-cover changes in albedo
Cumulative	Impact through worldwide distribution of change	(a) Groundwater pollution and depletion (b) Species depletion/genetic alteration (biodiversity)
	Impact through magnitude of change (share of global resource)	(a) Deforestation (b) Industrial toxic pollutants (c) Soil depletion on prime agricultural lands

Source: Turner et al. (1990b: 15).

change refers to the accumulation of regional and localized changes that are distributed widely throughout the world. Such changes not only involve the degradation of ecosystems such as coral reefs, groundwater resources, or rain forests but also entail the accumulating contamination of air, water, and land as well as mounting deforestation and soil loss under the pressures of increasing population, economic growth, and the global embrace of values of high consumption. Such cumulative changes represent many of the severe environmental problems currently confronting human (and particularly developing) societies. Table 1.1 summarizes these two types of change.

Both types of global environmental change pose distinctive challenges to the creation of an adequate knowledge base concerning both the environmental perturbations and the vulnerability of human and ecological systems, to the establishment of causal linkages in the structure of the risk, to the shaping of appropriate societal responses, and to the enhancement of the capability of the current global state system to respond effectively. The risks associated with systemic global environmental change confront assessors with difficulties that, although present in other types of risks (e.g. acid deposition or low-level toxins in aquatic systems), combine complexities in new configurations or pose new levels of analytic difficulty. For its part, cumulative global environmental change often lies beyond the pale of existing databases while its physical expression occurs in particular ecosystems and landscapes and its human consequences remain embedded in local cultures, religion, and indigenous knowledge systems.

The sources and effects of both systemic and cumulative global environmental risks are widely diffuse, often unmeasured, sometimes unidentified, and cumulative over long time periods. *Risk*, in our usage in this volume, is used synonymously with hazard to refer to "threats to human beings and what they value" (Kates, Hohenemser, and Kasperson 1985: 21). *Global environmental risk*, drawing on the discussion above, refers to threats (to human beings and what they value) resulting from human-induced environmental change, either systemic or cumulative, on the global scale.

Almost all human activities modify the environment in some way and, over sufficient time periods, may significantly degrade ecosystems. The enormous variety of land-use changes, for example, causes global environmental change in highly diverse ways – by disturbing carbon storage, nutrient cycles, the hydrological cycle, atmospheric concentrations, and the reflection of solar energy from the earth's surface (Malone 1986). Data concerning human activities are very meagre, the effects are often apparent only over the long term, and the various interactions in environmental processes and cycles are often poorly understood and unmeasured. Characterizing the risks, therefore, is intrinsically difficult because many of the risks are *elusive* – incrementally and almost imperceptibly they alter global life-support systems and the basic conditions of human welfare – or even *hidden* – they are shielded from societal awareness and response by ideology, competing societal priorities, political marginality, and cultural bias (Kasperson and Kasperson 1991).

Throughout this book five themes predominate:

1. *Global environmental risk is the ultimate threat.* This is true both because what is at stake is the life-support capabilities of the planet and also because such risk is only partially knowable, controllable, and manageable. Humans, it bears emphasizing, are not invulnerable and their actions carry long-term potentials for developing or destroying civilization as we know it. When the seemingly invulnerable realize (as now seems to be occurring) that they or their descendants cannot exploit the environment with impunity, then real crisis occurs, as George Perkins Marsh (1864) made vividly clear in the nineteenth century. We may be on the verge of that recognition on a planetary scale.

2. *Uncertainty is a persistent feature both of understanding process and causation as well as of predicting outcomes.* Anticipating and responding to global environmental risk means coping with the uncertain, the unknowable, and the inherently indeterminable. Therefore, human responses require a scientific language and social behaviour that embrace what we do not know and that incorporate a search through possibilities that prizes not only anticipation but adaptation and social learning as well.

3. *Global environmental risk manifests itself in different ways at different spatial scales.* Risk, responses, variability, vulnerability, and negotiation do not occur in the abstract or in a void. They interact in specific space, place, time, landscape, and culture – in other words, in environmental and sociopolitical context. Also, different elements of causation, driving forces, and impacts become apparent at household, village, regional, national, and international scales. So this book also addresses the regional variety and expression of environmental risks and response strategies through the richness of case studies drawn from various environments, economies, and societies. Only through context can we fully uncover the subtlety and complexity of global environmental risk and sensitively address the possible interventions.

4. *Vulnerability is a function of variability and distribution in physical and socio-economic systems, the limited human ability to cope with additional and sometimes accumulating hazard, and the social and economic constraints that limit these abilities.* Vulnerability can often be overcome by science, technology, political control, and the application of indigenous knowledge. But it is equally possible that the misapplication of these tools may increase vulnerability or create new and unanticipated environmental risks. Global environmental risk may also stimulate responses that seek to insulate the few temporarily at the expense of the many permanently. Although the Intergovernmental Panel on Climate Change has recently emphasized the importance of vulnerability (IPCC 1997, 1998, 2001), the concept does not yet occupy centre stage in discussions of global environmental risk. In this volume, vulnerability occupies a prominent place in the essential conception of risk, how risk occurs, the forms it takes, and the response strategies that may be possible or that provide greatest leverage for intervention.

5. *Futures are not given, they must be negotiated.* Enormous scope exists for channelling the energies of imagination, expectation, fear, and hope into participatory forms of creating global futures that are more equitable, less vulnerable, and less intolerably uncertain. This will mean not only emancipating the creative force of choice but also empowering people for their own safer and more fulfilling worlds. Ulrich Beck (1992) may well be prescient in his prediction that the hallmark of the future will be the "risk society," in which social processes centre on the allocation and avoidance of risk rather than on the distribution of wealth.

Broad and emancipatory risk assessment

The purpose of this book is to develop these five themes through a variety of perspectives. We begin by examining the role of global risk assessment as an approach for defining and characterizing environmental threats in order to test the power – and limits – of its philosophies and means of analysis. We shall see that, through its evolutionary search for better understanding, risk assessment is already well along in effecting a shift from a bias favouring expertise and top-down assessment towards an approach that accommodates more subtle and complex combinations of ecological and human harm, justice, social and civil rights, political accountability, and the building of social capital and trust (UK Interdepartmental Liaison Group on Risk Assessment 1998). In addition, risk assessment is actively incorporating ecological concepts and social and economic processes to supplement its long-standing attention to environmental events and technological hazards. Uncertainty, probability, trajectory analysis, equity assessment, and negotiated bargains are enriching and extending risk assessment. Inevitably, the theoretical bases and conceptual models needed to support this broader analysis have become more diverse and synthetic.

This layering of broader interpretative approaches, however, is not sufficient by itself to cope with the more problematic issues of uncertainty, vulnerability, value conflict, and political control. Hence, we also seek a more emancipatory form of risk analysis – an arrangement of societal transformation that creates political opportunities that allow people to identify preferred choices and to take alternative pathways to sustainability. Also, we argue that the science of risk analysis needs to be extended to be interdisciplinary at core, to embrace more diverse knowledge, and to identify multiple theoretical perspectives in order to embrace people's hopes and fears in terms that they can comprehend and to which they can contribute and respond. These are the forefronts of the new work now emerging in risk theory and praxis.

In essence, this new kind of global environmental risk analysis we seek is democratic in principle, humanistic in concept, and contextualized in assessment. Of course, this will not be easy to attain. It requires approaches – as recognized early in *Our Common Future* (WCED 1987) and by the Science Advisory Board of the US Environmental Protection Agency in its *Reducing Risk* (USEPA 1990) and *Future Risk* (USEPA 1988) and the US National Research Council in its *Understanding Risk* (Stern and Fineberg 1996) – that see risk as fully entwined with development, social processes, operations of the economy, and national policy. We therefore maintain that we need to locate risk squarely in its broader context of societal change, as Smith argues in chapter 3. We also need to provide better taxonomies of risk set sensitively in the context of different cultural traditions. This forms a second major task that is addressed by Norberg-Bohm and colleagues (chapter 2), whose important contribution suggests that the very character of global environmental risk is the product of both the physical character of locality and the pattern of physical, biological, and human processes that arise in any given locality, together with the particular history of human response to those variables. Clark University researchers pick up this theme in their analysis of environmental criticality (chapter 8), and the theme is explored concretely by N.S. Jodha in his analysis of environmental risk in the middle mountains of Nepal (chapter 9) and Diana Liverman in her assessment of climate change in Mexico (chapter 10).

Risk

Throughout this book, imagining sustainable futures is not a matter of wishful thinking but an integral part of risk analysis. Sustainable futures involve centrally issues of emancipation and empowerment. Human futures are not inevitable, as the global models of the 1970s seem to suggest (chapter 16), though it is in the interests of certain constituencies to make them seem obvious, already decided, or inevitable outcomes of "the way things are." Global environmental risk analysis "puts the future on the table" for negotiation, since it calls into fundamental question the trajectories of our current orthodoxies and notions of "business as usual."

The idea that the future is negotiable, and that affected parties are now differentially involved (or not involved) in the negotiations, brings forward considerations of power, equity, and social justice – and equitable outcomes and equitable processes for getting to those outcomes (see chapters 2, 7, and 15). The obstacles to broad public participation in creating global futures are many, ranging from lack of access to information and expertise all the way to brute exclusionary force. Equity and the future are linked not just through reference to "responsibilities for future generations" but by questions of who controls access to the future and who chooses the trajectories of change. Those who live in the present live, after all, in the layered remnants of a series of failed former utopias – concretized versions of earlier visions of how things might be.

Imagining futures, as developed in this book, is a process that relates in part to science (so that we can understand and anticipate change) but that also involves the search for human values, economic systems, and social structures by which high-risk pathways can switch to more sustainable tracks (see the compelling discussion in chapter 16). Accordingly, O'Riordan and Timmerman argue (chapter 14), imagining futures cannot be separated from fundamentals of social justice and civil rights, effective power, and collective self-reliance. It is bound up in the very essence of what will be necessary to realize sustainable development, as explored by Robinson (chapter 15) for the case of Canada. The quest for sustainability, it is clear, may reflect very different social visions of the future and colour how societies may best engage alternative risk situations (chapter 16).

This is an area of research and endeavour that is only beginning to command recognition. Past efforts to think about the future suggest some useful directions but also limits. This book argues for a series of approaches to risk anticipation in a number of carefully selected communities. These areas are high risk and highly vulnerable, for the reasons already stated, and have cultural settings and political structures that provide the richness of real variety. We make a start by exploring the conceptual and methodological issues in analysing trajectories of threat across regions (chapter 8). Then Jodha examines case studies in the mountain regions of northern India and Nepal (chapter 9), Liverman focuses on the arid areas of Mexico (chapter 10), Broadus analyses potential sea-level rise for the coastal regions of the southern coasts of Bangladesh and the Nile Delta (chapter 11), O'Riordan evaluates sea-level rise in the North Sea (chapter 12), and Ikeda does the same for the Sea of Japan (chapter 13).

We make no unrealistic claims for answers in this book. Confronting global environmental risk is not a matter for a guidebook or a blueprint: it must avoid undue or premature prescription or programmatic approaches. The interconnections between physical process and human change, complicated by history, injustice, and ignorance, render any simple formulation of the problems or prescriptions for solutions unrealistic. At this stage, we seek to conceptualize the issues more clearly, to apply the best traditions of risk analysis and policy formulation in characterizing global risk challenges, to extend the risk tradition through new theoretical perspectives and a heightened appreciation of social and political context, and to illuminate a richer analysis through examples that provide an encapsulation of diverse experience in nature–society encounters. In short, we seek to take up and contribute to the challenge for the global sustainability transition: to *anticipate* the unfolding environmental crises, to *conceive* an alternative paradigm for development, and to *actualize* the political will for equitable change (Raskin et al. 1996: 1).

Global environmental change is, therefore, among the murkiest of all environmental risks because, unlike threats where it is possible to establish clear causal linkages and effects, such change involves risk bundles that are composed of interactions among streams of fundamental human processes – public policies, economic activities, technological applications, and varying lifestyles. Thus, just as the human threat at hazardous waste sites is the product of synergism among hundreds of unidentified chemicals in a common risk brew, global environmental risks are *interactive phenomena* – interactive among human processes, between human and environmental processes, between environmental perturbations and vulnerability, and between environmental changes and adverse consequences. Similarly, though on larger scales, causes and effects are often separated over time and space. The self-assessments of the US Environmental Protection Agency (USEPA 1987, 1990) of its "unfinished business" after 20 years of the agency's existence explicitly recognizes that risk generation is embedded in fundamental economic processes and state policies. Significant further progress in reducing environmental risks (the agency recognizes) requires intervention into the basic structure of economy; state sponsorship of agriculture and industry; and the process of technology development and deployment, international trade and finance, with all the associated political ratifications. Jiang et al. (1995) and Jiang (1999) note that this is as true for environmental degradation in China as for the US economy.

Even where basic paths of causality may be evident, the relationship among human causes, environmental perturbations, and effects may be non-linear or even chaotic. Certainly, scientific research has advanced our understanding of the environmental processes and human driving forces of change; however, it has also brought society within sight of the limits of its understanding of the environment and of the complex character of human-induced change. The traits of such change are often unknown. The increase in CO_2 concentrations recorded at the Mauna Loa Station displays a rather consistent increase in mean levels and regular seasonal fluctuations. Meanwhile, nearly all studies of potential climate change impacts have been based on linear impact models. It is clear, however, that the reaction of ecosystems, climatic processes, species, and human systems may exhibit thresholds and sharp punctuation in changes, non-linear responses, or even chaotic behaviour. The rate, magnitude, and timing of such change; the interconnections among biophysical and social processes; and the spatial distribution of effects, involve broad uncertainties, ignorance and indeterminacies, and (often) vigorous scientific disagreement. The development of anticipatory assessment strategies – such as the Famine Early Warning System (FEWS) supported by the United Nations Agency for International Development (USAID) – and more resilient societies and institutions are both needed to address these issues. This remains as one of the pressing global tasks in the search for more sustainable futures, as clearly indicated in experience in implementing Agenda 21 in the years following the Earth Summit (United Nations 1992). Current environmental monitoring systems were often not created with a societal warning purpose in mind; rather, they were designed to provide information on the changing "noise" level in the environment. Effective early-warning systems are needed to address three related issues: these are (1) the existence of irreversible changes; (2) the chance of overshoots, flip-flops, and other discontinuities in ecological, physical, and socio-economic systems; and (3) the effectiveness of proposed management approaches in avoiding (or at least minimizing) the possibility of irreversible trends or discontinuities (Izrael and Munn 1986).

Scale is an important consideration in alerting systems for global environmental risks. For institutions at all scales to act on such risks, it is often necessary to have unambiguous evidence, an imminent threat, and a constituency of concern. States tend to pursue development goals, to delay in the face of uncertainty, and to react to aggregated problems or crisis events. They characteristically err toward tardy and costly interventions, as chapter 8 makes abundantly clear. If early-warning systems for global risks are to be effective, two types of alerting systems are likely to be needed. One type – a "trigger-related system" – would allow the state and international institutions (and, by extension, environmental managers at other scales) to react promptly to potential large-scale natural and technological threats. The second type – the vulnerability-related alerting system – will focus upon changes in the situations of potential victims, often at the subnational and local scale. These alerting systems would recognize that the state is often part of the risk problem rather than the solution and would disaggregate effects to focus at varying scales on differential risk and impact. Knowledge supporting early warning and intervention will need to be localized and to be sensitive to specific livelihood and cultural conditions (Cohen et al. 1998), whereas "place-based" strategies that integrate resource management efforts across entire ecosystems or watersheds will be required (Enterprise for the Environment 1998).

Although a key term in any risk calculus, vulnerability is typically a shadowy element in risk analysis generally and in assessing global environmental risks in particular. Past risk studies are quite definitive in demonstrating that the consequences of environmental change, and the extent of associated human harm and disruption, occur unevenly across regions, countries at different levels of development, social groups, and generations (Baird et al. 1975; Blaikie et al. 1994; Burton, Kates, and White 1978, 1993; Claussen and McNeilly 1998; Kasperson 1983; Liverman 1990c; Low and Gleeson 1998). In her review of the literature, Dow (1992) notes that vulnerability arises from diverse sources, including social relations, technology, biophysical conditions, economic relations, and demography. This vulnerability assumes different forms and magnitudes at national, regional, local, and household levels. Yet vulnerability is often neglected or poorly incorporated into assessment studies, despite the IPCC's substantial effort to address vulnerability in its second and third assessments (IPCC 1996b, 1998, 2001). Sensitive biophysical systems and vulnerable social groups often fail to coincide, thereby complicating even well-intended analyses. The social sciences have generally failed to develop a robust theory of vulnerability, much less to relate it to environmental change. This is well illustrated by the use of very general pressure–release or pressure–response schema, which shed little light on the structure of causal relations in risk reduction or sustainability discussions (e.g. Blaikie et al. 1994). This limited state of knowledge and theory greatly complicates assessments of global environmental change and its distributional effects; as a result, equity, justice, and security issues (and who will adjudicate these matters) remain opaque, ill-defined, and contentious (Parikh 1992; Parikh and Painuly 1994).

A further challenge to creating a robust knowledge base for global environmental risks is, as Clark (1990: 6) notes, their simultaneous local and global character. Indeed, the human and environmental processes of change and their impacts are both scale and place specific. Recognition of this fact has been a great strength of Russian studies of environmental change that have related human-induced perturbations to the evolutionary dynamics of particular landscapes (Mather and Sdasyuk 1991). Not only do analytic "cuts" at different scales and at different places reveal varying "faces" of environmental transformation and societal impacts, they also focus attention on the interactive processes that link phenomena at different scale levels. Such cross-scale assessments, as Moser (1998: 6–7) points out, can provide key insights into the structure of risk causation; build bridges for management and policy; strengthen assessment capacities of different scales; create alternative framings of the environmental problem; and produce knowledge that is more integrative, legitimate, and usable. Figure 1.1 suggests how complicated these relationships are in global environmental change. The long traditions in geography of studies of the "hazardousness" of place (Hewitt and Burton 1971) and cultural ecology of small-scale areas (Blaikie and Brookfield 1987; Turner et al. 1990b), as well as small-scale studies of common-property governance (Jodha 1992a; Lane 1998; Ostrom 1990; Ostrom et al. 1999), speak powerfully of the multitude of ways by which humans transform the earth and by which cultural values and institutions shape risk experience.

It is also apparent that the regional (or meso) scale offers a powerful entry point for connecting global and local scales (Easterling 1997; Schmandt and Clarkson 1992). Since environmental degradation is often "exported" from one region to another, regional and local studies of global environmental change need to operate in an "open systems" context (Kasperson, Kasperson, and Turner 1995, 1996). The use of concepts such as "carrying capacity," which characteristically treat the region as a closed system, fundamentally limits useful analyses and often produces

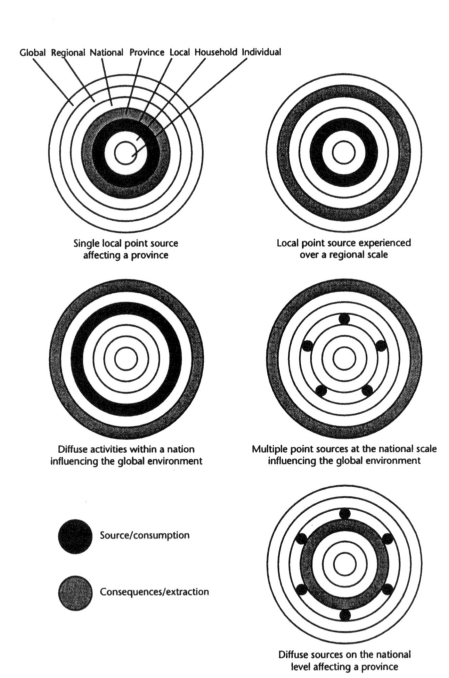

Fig. 1.1 **Space and scale relationships in global environmental change**

misleading inferences (Sagoff 1995). Effective response systems and institutions will require political coordination of institutions and interventions across multiple scales and actors in a stateless context, or the creation of new international regimes (Young 1989, 1994, 1997), whereas credible analyses of the causes and effects of environmental change will need to be multilayered and spatially structured (Wood 1990).

Although global environmental risks make their way on to the societal agendas one by one, according to eclectic and imperfect societal processes that filter what is most pressing or most tractable, each risk problem is in fact a microcosm of broader process of environmental and social transformation. Thus the *risque du jour* may be viewed as a "moment" in a constantly changing societal risk portfolio or relief diagram or, as Palmlund (1992) would argue, a fragment of drama in the risk theatre. The isolation of an environmental risk from the fabric of competing risks and burdens that make up the riskiness of human experience at particular places and in particular societies carries its dangers.

Smith in chapter 3, for example, argues that developing countries pass through a "risk transition" in which the historically high "traditional" risks associated with rural poverty (e.g. infectious diseases) decrease with economic development (although at different rates in different regions, places, or groups), whereas industrialization, agricultural changes, and urbanization increase the number and severity of "modern" risks. As society's risk portfolio ebbs and flows in internal composition (and may even decline in aggregate), social groups and regions shift in the extent to which they are more or less vulnerable and threatened. Furthermore, particular periods occur when traditional risks are still significant but new and unfamiliar risks are also unfolding. These new risks may not be additive but may interact in perplexing and novel ways with existing risks to produce new risk syndromes, thereby confounding established means of coping; introducing new winners and victims; and greatly complicating the alerting, assessment, and management tasks.

Risk scholars and analysts have yet to turn their attention to the multiple types of transitions and changing societal risk portfolios that characterize the former socialist societies of Eastern Europe in their transition to democratic polities and capitalistic economies. Interlocking transitions are fundamentally restructuring the nature of production and consumption systems and, by extension, the array of natural and technological risks facing society. Old vulnerabilities are disappearing as new ones are being created. Values for the future are being negotiated, often implicitly. How nature–society relations are being altered, risk distributions and vulnerabilities reshaped, coping systems altered, and overall resilience to environmental change affected, are only beginning to become apparent (Victor, Raustiala, and Skolnikoff 1998).

Problems of such complexity, it is apparent, are only partly knowable. Much current discussion of climate change carries the assumption that, by working harder – by improving our scientific understanding, by creating more complex models, by constructing faster computers, and by searching more assiduously the relevant historical experience – scientists will unlock the needed remaining truths (USNRC 1983b). Well-conceived and properly mounted assessment programmes, it is believed, will progressively whittle back uncertainties so that well-founded public policy choices will become evident. This optimism reflects a dangerous hubris about human capabilities to comprehend the universe, to predict the future, and to create needed knowledge through the applications of science. Such optimism is certainly not based on risk-analysis experience, where assessments of complex problems and technologies (and particularly those in which scientific questions and sociopolitical and moral issues mingle) have revealed that as assessors narrow some uncertainties they typically uncover other new ones (Kasperson and Kasperson 1987). Global environmental risks, and particularly those of the systemic type, appear almost certain to surpass in complexity even the most perplexing of what Edwards and Von Winterfeldt (1986) term "technological mysteries" and what Weinberg (1972a) labels "trans-scientific problems."

Understanding past and ongoing global environmental risks, much less anticipating future ones, clearly involves diverse ways of knowing, as chapter 16 discusses at length. Combined properly with other knowledge systems, risk analysis has a potentially valuable role. The concept of risk – threats to people and to what they value – is inherently interdisciplinary and unusually well situated to joining enquiry into the natural and life sciences with insights emerging from the social sciences and humanities (Committee on Shipborne Wastes 1995; Kates, Hohenemser, and Kasperson 1985). It also can be (indeed, should be) explicitly democratic in process, so that scientific knowledge, professional judgement, "folk" knowledge, and cultural perspectives all enter into a broad integrated risk analysis. At the same time, as Blaikie and Coppard (1998) have recently noted for Nepal, the current celebration of the diversity and quality of local environmental knowledge can quickly produce diminishing returns and become misleading, so that integrative knowledge bases that treat both scientific and indigenous knowledge are needed (Agarwal 1995). The focus on threats and possible consequences provides extraordinary potential power for informing policy and societal response. Since the birth of risk analysis lay in the systematic appraisal of highly complex problems involving large uncertainties associated with human interactions with nature and technology – with such diverse risks as hurricanes, famines, aircraft accidents, and nuclear power – risk research has established a knowledge base and methodological capability well suited to engaging the challenging issues of global environmental change. It is not surprising, therefore, that the World Commission on Environment and Development (WCED 1987) singled out the establishment of a global risk-assessment programme as a particularly urgent priority in global change research. Specifically, such an assessment programme could, as the Commission concluded, powerfully address the following:

- identifying critical threats to the survival, security, or well-being of all or a majority of people, globally or regionally;
- assessing the causes and likely human, economic, and ecological consequences of those threats, and reporting regularly and publicly on their findings;
- providing authoritative advice and proposals on what should or must be done to avoid, reduce, or (if possible) adapt to those threats; and
- providing an additional source of advice and support to governments and intergovernmental organizations for the implementation of programmes and policies designed to address such threats (WCED 1987: 325).

The potential, alas, is well short of being fulfilled. A comprehensive global risk-assessment programme is yet to emerge to guide global environmental initiatives. Beyond the climate-change arena, and perhaps the use of integrated assessment models (see below), the potential for global risk assessment has yet to be extensively tapped. At the same time, risk-analysis methodologies, especially in the emancipatory form we envision, continue to hold high potential for advancing our understanding of complex, global-scale risks.

In chapter 4, Funtowicz and Ravetz argue that a new sort of risk assessment is emerging to deal with the most difficult of assessment problems – problems that combine "high systems uncertainty" (as described above) with "high decision stakes." When both systems uncertainty and decision stakes are low, solving puzzles in the Kuhnian sense generally works (Kuhn 1962). When either factor reaches a medium level, then the "professional consultant's" skill and judgement become necessary. But in cases of the outermost region of the Funtowicz/Ravetz model (see chapter 4), such as those characteristic of global environmental change, traditional science becomes substantially inadequate and a new type of science, which they term "second-order" or "post-normal" science, becomes necessary. This new science, in their view, will involve not merely new types of expertise but a new conception of the objects, methods, and social functions of the material world and the interactions of this material world with structures of power and authority. Accordingly, the three types of risk problems will each require a different type of knowledge as well as a different type of science.

If risk analysis and other forms of environmental assessment are to enrich the knowledge base on which to base actions aimed at securing the well-being of the planet, then it cannot be isolated from moral and ethical questions. Although science and risk assessment have become highly secularized, they are intended to serve moral purposes. Earlier calls by some (e.g. USNRC 1983b) to distinguish sharply between risk assessment (the scientific job) and risk management (the value-setting job) illustrate this division but have turned out to be ill conceived. All the truly difficult risk problems, including those posed by global environmental change, are laden with basic value issues: whether humans will survive and (if so) with what kind of existence; whether the future can be ensured; whether harmony or disharmony with nature will prevail, and what type of future it will be; and whether existing humans will be responsible for distant peoples, generations, and species. The planetary crisis is thus both a scientific and a spiritual crisis. Progress in risk analysis, it is clear, must go hand in hand with the creation of a new global ethic, and means must be found to integrate them in the pursuit of alternative futures – a fundamental insight set forth by the World Commission on Environment and Development (WCED 1987) and elaborated since then (e.g. Barney 1999; Hammond 1998).

The history of risk management in the United States has, as Kates (1986) has suggested, involved three generations of ethics. The first generation proceeded on the principle of non-maleficence – do no readily avoidable harm to persons or nature. These ethics underlie the definition of corporate responsibility and much of the regulatory system enacted over the decade following the first Earth Day in 1970. The second generation, the child of cost–benefit analysis, involved the weighing (or balancing) of aggregate risks and benefits, the calculus of how to "optimize" risk reduction for society. The US Environmental Protection Agency's clarion call (USEPA 1990) for bringing its own programmes and resources into closer accord with expert assessment and opportunities in order to reduce more efficiently overall societal risk is a clear indication that it is operating with second-generation risk ethics. The third generation of risk ethics, by contrast, focuses on equity, fairness, and social justice. These ethics are the ethics of the present century and the ethics of global environmental change. They are the ethics that are at the heart of the current environmental justice movement in the United States and underlie advocacy of the precautionary principle in Europe and elsewhere. They are concerned not only with balancing overall benefit and harm but also with the export of risk to developing countries and to unborn people, with allocation of risks between workers and publics, with the commitment of toxic wastes to the biosphere, with protection of the most vulnerable, and with fairness of process as well as outcomes, as we note below. Risk-assessment studies (e.g. Cothern 1996; Kasperson 1983; MacLean 1986) and global change studies (Agarwal and Narain 1991, 1999; Claussen and McNeilly 1988; Kasperson and Dow 1991; Weiss 1989, 1990, 1995) are only now beginning to address in depth third-generation ethics. Meanwhile, second-generation ethics and second-order science continue to hold sway in most scientific and political assessments of global change.

A final social issue concerning the knowledge base needed for addressing global environmental risks is the question of who is creating the knowledge and for what ends. How the world community learns about global environmental change may be as important as the level of knowledge achieved. Little is known about the distribution of scientific knowledge within and among the different states in the global community, international institutions, and non-governmental organizations (NGOs), or among the political élite, scientists, policy makers, and publics. It is clear, however, that Western scientists are creating much of the formal knowledge about global environmental change, as in the Intergovernmental Panel on Climate Change (IPCC) programme, the International Geosphere–Biosphere Programme (IGBP), and the International Human Dimensions Programme (IHDP), for example. Oran Young's study (Young 1989) of how scientists and their knowledge affect decision-making in international resource regimes suggests the key role that the creation of science plays in at least four different policy-making contexts – agenda setting, regime formation, social choice, and compliance. The degree of influence that scientific knowledge has on state behaviour and international policy-making institutions depends on such factors as the degree of scientific consensus or disagreement, the extent of social conflict over these proposals, and the goals and ideology of the state.

Peter Haas's study (Haas 1990b) on the politics of environmental co-operation in the Mediterranean Sea and an analysis by Ronnie Hjorth (1994) of international cooperation on the Baltic underline the significant impact that expert-based consensus can have on achieving political agreement on regional action plans. The need of decision makers to rely on "epistemic communities" (experts and knowledge-based groups) to reduce uncertainty while protecting their own autonomy often extensively shapes governmental learning and the development of new national objectives (Haas 1990a). Individual nations need effective tools to cope with their own perceptions of the problem and yet enter into regional or global responses.

Thus far, scant attention has been paid to the political implications (and economic feasibility) of various methods of assessment by which developing countries can tackle issues such as climate change. How well

the design, financing, and implementation programmes of the environmental-monitoring programmes of states and international institutions
such as the United Nations Environment Programme (UNEP), the United
Nations Development Programme (UNDP), and the Food and Agriculture Organization of the United Nations (FAO) fare in providing international access to the information bases created will be key in determining the degree of self-reliance and autonomy achievable by various
countries. The Haas (1990a, b) and Hjorth (1994) studies offer insight on
negotiating the problems of cooperation between lesser and more developed countries. In the case of the Mediterranean, owing to wise planning
by the UNEP, an equal distribution of laboratories ensured that developing countries had access to resources to support indigenous research
and to deal more effectively with regional marine pollution. Such a process (geared, as it was, to equity considerations) helped to ensure that
the pressure from economically dominant countries could be minimized
through the intervention of the UNEP's planning secretariat. This stands
in sharp contrast to the experience in climate change, where the great
majority of analysts are from industrialized countries and scientific work
has focused on issues of primary concern to the developed world, where
little political attention or funding has existed in the south, and where a
widening gap between the capacities of scientists from the north and
those from the south is increasingly apparent (Kandlikar and Sagar 1999;
Sagar and Kandlikar 1997). Both cases suggest the importance of distributional issues in social access to knowledge and scientific and political
capabilities to act on national and global environmental change issues.

These risk knowledge and capability issues suggest that both the causes
and potential political responses to environmental changes are deeply
embedded in the human driving forces that create these risks in the first
place. The global economy and the international finance system often
seek ends in basic conflict with environmental values. Global patterns of
inequality in wealth between developed and developing societies are
growing rather than narrowing. State systems associated with traditional
notions of military security and armaments systems still structure much of
the global consumption of resources, but the overall system in a post-cold
war age is highly fractured politically (Sagasti and Colby 1995). National
and international institutions are only now beginning to address how development and market economies may internalize environmental values.
Meanwhile, corrupt and inept governments in many parts of the third
world siphon off the limited social investments that do occur. It is these
global driving forces, pervasive inequalities, ingrained patterns of political corruption, and power structures that must change if global environmental risks are to be confronted effectively and more sustainable global
futures negotiated.

A general model of societal response

With a keen appreciation of the limits of science and modelling, we offer figure 1.2 as a schematic representation of the processes involved in societal response to global environmental risk. Although intentionally broad, it is much more focused than the Bretherton "wiring diagram" (ESSC and NASA 1988: 58; Rayner et al. 1991) or the "social process diagram" (Miller and Jacobson 1992). This model is designed to mesh with the IGBP diagram of biogeophysical processes and efforts to model human driving forces. The underlying "driving forces" identified here correspond to those broadly identified in previous analyses of the human dimensions of global environmental change (see, for example, Stern, Young, and Druckman 1992), namely population growth, industrial ecology, land-use change, global political economy, and societal values. The starting point for analysing societal responses is, therefore, the "Driving Forces" box on the left-hand side. The other boxes identify issues at key junctures in societal response.

This is a process model, intended to identify cyclical and iterative processes and feedback loops. Scale and time are obviously important factors, but it is not feasible to represent them graphically here. Therefore, in order to use this model appropriately, it is necessary to conceptualize different scales and changing conditions through time, using multiple cycles of the model. Relationships among different scales need to be explicitly considered in applying the model. Separate cycles can also be distinguished for various social groups, or for a higher level of aggregation, such as "society."

The basic structure of the model purposely shows that failures to address environmental degradation can occur at various points in the system and often eventually feed back into the driving forces. In other words, responses mitigate the effects of the driving forces but they also often aggravate them. To call explicit attention to the role of political and economic power as constraints on the response system, a "political influence and resources" component, largely shaped by political economy and the complex of driving forces, influences all major components of the response system. Each major section of the model oversimplifies highly complex structures and processes that exist internally in the components. More elaborate and sophisticated representations of relationships denoted by each box are needed and can be developed for the various components. A model for the element of "social amplification and attenuation," for example, illustrates this need (see below). As in integrated assessment, linked models are one profitable approach to analysing social response. The discussion that follows depicts selected major issues and processes that need attention in analysing sociopolitical responses to

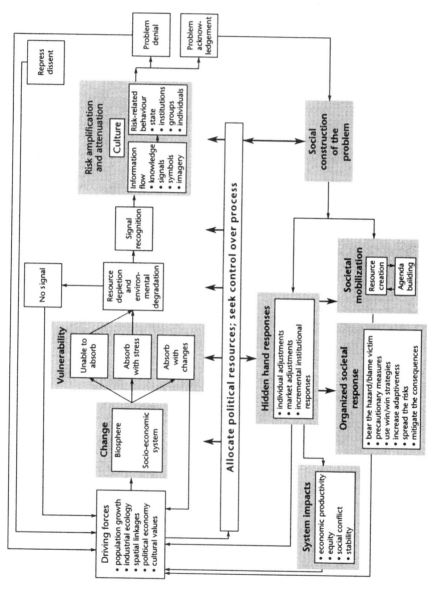

Fig. 1.2 **Societal response to global environmental risk**

global environmental change. It seeks to be suggestive rather than comprehensive, for the range of issues and theoretical approaches defy succinct treatment by any single model.

More detailed discussions typically view human driving forces as feeding into the "proximate" causes of environmental change (Stern, Young, and Druckman 1992). In this model, driving forces directly cause changes in both the biospheric and the socio-economic system, which also feed into each other. The risks of global environmental change are related not only to the human forces of change but also to the degree of fragility of the biosphere and the vulnerability of the socio-economic system. The varying degrees of vulnerability are, in this model, arbitrarily divided into three outcomes. Fragile ecological systems are unable to absorb further changes, and environmental degradation results. Less fragile or vulnerable systems may absorb changes for a time, until some threshold level is reached in the environmental or socio-economic system; then, serious declines occur. A highly resilient system is able to absorb a greater magnitude of changes. All instances of absorption of change are themselves agents of change: that is, the system that absorbs new stresses is not homoeostatic but dynamic. Some absorption may be made through spontaneous, incremental, or even unconscious adjustments (and become "hidden hand" responses) that also feed back into the driving forces, as either mitigating or aggravating influences.

Accompanying degradation will be events and consequences that are potential "signals" to society. Such signals occur with or without societal recognition. If no recognition occurs, there is no (or limited) social learning or stimulated action by the societal response system and the problem feeds back into the driving forces (perhaps as positive feedback, thereby sustaining or enlarging the forces of degradation). If the degradation is recognized – whether by individual farmers, the media, or government officials – and elicits recognition and concern, then the response mechanisms of society receive the signal.

Once the signal is received and recognized that an environmental change poses a threat, society will move to interpret the threat, assess its social meaning and significance, and locate it on its agenda of concerns. This may occur at many different scales and with varying degrees of formality in assessment. We refer to this process as the "social amplification or attenuation" of risk. It is this process by which those at risk, the mass media, societal groups, institutions, and the state itself, gauge whether a change constitutes a risk worthy of attention and how serious any risk may be. Key elements in this social processing of the risk include information flow; communication channels through which individuals, groups, and institutions interact; the formation of public perceptions and concerns; the role of "social stations;" the presence or absence of social capital and trust; and the risk-related responses undertaken. This model construes risk communication very broadly to include mass-media coverage, traditional knowledge, whistle-blowing, scientific knowledge, and propaganda. Interest groups and power factions disseminate information for the purpose of moulding societal opinion, building alliances, and creating political resources. On the basis of this information and its societal manipulation and interpretation, individuals and groups champion or respond to the risk, according to their cultural biases, values, differing societal contexts and agendas, and individual perceptions. The direct biophysical and economic assets interact with more subtle qualitative characteristics of the risk and social processes in ways that enlarge or attenuate both the risk consequences themselves and the associated signals to society.

Eventually, various societal assessments (formal or ad hoc) as to whether the change indicates a phenomenon requiring action or attention (i.e. response) will occur. This "processing" of the risk will often be ephemeral and context dependent, indicating different "snapshots" and urgency of risk at different times. It will also depend upon the confluence of streams that occur at critical junctures, in which policy windows open and action becomes possible (Kingdon 1995). The two major possible outcomes from this process are problem denial or problem acknowledgement. Problem denial moves the risk to the lower rungs of society's agenda, thereby potentially feeding back positively into the driving forces. Problem acknowledgement, by contrast, leads to a location of the risk on the agenda of societal concerns and to some activation or mobilization of a constituency of advocates, and prompts the response system to assess further and to define (or, more likely, choose within a pool already defined) an options pool for risk management.

Once social acknowledgement and initial assessment of the problem have occurred, it is structured into an "issue." How the issue is conceived and socially constructed is critical; political ideology and cultural values are major factors influencing this social construction. The power structure of society directly influences this process, of course, and particular values or ideologies will be dominant in a given society or set of constituencies. In the regional studies appearing in *Regions at Risk* (Kasperson, Kasperson, and Turner 1995), deforestation in Borneo was constructed by political authorities as a problem of "shifting cultivation," whereas deforestation and associated fires in Amazonia were seen as the inevitable price of frontier development or hamburgers for US markets. If the risk problem is constructed as one of concern only to those most directly involved with the production or utilization process (the manager closest to the environmental source), adjustments may be largely cosmetic or of the "hidden hand" variety. If, however, a problem achieves sufficient salience

as to require concerted social action and comes together with other political confluences needed for policy action, interventions to ameliorate
the risk may be undertaken. Building broad participation into decisions
about the definition of the environmental problem is one of the most
challenging tasks for a more democratic risk assessment, probably more
important than the current preoccupation with assessment and characterization (Fischer 1993).

Societal mobilization often must occur to set the terms of debate, to
generate support for action, to influence opinion, and to channel the
problem to particular institutional structures. Agenda building and resource creation (as through media attention) are major activities in
societal mobilization over risk. Actions loop back into information and
management processes at this point. Much is known about agenda
building, societal mobilization, and "policy windows" (Kingdon 1995).

Activation of organized societal response – some kind of organized
action to address the problem – often follows. Societies respond to risk in
myriad ways. This model (fig. 1.2) centres on the dominant risk strategy
chosen. Such management strategies are familiar in the risk literature and
typically include such options as bearing the risk, preventing or reducing
the risk, initiating precautionary actions, spreading or reallocating the
risk, fostering greater adaptiveness, or ameliorating the consequences
after they have registered their initial impacts. Among these risk-
response strategies, *avoidance* refers to action to evade the problem
or exposure to it (as in migration from degraded areas); *consequence
amelioration* refers to dampening the impacts (as through medical programmes or food relief); *bearing the risk* means deciding to absorb the
stress caused by the degradation (although the deciders and the bearers
are often not the same people); *risk spreading* refers to strategies, such as
insurance, that reallocate the risk over larger populations without changing the risk itself; and *adaptation* involves strategies by which society
changes itself in order to reduce further impacts of the risk. Each of these
approaches can make use of mechanisms that are often specific to particular political cultures – such as societal norms, regulations, markets,
liability or legal resources, or social and economic incentives. To these we
add structural changes – those responses that seek to address directly the
basic nature of the driving forces. Technology development; redistribution of land, wealth, power; and fundamental reorientation of social and
economic priorities are responses with substantial power to alter patterns
of regional and global environmental change. Other important responses
in this category include enhanced status for women, family planning, debt
cancellation, appropriate aid, land reform, support for the traditional
rural sector, and inclusion of environmental (and social) costs in national
accounting systems.

The specific character of the political system strongly influences the institutional structure and organized societal responses and the degree of influence emanating from the driving forces. The responses themselves feed directly back into the driving forces, either reproducing existing social and economic systems or inducing restructuring.

Critical issues in societal response to environmental risk

Figure 1.2 suggests, at least implicitly, a number of critical nodes or issues that merit particular attention in the evolution of societal responses to global environmental change. To illustrate the use of this model we select several of these issues for more detailed discussion, noting that others contained in the model but not subjected to detailed examination here require explicit treatment in any systematic and robust analysis.

Vulnerability

We have resolved to put vulnerability centre stage in the analysis of global environmental risk in this volume. The concept of *vulnerability*, defined here simply as "the differential susceptibility to loss from a given insult," is a central but still incompletely developed concept, as we elaborate below in chapters 5–7. Indeed, the variety of terms is useful to convey the breadth or various dimensions of loss encompassed by our definition of vulnerability. Yet it is abundantly clear that the impacts of environmental change or stress lie partially with the character of the environmental change or perturbation – its magnitude, suddenness, spatial extent, and exposure – and partially with the vulnerability of the regions, ecosystems, people, and social organizations affected. Sen (1981) has demonstrated powerfully that food crises can occur without a decline in absolute food availability, finding their roots instead in changing patterns of social and economic entitlements. Environmental risk, as noted above, is an interactive phenomenon involving both nature and society, and particularly inequality and a lack of buffering against environmental threats.

The variety of efforts across disciplines to address vulnerability has generated a pot-pourri of terms, differing approaches, and concepts. The related concepts of *resistance* (the ability to absorb impact and maintain functioning), *resilience* (the ability to maintain a system and to recover after impact), and *exposure* (the presence of a threat to a group or region), provide some guidance (also see Bohle, Downing, and Watts 1994: 38 on exposure, capacity, and potentiality). These three dimensions account for most of the discussion included under the general rubric of vulnerability. Here we examine the meaning and role of vulnerability in risk incidence, leaving in-depth exploration of issues for chapters 5–7 and 10.

Beginning with the early work on stability and resilience in ecology, the impact of a perturbation on a system has been linked with other system characteristics, such as diversity or complexity of food webs. Holling (1973: 17) defines *resilience* as "the ability of these systems to absorb changes ... and still persist." Resistance and persistence have long been used in ecology in the sense of ability of a system or population to resist movement away from an equilibrium point (Connell and Sousa 1983; Margalef 1968). Rappaport (1984), in his *Pigs for the Ancestors*, conceives adaptation as sequences of responses that have structural properties.

Timmerman (1981: 21) reviews the concepts associated with vulnerability, looking towards explanations of societal collapse. He offers his own definitions:

Vulnerability is the degree to which a system, or a part of a system, may react adversely to the occurrence of a hazardous event. The degree and quality of that adverse reaction are partly conditioned by the system's *resilience*, the measure of a system's, or part of a system's, capacity to absorb and recover from the occurrence of a hazardous event.

Frederick (1994) sees the vulnerability of water-use systems to climatic variability as dependent, in part, on the *slack* between supply and demand and characteristics of the system. The characteristics he identifies are *robustness* – the ability to respond to the spectrum of uncertainties associated with factors of climatic variability – and *resilience* – the ability to function under a range of conditions and to return to design performance levels quickly after failure. Definitions addressing natural hazards focus on the interface between everyday socio-economic circumstances and the physical environment. The IPCC (1997: 1) sees vulnerability as "a function of the sensitivity of a system to changes in climate (the degree to which a system will respond to a given change in climate, including both beneficial and harmful effects) and the ability to adapt the system to change in climate (the degree to which adjustments in practices, processes or structures can moderate or offset the potential for damage or take advantage of opportunities created, due to a given change in climate)."

Although the specific terms, concepts, and criteria used to address vulnerability differ, basic concepts are consistent – the ability to continue to function within a normal range despite perturbation and the ability to recover from perturbations that substantially disrupt the normal functioning of the system. These two concepts address internal characteristics of the affected system. Economic, political, or biophysical changes may, however, interact to alter the resistance or resilience, or – significantly – the characteristics of exposure experienced by a system.

These efforts at defining vulnerability reflect the character of the concept. Vulnerability is a useful general concept that depends on precise definition of event and context to provide insight into specific situations. Key considerations in that precision include the following: (1) vulnerability is viewed as comprising internal characteristics of the subject and external processes contributing to differential exposure; (2) vulnerability is understood as relative to a given hazard; thus, the degree of vulnerability will differ among hazards of varying kind and magnitude; and (3) the unit of analysis varies *in scale* as suggested by the range from "system" in Timmerman's definition (Timmerman 1981), to "class," as emphasized by Blaikie et al. (1994), to "household," addressed by Chambers (1989), and to the "nutcracker" of increasing pressure from both rising vulnerability and the severity of hazard that Blaikie et al. (1994: 22) envision. The factors influencing vulnerability, in short, vary widely among scales of analysis, so that scale-sensitive analyses of global environmental risk remain a pressing need (Ghan et al. 1993).

The structure of global environmental risk, then, reflects human driving forces, patterns of environmental change, the fragility of ecosystems, the vulnerability of social groups, and the adverse effects that result (fig. 1.3). The stages in the evolution of risk consequences are highly interrelated and interactive. The magnitude of the impact will depend not only on the *extent* of the environmental change – decreased length of the growing season or number of hectares of cultivable land inundated by increased flooding or sea-level rise – but on the ongoing processes and existing factors contributing to the resilience or capacity of a group or a region to be harmed (i.e. its *vulnerability*). Five broad categories of

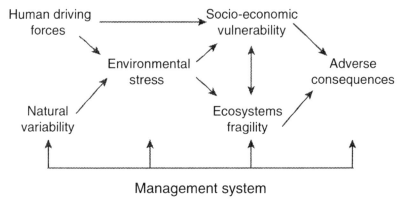

Fig. 1.3 **Simple model of the structure of environmental risk**

factors and processes are important in determining the vulnerability of affected populations: these are *ecosystem fragility, economic sensitivity, social-system sensitivity, individual decision-making,* and *demographic characteristics.* We discuss these at length in chapter 7; here, we present only a brief treatment.

Social-system sensitivity is broadly defined to include institutional characteristics, technological dependencies, and social relationships as they affect capacities, constraints, and decision-making processes within and across social scales of analysis. *Ecosystem fragility,* by contrast, addresses the spatial shifts as well as the regional changes accompanying environmental change. Much of the research on the impacts of climate change has focused heavily on this ecosystem dimension of impact. General circulation models (GCMs) predict with low resolution different levels of warming across latitudes. In some areas, warming may increase the growing season; in others, it may require change in cropping or farming practices. Other analyses focus on the *rate* of climate change and the *ability* of species to keep pace with shifts in favourable habitat conditions, as Ezcurra and colleagues discuss in chapter 6. Predictions of sea-level rise note particularly the vulnerability of island states, coastal regions, and their populations. Depending on the sensitivity of the ecosystem, relatively small environmental changes may have large impacts, a finding well established in the climate-impacts literature (Downing 1996; Kates, Ausubel, and Berberian 1985; Parry, Carter, and Konijin 1988; Rosenzweig and Hillel 1998). Marginality in an ecosystem suggests that the greatest impacts may arise outside the areas experiencing the greatest quantitative climate change, in areas on the margins of an ecosystem's favourable set of conditions; the primary influence of anthropogenetic climate change on ecosystems is likely to occur through the rate and magnitude of change in climate means and extremes and through the direct effects of increased atmospheric CO_2 concentrations (IPCC 1997: 2).

Sensitivity is often used to measure the magnitude of impacts of environmental change. Change can have positive and negative impacts on human societies. Blaikie and Brookfield (1987) use the terms *sensitivity* and *resilience* to describe the quality of land systems: *sensitivity,* in their view, refers to "the degree to which a given land system undergoes changes due to natural forces, following human interference" and *resilience* refers to "the ability of land to reproduce its capability after interference, and the measure of the need for human artifice to that end" (Blaikie and Brookfield 1987: 10). For its part, the closely related concept of *fragility* combines two dimensions – the capacity to be wounded by a particular environmental perturbation (either nature- or human-induced), and the ability to maintain structure and essential functions and to recover (Turner and Benjamin 1994).

Obviously, these different attributes of an ecosystem are highly relevant to analyses of human-induced environmental stresses. Fragile environments degrade more readily under mismanagement and exact higher societal costs for risk management or for substitution in production systems. More robust environments do not degrade so rapidly and respond to substitutes and mitigation interventions more economically. It should be noted, on the other hand, that less-fragile environments are often those under the most human stress and hence are those most subject to mismanagement. Meanwhile, as the IPCC (1997: 3; 2001) has concluded, the projected rapid rate of change relative to the rate at which species can re-establish themselves, the isolation and fragmentation of many ecosystems, the existence of multiple stresses, and limited adaptation options suggest that some ecosystems (e.g. forests, montane systems, and coral reefs) are particularly vulnerable to global climate change.

In chapter 5, Liverman points out that, relative to global environmental change, the most vulnerable people are often believed to be those living in the most ecologically fragile areas. This tendency to equate the probability and magnitude of impacts with the vulnerability of populations, however, fails to capture the richness and subtlety of differences in coping and recovery abilities noted in broader definitions of vulnerability (Chambers 1989; Dow 1992; Dow and Downing 1995; Liverman 1990c; Watts and Bohle 1993). That the population of environmentally fragile areas is also among the world's poorest is a frequent observation, but one that Kates and Haarmann (1992) found difficult to support strongly, on the basis of existing national data and case-study reports.

The *vulnerability of economic systems* is a key issue in impacts. The economic marginality of an activity in relation to first- and higher-order impacts of change that cascade through the economic system becomes an important consideration. Economic marginality describes a situation in which the returns to an economic activity barely exceed the costs or barely provide subsistence. Again, it may be that smaller changes in physical conditions affecting economically marginal practices will result in greater changes. The importance of economic assets and knowledge of the probability and magnitude of events are important, for example, to small-scale farmers' strategies for coping with climatic variability in drought-prone areas of Africa (Akong'a and Downing 1988; Chambers 1989; Homewood 1996) or in flood-prone areas of Bangladesh (Karim 1996). Uncertain changes in environmental patterns may remove some of the buffering offered by established coping strategies and institutions based upon "peasant science" and the local tradition of indigenous agronomic knowledge and experimentation (Watts 1987: 179). In an economic vein, Sen's concept of "entitlements" – access to resources, food, and assistance – suggests a wide range of circumstances surrounding economics and everyday life that may influence vulnerability and points to some of the limits of income as a measure of vulnerability (Sen 1981). Individual decisions about risk-management issues at the moment of disaster or in anticipation will also depend on the social context, such as information channels, types of entitlement available, and other factors as well as the psychological factors of risk heuristics and biases (Kahneman, Slovic, and Tversky 1982).

In his classic *The Moral Economy of the Peasant*, Scott (1976) argues that risk-avoiding behaviour has coloured how South-East Asian peasants manage subsistence agriculture. Generally, cultivators prefer to minimize the probability of a disaster rather than to maximize average return. Accordingly, a "safety-first" principle guides choices in the production process – through a preference for crops that can be eaten over crops that must be sold, an inclination to employ several seed varieties in order to spread risks, and a preference for varieties with stable (if modest) yields (Scott 1976: 23). The growth of the colonial state and the commercialization of agriculture have enlarged the vulnerability and complicated the traditional risk security system of the peasants through the pervasive impacts of changes in the character of social and economic relations.

Social-structure sensitivity is no less important, although often more neglected, than ecological fragility and economic sensitivities. It is broadly concerned with the response capabilities of particular social groups within a threatened region. Many researchers point out that vulnerability does not lie with extreme events or situations but is embedded in the routine or "normal" functioning of social relationships (Dow 1992; Hewitt 1984a; Jeffery 1981, 1982; Torry 1979). Political-economy perspectives stress the importance of social divisions, often class, in accounting for differential access to resources and assets and differences in the magnitude of impacts experienced. Among social relations, class, gender, and ethnicity have emerged as significant elements in differences both in perception, as above, and in limiting the options available to different groups (Blaikie et al. 1994; Cannon 1994). Social relations that restrict women's options and abilities to respond to famine result in more severe impacts on women (Ali 1984; Schroeder 1987).

The nature of social and economic institutions is yet another factor that may contribute to vulnerability. The infrastructure of social institutions and resource management pervasively shapes the breadth and depth of the response capability. Whether institutions are highly adaptive or more narrowly prepared to respond to only a limited set of opportunities, the potential to benefit (or to avoid harm) rests largely with the ability to identify and react to opportunities. Increasing the reliance on management systems to reduce the variability of an ecosystem may erode the

overall resilience of the ecosystem and its associated social and economics system (Holling 1986). In the context of global climate change, with its
potential shifts in resource availability and environmental stresses, the
ability of the legal system to adjust to changes in the resource base will
obviously affect the magnitude of potential disruptions. The legal and institutional rigidity of water-resources management throughout the world
is a major obstacle to efficient use of freshwater sources. To that end, the
Global Water Partnership (1996), an action-oriented network of organizations, has resolved to create practical tools for solving water problems
at the regional and local level, while Gleick (1998: 18) has proposed sustainability criteria for water planning. The ability of social institutions to
provide warnings, diagnosis, insurance, and planning support is a key
element in shaping successful societal coping.

Demographic and health considerations among human populations
represent an area of specific concern and investigative depth in relation
to a number of hazards. Population growth in marginal areas – such as
coasts, drylands, and degraded areas – may expose more people to additional stresses. Other demographic and social characteristics (such as age
and occupation) of those populations will also be important in evaluating
overall levels of vulnerability. Special-needs populations (elderly, dis-
abled, lactating women, children under five) are among the most vulnerable persons in natural disasters. Vulnerability, in short, leaves a finger-
print that becomes increasingly complex as more social relations and
scale disaggregation are integrated into the analysis.

Alerting systems

Society uses a variety of means to inform itself of potential environmental
and social hazards. *Monitoring* implies an active and often systematic
search by institutions or individuals for information or signals. *Alerting*
refers to the warning function associated with the identification and assessment of risk. These monitoring and alerting functions play key roles
in identifying social problems and initiating societal responses. As Rappaport (1988) points out, the perception of a hazard begins with the first
sense of danger and *changes with increased information*. The identification of risks is closely tied to social perceptions and occurs in a variety of
public arenas – scientific research, the mass media, environmental groups,
and the dissemination of industrial products. Such perceptions arise according to principles of cultural and political selection, including agenda
competition among problems, cultural preoccupations, and political
biases, as well as more formal risk screening.

Global environmental risks pose distinctive challenges to established societal monitoring and alerting practices. As already noted, a high degree of scientific and political uncertainty often pervades such issues. Past experience of interactions, rates, or magnitudes of change may be unreliable as a guide to future risks on the global scale; instead, societal alerting and assessment systems will need to take on a larger burden of anticipation.

Such alerting occurs in a wide diversity of institutions, roles, and contexts. Knowledge of change does not always arise from purposeful monitoring of the environment, or even of one component of that environment, but often by serendipity or by chance, from monitoring a related societal process. In many cases, the motivation for monitoring is aimed at the realization of other societal goals – such as health promotion, economic growth and development, or policy effects. Additionally, institutions and groups experience a change that suggests a departure or discontinuity not previously experienced. Mountain dwellers of Nepal, for example, learn of environmental change through increased mudslides, hazards, declining crop yields, and growing scarcity of fuelwoods, whereas the government learns about them through accounts of local crises trickling up from remote areas or through reports of failures of development or aid programmes (Jodha 1995).

Table 1.2 presents an overview of societal alerting mechanisms related to environmental concerns. Many elements of society are potential sources of alerts. Involved are the daily checks on the weather or the level of a river in midsummer, the conduct of formal science, the activities of citizen groups' watching and reporting on their concerns, as well as the media's scrutiny of events and editorial determination of "what's news." From this diverse array of perspectives and capabilities emerges a complicated web of interactions – scientists monitoring atmospheric composition, journalists looking over the shoulders of scientists and government, and individuals searching for information directly through their own experience and indirectly through interaction in social and information networks.

Scientific investigation is likely to play a pre-eminent role in identifying global environmental risks, especially those of the systemic type. Increasing sophistication in technical capabilities to detect hazard precursors or effects will be needed to identify new environmental threats at the global scale. Debates within science also stimulate the search for information. Perhaps the most massive (if diffuse) monitoring system exists at the governmental level in efforts to track emerging political issues and to assure "proper functioning" of existing programmes and policies.

Building understanding of the connections among monitoring data is part of societal learning about environmental change. The recent advances in climatology provide an interesting example. The range of theoretical

Table 1.2 **Societal monitoring and alerting mechanisms**

Monitoring/alerting group	Focus of monitoring	Monitoring system	Alerting system	Constituency
Individuals	Direct individual or community impacts	Direct and indirect experience	Social networks	Peer groups, friends, neighbours
Individual resource managers (e.g. farmers)	Production systems, resource inputs	Direct observation; market conditions	Petitions to government	Interest group, neighbours
Citizens' groups	Major special-interest topics; activities of other organizations or arenas	"Networking" – sharing or seeking information from other groups	Social networks; media contacts; direct mail; newsletters	Organizational network; members; subscribers
Mass media	Regular defined "beats" on selected topics or by style of monitoring (i.e. investigative journalism)	Professional standards of "what's news"; "gatekeeping"	Timing, duration, and level of coverage; media "signals" and imagery	Subscribers, viewers, advertisers
Corporations; trade and industry groups	Self-interest areas, including regulations, science, and public opinion	Media coverage, legislative and regulatory actions; product-related outbreaks	Newsletters; mandatory notifications to government	Stockholders, consumers, employees
Scientists	Research interests, self-defined and socially defined	Research; direct observation; technology	Debates; dissent; public testimony; research reports	Peer scientists, the public; funding sources
NGOs	Institutional agenda; "mission" of organization	Public concerns; outbreaks; media reports; public opinion	Reports; press releases; testimony	Legislators, public-opinion leaders, members
Government agencies; legislature	Issues on the public agenda; congressional actions; expert-defined risk	Screening systems; rules and regulations; public opinion; agency-sponsored research	Reports; laws; regulations; action	Political supporters; élite publics

understanding within climatology has greatly increased, owing to new satellite capability to observe the entire planet and to detect relationships across altitudes and across regions. The Earth Systems Science Program plans to incorporate its observations into its ongoing activities in an impressive range of ways. Meanwhile, physical scientists concerned with global environmental change and global processes of change are seeking to extend their range of observation further back into history, as well as to expand the scope of current efforts. Longer time periods will allow statistical techniques to identify trends or bands in data sets that may be difficult to glean through other means.

Other researchers have directed their attention to the question of *what monitoring systems should look for*. R.E. Munn (1988) examines the potential of integrated monitoring to provide early warnings through monitoring *stresses*, *feedbacks*, and *lags* in ecological systems. The possibility of surprise accompanying global environmental change is constantly present and suggests that current monitoring systems may be inadequate to provide the needed warnings. According to Munn (1988: 216), a diversity of methods is needed, including the following:

1. Studies of historical surprises;
2. Biological early indicators;
3. The mapping of signal-to-noise fields;
4. Identification and quantification of stresses, feedbacks, and lags; and
5. Creative scenario writing.

Schneider and Turner (1995) identify specific measures that can be taken to increase society's resilience and its ability to adapt to surprise (table 1.3).

The FEWS is an active early warning system at the international level that suggests possibilities for integrated physical–social alerting systems for global environmental change. It selects both ecological and social indicators of pre-famine conditions based on research linking indicators of vulnerability to famine impacts. These indicators include agrometeorological measures of rainfall, pest damage, flooding, growing conditions, animal-carrying capacity, brush fire damage, social measures (including level of food reserves), existing food stress, contribution of cash crops, changes in cereal prices, food production trends, health and nutritional data, and conflict/civil disruption (FEWS 1999). At the same time, the severity of the 1992 famine in Somalia and the 1997–1999 famine in North Korea indicate the limits of what even early warning can provide in the face of more general political disintegration.

Proposals have also been made for Drought Early Warning Systems (DEWS) designed to forewarn governments, those likely to be directly affected by drought episodes, and the donor community (Glantz and Degefu 1991: 258–260). Such a system would be based on both quantita-

Table 1.3 **Increasing resilience and adaptability to surprise**

1. Diversifying economic productive systems: the tendency towards increased economic specialization carries the risk of vulnerability to controls (e.g. markets or absentee landlords) well beyond the local area, which can have both positive and negative impacts on local resilience to environmental perturbations.
2. Avoidance of technological monocultures: reliance on a single technology, such as the extensive use of a single crop variety, may be vulnerable to environmental or other perturbations with negative impacts on the economy. (In this case, long-term impacts might follow without a backup system, such as a "crop variety bank.")
3. Strengthening "entitlements" or access to resources broadly defined: providing robust safety nets to respond to unforeseen events is a critical part of resilience.
4. Adaptive management systems: organizational theory suggests that different management systems have different capacities for dealing with surprise: those characterized by openness, participation of all parties, and flexibility may be more adaptive than those characterized by command-and-control (e.g. IPCC embedded into the Conference of Parties, Climate Convention by means of rolling re-assessments).
5. Disaster-coping systems: improving designs of early-warning, monitoring and altering systems, and strengthening the capability of private and public sectors to respond rapidly to potential disasters should be encouraged.
6. Organizational memory and social learning: measures that improve memory and the ability to learn from surprises improve overall resilience and reduce vulnerability to surprise.

IPCC, Intergovernmental Panel on Climate Change.
Source: Schneider and Turner (1995: 138–139).

tive and qualitative information, including anecdotal information about sightings of "famine foods" in the market place. Many limitations to existing scientific practice need to be addressed. Research on indigenous knowledge and ethnoscience suggests that indigenous people often "see" nature differently owing to its centrality in their everyday practices (Thompson and Rayner 1998; Wynne 1996). In chapter 4, Funtowicz and Ravetz underline the greater attention that is needed to the limits of scientific knowledge and ignorance in confronting environmental change. Chambers (1984) notes key types of bias that operate against contact with and learning from poorer people and that contribute to an ongoing mis-representation of rural poverty, namely:

1. *Spatial* – research tied to urban area, tarmacs, roadsides;
2. *Project* – associated with places where there are projects;
3. *Person* – towards those who are better off, i.e. users versus non-users, men versus women, healthy and active versus sick;
4. *Season* – avoiding the bad times of the year;
5. *Diplomatic* – not seeking out the poor for fear of giving offence;
6. *Professional* – confined to the concerns of an outsider's specialization.

The FEWS's attempt to incorporate rural poverty into its assessments may prove difficult to replicate with other research and monitoring strategies.

All monitoring methods, it must be appreciated, have their biases and blind spots. Farhar (1979), for example, discusses the Metromex study of the St Louis urban weather anomaly in which a 30 per cent increase in summer precipitation over a 30-year period passed virtually unnoticed (Changnon 1981). The farmers (who might be expected to be more sensitive to the change) attributed higher crop yields to better agricultural technology – mainly fertilizers and better seed varieties. Perhaps they were monitoring yields and the impact of management efforts more closely than rainfall; in any case, they did not include the role of increased rainfall in their attributions of the source of the higher yields. An array of case studies organized by Glantz (1988) bears out the common failure of alerting systems, whether based on scientific or indigenous knowledge, to heed potential signals.

The complex array of societal efforts to collect information about the environment provides some level of ongoing awareness and some alerts of unfolding environmental hazards. Such systems comprise diverse mechanisms (official and unofficial) and a blend of knowledge, ignorance, and bias. How to design the alerting function so as to increase society's anticipatory capabilities for confronting surprises requires both greater understanding and more adaptation of mechanisms to particular political cultures.

Social amplification and attenuation

As figure 1.2 suggests, once society has recognized signals depicting an environmental change and identified a potential threat, social institutions and groups must "process" the risk information and assess the threat. This social processing can amplify or attenuate the risk signal and shape the interpretation of risk and the ways by which society will respond. The notion of risk amplification and attenuation is based on the recognition that psychological, social, institutional, and cultural processes interact in ways that can heighten or dampen social perceptions of risk and shape individual, group, and institutional behaviour (Kasperson 1992; Kasperson and Kasperson 1996; Kasperson et al. 1988).

Behavioural responses, in turn, generate secondary social or economic consequences. These consequences extend far beyond direct harm to human health or the environment to include significant indirect impacts such as liability, insurance costs, loss of confidence in institutions, stigmatization, or alienation from the political system, as shown in figure 1.4. Such secondary effects often (in the case of risk amplification) trigger

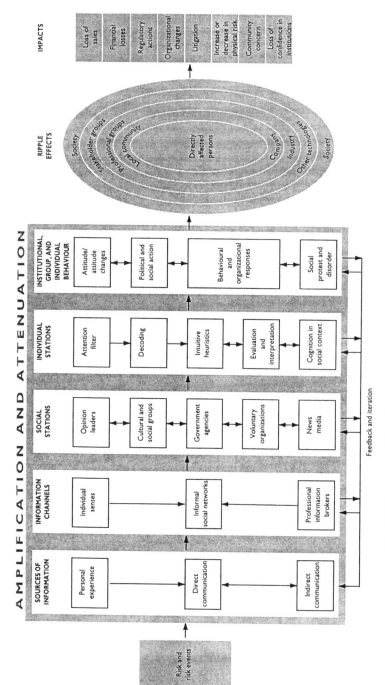

Fig. 1.4 **Framework for the analysis of risk amplification and attenuation**

demands for additional institutional responses and intervention, or, conversely (in the case of risk attenuation), place impediments in the path of needed protective actions. In our usage, "amplification of risk" includes both intensifying and attenuating signals about risk. Thus, alleged "overreactions" of people and organizations receive the same attention as alleged "downplaying."

In our conception, environmental risks are viewed as partly biophysical threats of harm to people and partly as "socially constructed," a product of culture and social experience. Hence, hazardous events typically involve transformations of the physical environment (through the release of energy, materials, or information) but also involve perturbations in social systems and value structures. Impacts of such changes and events remain limited in the social context unless they are observed by human beings and communicated to others (Luhmann 1986: 63; 1989: 28–31). The consequences of this communication and other social interactions may lead to additional physical transformations – such as changes in technologies; changes in methods of land cultivation; or changes in the composition of water, soil, and air. The experience of risk is, therefore, both an experience of physical harm and the result of cultural and social processes by which individuals or groups acquire or create social interpretations of hazards (UK Interdepartmental Liaison Group on Risk Assessment 1998). These interpretations provide rules of how to select, order, prioritize, and explain signals emanating from nature–society and technology–society interactions. Additionally, each cultural or social group selects certain risks and adds them to its strand of "worry beads" to rub and burnish even as it rejects other risks as not meriting immediate concern or not fitting the current agenda of concern (Thompson and Rayner 1998).

The amplification process, as we conceive it, often starts with either a process or physical change; an event; or a report on environmental or technological events, releases, exposures, or consequences. Alternatively, environmental groups or individuals who are actively monitoring the experiential world and searching for hazard issues related to their political agenda may elevate routine incidents into events or even crises. Individuals or groups select specific characteristics of these events and interpret them according to their values, perceptions, and world views. They also communicate these interpretations to other individuals and groups (and receive interpretations in return). Social groups and individuals process the information, locate it in their agenda of concerns, and may feel compelled (or see opportunities) to respond. Some may change their previously held beliefs, gain additional knowledge and insights, and be motivated to take action; others may use the opportunity to compose new interpretations that they send to the original sources or other interested parties; and still others find the added information as confirming long-held views of the world and its order.

The individuals or groups who collect information about risks communicate with others, and through behavioural responses act as *amplification or attenuation stations*. Such stations can be individuals, groups, or institutions. It is obvious that social groups or institutions can amplify or attenuate signals only by working in social aggregates and participating in social processes. However, individuals in groups and institutions do not act or react merely in their roles as private persons; rather, they act according to the role specifications associated with their positions. Amplification may, therefore, differ among individuals in their roles as private citizens and in their roles as employees or members of social groups and organizations.

Role-related considerations and membership in social groups shape the selection of information and communications that the individual regards as significant. Interpretations or signals that are inconsistent with previous beliefs or that contradict the person's values are often ignored or attenuated; they are intensified if the opposite is true. The process of receiving and processing environmental risk-related communications by individuals is well researched in the risk-perception literature (Freudenburg 1988; Slovic 1987; Stern and Fineberg 1996). However, this is not sufficient: individuals act also as members of cultural groups and larger social units that co-determine the dynamics and social processing of risk. In this framework, we term these larger social units *social stations of amplification*. Individuals in their roles as members or employees of social groups or institutions not only follow their personal values and interpretative patterns but also perceive risk information and construct the risk "problem" according to cultural biases and the rules of their organization or group (Jasanoff and Wynne 1998). Cultural biases and role-specific factors are internalized and reinforced through education and training, identification with the goals and functions of institutions, beliefs in the importance and justification of social outcomes, and rewards (promotions, salary increases, symbolic honours) and punishments (demotions, salary cuts, disgrace). Meanwhile, conflicts between personal convictions and institutional obligations may become apparent.

Both the communication processes depicting the risk or risk event and the associated behavioural responses by individual and social-amplification stations generate secondary effects that extend beyond the people directly affected by the original hazard event or report. Secondary impacts include the following effects:

- enduring mental perceptions, images, and attitudes (e.g. antitechnology attitudes, alienation from the physical environment, social apathy, stress, or distrust of risk-management institutions);

- impacts on the local or regional economy (e.g. reduced business sales, declines in residential property values, and drops in tourism);
- political and social pressure (e.g. political demands, changes in political climate and culture);
- social disruption (e.g. protesting, rioting, sabotage, terrorism);
- changes in risk monitoring and regulation;
- increased liability and insurance costs; and
- repercussions on other technologies (e.g. lower levels of public acceptance) and on social institutions (e.g. erosion of public trust).

Secondary impacts that are, in turn, perceived by social groups and individuals may generate additional stages of amplification and produce higher-order impacts that may carry over, or "ripple," to other parties, distant locations, or future generations. Each order of impact not only will disseminate social and political impacts but also may trigger (in risk amplification) or hinder (in risk attenuation) positive changes for physical or material reduction of risk. The concept of social amplification of risk is hence dynamic and iterative, taking into account the continuing learning and social interactions resulting from society's experiences with risk.

The analogy of dropping a stone into a pond (see fig. 1.4) illustrates the spread of these higher-order impacts associated with the social amplification of risk. The ripples spread outward, first encompassing the directly affected victims or the first group to be notified, then touching the next higher institutional level (a company or an agency), and (in more extreme cases) reaching other parts of the industry or other social arenas with similar problems. This rippling of impacts is an important aspect of risk amplification since it suggests that the social processes can enlarge (in risk amplification) or shrink (in risk attenuation) the temporal, geographical, and sectoral scales of impacts. It is important to recognize that the responses may become embedded in the structure and memory of institutions, policy, and knowledge and thereby form the historical legacy affecting future decisions and public reactions.

Integrated assessment

During the past decade of the 1990s, no area of assessment or methodology development has received greater attention in global change studies than what is termed "integrated assessment" or "integrated assessment modelling." Indeed, in both Europe and the United States, funding agencies have accorded high priority and substantial financial support to the development of large-scale models for integrated assessment, and surely the largest venture in risk assessment has been the Intergovernmental Panel on Climate Change, or IPCC.

Integrated assessment refers to a set of formal models or studies "that are combined into a consistent framework to address one or more issues in the problem of global environmental change" (Toth 1994: 11). Asselt and Rotmans (1996: 121), for their part, define it as "an interdisciplinary process of combining, interpreting and communicating knowledge from diverse scientific disciplines in such a way that the entire cause–effect chain of a problem can be evaluated from a synoptic perspective with two characteristics: (1) it should have added value compared with a single disciplinary oriented assessment, and (2) it should provide useful information to decision makers." Such assessment addresses several central problems in risk analysis generally – how to include quite comprehensively the important elements or components of a risk problem, how to identify linkages among these components, and how a complex problem can be decoupled so that discrete parts of the problem can be analysed in context (Shlyakhter, Valverde, and Wilson 1995). Such assessments, drawing on the definitions noted above, need to begin with the objectives and concerns of those who will use the assessments; clearly, different outcomes will matter to different users and decision makers. Many kinds of integrated assessment are possible, including a simple diagram or schematic representation, simulation gaming, scenario construction, qualitative interpretation (as in works on environmental history), or a complex of linked computer models; however, it is the last type that has dominated (Dowlatabadi 1995). Such integrative assessment, whether qualitative or computer based, is particularly useful in facilitating the following:

- the placement of environmental risks in the broader context of global environmental change;
- the identification and assessment of potential interventions to prevent or adapt to risk;
- the construction of a framework to structure the existing knowledge base;
- the identification and assessment of major uncertainties; and
- the demarcation of priorities for further research (Rotmans and Dowlatabadi 1998: 295–296).

Several such large computer-based models are well known. IMAGE (Integrated Model to Assess the Greenhouse Effect), developed in the Netherlands by Rotmans and colleagues, provides a rich description of physical and biogeochemical processes involved in climate change. This model itself has recently become a part of ESCAPE (the Evaluation of Strategies to Address Climate Change Adapting to and Preventing Emissions), a much larger framework of analysis that relates three principal subsystems – energy–industry, terrestrial environment, and atmosphere–ocean (Alcamo et al. 1994), and one that has also been superseded by the integrated assessment model TARGETS, the acronym for Tool to Assess Regional and Global Environmental Health Targets for Sustainability (Rotmans et al. 1994; Rotmans and de Vries 1997). Some benefit–cost models, such as the PRICE model (Nordhaus and Popp 1997) and that of Peck and Teisberg (1992, 1995), assess benefits, the value of scientific knowledge, and the costs of climate change; they conclude that actions intended to secure early abatement of CO_2 emissions are not economically justified. The ICAM (Integrated Climate Assessment Model) developed at Carnegie-Mellon structures the climate-change problem in terms of the perspectives of different decision makers, gives in-depth attention to the types and propagation of uncertainties, and concludes that subjective perceptions of different participants are more important in determining policy objectives than the scientific uncertainties (Dowlatabadi and Morgan 1993: 1813).

Despite the extensive use of large-scale formal modelling and computer-based analysis, integrated assessment, at its core, impinges on three themes that thread their way through this volume – how to draw upon diverse knowledge to inform the decision process, how to create an emancipatory risk analysis, and how to negotiate alternative futures for societies and for the planet as a whole. It is clear from the integrated assessment work thus far that current understanding of the natural and social systems is so incomplete that it is not possible to build models that meet any test of comprehensiveness, much less treat all major interactions and feedbacks. The general response in the policy research community has continued to be – as Dowlatabadi and Morgan (1993: 1932) note early on – to model what was understood and "wave their hands" at the rest, a problem discussed at length in chapter 16. None the less, it is quite remarkable how integrated assessment, as one rather specialized assessment tool, has dominated the funding and first generation of global environmental change research. The authors note, for example, that integrated assessment has commanded the greatest space of any topic in the first two open meetings of the IHDP at Duke University in the United States and at IIASA in Austria. Although it is clear that integrated assessment provides one important avenue to creating the knowledge base needed for anticipating and responding to global environmental changes, it is also important that we maintain realistic expectations about what can be gained and that we continue to search for means to make such methodologies more participatory and more sensitive to social-choice systems.

Other efforts have also sought a more integrative approach to nature–society relations, including societal response. Environmental historians, such as William Cronon (1991), have provided a powerful interpretive capability to describe and analyse the emergence of regional environments over long time periods. The monumental work *The Earth as*

Transformed by Human Action (Turner et al. 1990a) has provided 300-year environmental histories of such diverse regions as the Ordos Plateau, Amazonia, Borneo, Caucasia, the Basin of Mexico, the Russian Plain, and the US Great Plains, drawing upon ecology, anthropology, history, and demography. Palm (1990) develops a conceptual framework that focuses upon the complexity and interactions between societal structures and behaviours at different levels or scales. Chapter 8 of the present volume reports on an international collaborative effort to provide comparative analyses using a common research design to analyse environmental change and societal responses in nine different environmentally threatened regions across the globe.

Assessing the state of social science research on global environmental change early in the 1990s, Miller and Jacobson (1992) argue that a substantial theoretical base exists in (1) the theory of the demographic transition, (2) theories linking resource use with levels of economic development and prices, (3) theories about the pace and direction of technological innovation, (4) theories about attitude formation and change, and (5) theories about the operation and responsiveness of economic, political, and social systems. What was needed, in their view, was a model exploring linkages among the concepts and theories. Accordingly, a working group sponsored by the Consortium for International Earth Sciences Information Network (CIESIN) and the Aspen Global Change Institute constructed a "social process diagram." This diagram aimed to do for the social sciences what the "Bretherton" diagram (which identifies major topical areas and depicts linkages among the biological, chemical, and physical sciences in global environmental change) did for organizing the study of global change by the physical sciences. Unfortunately, whereas the social process diagram suggests a multitude of components and linkages among social and economic phenomena, its high level of generality precludes any significant explanatory power and shows the limitations of seeking to capture generic socio-economic processes that are strongly rooted in particular regions, places, and cultures.

Thus, the scope of available approaches to integrated environmental risk analysis is broad and variegated. The hazard and climate-impact models are particularly powerful in their depiction of the causal structures that link human wants and technology choices with environmental perturbations, relating them to societal vulnerabilities and eventually with the diverse consequences to human health and society. These relatively generic models have been applied across a wide diversity of hazards and societies. Their strengths lie in their ability to link environmental changes with ecological and societal impacts and to construct causal pathways involved in such change, and the direct power that they bring to policy and management analysis. Their principal limitations stem from a lack of attention to the "upstream" portions of human-induced environmental change – the underlying economic and social forces that alter the environment and how such forces operate through political economy and culture to shape and constrain societal responses. A primary task for applying risk concepts to global environmental change is to broaden current risk models into these areas or to supplement risk thinking with complementary "upstream" social science theory. It is also abundantly clear that risk analysis provides a powerful means for linking theory on social and policy change with patterns and processes of environmental degradation.

Adaptive management

The three chapters that follow make it clear that global environmental risks often have distinctive properties – including the large scale of impacts, the time and space separation between risk sources and affected areas and peoples, and the synergistic ways that new and traditional risks interact in many developing societies. Funtowicz and Ravetz point out, in particular, that global environmental risks combine two properties – high uncertainty and high decision stakes – that pose great challenges for both assessment and management.

The nature of global environmental risks inevitably raises the question of what type of management systems, institutions, and approaches are likely to be successful in dealing with them. The effectiveness of institutions, it has been argued in the *Science Plan* of the Institutional Dimensions of Global Environmental Change Planning Committee (Young et al. 1999: 45–48), is a function of the match between the characteristics of the institutions and the nature of the biogeophysical systems with which they interact. Institutional failures arise owing to mismatches, arising from such sources as ignorance of relevant ecological processes, the power of special interests, the influence of dominant paradigms, reliance on false analogies, the operation of inappropriate norms or principles, the influence of institutional path dependence, and jurisdictional problems. In this context, conventional environmental management characterized by detailed scientific assessment aimed at creating relatively high certainty and anticipation, effective "rational" choice among options, and a "command-and-control" approach to implementation is unlikely to succeed. Elsewhere, we and our colleagues (Cook, Emel, and Kasperson 1990) have argued in the context of radioactive waste management, as Lee (1993) has done more generally, that an adaptive-management approach is likely in many cases to be much more robust, resilient, and effective over time with these risks. Such a management style proceeds from the recognition that global environmental risks cannot be comprehensively assessed (as

we note above in our discussion of integrated assessment), that large uncertainties and even ignorance typically characterize some major elements of the risk problem (see the treatment by Norberg-Bohm and colleagues in the next chapter), that surprises are certain to occur, that a
complex sociopolitical and value-laden environment will continually interact with environmental risks to shape the complexity of decision, that
flexibility to respond to the unexpected is imperative, that decision-
making will need to be iterative, and that social learning from experience
will be a paramount need. It is also a management style that is highly
congruent with a democratic, broad, humanistic approach to risk analysis.

Climate change illustrates these issues well. Uncertainties beset the
causal structure of climate change risks at all stages – including the particular types, timing, and patterns of consequences (good and bad) that
may occur – and these uncertainties are not likely to be narrowed significantly within at least the next decade. These uncertainties interact
with political structures in ways that touch upon a wide array of interest
and idealogies. As Skolnikoff (1999) has recently pointed out, nearly every
US agency in the executive branch has some legitimate interest in the
climate issue while most congressional committees will be drawn in, with
turf to defend or expand and particular visions of the national interest to
pursue. In this setting, Skolnikoff (1999: 19) argues, scientific evidence
has "a long row to hoe" in order to have a decisive impact on public
policy.

Successful institutions charged with managing global environmental
risks are likely to be those that develop a management approach and
culture in which learning and adaptation thrive. They are likely to be
institutions with greater openness and permeability in their boundaries
so that greater information sharing occurs with constituents, clients, and
critics. The locus of institutional learning and innovation will expand to
include lower as well as higher levels of the organization. Given the uncertainties and complexities surrounding the risk, it will be necessary to
conceptualize the management task as experimental in nature. Such an
approach allows the explicit acknowledgement of uncertainties, rather
than their suppression or distortion. Currently, a common organizational
response to uncertainty is to convert it into risk, in which probabilities,
outcomes, and causal relationships are purported to be known. Adaptive-
management approaches to uncertainty and conflictual problems recognize that knowledge is both incomplete and contested and that elimination of uncertainties is not achievable. Thus, a management process using
some trial and error, with continuing corrections and alterations from
new understandings or unexpected outcomes, will be better positioned to
address the challenges posed by this class of risk problems.

Radioactive-waste management affords insights into the institutional challenges that emanate generally from global environmental change. The complex nature of radiological hazards, the very long time-scales involved in isolating wastes from the biosphere, the conflicting values surrounding the use of nuclear power, and the fact that siting a high-level nuclear waste disposal facility is a first-of-a-kind venture have proven highly challenging to management authorities in various advanced industrial societies. Most began with command-and-control *modus operandi* developed for other waste and energy problems; however, a multitude of failures, underestimation of complexity and difficulty, commingling of technical and social problems, and widespread public opposition have mired many of these programmes in technical debate, public contention, and dramatically escalating programme costs. Most societies have been forced to reinvent (or create entirely new) institutions and management strategies.

The radioactive-waste experience suggests how critical it will be to fit management institutions and strategies to the nature of the risks and their attendant social considerations. Global environmental risk covers a highly variegated array of hazards, as Norberg-Bohm and colleagues demonstrate so well in the next chapter (2), so that no single management approach or strategy will be optimal. However, many with the highest global stakes will fall within the domain of highly problematic risks that Funtowicz and Ravetz identify in chapter 4. For these risks, assessment will need to be more participatory, futures will need to be negotiated as part of decisions, knowledge bases made more diverse, and management more adaptive and learning based.

Environmental change and international security

Global environmental risks and social and economic inequalities threaten international security and peaceful relations among states by limiting economic development; contributing to differential wealth and well-being; and increasing competition, tensions, and conflict (Dalby 1992; Homer-Dixon 1999; Mathews 1993). A variety of risks and risk sources are involved, including the following:

- disputes arising directly from human-induced local environmental degradation;
- ethnic clashes arising from population migration and deepened social cleavage due to environmental scarcity;
- civil strife caused by environmental scarcity that affects economic productivity and, in turn, people's livelihoods, élite groups, and the ability of states to meet changing demands;
- scarcity-induced interstate war over, for example, water; and

- North–South conflicts over mitigation of, adaptation to, and compensation for global environmental problems (Homer-Dixon 1999: 5).

Lonergan (1997) who has done important work in developing indicators and databases relevant to environmental insecurity, has set forth a typology to categorize links between environment and population movement, using the five components natural disasters, "slow-onset" changes, industrial accidents, development projects, and conflict and warfare.

Resource depletion and environmental degradation, it is increasingly appreciated, can be major sources of tension in international relations. Population growth is a significant factor limiting economic growth and restricting improved standards of living in developing countries. Low standards of living, in turn, aggravate global inequalities by increasing the gap between rich and poor. Meanwhile, urbanization in third world countries juxtaposes urban poor and migrants from rural poverty with the rich élite. Environmental refugees in the wake of environmental breakdowns may well become more commonplace in a more densely populated and environmentally inequitable world (Kavanagh and Lonergan 1992; Lonergan 1998). Declines in availability and quality of land resources may lead to rising prices for food and increased dependency on outside sources. Many countries are facing serious deficiencies in water supply and sanitation, and water disputes are apparently increasing among water-short countries. Acid rain signals a growing class of transboundary problems involving environmental pollution that may lead to international tensions. The depletion of non-renewable energy resources has been an underlying dimension of international conflict in the Middle East. The agricultural and economic implications of dislocations connected to altered temperature and rainfall distributions associated with climate change could engender tensions between "winners" and "losers."

Perhaps no environmental problem is more acute than the growing global scarcity of high-quality freshwater resources. Several factors escalate political rivalries and potential conflict, including (1) the degree of scarcity, (2) the extent to which water supply is shared by more than one region or state, (3) the relative power of the basin states, and (4) the ease of access to alternative freshwater resources (Gleick 1992, 1998). The global water situation reveals a growing inequality and potential for conflict. More than a billion people lack access to safe, clean water; 1.7 billion (1.7×10^9) lack appropriate sanitation facilities; waterborne diseases are still widely prevalent; and irrigation development remains highly unequal (Falkenmark 1986; Falkenmark and Lindh 1993; Gleick 1998, 2000).

In his assessment of potential water scarcity, Gleick (1998: 113) lists countries in which annual renewable water availability in the mid-1990s was below 1,700 m^3 per person, a minimal water requirement that Postel

(1997) describes as a level below which domestic food self-sufficiency becomes almost impossible (see also Falkenmark and Lundqvist 1997: 4). Gleick (1993a, b) previously estimated that the annual availability of water would fall below 500 m^3 per person by 2025 for 19 of these countries, mostly located in Africa and the Middle East. The number of countries at risk may well increase as a result of population growth and economic development. The World Health Organization (WHO 1996) estimated that nearly 2.6 billion people (nearly a billion more than 5 years earlier) lacked access to sanitation and 1.3 billion people lacked access to basic water sources. Water inequities, it is clear, will add to and interact with other cumulative inequities already apparent or growing throughout the globe, and a substantial list of potential regional water conflicts now exists (German Advisory Council on Global Change 1999: 205–214).

Much debate currently exists over the relationship between environmental degradation and international security. The second and third Open Meetings of the Human Dimensions of Global Change Research Community in 1997 and 1999, respectively, debated at length progress over the past five years in analytic frameworks for assessing environmental security, suggesting to the authors that such frameworks and models remain very limited in providing satisfying interpretations. The problems are several. As yet, there is no general agreement as to the meaning of environmental security, with definitions ranging from the security of nation-states to others emphasizing human security and individual well-being. The causal linkages between environmental change and attributes of environmental security are as yet poorly defined and understood. As Myers (1989: 39) observes, policy makers have a limited capacity to think methodologically about matters long outside their purview. Moreover, the "new security" studies have yet to create robust models that demonstrate the causal links between environmental degradation and civil strife, or international security more generally. Observers (Deudney and Matthew 1999; Levy 1995) correctly point out the dangers of muddled thinking in making arguments and claims. This topic appears well suited to the types of careful modelling of causality linkages, as well as uncertainty, that have characterized the best of risk analysis, and has recently been stressed in the *Science Plan* for the Global Environmental Change and Human Security Planning Committee (Lonergan 1999) of the IHDP.

As with other risk-analysis experience, such studies may suggest implications different in kind from those currently emerging from the highly politicized suggestions that characterize the discussion of this topic. To what extent should interventions focus on human driving forces, such as population increases? How much can be gained from global transfers of funds from military programmes to society–nature interactions? Do the various programmes in progress under Capacity 21, coordinated by the

United Nations Development Programme (UNDP 1996), really hold significant potential for enlarging national risk-management capacities and reducing international insecurities? And what about the overall fabric of existing and envisioned global environmental risk management institutions – UNEP, WHO, the World Bank, the Global Environmental Facility – how well will such institutions be able to anticipate and ameliorate emerging environmentally based international tensions? How should they be changed to overcome current limitations, and what new institutions will be needed?

Looking to the future

This chapter is intended to suggest a structured overview of issues that must be confronted if society is to deal more adequately with the ongoing environmental changes that threaten global well-being. It is abundantly clear at the dawn of the twenty-first century that the past is an imperfect guide to the future. Accumulated and growing environmental stress at the global scale is increasingly dominating risk portfolios at national, regional, and local scales, signalling a mismatch between institutional jurisdictions and future risk challenges. The burden on society to identify and anticipate such changes in advance of the experience of their impact has grown enormously, and it is apparent that future surprises will be numerous and wrenching. At the same time, it is also clear that the international political economy and existing governance arrangements, to which the various international conventions and institutions look for answers, are deeply implicated as sources of the problems. The changing role of the state and the growing power of NGOs and scientific "epistemic" networks point to the possibility of major changes in the structure of international regimes addressing environmental problems. Such regimes will need to overcome the continuing, perhaps growing, divergence between the rapid pace of global environmental change and the much slower rate and desultory nature of societal response outlined in chapter 8 if we are to succeed in anticipating global risks and sustain the planetary environment.

Note

1. The debate over appropriate usage of the terms "hazards" and "risks" has subsided, but it is unresolved. In this chapter, we use the term "risk" throughout in its broadest sense, encompassing both the recognition of a threat, often distinguished as a "hazard," and the more precise knowledge of the probability of occurrence denoted by the specific usage of risk.

Part One

Characterizing global environmental risks

Editors' introduction

Are global environmental risks really new risks? If so, in what sense are they new, or how do they differ from risks that societies have faced throughout history? Or is it the global political context in which such risks occur, the fragmented management setting, and the weak international institutions that differ so markedly from other risk situations?

This section addresses the nature of global environmental risks and how they may best be characterized. In the field of risk analysis, the 1990s have been an active period of exploring ways of characterizing and comparing risks. Rightly so, because progress in our ability to identify the essential properties of a given risk (including its perceptual, social, and economic dimensions) is the basis for more systematic and well-founded social engagement of risk. Nearly three decades after the United Nations Conference on the Human Environment in Stockholm, the extent to which societies throughout the world still deal with environmental and health threats in a *risque-du-jour* fashion is striking. The conclusion from the self-examination by the US Environmental Protection Agency (USEPA 1987) – that public outcries, media attention, and legislative concerns drive risk-management efforts more than the agency's own assessment of what problems most need attention and are most amenable to control – is a telling result. No less striking has been the self-assessment of the UK Interdepartmental Liaison Group on Risk Assessment (1998) that broader approaches are needed to capture the concerns of the public in risk matters. How best to manage risks is a matter for societal debate and introspection and undoubtedly varies with economic and cultural settings, but it is clear that global environmental risks will rapidly eclipse human capacity to respond in a timely and effective way unless such risks are anticipated (to the extent possible), characterized for the benefit of those who must make decisions, compared with other risks competing for limited societal resources, and accorded priorities in the allocation of public and private efforts across the globe. Indeed, the newly established Institute for Global Environmental Strategies (IGES 1999; IGES and Environmental Agency of Japan) in Japan has identified the need for a clearer picture of the future and for planning anticipatory counter-measures to risk among the major environmental messages of the twentieth century.

But what are the distinctive properties, if any, of these risks? What are the risk attributes that need to enter into characterization and priority-setting activities? Can indicators be found by which characterization can proceed? And what are the limits and areas of uncertainty and ignorance that risk analysts and decision makers need to observe if they are to avoid inflicting more harm and generating unintended effects rather than securing a safer and healthier planet? And, finally, with what other risks should we compare global environmental risks to learn from past experiences? The three chapters that follow take up these tasks and shed important light on directions for the improved risk analysis to which this volume aspires.

In chapter 2, Norberg-Bohm and her colleagues take on a most difficult challenge – can a framework be found, based upon sound and replicable methods, by which various environmental risks can be compared and ranked? It will need to be an approach applicable to developing countries, where the need for priority setting is greatest but where securing needed data is most problematic. It will also need to be robust in facilitating international comparisons despite the enormous differences among countries. Drawing upon a causal model of hazard and a taxonomy of technological hazards (Kates, Hohenemser, and Kasperson 1985), and adding the new element of "valued environmental component" to this past work, they recognize 18 descriptors of risk which they apply to 28 different environmental problems. The system for characterizing hazards developed from this exercise is designed to compare risks *within* and *between* countries, and has the virtue of doing this with high transparency (i.e. value judgements are apparent). Four country studies (treating India, Kenya, the Netherlands, and the United States) demonstrate the applicability of the risk-characterization system. Since this work is intended as a methodological contribution aimed at stimulating other efforts, we include in the appendices extensive detail on the methods used and guides for the analyst.

Smith takes a different approach in chapter 3 where he focuses on what he terms "risk transitions" in developing countries. The panoply of risks confronting a developing country at any point in time is an amalgam of risks arising in three transitions – a demographic transition, an epidemiological transition, and a risk transition. The demographic transition involves the shift from high birth and death rates to low birth and death rates; the epidemiological transition entails a substantial reduction in traditional diseases and effects of natural hazards and a rise in modern diseases; and the risk transition, driven by the other two transitions, involves a decoupling of risk from ill-health and a growth of risk latency periods and uncertainties. The transition risks typically include the traditional risks that have long confronted such societies – poverty, natural disasters, infectious disease, and epidemics – but they also include the modern risks, such as degenerative diseases, risks with long latency periods, technologies with low-probability/high-consequence risks, and the risks associated with transboundary activities. Since these risks often overlap and interact in new synergistic combinations, they confront developing societies with perplexing new management challenges and new needed capabilities. Risk analysis in such contexts, Smith argues, needs to focus on *net risk*, in order to capture the arrays of changes contributing to risk lowering and risk raising. However, scales and risk distributions will also need close scrutiny, as risk transfer and risk redistribution are capable of producing unforeseen and nasty surprises. None the less, risk assessment aimed at characterizing risks during transitions and identifying net-risk-lowering strategies holds promise not only for more robust anticipatory analyses of risk but for charting comparative developmental pathways for developing countries.

Funtowicz and Ravetz take yet a different tack in chapter 4. Appreciating that global environmental risks often appear with non-linear or chaotic characteristics, scale interactions, and long time horizons over which they unfold and register consequences, they centre upon levels of systems uncertainty and intensity of decision stakes as defining attributes for affording insights into the special challenges of global environmental risks. Along with the editors of this volume, they seek a risk analysis geared to the distinctive problems posed by the nature of the risks. In particular, they argue that the customary use of traditional scientific methods is unlikely to capture the essential elements of these risks, primarily because such methods have limited capacity to provide answers or guidance for problems that are highly uncertain and politically contentious. These are key characteristics that decision makers in various countries have found very troublesome and which appear to be particularly prone to the social amplification of risk discussed in chapter 1. Their chapter explores, therefore, the use of what they term "post-normal science" to characterize and assess this special class of risk.

Within the context of this volume, several issues are particularly note-worthy concerning the Funtowicz and Ravetz discussion. First, it would be highly instructive to relate global environmental risks using the di-mensions and aggregation schemes set forth by Norberg-Bohm and col-leagues to the different types of risk assessment recognized by Funtowicz and Ravetz. It would appear that certain of Norberg-Bohm and col-leagues' aggregations, such as "disruption" or "manageability," might provide more powerful defining dimensions than systems uncertainty. Secondly, it is apparent that many of the difficult global environmental risks fall in the domain of high uncertainty and high decision stakes where conventional risk assessment, in the Funtowicz and Ravetz view, is most limited. So they argue for a kind of risk assessment – with broader participants, a new role for risk bearers, diverse peer review – entirely in keeping with the more democratic and humanistic risk analysis envi-sioned in chapter 1 for global environmental risk but also for assessing contentious and socially amplified risks more generally.

Taken together, these chapters provide important directions for the future development of approaches and methodologies for characterizing global environmental risks. It is clear that assessing global environmental risks will require full and creative use of the array of sound methods and techniques that currently exist in risk analysis and, as chapter 1 makes clear, these are considerable. At the same time, the three chapters pro-vide a powerful critique of any approach based on a narrow, technical conception of risk. "Business as usual" will not suffice for elucidating global environmental risks and may serve only to mislead well-intentioned management initiatives. Developing the new risk analysis, which builds upon the directions pointed to in this section, is a high priority for the next generation of risk analysts.

2

International comparisons of environmental hazards

*Vicki Norberg-Bohm, William C. Clark,
Bhavik Bakshi, Jo Anne Berkenkamp,
Sherry A. Bishko, Mark D. Koehler,
Jennifer A. Marrs, Chris P. Nielsen, and
Ambuj Sagar*[1]

The last decades of the twentieth century have witnessed a growing awareness of not only the severity but also the diversity of environmental problems. Global environmental concerns such as ozone depletion, climate change, and biodiversity loss have stimulated international environmental actions. Regional and local problems – such as acid rain, depletion of renewable resources, and drought – have also been the subject of public debate and governmental action. An increasing number of environmental problems compete for places on political agendas, for the attention of regulatory agencies and international governmental bodies, and for the limited resources available for environmental management.

This rapid proliferation of environmental concerns, as noted in the Introduction to this volume, poses a number of challenges for the strategic design of good public policy. Most obviously, to quote a former administrator of the US Environmental Protection Agency (USEPA), "In a world of limited resources, it may be wise to give priority attention to those pollutants and problems that pose the greatest risk to our society" (USEPA 1987: ii). More broadly, as senior officials of the United Nations Environment Programme (UNEP) have long argued, we need a systematic understanding of how the severity and nature of environmental problems vary from place to place around the world. This need can only become more pressing as bilateral and international negotiations increasingly take on environmental dimensions.

By and large, today's world is poorly equipped to meet these strategic challenges. True, we have more data on environmental problems than ever before. A variety of international and national compendia (e.g. *Environmental Almanac, Global Environmental Outlook, OECD Environmental Data, Environmental Data Report, State of the World, World Resources*) are now available and are being updated on a regular basis. The advance of the Internet has swelled the availability of on-line environmental information. In the United States, the National Council for Science and the Environment (formerly the Committee for the National Institute for the Environment, CNIE), has posted numerous international, national, regional, and local environmental reports on its web site (http://www.cnie.org). Many of these reports are electronically accessible via the likes of UNEP's Global Resource Information Database (GRID) in Arendal, Norway (http://www.grida.no). In 1991, the Organization for Economic Co-Operation and Development (OECD) launched its series of country-level "environmental performance reviews" as a means of assisting member nations "to improve their individual and collective performances in environmental management" (OECD 1997a: 5). The 1992 *World Development Report, Development and the Environment*, focused attention on the environmental priorities that directly affected the welfare of the largest number of people in the developing world. Moreover, a great deal of progress has occurred in identifying policy instruments that should improve the effectiveness and efficiency with which specific environmental problems are managed (e.g. O'Riordan et al. 1998; OTA 1995). In practice, however, these important advances have left the strategic task of setting priorities among environmental problems largely unaddressed. Indicators vary from country to country: some focus on sources of environmental degradation, some on effects, and some on the state of the environment *per se*. As a result, meaningful comparisons are generally impossible. Resource allocation sporadically tracks successive "problems of the month," responding more to media attention and political grandstanding than to any more fundamental criteria. The lack of common definitions and metrics left the United Nations 1992 Conference on Environment and Development (UNCED) hardly more able to conduct systematic international comparisons of environmental problems than was its forerunner in Stockholm 20 years earlier.

Much needs to be done to mitigate these shortcomings of current practice. The methodological framework proposed in this chapter is intended as one modest contribution to the required effort. In brief, we have attempted to design a framework that can help both to define the data most relevant for priority setting and international comparisons among environmental problems and to structure those data in a way that makes them usable to a variety of participants in policy making. We have used data from a study conducted in 1988–1991. Experience in a number of other policy fields, most notably macroeconomics, suggests that such strategically defined data frameworks constitute essential foundations for long-term improvements in management performance (Hutchison 1977).

Reflecting the general recognition of this fact, several efforts have already been made to develop strategic frameworks for evaluating priorities in environmental policy. As part of a retrospective assessment of the world's environment ten years after the United Nations Conference on the Human Environment in Stockholm, Gordon Goodman developed a set of indicators for characterizing and ranking, at an aggregate global level, environmental problems around the world (Goodman 1980). At roughly the same time, scholars at Clark University's Center for Technology, Environment, and Development (CENTED) began their seminal research on the characterization of health and environmental hazards arising from the use of technology (Hohenemser, Kates, and Slovic 1985; Kates, Hohenemser, and Kasperson 1985). At a national level, in 1986 USEPA began an effort to compare the risks posed by more than 30 environmental problems faced by American society (USEPA 1987). In a review and follow-on study, the Science Advisory Board of USEPA highlighted and worked to overcome several of the shortcomings of the original study (USEPA 1990). Subsequent studies by the United States National Research Council (Stern and Fineberg 1996; USNRC 1994) have addressed means of improving risk characterization and decision-making.

Although our work proceeded simultaneously, we have found that, in developing the method presented in this chapter, we also identified and tried to overcome many of the same shortcomings and concerns that have been identified by the Scientific Advisory Board of USEPA. These included the inconsistencies created by basing definitions of an environmental problem on legislation and programmatic organization; limiting hazards to those under the jurisdiction of USEPA and thus not addressing the full range of environmental hazards; the need to emphasize the important role that values play in risk comparisons; and the importance of considering temporal and spatial dimensions of hazards in an analysis of relative environmental risk.

Research over the past several years has shed much light on the problem of establishing priorities. A debate has ensued on the role of comparative hazard assessment in policy-making (Finkel and Golding 1994). Some efforts have been under way to apply comparative hazard assessment at the state and local level (Davies J. C. 1996) and to improve the methodology for comparative hazard assessment (Morgan et al. 1996, 1997). Furthermore, a growing interest in life-cycle assessment has generated new efforts to develop methods for weighting and comparing environmental hazards (Baumann and Rydberg 1997; Ehrenfeld 1997;

Owens 1997). However, none of the existing approaches is directly applicable to developing countries where the needs for priority setting are arguably greatest, whereas the scarcity of data is the most problematical. Moreover, none allows international comparisons to be carried out in the face of the vast differences among nations – both in the importance of specific environmental hazards and in the values relevant to their evaluation. Smith, in the chapter to follow, shows how risks change with the development process. The approach presented here is an effort to begin constructing a framework for characterizing environmental hazards that addresses these unmet needs and thus complements existing efforts.

The remainder of the chapter is organized as follows. We first outline our framework for comparing hazards, with emphasis on its logical structure, the set of environmental hazards covered, and the indicators used to characterize them. We then present an application of the framework to the characterization and comparison of hazards in four countries (India, Kenya, the Netherlands, and the United States). Here the emphasis is on alternative ways of incorporating values in the aggregation and comparison of indicator values. Our tentative conclusions regarding the framework, its strengths and limitations, and its possible applications follow. Presenting this approach in its present preliminary state of development is not to suggest definitive rankings among problems or countries; rather, our principal purpose is, in the spirit of this volume, to explore and encourage the use and further development of the comparative-hazard approach by others. Put somewhat differently, we are interested more in producing a useful tool and an illustration of what can be built with it than in the particular demonstration piece we construct along the way. We include with this chapter, therefore, a set of appendices that provide sufficient details to allow independent application and evaluation of this comparative-hazard approach.

The method

An overview

Early efforts at comparative-hazard assessment were often based on single measures of hazardousness, such as human mortality. Such simple characterizations proved to be of limited use in promoting productive debates on public policy for at least two reasons. First, data on consequences in general, including mortality, are often much less available and more controversial than data about other dimensions of hazard – for example, emissions or exposure. Second, empirical studies have shown that, even when consequence data are available, many people give strong weight to other aspects of hazardousness in their informal assessments. Analytic approaches that neglect these "other aspects" have often found it difficult to establish their relevance to (or legitimacy in) the formulation of public policy. In recent years we have, therefore, witnessed a trend towards the adoption of a more varied and complex set of measures for comparative-hazard assessment (Covello 1991; Hohenemser et al. 1983). Unfortunately, the potential benefits of this variety have been offset by its apparent arbitrariness – complex sets of indicators have had little obvious rationale or cohesion.

A rich but systematic approach that largely avoids these difficulties was developed for comparing technological hazards by researchers at CENTED at Clark University (Kates, Hohenemser, and Kasperson 1985). We have adapted CENTED's approach, particularly its "causal taxonomy" of hazard, for the comparative environmental assessment presented here. In the following discussion we present a detailed discussion of our causal taxonomy for environmental hazards and its advantages over simpler approaches to hazard assessment. First, however, it may be helpful to introduce our approach by providing a broad outline of its basic components.

We began with a simple linear model of hazard causation, anchored at one end by the human activities or natural events that perturb the environment and at the other by consequences of those perturbations for people and things they value. This "causal model," derived from the causal model of Hohenemser, Kasperson, and Kates (1985), is presented in figure 2.1. It provides the "backbone" against which we define a number of key characteristics of environmental hazard – for example, persistence of a pollutant in the environment, the population exposed, and health consequences. For each of these characteristics, we next defined specific measurable "descriptors," and normalized them on a scale of 1 to 9. We have developed a total of 18 descriptors that are used to characterize each environmental hazard. The same model structure and descriptor definitions are applied to all hazards characterized through the framework (see appendix A for a full discussion and for several technical exceptions).

Next, we identified the specific set of potential environmental problems to be characterized. These were selected according to three criteria: (1) that they be understandable and meaningful to policy makers and other non-specialists; (2) that they be comprehensive, so as to include virtually

Fig. 2.1 **The causal structure of environmental hazards**

all the potentially significant environmental problems faced by any nation; and (3) that they be defined at a level of resolution yielding a list neither too general nor too complex for practical utility. A group of 28 problems, ranging from biological contamination of fresh water to depletion of stratospheric ozone, was ultimately selected for characterization in this study.

Finally, for each hazard, in each country or other region evaluated, we used data available at the time (1988–1991) to assign each descriptor a score. The method, therefore, results in a matrix of 18 by 28 scores – 504 in all – for each country studied. Such a large set of numbers is not particularly informative by itself; rather, useful interpretation requires aggregation of the descriptors. The choice of how to weight or group descriptors is essentially a choice about what aspects of hazardousness will be used for comparing and prioritizing environmental hazards. The crucial input to the aggregation process is not scientific data but human values. How should future consequences be weighted relative to current ones? How should degree of perturbation to natural cycles be weighted relative to known impacts on human health? Although our approach does not attempt to answer such value-laden questions, it does provide a transparent and flexible method that can help users to evaluate the implications of alternative value assumptions for the ranking of environmental hazards. Examples of several value positions often adopted in debates on environmental priorities are explored below.

The causal structure of environmental hazards: A framework for analysis

Based on the causal structure of environmental hazards illustrated in figure 2.1, any environmental problem is characterized by a sequence of events that can be summarized as follows: natural factors or human activities (e.g. energy production) lead to changes in material and energy fluxes (e.g. release of sulphur dioxide), which in turn lead to changes in components of the environment that society values sufficiently to single out for concern (e.g. acidity of precipitation), which in turn leads to exposure of, and consequences for, humans and things they value (e.g. forest die-back). The actual ontogeny of real environmental problems is, of course, a much more complex affair, replete with feedbacks and other non-linearities. For purposes of the present framework, however, the simpler model suffices.

The model starts with natural phenomena or *human activities* that are the initial sources of changes in the environment. The human activities are defined by both the choice of technology and level of its use.

Human activities or natural phenomena lead to measurable changes in chemical flows or in other physical or biological components of the environment, which are together described as *changes in the material fluxes*. In the case of pollution-based environmental problems, this refers to an increase (or decrease) in chemical constituents; in the case of renewable-resource depletion, this describes a decrease (or increase) in the stocks of plants or animals; for natural hazards, this describes a change in the accustomed or usual flow of materials or energy, noting that the accustomed flow may be zero.

Changes in these material fluxes then lead to changes in *valued environmental components (VECs)*, which are most simply described as those attributes of the environment that humans choose to value. The important concept of VECs was articulated by Beanlands and Duinker (1983) in their retrospective review of environmental impact assessment experience in Canada. They argued that successful assessments had forsaken the early *checklist* objectives of assessing *everything important* about the environment in favour of much shorter lists of especially valued components of the environment. They emphasized that these VECs were ultimately social, rather than scientific, constructs: that is, they reflected particular aspects of the environment – sometimes, but not always, highlighted by scientific research – that people choose to value. Where society has been unwilling to choose a modest number of such environmental components on which to focus its management attention – or where environmental-assessment procedures have made such focus impossible – Beanlands and Duinker show that effective policy has seldom evolved. In practice, VECs turn out to be the largish categories of environmental problems that appear in the table of contents of national *state of the environment* reports, or as the subject of debates on the floors of parliaments. In general, we value these components not in and of themselves (although this point would be debated by the deep ecologists), but because changes in them might lead to downstream consequences more directly bearing on human or ecosystem welfare.

Exposure is the pathway by which changes in VECs inflict downstream consequences. For example, pathways of exposure to pollution that cause human health consequences include inhalation, ingestion, and dermal contact.

Consequences for humans and things we value are defined here as increased risks to human life and human health, to ecological systems, and to productivity and the physical infrastructure. In this taxonomy, we have included loss of species as a consequence of value to humans and thus will not define it separately as an environmental problem. As discussed below, we have defined our list of hazards in terms of changes in VECs. Because changes in many different VECs can lead to the loss of species, within the framework of this study loss of species is not considered to be a single environmental problem.

As an example, figure 2.2 shows the causal structure of stratospheric ozone depletion. This figure demonstrates that, in practice, the structure may be expanded to include more than one stage of any of the components of the taxonomy. In the case of ozone depletion, a change in one VEC (increased CFCs in the stratosphere) leads to a change in another (decreased ozone in the stratosphere). The example of ozone depletion also shows that, in practice, the taxonomy may look more like a pitchfork than a simple chain. In this case, changes in the concentration of several ozone-depleting gases lead to a decrease in ozone in the stratosphere, and the change in the VEC of UV radiation on the earth's surface leads to several different consequences.

One of the advantages of using a causal structure as a framework for analysis is that it provides a valuable link between the functions of hazard assessment and hazard management (Hohenemser, Kasperson, and Kates 1985; Kates, Hohenemser, and Kasperson 1985). The causal taxonomy clearly delineates the sequence of events that create a hazard, thus providing a guide to available points for intervention. For demonstration purposes, we look once again at the case of ozone depletion. Figure 2.3 shows that environmental problems can be managed by modifying human activities, altering environmental processes, preventing exposure, or preventing consequences. Although each of these points of intervention may not be equally viable, the causal taxonomy is a valuable aid in conceptualizing the options. We view the further development of this tool to provide more information relevant to management as one of the key next steps (see, for example, Clark 1991).

The set of descriptors

For each stage in this causal structure, we have developed several "descriptors" that characterize environmental problems. The causal structure and descriptors together form a causal taxonomy for the evaluation of environmental hazards, as presented in figure 2.4. The descriptors are summarized in table 2.1. More in-depth definitions, measurement scales, and applications of these descriptors are provided in appendix A. Many of our descriptors have logarithmic scales, in some instances in base 2 and in others in base 10. As noted in the CENTED study, this is justified on the basis of both the nature of human perception and the uncertainty of measurements (Hohenemser, Kasperson, and Kates 1985; Kates, Hohenemser, and Kasperson 1985).

This causal structure is a particularly valuable framework for hazard

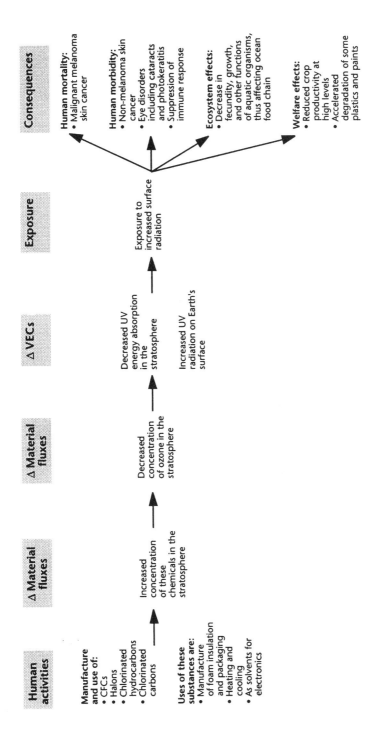

Fig. 2.2 **The causal structure of stratospheric ozone depletion**

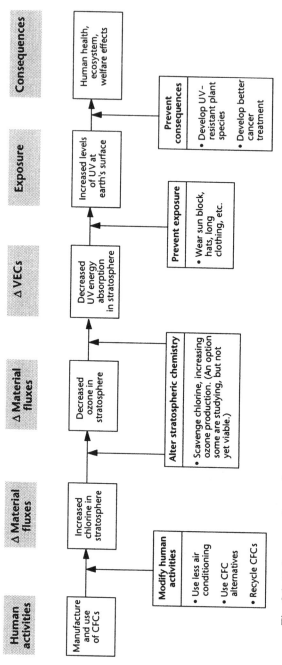

Fig. 2.3 **Relevance of causal structure for hazard management: The example of stratospheric ozone depletion**

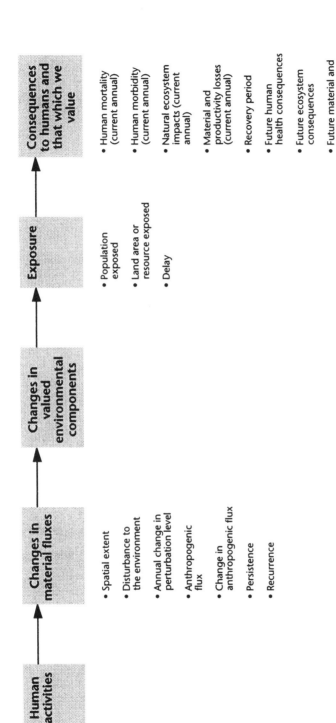

Fig. 2.4 Causal taxonomy for comparative assessment of environmental hazards

Table 2.1 **Descriptors**

Material flux descriptors

1. *Spatial Extent*. Measures the spatial extent of a single release for which there is a significant change in the material levels in the environment.
2. *Disturbance to the Environment (Magnitude of Perturbation)*. Measures the degree to which a change in the material of interest (pollutant or resource) is above or below the background, normal, or long-term average levels.
3. *Annual Change in Perturbation Level*. Measures the annual change in the perturbation level.
4. *Anthropogenic Flux*. Measures the extent to which human activities contribute to the total flux of materials into or out of the environment, i.e. compares the anthropogenic flux with the natural flux.
5. *Annual Change in Anthropogenic Flux*. Measures the annual rate of change in the anthropogenic flux.
6. *Persistence*. Measures the time period required for the material most directly perturbed in the environment to return to the level existing before the event that caused a flux of that material.
7. *Recurrence*. Measures the frequency of the event that causes a flux.

Exposure descriptors

8. *Population Exposed*. Measures the percentage of current population within a country that is exposed to the change in a valued environmental component (VEC).
9. *Land Area or Resource Exposed*. Measures the percentage of land area or resource within a country that is currently exposed to the changed VEC.
10. *Delay*. Measures the delay time between the release or altering of materials and the earliest onset of serious consequences.

Consequence descriptors

11. *Human Mortality (Current Annual)*. Measures the average annual percentage of population that dies as a direct result of the hazard.
12. *Human Morbidity (Current Annual)*. Measures the average annual percentage of population that becomes significantly ill from the hazard.
13. *"Natural" Ecosystem Impacts (Current Annual)*. Describes the impacts to "natural" elements within ecosystems.
14. *Material and Productivity Losses (Current Annual)*. Measures the average annual losses of material and non-labour productivity.
15. *Recovery Period*. Measures the time period required to recover from the effects of the relevant material or energy flux to the background state of human health, ecosystem, and productivity, assuming that the causative human activity is stopped.
16 *Future Human Health Consequences*. Describes the severity of harm to future health caused by today's human activities.
17. *Future Ecosystem Consequences*. Describes the severity of harm to future ecosystems caused by today's human activities.
18. *Future Material and Productivity Losses*. Describes the severity of harm to future material and non-labour productivity caused by today's human activities as a percentage of gross national product (GNP).
19. *Transnational*. Describes the nations in which human activities lead to the consequences.

assessment because it reflects the multidimensional nature of environmental hazards. Considering, first, only consequences, we are concerned about human mortality and morbidity, ecosystem damages, and material losses. These different types of consequences, however, cannot be adequately measured in a common metric. Furthermore, when considering consequences, we care both about current effects and future effects, but the correct valuation of effects occurring now versus those expected in the future is hotly debated. Much evidence indicates that policy makers would rather have information about several relevant aspects of a problem than a single number (e.g. a cost–benefit ratio) composed of either detailed or unspecified assumptions (Ezrahi 1990). It is, therefore, useful to have a method such as this, in which these different aspects of consequences are disaggregated and in which any aggregation scheme is transparent.

The foregoing discussion explains the need for a range of consequence descriptors but still leaves open to question the usefulness of descriptors at each stage in the causal chain. There are three major reasons for including this entire set of descriptors. First, considerable uncertainty and ignorance exist regarding the consequences of environmental hazards. Comparatively speaking, we have a much better understanding of the magnitude of human perturbations of the environment. Thus, we generally have better and more accurate information for the descriptors to the left of our causal chain (the material flux descriptors) than for those to the right (the consequences). In fact, estimates of consequences depend on estimates at each stage of the causal chain, and therefore the errors and uncertainties in each stage are compounded in the measure of consequences (see also Schneider 1983). Owing to this uncertainty and ignorance, many environmental problems are in fact managed and even assessed in terms of the disruption that they cause to natural processes. One strength of this method is that it allows comparison of problems based both on perturbations to the environment and on known consequences.

A second reason for using this full causal taxonomy for comparative hazard assessment is that public perceptions of risk depend not only on consequences but also on factors such as controllability, knowledge, and dread (Slovic 1987; Slovic, Fischhoff, and Lichtenstein 1985). Although the method does not describe the full range of variables important in risk perception, several of our descriptors – such as persistence, frequency of occurrence, and spatial extent – have a direct bearing on these factors.

A third reason reflects an interest in characterizing as well as ranking environmental hazards, in order to ascertain whether particular patterns in the types of environmental hazards exist that cause the greatest consequences. Characterization is also important as one moves from hazard assessment to hazard management. The causal structure and its associated descriptors provide a method for the full characterization of environmental hazards.

The set of environmental problems

In principle, environmental problems can be defined in terms of their manifestation at any point along the causal chain described above. The most conventional approach focuses on important environmental changes *per se* (e.g. all activities leading to, and consequences resulting from, the depletion of stratospheric ozone). An alternative approach, which in recent years has been advocated by many concerned with environmental management, is to define problems in terms of the sectors of human activity that lead to environmental disruption (e.g. the environmental problems of energy use). The advantages of this approach have been clearly articulated by the Brundtland Commission, which emphasized that, in order to reduce environmental damage, one had to consider environmental consequences when undertaking activities in all sectors of the economy (WCED 1987). For managing a well-defined set of priority environmental problems, we side with the Brundtland Commission in advocating the second, or source-oriented, approach; however, for the more fundamental comparisons and priority-setting tasks that are our principal concern here, we have determined that the conventional "environment-centred" approach is most useful.

As mentioned above, we wanted our set of environmental hazards to be meaningful to the environmental policy debate, to do justice to the great variety of environmental concerns expressed around the world, and to be defined at a level of resolution that is neither too general nor too complex. Two other criteria were that they be internally consistent and justifiable on the basis of both theory and experience. We met these criteria by defining our list of problems by the change in a VEC, and not by some other point in the causal chain. This key decision is best explained by walking through an example such as climate change, as illustrated in figure 2.5. It is not the human activity of fossil-fuel burning that is the problem. In fact, such activities represent benefits to society as well as sources of environmental problems. Likewise, the change in the carbon cycle does not concern us except as it changes a component of the environment we care about – in this case, the thermal radiation budget. Moving further down the causal chain, exposure is the link between changes in environmental components and the consequences we care about. It holds no promise as a unique definition of an environmental problem. Similarly, consequences for humans and the things we value are not unique definitions of a single problem.

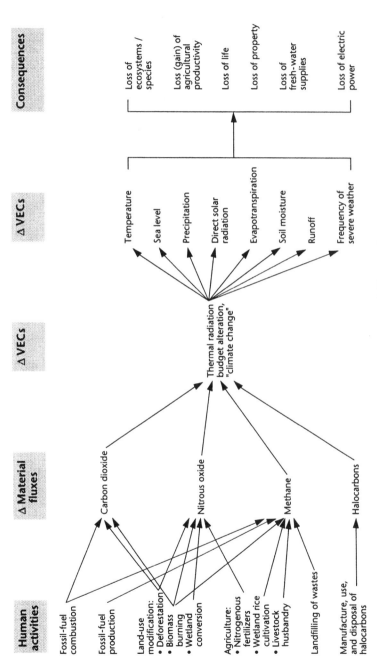

Fig. 2.5 **The causal structure of climate change**

69

In sum, defining environmental hazards by the VEC allows us to develop a comprehensive yet non-overlapping set of environmental hazards. Of equal importance in our choice to define the environmental hazards by VECs was practical experience in environmental-impact assessment. A review of environmental-impact assessments in Canada, for example, concluded that the most effective assessments were based on an agreement on which aspects of the environment were to be valued, that is on a clearly defined set of VECs (Beanlands and Duinker 1983).

The list of environmental problems finally chosen for this analysis appears in table 2.2. This list includes pollution-based hazards, depletion

Table 2.2 **Environmental hazards**

Pollution-based hazards
Water quality
1. Fresh Water: Biological Contamination
2. Fresh Water: Metals and Toxic Contamination
3. Fresh Water: Eutrophication
4. Fresh Water: Sedimentation
5. Ocean Water
Atmospheric quality
6. Stratospheric Ozone Depletion
7. Climate Change
8. Acidification
9. Ground-Level Ozone Formation
10. Hazardous and Toxic Air Pollution
Quality of the human environment
11. Indoor Air Pollutants: Radon
12. Indoor Air Pollutants: Non-radon
13. Radiation: Non-radon
14. Chemicals in the Workplace
15. Accidental Chemical Releases
16. Food Contaminants

Resource-depletion hazards
Land/water resources
17. Agricultural Land: Salinization, Alkalinization, Waterlogging
18. Agricultural Land: Soil Erosion
19. Agricultural Land: Urbanization
20. Groundwater
Biological stocks
21. Wildlife
22. Fish
23. Forests

Natural hazards
24. Floods
25. Droughts
26. Cyclones
27. Earthquakes
28. Pest Epidemics

of natural resources, and natural disasters. We recognize that these are often considered to be quite different types of problems that are managed under different legislation or by different agencies. None the less, it is crucial to include all three types of environmental problems, because each is likely to be among the most hazardous in some countries. Appendix B describes these problems in terms of their causal structure.

Interpreting the data: Aggregation schemes

As noted above, the application of this method to a single country results in 504 scores – 18 descriptors scored for each of 28 environmental problems. In order to render this information useful for comparing, ranking, and characterizing hazards, data reduction is necessary.

Various sophisticated data-reduction methods are available, some attempting statistically efficient representations of information content, others seeking accurate representations of particular users' concerns and preferences. We used one such method, principal-component analysis (PCA), which is often helpful in interpreting multidimensional data sets. PCA sometimes reveals variable groupings that are not intuitively obvious but which, upon further reflection, provide meaningful results. In our case, to the limited extent that the PCA pointed toward tight groupings of descriptors, it largely validated the intuitively meaningful approaches to data reduction that we had previously developed. This result, along with our ultimate concern for usefulness and transparency of the method, has led us to adopt some very simple aggregation schemes in the work reported here. Appendix E presents the PCA.

In the final analysis, all data-reduction or -aggregation schemes incorporate value judgements. Aggregating consequence descriptors, for example, requires one to decide how to value human health, ecosystem, and economic effects in relation to each other. In our interpretation of the country studies below, we present several aggregation schemes that we found particularly informative. Interpretation of the data, however, is not limited to these particular aggregations, nor does the method prescribe particular aggregation schemes. One of the strengths of this method is that it allows other users to explore prioritizations based on their own sets of values and to communicate clearly those value judgements and their implications along with the conclusions. We emphasize that this can and should be a two-way process. Learning what our environmental values are and ought to be requires a means of examining and reflecting upon their implications for the ordering of specific real-world priorities (Wildavsky 1979b). This method can contribute to the necessary dialectic between general environmental values and specific environmental concerns.

Having pleaded for useful simplicity, however, it is clear that a fine line exists between the usefully simple and the misleadingly simplistic. With data such as those generated in the scoring of the case studies, it is tempting simply to add the scores for all 18 descriptors of each problem, creating a "total hazard score." We discourage such an approach: a simplistic summation implies that each descriptor merits equal weighting in the determination of rank, an assumption for which no basis exists. There is another fundamental reason for avoiding a simple summation. For a majority of the hazards, a normalized score of 9 will imply a more hazardous problem than a score of 1, but this is not necessarily the case. This is particularly true for Recurrence and Delay. For example, there is no inherent reason to believe that a problem that recurs more frequently is more hazardous than one that recurs less frequently. In our interpretation of the case studies, we show how these particularly tricky characterizing descriptors can, none the less, be useful components of aggregation schemes. The representative aggregation schemes we present seek to retain simplicity and transparency without falling into such simplistic traps.

Throughout the analysis presented below, the method is used to rank problems in three categories – high, medium, or low hazard. Although the problems could be ranked numerically in each category, such a detailed ranking demands of the method more precision than it was designed to deliver and places more confidence in numerical scores than the data generally justify. Rather than developing an ordinal ranking, the method best serves its purpose by distinguishing the problems that present the greatest hazard from those presenting intermediate or low hazards. "High" and "Low" ranked problems are defined, respectively, to reflect approximately the top and bottom quartiles of the 28-item list of potential hazards.

Some details are necessary. In defining quartiles, if several problems had tie scores at the 25 per cent cut-off, we used the following rules. For the upper quartile (problems ranked "high") we included all hazards tied at the 25 per cent cut-off line; for the lower quartile (problems ranked "low") we excluded all hazards tied at the 25 per cent cut-off line. One exception to this is the aggregation scheme **Total Ecosystem Consequences**, for which the aggregate scores fall into three groups. For example, in the US case, three problems have aggregate scores of 14, fourteen problems scored 10, one problem scored 6, and ten problems scored 2. In this case, the above-defined rule for "cut-off" points for quartiles would lead to identifying 17 problems as highest ranked, a rather meaningless categorization. Similar results were found in the other case studies. Therefore, for the aggregation scheme **Total Ecosystem Consequences**, we chose to define high, medium, and low by the break points in the scoring: a score of 14 or above indicates a high hazard, a score of 6 or 10 is a medium hazard, and a score of 2 is a low hazard. This points to one of the shortcomings of the method – the descriptors for ecosystem consequences simply could not differentiate with the degree of sensitivity we would like. Further development of ecosystem descriptors that can both provide greater differentiation and remain scorable would contribute greatly to efforts for comparative hazard assessment.

Limitations of the method

Because this method was developed for hazard assessment rather than hazard management, it is important to stress that it provides only one of the key inputs into the process of allocating resources for managing environmental hazards. In this sense, our effort is similar to the USEPA study *Unfinished Business* (USEPA 1987), in that it focuses on characterizing the hazard or harm and does not perform the following tasks that are also required for setting policy agendas and managing hazards: (1) to evaluate the economic costs or technical potential for controlling the risks; (2) to quantify or list the benefits to society from the activities that cause the environmental risk; (3) to evaluate existing governmental efforts that have ameliorated or exacerbated an environmental problem; (4) to evaluate the full range of the qualitative aspects important to the public's perception of risk, including voluntariness, familiarity, or equity. Rather, the contribution of the method is its ability to highlight the environmental problems that may be "most" hazardous and to characterize the key differences in the set of environmental hazards that we are facing.

We also emphasize our awareness that the task that we have addressed – comparing environmental problems within and between countries – is of such breadth and complexity that efforts to accomplish it can probably never reach the standards of clarity, certainty, and accuracy to which we aspire. Our intent is only to contribute to this end. In the course of this project we made many difficult choices, and the method is neither as simple to apply as we had originally hoped nor as precise as we might like. Although the method employs a quantitative scoring process based on a seemingly clearly defined set of characteristics of hazards, informed judgement plays a significant role in the method's application. Indeed, assumptions and interpretations enter the process at almost all steps, from scoring the descriptors to choosing aggregation schemes. We want, therefore, to make certain that potential users of this method understand that it remains an imprecise tool. In order to pass on some of our experience to future users, we have included "tips on applying the method" in appendix C.

Application of the method: Four country studies

Choice of case studies

In order to demonstrate, evaluate, and refine the method outlined above, we have used it to characterize and prioritize the environmental problems in four countries – India, Kenya, the Netherlands, and the United States. The United States was chosen as our first case, owing to the large amount of information available on environmental hazards and environmental risk assessment. If the descriptors could not be scored well for the United States owing to lack of information, then we would have little hope of being able to score them for most countries around the world. The other countries were chosen on the basis of two factors – diversity and availability of information at the time of our research. In terms of diversity, these four countries represent a range of climates, population density, levels of economic development, and predominant economic activities. These variables are likely to have significant effects on the type of environmental problems facing a country. In addition, the United States and India are both large countries that have tremendous diversity in these variables within their own borders. Applying the method to four countries with major differences in key variables has tested both its scorability and its usefulness as a tool for comparative hazard analysis. Appendix C presents the data for each country.

Comparing hazards within a country: Results from the US case

We begin by presenting several different aggregations based solely on the consequence descriptors. We start here because consequences are widely used in hazard assessment, are easily understood, and are clearly critical variables for comparative hazard assessment. As argued in our presentation of the method, however, there are several characteristics of hazards other than consequences that provide important information for comparative hazard assessment. We end this section, therefore, by presenting several aggregation schemes involving other descriptors in the causal chain, demonstrating how these can also aid in priority setting. To make the discussion easier to follow, throughout the remainder of this chapter, descriptors are written in *italics* and aggregation schemes are in **bold type**. Table 2.3 provides a summary of all of the aggregation schemes.

Figure 2.6 shows the rankings of environmental hazards based on three different aggregations of consequences. The first column is the aggre-

Table 2.3 **Aggregation schemes used in interpreting data**

Aggregation scheme	Sum of the following descriptors
Total Human Health Consequences	11. *Human Mortality (Current Annual)* 12. *Human Morbidity (Current Annual)* 16. *Future Human Health Consequences*
Total Ecosystem Consequences	13. *"Natural" Ecosystem Impacts (Current Annual)* 17. *Future Ecosystem Consequences*
Total Material and Productivity Consequences	14. *Material and Productivity Losses (Current Annual)* 18. *Future Material and Productivity Losses*
Total Consequences	11. *Human Mortality (Current Annual)* 12. *Human Morbidity (Current Annual)* 13. *"Natural" Ecosystem Impacts (Current Annual)* 14. *Material and Productivity Losses (Current Annual)* 16. *Future Human Health Consequences* 17. *Future Ecosystem Consequences* 18. *Future Material and Productivity Losses*
Total Current Consequences	11. *Human Mortality (Current Annual)* 12. *Human Morbidity (Current Annual)* 13. *"Natural" Ecosystem Impacts (Current Annual)* 14. *Material and Productivity Losses (Current Annual)*
Total Future Consequences	16. *Future Human Health Consequences* 17. *Future Ecosystem Consequences* 18. *Future Material and Productivity Losses*
Pervasiveness	1. *Spatial Extent* 6. *Persistence* 8. *Population Exposed* 9. *Land Area or Resource Exposed* 15. *Recovery Period*
Disruption	2. *Disturbance to the Environment (Magnitude of Perturbation)* 3. *Annual Change in Perturbation Level* 4. *Anthropogenic Flux* 5. *Annual Change in Anthropogenic Flux*
Persistence in Time	6. *Persistence* 15. *Recovery Period*
Pervasiveness in Space	1. *Spatial Extent* 8. *Population Exposed* 9. *Land Area or Resource Exposed*
Manageability	7. *Recurrence* 10. *Delay* 19. *Transnational*

Environmental Problems	(a) Total Human Health Consequences	(b) Total Ecosystem Consequences	(c) Total Material/ Productivity Consequences
1. Fresh Water: Biological Contamination	●●●	●●●	●●●
2. Fresh Water: Metal and Toxic Contamination	●●●●●	●●●	●●●●●
3. Fresh Water: Eutrophication	●●●	●●●	●●●●●
4. Fresh Water: Sedimentation	·	●●●	●●●●●
5. Ocean Water	●●●	●●●	●●●●●
6. Stratospheric Ozone Depletion	●●●	●●●	●●●
7. Climate Change	●●●	●●●	●●●
8. Acidification	●●●	●●●	●●●
9. Ground-Level Ozone Formation	●●●	●●●	●●●
10. Toxic Air Pollution	●●●●●	●●●	●●●
11. Indoor Air: Radon	●●●●●	·	●●●
12. Indoor Air: Non-radon	●●●●●	·	·
13. Radiation: Non-radon	●●●●●	·	·
14. Chemicals in the Workplace	●●●●●	·	·
15. Accidental Chemical Releases	●●●●●	·	·
16. Food Contamination	●●●●●	●●●	●●●
17. Agricultural Land: Salinization, Alkalinization, Waterlogging	·	·	●●●
18. Agricultural Land: Soil Erosion	·	·	●●●●●
19. Agricultural Land: Urbanization	·	●●●	·
20. Groundwater	·	●●●	●●●●●
21. Wildlife	·	●●●●●	●●●
22. Fish	·	●●●●●	●●●
23. Forests	·	●●●●●	·
24. Floods	●●●	·	●●●●●
25. Droughts	·	●●●	●●●
26. Cyclones	●●●	·	●●●●●
27. Earthquakes	●●●	·	●●●●●
28. Pest Epidemics	●●●	●●●	●●●

High Hazard ●●●●●
Medium Hazard ●●●
Low Hazard ·

(a) **Total Human Health Consequences** consists of the descriptors (11) *Human Mortality (Current Annual)*, (12) *Human Morbidity (Current Annual)*, and (16) *Future Human Health Consequences*.

(b) **Total Ecosystem Consequences** consists of the descriptors (13) *Natural Ecosystem Impacts (Current Annual)*, and (17) *Future Ecosystem Consequences*.

(c) **Total Material and Productivity Consequences** consists of the descriptors (14) *Material and Productivity Losses (Current Annual)*, and (18) *Future Material and Productivity Losses*.

Fig. 2.6 **Hazard rankings by total consequence aggregation for the United States**

gation scheme **Total Human Health Consequences**. This aggregation is the sum of the descriptors (*Human Mortality*, *Human Morbidity*, and *Future Human Health Consequences*) relating to current and future health effects. This is an important aggregation scheme because human health concerns have been, and will continue to be, one of the main reasons for enacting environmental-protection measures.

Few would argue, however, that health concerns alone should determine environmental priority setting. We, therefore, present comparable aggregation schemes for ecosystem impacts and material and productivity losses. The second column in figure 2.6 presents the aggregation scheme **Total Ecosystem Consequences**, which is the sum of the descriptors *Natural Ecosystem Impacts* and *Future Ecosystem Consequences*. The third column presents the aggregation scheme **Total Material and Productivity Consequences**, which is the sum of the descriptors *Material and Productivity Losses* and *Future Material and Productivity Losses*.

Examination of these three aggregation schemes leads to conclusions similar to those found by CENTED researchers in their study of technological hazards (Kates, Hohenemser, and Kasperson 1985) and by the USEPA (1987) and its Scientific Advisory Board (USEPA 1990) in their studies of environmental problems: how a problem is ranked in relationship to other problems depends on what dimensions of the problem or types of effects are being considered. Specifically, when looking only at consequences, what one considers to be the most pressing environmental problems depends on whether one focuses on health or ecosystem or material losses. Most problems rank high in one – and only one – of the consequence aggregations. Only one problem (Fresh water: Metals and Toxics) ranks high in all three categories of consequences. Four problems (Fresh water: Eutrophication; Fresh water: Sedimentation; Ocean Water; and Hazardous and Toxic Air Pollution) rank high in two of the three consequence aggregations, and only two problems (Fresh water: Biological Contamination; Agricultural Land: Salinization) do not rank high in any category.

An alternative treatment of the method's consequence descriptors is to construct two temporally based aggregations – **Total Current Consequences** and **Total Future Consequences**. **Total Current Consequences** is composed of *Human Mortality*, *Human Morbidity*, *Natural Ecosystem Impacts*, and *Material and Productivity Losses*, whereas **Total Future Consequences** includes *Future Human Health Consequences*, *Future Ecosystem Consequences*, and *Future Material and Productivity Losses*. These aggregations allow us to compare those environmental hazards that pose the greatest hazard currently with those that are expected to become significantly worse in the next 25 years. This comparison is of particular interest, since determining how to weigh present versus future

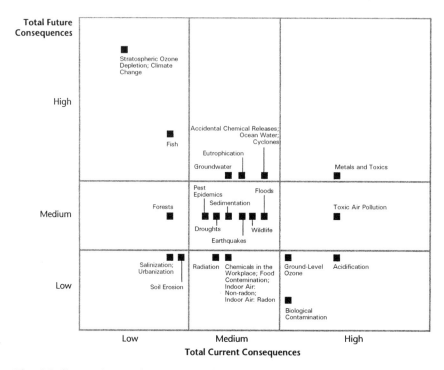

Fig. 2.7 **Comparisons of hazard rankings for the United States: Total current consequences versus total future consequences**

consequences is one of the most difficult tasks facing policy makers (Weiss 1989).

Figure 2.7 presents a comparison of these two aggregation schemes for the US case: the horizontal axis represents **Total Current Consequences**; the vertical, **Total Future Consequences**. The lines on the graph correspond to our quartile divisions, used to distinguish the problems ranked high, medium, and low. The "corners" of the graph represent combinations of current and future consequences that are of particular interest. A problem falling into the upper right-hand corner of the graph ranks high in both these aggregation schemes. Such problems are worthy of special attention, for not only are they among the worst current problems but also they are expected to become significantly more severe in the future. In other words, they are problems that are not yet being effectively managed, even to the minimal level of preventing increasing harm. In the United States, Fresh Water: Metals and Toxics; Accidental Chemical Releases; Ocean Water; and Cyclones fall into this corner. In contrast, the lower right-hand portion of the graph contains those problems that are currently severe but which, unlike those in the upper right quartile, are expected to improve in the future. This is due to projections that management efforts already under way will successfully reduce the future consequences of this set of problems. For the United States, Biological Contamination of Fresh Water; Acidification; and Ground-level Ozone fall into this quartile, owing to expectations that these will be controlled under the new Clean Water and Clean Air Acts. A final portion of this graph that is of particular interest is the upper left-hand corner: it represents hazards that are not known to have serious current consequences but which are expected to have significant future consequences. Not surprisingly, Climate Change and Stratospheric Ozone Depletion display this characteristic.

Two cautions are in order regarding the use and interpretation of the aggregation **Total Future Consequences**. First, the future-consequence descriptors were scored in relation to the current consequences, and thus measure whether a problem is expected to become significantly better or worse, with "significant" defined roughly as a doubling or halving of consequences. Thus, the future-consequence descriptors and the aggregation **Total Future Consequences** are measures that combine trend and magnitude information. Second, the expectation of future harm is based roughly on a "business as usual" scenario, which takes into account current management activities, including those recently enacted that may not yet have produced results. Current policy debates highlight the contentiousness of what constitutes "business as usual," and past experience has shown how often "business as usual" scenarios prove to be incorrect. Despite our recognition of the limitations of a "business as usual" approach, the alternative (which is to use several future scenarios) would have created too much complexity for a method of this type. It would have defeated one of our key goals – to develop a method that could be applied in a fairly short period of time by an informed generalist.

Thus far, the aggregation schemes we have discussed have been based on the consequence descriptors. As argued previously, one of the reasons why it is problematical to depend solely on current or projected effects to identify priority environmental problems is that much uncertainty and ignorance exist regarding the consequences of actions that perturb the environment. We are, therefore, interested in aggregation schemes that can give us an indication of the degree of perturbation caused by various environmental problems. The two aggregation schemes presented next involve descriptors from other points in our causal chain that capture this.

The first, **Pervasiveness**, seeks to capture both the spatial spread and the natural longevity of pollutants or other environmental perturbations. It is composed of the descriptors *Spatial Extent, Persistence, Population Exposed, Land Area or Resource Exposed*, and *Recovery Period*. In figure 2.8, we compare scores on **Pervasiveness** with scores on **Total**

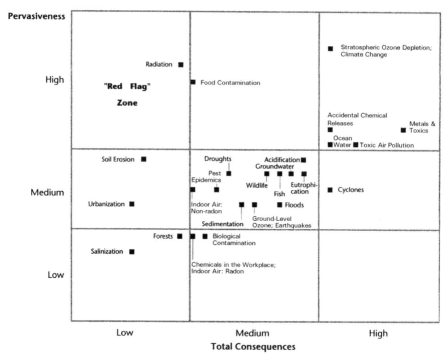

Fig. 2.8 **Comparisons of hazard rankings for the United States: Total consequences versus pervasiveness**

Consequences – an aggregate indicator created by summing the scores of the **Health**, **Ecosystem**, and **Material Consequence** aggregations shown in figure 2.6. Not surprisingly, all problems that rank high on **Total Consequences** have high or medium scores for **Pervasiveness**. Of particular interest in this graph are the hazards occupying its upper left-hand portion: they have relatively low known consequences but high pervasiveness. Both Radiation and Food Contamination fall into this zone for the United States. Radiation ranks high for the aggregation **Pervasiveness** for the following reasons: *Spatial Extent* scores 9 for all countries, owing to the potential for nuclear accidents such as that at Chernobyl to cause radioactive particles to be carried for long distances. The two temporal descriptors, *Persistence* and *Recovery Period*, scored 9 for all cases because of the longevity of isotopes in nature and the generational time required for carcinogenic effects to run their course. *Population* and *Land Area or Resource Exposed* had high scores based on extrapolations from Netherlands data that showed that external radiation from building materials accounted for 40 per cent of the anthropogenic exposure.

Scorers assumed this to be a significant source of radiation in the United States as well.

Although food contamination has only a moderate *Spatial Extent*, it ranks high for the aggregation **Pervasiveness** because of high scores for the other four descriptors that compose this aggregation scheme. *Persistence* and *Recovery Period* both score 9 owing to the longevity of some pesticides in the environment. *Population Exposed* and *Land Area or Resource Exposed* both score 9 because pesticides and other chemicals are used in most agriculture and livestock production, and thus most of the population is exposed to pesticides, hormones, or other contaminants through the food they eat. In addition, these agricultural chemicals leach into water supplies. We have dubbed such hazards "Red Flag" problems because their high pervasiveness suggests that, if it turns out that currently unknown effects of these changed VECs actually do exist, we could experience unexpectedly high consequences in the future. These are thus problems that call for extra monitoring, and possibly policy action as insurance against surprises.

Two environmental problems that are currently receiving much attention – Climate Change and Stratospheric Ozone Depletion – are good examples of the relevance of the **Pervasiveness/Total Consequences** analysis. Twenty years ago, they would have fallen into the red-flag zone, as their consequences were poorly understood or altogether unsuspected. Now it is clear that these hazards pose grave environmental risks precisely because the pollutants that cause them are globally pervasive and temporally persistent.

A second aggregation scheme that describes perturbation to the environment is **Disruption**. This aggregation consists of the descriptors *Disturbance to the Environment*, *Change in Perturbation*, *Anthropogenic Flux*, and *Change in Anthropogenic Flux*. The **Disruption** scheme identifies those hazards that are causing drastic additions (as in pollution) or extractions (as in harvesting of renewable resources) of materials into or out of the natural environment. A graph of **Disruption** versus **Total Consequences** is shown in figure 2.9. Another red-flag zone appears in the upper left-hand corner, identifying problems that have high disruption but low known consequences. Salinization falls in this zone. Salinization ranks high for the aggregation **Disruption** for several reasons. Some United States farmland has a salinity that is three to four times the natural amount, leading to a score of 7 for the descriptor *Disturbance to the Environment*. Furthermore, salinization from irrigation is a completely anthropogenic phenomenon, although it is somewhat ameliorated by rainfall. This leads to a score of 9 for the descriptor *Anthropogenic Flux*. Salinity tends to increase on a given land with repeated irrigations, and the hectarage of land that has been salinized in the United States is

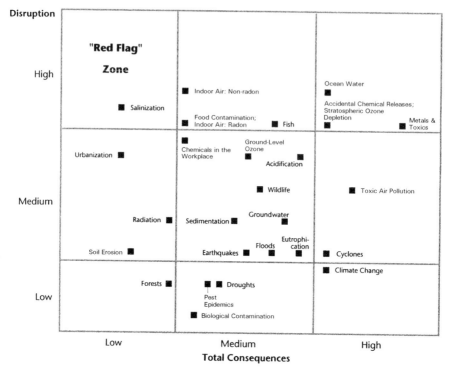

Fig. 2.9 **Comparisons of hazard rankings for the United States: Total consequences versus disruption**

also growing. Specific quantitative data on the rates of growth are limited. Assumptions based on the rate of growth of irrigated acreage led to scores of 6 and 7 for the descriptors *Change in Perturbation Level* and *Change in Anthropogenic Flux*, respectively.

Again, owing to society's ignorance and uncertainty regarding the consequences of severe or rapid disruption of the natural environment, perhaps more attention should be paid to such hazards. The lower right-hand corner of this graph is also of interest, since it contains Climate Change, which scores high for **Total Consequences**, but is low for **Disruption**. Although human activities are disrupting the carbon cycle at a faster rate than ever before in history, it is still a rather minor disruption compared with the entire carbon cycle. The reason, of course, is that we believe that small perturbations to the carbon cycle will turn out to be magnified by the earth system into relatively large changes in the distribution of the energy received from the sun, in the climate, and in ecosystem function. This is a consequence of detailed feedback mechanisms missed by the simple **Disruption** index, and serves another warning that no single-dimensioned ranking of environmental problems is likely to be without major shortcomings.

As stated previously, the foregoing aggregation schemes are suggestive, not prescriptive. Several others that we believe can be useful for ranking and characterizing environmental hazards are described below.

1. *Persistence in Time:* this is the sum of the descriptors *Recovery Period* and *Persistence*. It describes the natural longevity of pollutants or perturbations in the environment and is the temporal component of the aggregation **Pervasiveness**.

2. *Pervasiveness in Space:* this is the sum of the descriptors *Population Exposed*, *Land Area or Resource Exposed*, and *Spatial Extent*. It describes the physical reach of pollutants or perturbations in the environment and is the spatial component of **Pervasiveness**.

3. *Manageability:* this aggregation scheme is the sum of the descriptors *Transnational*, *Delay*, and *Recurrence*. Problems that score high in *Transnational* will be more difficult to manage because the problem is exacerbated by the actions of other countries. Problems that score high in *Delay* will be difficult to manage because they will be out of the public eye for long periods of time and it will, therefore, be difficult to rally political support to remedy them. Finally, problems that score high in *Recurrence* will be difficult to manage, because the less frequently problems occur the less likely it is that the public will maintain an active awareness of them.

Comparing hazards across countries: India, Kenya, the Netherlands, and the United States

For this comparison, we use several of the aggregation schemes illustrated above. We begin in figure 2.10 with a comparison of environmental problems described in terms of their **Total Consequences**. The figure shows that none of our 28 environmental problems ranks high for **Total Consequences** in all four countries. Similarly, only three environmental problems – Fresh Water: Metals and Toxics, Ocean Water, and Hazardous and Toxic Air Pollution – rank high for three countries. This result serves as an important reminder for international cooperation that different nations are facing very different sets of severe environmental threats.

It is also instructive to examine shared problems across pairs of countries, since many environmental problems are ranked high across two countries for **Total Consequences**. The Netherlands and the United States share five out of seven of their top-ranked hazards; this is due to

Environmental Problems * indicates high hazard ranking	Netherlands	USA	India	Kenya
2. Fresh Water: Metal and Toxic Contamination	*	*	*	
5. Ocean Water	*	*	*	
10. Toxic Air Pollution	*	*	*	
6. Stratospheric Ozone Depletion	*	*		
7. Climate Change	*	*		
9. Ground-Level Ozone Formation	*			
18. Agricultural Land: Soil Erosion			*	*
23. Forests			*	*
1. Fresh Water: Biological Contamination			*	
3. Fresh Water: Eutrophication	*			
15. Accidental Chemical Releases		*		
21. Wildlife				*
22. Fish				*
24. Floods			*	
25. Droughts				*
26. Cyclones		*		
28. Pest Epidemics				*
4. Fresh Water: Sedimentation				
8. Acidification				
11. Indoor Air: Radon				
12. Indoor Air: Non-radon				
13. Radiation: Non-radon				
14. Chemicals in the Workplace				
16. Food Contamination				
17. Agricultural Land: Salinization, Alkalinization, Waterlogging				
19. Agricultural Land: Urbanization				
20. Groundwater				
27. Earthquakes				

(a) The **Total Consequences** aggregation consists of the descriptors (11) *Human Mortality*, (12) *Human Morbidity*, (13) *Natural Ecosystem Impacts*, (14) *Material and Productivity Losses*, (16) *Future Human Health Consequences*, (17) *Future Ecosystem Consequences*, and (18) *Future Material and Productivity Losses*.

Fig. 2.10 **Comparison of hazards ranked as most severe in terms of total consequences for the Netherlands, United States, India, and Kenya**[a]

their similar levels and types of industrialization, as well as similarities in their environmental-management programmes. Comparing India and Kenya, the two developing nations in our study, only two problems ranked high in both for **Total Consequences**. In fact, India shares more top-ranked problems with the Netherlands and the United States than with Kenya. This is due to the fact that India is a diverse country with major areas of concentrated industrialization, whereas Kenya has much lower levels of industrial activity. Thus, more of India's top-ranked problems are attributable to industrial activities whereas more of Kenya's result from resource depletion. These results serve as a warning against making the erroneous assumptions that all developing countries face similar environmental hazards, or that one set of problems faces industrialized countries and another confronts developing countries. Many more case studies are needed to determine meaningful ways of grouping countries in terms of the most severe environmental hazards that they face.

Our second set of cross-country comparisons involves the aggregation schemes **Total Current Consequences** and **Total Future Consequences**, as illustrated in figures 2.11 and 2.12. For both aggregation schemes there are significant differences in the hazards that are ranked high for the different countries, emphasizing that the greatest risks are different for the four countries. Comparing these two aggregations, however, we find that more environmental problems are highest ranked across three or more countries in terms of **Total Future Consequences** than in terms of **Total Current Consequences**. Of particular note, Stratospheric Ozone Depletion and Climate Change are ranked high for **Total Future Consequences** in all four countries studied. This suggests that, to the extent that attention can be focused on problems largely important for their future consequences, there is an internationally shared basis for focusing on these two global hazards.

These shared high future-consequence rankings for Stratospheric Ozone Depletion and Climate Change for all four countries also point to one of the limitations of our method: not all countries are likely to suffer the same magnitude of consequences from a problem such as Climate Change, yet our future-consequence descriptors are not sensitive enough to allow for such distinctions. The complex scenario development needed for that type of analysis, even for problems with better-understood consequences, was not compatible with the ease of application we were striving for in our descriptors.

The two figures also highlight distinct differences between highest-ranked problems for **Total Current Consequences** and **Total Future Consequences** in each country. Two factors contribute to these differences: first, many of the worst current problems, such as acid rain and ground-level ozone formation in the United States, are being managed and are expected, therefore, to become less severe in the future; second, problems that do not currently rank among the worst may be becoming more severe at a significantly faster rate. This difference between hazards

Environmental Problems * indicates high hazard ranking	Netherlands	USA	India	Kenya
1. Fresh Water: Biological Contamination		*	*	*
2. Fresh Water: Metal and Toxic Contamination	*	*	*	
5. Ocean Water	*	*		
8. Acidification	*	*		
9. Ground-Level Ozone Formation	*	*		
10. Toxic Air Pollution	*	*		
15. Accidental Chemical Releases	*	*		
18. Agricultural Land: Soil Erosion			*	*
24. Floods		*	*	
25. Droughts			*	*
28. Pest Epidemics			*	*
3. Fresh Water: Eutrophication	*			
4. Fresh Water: Sedimentation	*			
12. Indoor Air: Non-radon				*
17. Agricultural Land: Salinization, Alkalinization, Waterlogging			*	
21. Wildlife				*
23. Forests				*
26. Cyclones		*		
6. Stratospheric Ozone Depletion				
7. Climate Change				
11. Indoor Air: Radon				
13. Radiation: Non-radon				
14. Chemicals in the Workplace				
16. Food Contamination				
19. Agricultural Land: Urbanization				
20. Groundwater				
22. Fish				
27. Earthquakes				

(a) The **Total Current Consequences** aggregation consists of the descriptors (11) *Human Mortality*, (12) *Human Morbidity*, (13) *Natural Ecosystem Impacts*, (14) *Material and Productivity Losses*.

Fig. 2.11 **Comparison of hazards ranked as most severe in terms of total current consequences for the Netherlands, United States, India, and Kenya**[a]

that are top-ranked for current and future consequences is most pronounced for Kenya and India, where only two problems in each country were ranked high for both current and future consequences. This indicates the existence of many problems that are not currently of great concern that are expected to become significantly more severe in the future.

Our final cross-country comparison is **Total Consequences** versus **Per-**

Environmental Problems * indicates a high hazard ranking	Netherlands	USA	India	Kenya
6. Stratospheric Ozone Depletion	*	*	*	*
7. Climate Change	*	*	*	*
3. Fresh Water: Eutrophication	*	*		*
5. Ocean Water	*	*	*	
9. Ground-Level Ozone Formation	*		*	*
2. Fresh Water: Metal and Toxic Contamination	*	*		
8. Acidification			*	*
10. Toxic Air Pollution	*		*	
15. Accidental Chemical Releases	*	*		
22. Fish		*		*
23. Forests			*	*
1. Fresh Water: Biological Contamination	*			
18. Agricultural Land: Soil Erosion				*
20. Groundwater		*		
24. Floods			*	
26. Cyclones		*		
4. Fresh Water: Sedimentation				
11. Indoor Air: Radon				
12. Indoor Air: Non-radon				
13. Radiation: Non-radon				
14. Chemicals in the Workplace				
16. Food Contamination				
17. Agricultural Land: Salinization, Alkalinization, Waterlogging				
19. Agricultural Land: Urbanization				
21. Wildlife				
25. Droughts				
27. Earthquakes				
28. Pest Epidemics				

(a) The **Total Future Consequences** aggregation consists of the descriptors (16) *Future Human Health Consequences*, (17) *Future Ecosystem Consequences*, and (18) *Future Material and Productivity Losses*.

Fig. 2.12 **Comparison of hazards ranked as most severe in terms of total future consequences for the Netherlands, United States, India, and Kenya**[a]

vasiveness, already investigated for the US case in our first "red-flag" graph. Figure 2.13 presents this graph for each of the four countries. Two particularly interesting observations can be made from this figure. First, as discussed for the US case, problems that fall into the upper left-hand

88

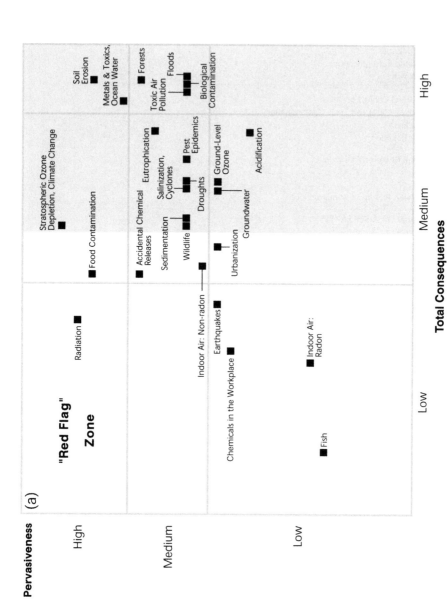

Fig. 2.13 **Cross-national comparisons of hazard rankings: Total consequences versus pervasiveness for (a) India, (b) Kenya, (c) the Netherlands, (d) USA**

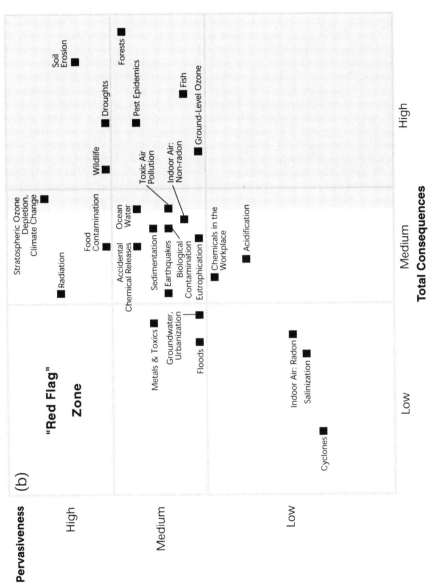

Fig. 2.13 (cont.)

Pervasiveness (c)

"Red Flag" Zone

High
- Radiation
- Food Contamination
- Accidental Chemical Releases
- Stratospheric Ozone Depletion, Climate Change
- Metals & Toxics
- Ocean Water
- Eutrophication

Medium
- Wildlife
- Groundwater
- Urbanization
- Indoor Air: Non-radon
- Soil Erosion
- Floods
- Sedimentation
- Pest Epidemics
- Droughts
- Acidification
- Biological Contamination
- Toxic Air Pollution
- Ground-Level Ozone

Low
- Forests
- Salinization
- Earthquakes, Cyclones
- Fish
- Chemicals in the Workplace
- Indoor Air: Radon

Low   Medium   High

Total Consequences

Fig. 2.13 (cont.)

90

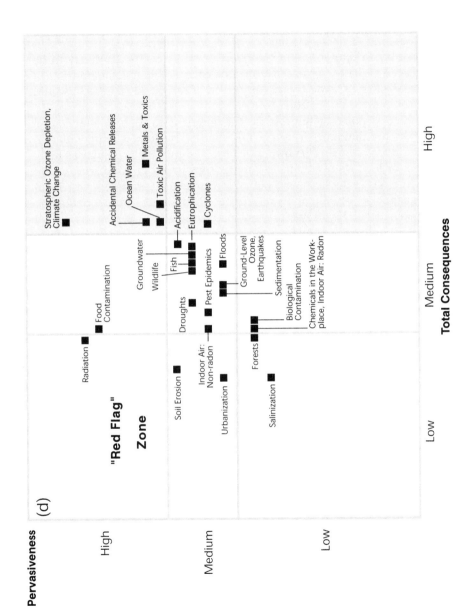

Fig. 2.13 (cont.)

91

corner of this graph (the red-flag zone) have high **Pervasiveness** and low **Total** (known) **Consequences**. Figure 2.13 demonstrates that each of the four countries of our study has two problems in common that are in or very near the red-flag zone: these are Radiation: Non-radon and Food Contamination. This across-the-board outcome serves as a warning that unknown but severe consequences from these problems are likely to be experienced globally. Again, they may be comparable to Climate Change and Stratospheric Ozone Depletion, which, if scored 20 years ago, would have been plotted in this zone.

The second point of interest on this graph is to look at the shaded portion of the diagram. So far in this chapter we have been comparing each country in terms of its own quartiles, and ranking problems as high, medium, or low; we now shift to comparisons based on total scores. The shaded regions in the figure are defined to include all problems with **Total Consequence** scores equal to or greater than those characterizing the upper quartile or high-ranked problems in the US case. For the US, a score of 30 or greater is needed to rank in the top quartile for the aggregation **Total Consequences**. The Netherlands has nearly the same number of problems with scores greater than 30, eight compared with seven. In contrast, there are eleven problems in Kenya and nineteen in India with scores greater than 30. This result emphasizes that the environmental challenges being faced by Kenya and India are in general much more hazardous than those faced by the United States and the Netherlands. Put somewhat differently, many of the "medium"-ranked problems of India and Kenya almost certainly pose greater threats to those countries' peoples and ecosystems than even the "high"-ranked problems of the United States and Netherlands pose to theirs.

Conclusions

The case-study evaluations presented here convince us that this method can successfully contribute to the difficult task of comparative environmental hazard assessment, both within and across countries. Our goal has been not to present one final set of rankings but rather to demonstrate how the method can be used to accomplish this task. The case studies have also demonstrated the complexity of an endeavour of this sort and emphasized the importance of the assumptions and judgements required to make comparisons in the face of limited and uncertain data. In addition, we have discussed our difficulties in developing and scoring descriptors for the important category **Future Consequences**.

A method of such admitted liabilities should lead one to ask the obvious question of why it should be used at all. Our justifications are twofold.

First, the arguments for conducting some sort of environmental-hazard comparison are compelling: thinking of environmental risks in isolation from each other and then drawing policy conclusions independently is extremely shortsighted. Despite the limitations of the USEPA report *Unfinished Business* (USEPA 1987), its great contribution to policy-making was its fundamental demonstration that effective environmental protection should not and need not be based on ad hoc responses to hazards.

As for the second justification, we simply do not know of any alternative mechanisms for international comparative assessment of environmental hazards that are any more satisfactory than the one we have sketched here. Despite its many shortcomings, the use of a causal taxonomy for comparative-hazard assessment permits a full characterization of environmental hazards. This allows for hazards to be ranked on the basis both of their consequences and of other important characteristics (such as disruption to the environment, pervasiveness, and manageability). By focusing on aggregation schemes that are chosen by the user and can be clearly communicated to others, the method makes transparent the value judgements inherent in any scheme for comparing and ranking hazards. This is of particular importance when ranking hazards that cause different types of consequences, that occur over different time-frames, and that appear in very different perspectives in different parts of the world. Lastly, this method provides clear links between the two distinct but interrelated tasks of hazard assessment and hazard management.

In short, we are at least as aware as any of our readers that the method presented here is only a beginning, and a highly imperfect one at that, for the urgent and complex task of comparing environmental hazards in international contexts. We can only hope and believe that it is a beginning by which others will be motivated and upon which they will be able to build.

Note

1. The study reported in this paper was conducted at the Center for Science and International Affairs at Harvard University's John F. Kennedy School of Government from 1988 to 1991. The work was carried out as part of the Center's larger programme on global environmental risks, led by Professor William C. Clark. Vicki Norberg-Bohm, then a doctoral candidate at the School, directed the study. The other authors contributed in the course of their graduate training at the institutions listed below: Jo Anne Berkenkamp, Marc D. Koehler, and Jennifer A. Marrs (Kennedy School of Government, Harvard University); Bhavik Bakshi, Chris P. Nielsen, and Ambuj Sagar (Massachusetts Institute of Technology); Sherry Bishko (Tufts University).

The study was supported by the Stockholm Environment Institute (SEI), The Rockefeller Brothers Fund through a grant to SEI, and the United Nations University. The early formulation of the study benefited greatly from meetings with Roger Kasperson, Christoph Hohenemser, Robert Kates, and their colleagues at Clark University's Center for Technology, Environment, and Development (CENTED). The grounding of our approach in CENTED's earlier work on hazard assessment is acknowledged in the text, but bears special mention here. CENTED contributed substantially to the subsequent development of our study through hosting a critical review meeting on an earlier draft and, through the Center's librarian, Jeanne X. Kasperson, making available their unparalleled collections of source materials. Other critical feedback was provided by members of the European Policy Unit at the European University Institute in Florence, by students in William Clark's classes on Comparative Environmental Policy at the Kennedy School, and by anonymous reviewers of an earlier version of the work presented by SEI to the Second World Climate Conference. To all, our sincere thanks.

Appendix A: The descriptors[1]

Summing

Any given environmental problem may have a variety of components that score differently for the same descriptor. For example, in the case of Stratospheric Ozone Depletion, persistence times of some CFCs are 10 years whereas others are 110 years. As a result, a summing rule has been chosen for each descriptor, as noted throughout and explained below:

1. *Total:* add up over all components;
2. *Highest Significant Score:* give the highest score for which a significant portion of the problem would score (significant portion is defined as about 20 per cent);
3. *Weighted Average:* take a weighted average over all components of the problem.

Material flux descriptors

1. Spatial Extent

This measures the spatial extent of a single release or event for which there is a significant change in the material levels in the environment. The quantitative scale is based on linear dimensions, whereas the categorical scale is on common geographical units (table 2.1A).

Table 2.1A **Scoring *Spatial Extent* of a single event**

Score	Distance scale (km)	Categorical definition
1	<10	Small region
3	10–100	Region
5	100–1,000	Subcontinental
7	10^3–10^4	Continental
9	>10^4	Global

- *Pollution-based hazards:* this measures the distance from a source for which a single release or event contributes to a change in material levels.
- *Renewable-resource depletion:* this measures the area (scored as the longest linear distance) in which local harvesting affects the quantity and/or quality of the resource.
- *Natural disasters:* this measures the area (scored as the longest linear distance) affected by a single occurrence.

Notes on scoring

The objective here is to distinguish between local, national, multinational within a region, and global problems. A score of 1 or 3 will be within a nation for all countries. Although theoretically a score of 5 could be within a nation for the largest nations, we have not been able to think of an example where this is the case (e.g. acid rain in the US, Canada, USSR, and China would be scored 5, and would be a multinational regional problem). Thus, it is safe to say that a score of 1 or 3 indicates that the problem occurs within a nation, whereas a score of 5 or over indicates that it is multinational. An exception to this is Australia, where a score of 5 is both continental and within a nation.

In general, scores for this descriptor are determined by the nature of the hazard and thus, for a given hazard, will typically score the same for all countries. The taxonomy includes four such "problem-specific" descriptors, each identified as such in its respective "Notes on Scoring" sections.

Summing rule

Highest significant score.

Notation for the material flux descriptors 2 through 5

The next four descriptors give a characterization of the material or energy changes due to human activities or natural forces. We are interested here in both *current levels or stock* (how much of the material is in the environment compared with background, normal or long-term average levels), and *flows* (how much of the material is being added to or extracted from the environment on an annual basis). We are also interested in how stocks and flows are changing over time. In other words, what are the trends? Are we perturbing the environment more or less than we were 5 to 10 years ago? This helps to assess whether the problem is getting better or worse.

Descriptor 2 characterizes the current level or stock, whereas descriptor 3 characterizes how the current level or stock is changing over time. Descriptor 4 characterizes the flow, whereas descriptor 5 characterizes how the flow is changing over time. In all four descriptors, the values are normalized in relationship to a background, normal, or long-term average in order to make scores comparable across the different hazards.

The notation below is used to explain how to calculate the scores for these descriptors.

- B = "natural" background levels.
- Mj = material or energy levels in year j.

$$= B + \int_0^j D\,dt - T\,dt$$

- Dj = annual anthropogenic flux of materials or energy (disturbance; discharge or removal) in year j.
- Tj = annual absorption/transformation of materials or of ecosystems toward the "natural" or undisturbed state.
- Aj = annual natural flux of materials in year j.
- y = current year.
- j = a given year.
- n = the number of years under consideration.

The measures defined below are used to score the next four descriptors. They are listed below as a summary, and described in more detail in the text that follows.

- P describes the disturbance to the environment or magnitude of perturbation (descriptor 2).

$$P = M_y/B$$

- ΔP: describes the change in P over time as a fraction of P; i.e. measures the change in the perturbation (descriptor 3).

$$\Delta P = dP/dt = (M_y - M_{y-1})/M_y \quad \text{or} \quad [(M_y - M_{y-n})/M_y]/n$$

- F: describes the anthropogenic flux in relation to the natural flux (descriptor 4).

$$F = D_y/A_y$$

- ΔF: describes the change in anthropogenic flux as a fraction of the anthropogenic flux (descriptor 5).

$$\Delta F = dF/dt = (D_y - D_{y-1})/D_y \quad \text{or} \quad [(D_y - D_{y-n})/D_y]/n$$

2. Disturbance to the Environment (Magnitude of Perturbation)

Measures the degree to which a change in the material of interest (pollutant or resource) is above or below the background, normal, or long-term average levels. In other words, this measures the degree to which the quantity of the material present in the environment has increased (where the environmental problem is caused by net material flow into the environment [e.g. CO_2 into the atmosphere]) or decreased (where the problem is due to net material flow out of the environment [e.g. loss of forests]) relative to the background levels. The perturbation is estimated by calculating the factor P, where P = current material levels/background levels for flow into the environment as in most pollutants, and P = background levels/current levels for flow out of the environment as in resource depletion. As a result, increased pollution concentrations and decreased forest cover, for instance, will both have P values greater than 1 (table 2.2A).

- *Pollution-based hazards:* this measures the level of the pollutant compared with the natural background level (i.e. the level that existed prior to human perturbation). Note that those pollutants that do not occur in nature will necessarily score 9.
- *Renewable-resource depletion:* this measures the long-term average pre-harvest level of the resource or the level to which the system would return in the long run if harvesting ceased, compared with the current stock of the resource.
- *Natural hazards:* this measures the level of the material or energy during the natural hazard, compared with background levels.

Table 2.2A **Scoring *Disturbance to the Environment***

Score	Scale[a]
1	$P < 1/2$
3	$1/2 < P < 1$
5	$1 < P < 2$
7	$2 < P < 4$
9	$P > 4$

a. For pollutants, P = current material levels/background levels, for flow into the environment; for resource depletion, P = background levels/current levels, for flow out of the environment.

Notes on scoring

When measuring perturbation, use the change in material quantity nearest to, or most affected by, the original human perturbation. For example, with Stratospheric Ozone Depletion, measure increase in atmospheric chlorine and not a reduction in stratospheric ozone.

Some hazards have bimodal distributions: for example, some average photochemical oxidant concentrations in urban areas are much higher than the average concentrations in rural areas, yet consequences to these elevated levels occur in both areas. Decisions on how to score such problems should be made on an individual basis, however, and should be scored for one of the distributions, rather than the average of both.

Summing rule

Weighted average of most significant effect.

3. Change in Perturbation

Measures the annual change in the magnitude of perturbation (i.e. the rate of change of descriptor 2). The annual change is measured by the factor ΔP. In technical terms, ΔP is the time derivative of M as a fraction of M (table 2.3A).

- *Pollution-based hazards:* this measures the change in the pollution level in the environment.
- *Renewable-resource depletion:* this measures the change in the stock of the resource.
- *Natural disasters:* this measures the degree to which the flows have changed in severity or frequency. This will usually score 5, except when human actions combine with natural factors to cause an increase or

Table 2.3A **Scoring annual rate of *Change in Perturbation***

Score	Halving/doubling time (years)	Percentage change per year
	(Halving time)	
1	<10	<−7.0
2	10–20	−7.0 to −3.5
3	20–40	−3.5 to −1.7
4	>40	<−1.7
5	No detectable change	0
	(Doubling time)	
6	>40	<1.7
7	20–40	3.5 to 1.7
8	10–20	7.0 to 3.5
9	<10	>7.0

decrease in the occurrence or severity of natural disasters, as can be the case in floods and droughts.

Notes on scoring

Given the way in which the scale is defined, a negative change in perturbation indicates a situation in which there is an improvement in the environment. A positive change in the perturbation indicates a situation where the problem is being exacerbated. The change in perturbation level should reflect an average over the last 5–10 years. If a significant change in trend has occurred during this time, score for the more recent trend (e.g. if levels in the environment were increasing from 1980 through 1985, but decreasing after 1985 owing to the implementation of pollution controls, score for the decreasing trend after 1985). Also, take care to eliminate biases due to extremely high or low values for a single year that are caused by random fluctuation.

Summing rule

Weighted average.

4. Anthropogenic Flux

This measures the extent to which human activities contribute to the total flux of materials into or out of the environment – i.e. it compares the anthropogenic flux with the natural flux). This is measured by the factor F, which measures the ratio of annual human flux to background flux (table 2.4A).

- *Pollution-based hazards:* this measures the annual anthropogenic flow of the pollutant compared with the annual natural flow.
- *Renewable-resource depletion:* this measures the annual net extractions of the resource compared with the sustainable yield (i.e. the level of extraction that would result in the highest average annual rate of extraction).
- *Natural disasters:* this measures the magnitude of the material flow due

Table 2.4A **Scoring *Anthropogenic Flux***

Score	Scale[a]
1	$F < 1/2$
3	$1/2 < F < 1$
5	$1 < F < 2$
7	$2 < F < 4$
9	$F > 4$

a. F, ratio of annual human flux to background flux.

to human factors compared with the magnitude of the material flow due to natural factors. These problems will generally score 1, again with the exception of problems where human actions have intervened to reduce or increase the severity of the material flow. (Note that this differs from human actions which reduce or increase the severity of consequences.)

Summing rule

Weighted average.

5. Change in Anthropogenic Flux

This measures the annual rate of change in the anthropogenic flux (i.e. the rate of change in descriptor 4). This is measured by the factor ΔF. In technical terms, ΔF is the time derivative of F as a fraction of F (table 2.5A).

- *Pollution-based hazards:* this measures the change in the annual anthropogenic additions of the pollutant to the environment.
- *Renewable-resource depletion:* this measures the change in annual extractions.
- *Natural disasters:* this measures the change in flow due to human factors, in other words, whether human factors are increasing or decreasing the severity or frequency of a natural hazard. This will generally score a 5.

Notes on scoring

The change in anthropogenic flux should reflect an average over the last 5–10 years. If a significant change in trend has occurred during this time,

Table 2.5A **Scoring annual rate of** *Change in Anthropogenic Flux*

Score	Halving/doubling time (years)	Percentage change per year
	(Halving time)	
1	<10	<−7.0
2	10–20	−7.0 to −3.5
3	20–40	−3.5 to −1.7
4	>40	<−1.7
5	No detectable change	0
	(Doubling time)	
6	>40	<1.7
7	20–40	3.5 to 1.7
8	10–20	7.0 to 3.5
9	<10	>7.0

score for the more recent trend (e.g. if emissions were increasing from 1980 through 1985, but decreasing after 1985 owing to the implementation of pollution controls, score for the decreasing trend after 1985).

Summing rule

Weighted average.

6. Persistence

This measures the time period required for the material most directly perturbed in the environment to return to the level existing before the event that caused a flux of that material (table 2.6A). (Note that this differs from the recovery period (descriptor 15), which reflects the time necessary for all subsequent effects including those on human health, ecosystem, or material welfare, to return to background levels.)

* *Pollution-based hazards:* this measures the time period that the material released remains in the environment (e.g. for acidification, persistence measures the atmospheric lifetime of acid-rain precursors as 3–10 days).
* *Renewable-resource depletion:* this measures the time period required to regenerate a specimen of the resource extracted (e.g. the lifetime of a fish caught or of a tree lost to deforestation).
* *Natural disasters:* this measures the duration of the event.

Notes on scoring

In general, this descriptor is defined by the hazard (i.e. it is problem specific), and thus for a given hazard it will score the same for all countries. In scoring, consider a 90 per cent return from the perturbed level to be sufficient. If a problem does not exist in a particular country, persistence must none the less be scored for the time that the environment remains perturbed for that type of hazard.

Summing rule

Highest significant score.

Table 2.6A **Scoring *Persistence***

Score	Time-scale
1	<1 week
3	1 week–1 year
5	1–10 years
7	10–100 years
9	>100 years

7. Recurrence

This measures the frequency of the event that causes a flux. Use the scale for *Persistence*.

Notes on scoring

Within the pollution and resource-depletion category, we will not expect to get a wide range of scores, since most of the relevant events are continual and will therefore score 1; however, accidental releases, chemical or nuclear accidents, and natural disasters will rank differently.

In general, this descriptor is defined by the hazard (except natural hazards), and thus for a given hazard will score the same for all countries.

Summing rule

Highest significant score.

Exposure descriptors

8. Population Exposed

Measures the percentage of current population within a country that is exposed to the change in a valued environmental component (VEC) (table 2.7A).

Notes on scoring

The "exposed" population should be broadly interpreted to mean all people who are currently exposed to a changed VEC. For example, for atmospheric problems such as acidification or smog, the percentage of the population exposed would include all inhabitants of an affected area, not just those individuals whose respiratory systems are unusually sensitive to such conditions. For accidental chemical spills, percentage exposed

Table 2.7A **Scoring *Population Exposed***

Score	Percentage of population[a]
1	<1
3	1–10
5	10–30
7	30–70
9	>70

a. Percentage of population within a country exposed to change in a valued environmental component (VEC).

Table 2.8A **Scoring** *Land Exposed*

Score	Percentage of land area[a]
1	<1
3	1–10
5	10–30
7	30–70
9	>70

a. Percentage of land area (or resource) exposed to changed VEC.

should include all inhabitants in the relevant vicinity of a potential accident. Only first-order effects should be considered: indirect effects, such as malnutrition from crop losses, should not be brought into the reckoning of the population exposed.

Summing rule

Total.

9. Land Area or Resource Exposed

This measures the percentage of land area or resource within a country that is currently exposed to the changed VEC (table 2.8A).

Notes on scoring

Interpret broadly, as described in the notes on scoring for descriptor 8.

Summing rule

Total.

10. Delay

This measures the delay time between the release or altering of materials and the earliest onset of serious consequences. The goal here is to measure the time delay between human actions and the first evidence that these actions are causing consequences (table 2.9A).

Notes on scoring

In general, this descriptor is defined by the hazard (i.e. it is problem specific) and thus for a given hazard will score the same for all countries. As with *Persistence*, if a problem does not exist in a particular country, delay must none the less be scored for the time between changes in material fluxes to the onset of consequences for that type of hazard.

Table 2.9A **Scoring *Delay* between human actions and first evidence of consequences**

Score	Time-scale
1	<1 week
3	1 week–1 year
5	1–10 years
7	10–100 years
9	>100 years

Summing rule

Earliest onset of first significant consequence.

Consequence descriptors

11. Human Mortality (Current Annual)

This measures the average annual percentage of population that dies as a direct result of the hazard (table 2.10A). Indirect effects, such as malnutrition from crop losses, should not be included.

Notes on scoring

Note that the scale is based on percentages of the population, not fractions.

Summing rule

Total.

Table 2.10A **Scoring *Human Mortality (Current Annual)***

Score	Percentage of population[a]
1	$<10^{-6}$
2	$10^{-6}-10^{-5}$
3	$10^{-5}-10^{-4}$
4	$10^{-4}-10^{-3}$
5	0.001–0.01
6	0.01–0.1
7	0.1–1
8	1–10
9	>10

a. Average annual percentage of population dying as a result of a specific hazard.

Table 2.11A **Scoring *"Natural" Ecosystem Impacts (Current Annual)***

Score	Categorical definition
1	No significant effect
5	Significant declines in productivity or local species extinctions
9	Regional or global extinction of significant species

12. Human Morbidity (Current Annual)

This measures the average annual percentage of population that becomes significantly ill from the hazard. Significantly ill is defined as a permanent injury or injury that interferes with normal activity.

Use the scale for human mortality.

Summing rule

Total.

13. "Natural" Ecosystem Impacts (Current Annual)

This describes the impacts on "natural" elements within ecosystems (table 2.11A). As a result, this descriptor does not take into account resources within ecosystems which are managed predominantly for the purpose of harvesting food or materials (e.g. crops or timber); these are captured in descriptor 14 below.

Summing rule

Total.

14. Material and Productivity Losses (Current Annual)

This measures the average annual losses of material and non-labour productivity as a percentage of national GNP (table 2.12A).

Damages to be included are material damage to both public and private capital stocks, crop losses, losses to recreation, and damage of natural resources.

Not to be included are reductions in labour productivity from lost days of work (or lost human lives), the costs of management or control measures (including equipment, managerial efforts, and health care), or aesthetic loss valuations.

Notes on scoring

Note that the scale is based on percentages of GNP, not fractions.

Table 2.12A **Scoring *Material and Productivity Losses (Current Annual)***

Score	Percentage of GNP[a]
1	$<10^{-6}$
2	$10^{-6}-10^{-5}$
3	$10^{-5}-10^{-4}$
4	$10^{-4}-10^{-3}$
5	0.001–0.01
6	0.01–0.1
7	0.1–1
8	1–10
9	>10

a. Average annual losses of material and non-labour productivity as a percentage of the gross national product (GNP).

Summing rule

Total.

15. Recovery Period

This measures the time period required for the VEC to recover from the effects of the changed material or energy flux to the prior state of human health, ecosystem, and productivity *assuming that the causative human activity is stopped* (table 2.13A). Assume no change in trends of management activities to ameliorate consequence.

Notes on scoring

It is important to note that the assumption here is not "no management change", but rather "no change in trends of management." This means, for example, that if a nation has been improving its sewage treatment at a certain rate and it appears likely that this will continue, then assume that rate of improvement when scoring. For management activities that have

Table 2.13A **Scoring *Recovery Period***

Score	Time-scale[a]
1	<1 week
3	1 week–1 year
5	1–10 years
7	10–100 years
9	>100 years

a. Time necessary for VEC to recover.

Table 2.14A **Scoring *Future Human Health Consequences***

Score	Categorical definition[a]
1	Significantly smaller than current consequences
5	Not differing radically from current consequences (more than half or less than twice as large)
9	Significantly greater than current consequences

a. Most likely consequences to future human health of today's human activities.

been holding steady, assume this will not change. To the extent that a nation has a credible action plan for ameliorating a problem, this should be taken into consideration. As a result, recovery periods for any given problem may be shorter in countries in which greater effort is currently apparent in aiding environmental recovery than in countries in which this is not the case.

As with *Persistence*, consider a 90 per cent recovery to be sufficient.

Summing rule

Highest significant score.

16. Future Human Health Consequences

This describes the severity of harm to future human health caused by today's human activities and assumes current trends in management activities to ameliorate consequences (i.e. use a "business as usual" scenario) (table 2.14A).

Notes on scoring

The term "future" means the next 25 years. Take the annual change in perturbation level and extrapolate out 25 years. Then examine such factors as the political and economic situations, and attempt to factor the non-quantifiable trends into the 25-year extrapolation. Use this to determine whether the effects of the considered activity will be more than one-half or less than twice as large as the current consequences.

If the problem does not cause significant consequences for a given country (and is not expected to do so in the next 25 years), score 1. In other words, if no future consequences are expected, score 1.

Summing rule

Total.

17. Future Ecosystem Consequences

This describes the severity of harm to future ecosystems caused by today's human activities. As with descriptor 15, harm to resources managed predominantly for harvesting of food or material should not be reflected here. Assume current trends in management activities to ameliorate consequences (i.e. use a "business as usual" scenario).

Score

Use scale for descriptor 16.

Notes on scoring

See notes under descriptor 16.

Summing rule

Total.

18. Future Material and Productivity Losses

This describes the magnitude of future material and productivity losses caused by human activities today. It assumes current trends in management activities to ameliorate consequences (i.e. use a "business as usual" scenario).

Score

Use scale for descriptor 16.

Notes on scoring

See notes under descriptor 16.

Summing rule

Total.

19. Transnational

This describes the nations whose activities lead to consequences (table 2.15A).

Notes on scoring

Interpret this in the narrow sense, as describing the nations whose release or alterations of the environment lead to the consequences. For natural disaster, always score 1 unless the consequences of the disaster are ex-

Table 2.15A **Scoring sources of *Transnational* consequences**

Score	Categorical definition
1	Consequences are caused mainly by a nation's own activities
5	Consequences are caused by activities of neighbouring countries and a nation's own activities
9	Consequences are caused by activities around the world

acerbated by activities of other nations (e.g. if flooding is worse owing to deforestation in a neighbouring country).

Summing rule

Highest significant score.

Note

1. Many of these descriptors were inspired by the work of Goodman (1980), the Center for Technology, Environment, and Development at Clark University (CENTED) (Kates, Hohenemser, and Kasperson 1985), and the US Environmental Protection Agency (USEPA 1987). Several descriptors are modifications in name, scale, or interpretation, but are none the less quite similar to those used by CENTED, including Spatial Extent, Magnitude of Perturbation, Persistence, Recurrence, Recovery Period, Population Exposed, Delay, and Human Mortality. The scale for Spatial Extent is the same as that used by Goodman.

Appendix B: Environmental problems

Pollution-based hazards

Water quality

Note: Fresh water includes both surface water and groundwater.

1. Fresh Water: Biological Contamination

Human activities
Human and animal waste disposal.

Changes in material fluxes
Levels of bacteria, viruses, and parasites in surface waters and/or groundwater.

Changes in valued environmental components
Reduction in quality of fresh-water supplies. This is particularly relevant for drinking water.

Exposure
Ingestion of water; ingestion of contaminated food; dermal contact.

Consequences

Results in relation to human mortality and human morbidity. Diseases carried include diarrhoea, cholera, sleeping sickness, and guinea-worm infestation. Loss of recreation through closing of waterways, beaches, etc.

2. Fresh Water: Metals and Toxic Contamination

Human activities

The use of pesticides, fossil-fuel combustion (deposition from the atmosphere into water, or onto land surfaces, and then runoff), industrial activities (including releases of chemicals, waste disposal, accidental releases, and underground storage), mining, consumer (municipal) waste disposal, urban runoff.

Changes in material fluxes

Increased pollutants in surface waters and groundwater. Toxic chemicals including heavy metals, inorganic compounds, and volatile organic compounds, herbicides, insecticides, fungicides.

Changes in valued environmental components

Reduction in quality of water supplies; increased toxicity of water.

Exposure

Human exposure through ingestion of contaminated water and food, dermal contact. Aquatic life exposed to increased pollutants in water.

Consequences

Human mortality and human morbidity. Some pollutants are carcinogenic, mutagenic, or both. Damage to aquatic ecosystems, such as reproductive deformities to animals (including birds and mammals) that depend on these aquatic ecosystems. Crop losses from decreased biological productivity due to contaminated irrigation water.

3. Fresh Water: Eutrophication

Human activities

Use of fertilizers in agriculture, animal husbandry, forest clearing, municipal waste disposal.

Changes in material fluxes

Increased nitrates and phosphates in water.

Changes in valued environmental components

Initial changed VEC is increased nutrient loading in water. This leads successively to increased algal growth, decreased clarity and increased particulate organic levels in the water, settling of particulate organic material into deep water, overabundant bacteria that consume oxygen and produce hydrogen sulphide, and a decrease in oxygen levels in deep water.

Exposure

To fish, through inhalation, lack of oxygen. For vegetation and other shellfish, ingestion of bacteria, reduced sunlight.

Consequences

Unsafe levels of nitrates can cause methaemoglobinaemia in infants, hypertension in children, gastric cancer in adults, and foetal malformations. Nitrates may be carcinogenic or mutagenic. Eutrophication can kill or displace fish and other aquatic life. Contamination of fish supply and losses to tourism.

4. Fresh Water: Sedimentation

Human activities

Agricultural practices leading to erosion from cropland; silviculture practices or deforestation leading to erosion from forestland (or formerly forested land); animal husbandry leading to erosion of rangeland; construction activities.

Changes in material fluxes

Increased sediments in river water; increased sedimentation in waterways.

Changes in valued environmental components

Reduction in quality of water supplies; increased toxicity of water; decreased navigability of waterways.

Exposure

Through ingestion, and use of rivers for transportation.

Consequences

Harm to aquatic life; decreased river navigation; destruction or decreased efficiency of hydroelectric projects.

5. Ocean Water

Human activities

Offshore oil drilling, oil transport, shipping in general. Disposal of industrial, consumer, commercial, and public wastes. Discharges of pollutants from rivers, direct coastal outfalls, and coastal urban and agricultural runoff.

Changes in material fluxes

Increased concentrations of oil, plastics, microbial/organic components, and toxic chemicals.

Changes in valued environmental components

Decreased quality and increased toxicity of ocean water; disruption of marine food chains.

Exposure

Entanglement by marine life. Ingestion by marine life including shellfish, fish, marine mammals, birds, plankton, and algae, sometimes several steps removed on the food chain. Ingestion of contaminated seafood by humans. Dermal contact. Visual contact.

Consequences

Human health and morbidity from exposure to toxins (especially through contaminated seafood), including carcinogenic and mutagenic effects as well as more immediate digestive disorders. Ecosystem damage, especially local effects from locally high concentrations. Loss of recreation, losses to tourism, and loss of food supplies.

Atmospheric quality

6. Stratospheric Ozone Depletion

Human activities

Manufacture, use, and disposal of halocarbons, including chlorofluorocarbons (CFCs), halons, chlorinated hydrocarbons, and chlorinated carbons. These substances are used in the manufacture of foam for insulation and packaging, as propellants, as heat-transfer fluids in heating and cooling systems, as solvents, especially in the electronics industry, and in fire extinguishers.

Changes in material fluxes

Increased concentration of CFCs, halons, chlorinated hydrocarbons, and chlorinated carbons. Through a series of chemical reactions this leads to decreases in the concentration of ozone (O_3) in the stratosphere.

Changes in valued environmental components

Increased ultraviolet (UV) radiation on the earth's surface. A reduction in the ozone shield leads to more radiation reaching the earth's surface.

Exposure

UV radiation contact with skin, eyes, ecosystems.

Consequences

Increased human mortality from malignant melanoma skin cancer. Increased human morbidity from non-melanoma skin cancer, and eye disorders including cataracts and acute photokeratitis (snow blindness). Suppression of immune-response system of humans and animals; slower growth and higher mortality among plants and animals. May aggravate nutritional deficiencies, infectious diseases, and autoimmune disorders. At high levels, may reduce crop productivity. Causes decrease in fecundity, growth, survival, and other functions of aquatic organisms, and thus affects ocean food chain. Accelerated degradation of some plastics and paints. Crop productivity losses due to UV radiation and secondary losses due to increased tropospheric smog.

7. Climate Change

Human activities

Fossil-fuel production, distribution, and combustion. Production, consumption, and disposal of halocarbons (CFCs, halons, and chlorocarbons). Wetland rice cultivation. Livestock husbandry. Use of nitrogenous fertilizers in agriculture. Landfilling of wastes. Land-use modification including deforestation, biomass burning, and wetland conversion.

Changes in material fluxes

Increases in several chemical constituents in the stratosphere including carbon dioxide (CO_2) from fossil-fuel consumption, deforestation and biomass burning; halocarbons; methane (CH_4) from landfills, fossil-fuel production and distribution, wetland rice cultivation, livestock husbandry, and biomass burning. Nitrous oxide (N_2O) from fossil-fuel consumption, use of nitrogenous fertilizer, deforestation, and biomass burning.

Changes in valued environmental components

Alteration in thermal radiation budget leading to climate change, which, in turn, will lead to other changes in key environmental components including temperature, sea level, precipitation, storm patterns (including frequency and severity), direct solar radiation, evapotranspiration, soil moisture, and runoff.

Exposure

Humans and ecosystems exposed to changed climate.

Consequences

Consequences, likely to vary significantly from one region to another, include severe disruptions of natural ecosystems with species loss; losses (or gains) in agricultural productivity; disruptions to human settlements and infrastructures, including property losses and loss of electric power; loss of life; loss of supplies of fresh water.

8. Acidification

Human activities

Fossil-fuel combustion and use. Industrial activities including use of smelters and paper manufacture.

Changes in material fluxes

Increased sulphur oxides (SO_x) and nitrogen oxides (NO_x) in the troposphere.

Changes in valued environmental components

Acidity of atmosphere. Through dry and wet deposition, acidity of soil and fresh water (lakes and streams).

Exposure

Inhalation of SO_x and NO_x by humans. Ecosystems exposed to lower pH.

Consequences

Fish kills and loss of aquatic life in acidified lakes. Forest dieback (from combined problems of acidification and elevated ozone). Possible human health effects include reduced lung function and possible water contamination. Premature mortality for sufferers of cardiac and respiratory problems. Materials damage including degradation of iron, steel, zinc, paint, and stone.

9. Ground-Level Ozone Formation

Human activities

Fossil-fuel combustion, biomass burning, industrial processes including use of organic solvents, surface coatings, chemical manufacture, and petroleum refining.

Changes in material fluxes

Increased nitrogen oxides (NO_x) from fossil-fuel combustion. Increased volatile organic compounds (VOCs, also called reactive hydrocarbons) from solvents and gasolines, highway vehicles, surface coating, organic solvents, solid-waste disposal, chemical manufacturing, and petroleum refining.

Changes in valued environmental components

Photochemical oxidant formation (i.e. increased ozone). This is formed through the reaction of VOCs and NO_x in the presence of sunlight.

Exposure

For human health, through inhalation.

Consequences

Damage to crops. Eye irritation. Decreased lung function including coughing, shortness of breath, possibly long-term lung damage, such as premature ageing of lungs. Deterioration of works of art. Forest dieback (in conjunction with acidification).

10. Hazardous and Toxic Air Pollutants

Human activities

The full spectrum of industrial activities involved with the manufacture, use, and disposal of chemicals including petroleum handling, pesticide application, waste-disposal sites, waste incineration, and metallurgical industries. Combustion of fossil fuels. Use of motor vehicles. Municipal sewage disposal, waste-water treatment.

Changes in material fluxes

Level of toxic chemicals in the atmosphere (hundreds of different toxic chemicals are released to the atmosphere).

Changes in valued environmental components

Toxicity of air.

Exposure

Inhalation by human and non-human life.

Consequences

Increased human morbidity and mortality, including both cancer and non-cancer health effects. Ecosystem effects.

Quality of the human environment

11. Indoor Air Pollutants: Radon

Human activities

Radon is a type of naturally occurring radiation that enters surrounding air, water, or both in human structures. The nature of some structures and ventilation systems allows for the accumulation of radon.

Changes in material fluxes

Uranium-238 and radium-226 are present in most soils and rocks in widely varying concentrations. Radon gas forms from the decay of radium-226 (the fifth daughter of uranium-238).

Changes in valued environmental components

Increased radiation level in human habitats.

Exposure

Inhalation by humans.

Consequences

Lung cancer.

12. Indoor Air Pollutants: Non-radon

Human activities

Combustion of fuels inside buildings. Use of chemicals, including cleaning solutions, pesticides, and office supplies, in buildings. Materials used to construct buildings. Level of ventilation in buildings is a key factor in determining levels of indoor air pollutants.

Changes in material fluxes

Increased levels of nitrogen oxides from combustion of fuels. Increased levels of chemical contaminants.

Changes in valued environmental components

Quality of air inside buildings.

Exposure

Inhalation by humans.

Consequences

Increased morbidity and mortality.

13. Radiation: Non-radon

Human activities

Medical exposure, operation of nuclear power plants, manufacturing of nuclear weapons, disposing of nuclear waste. Only anthropogenic sources of radiation are included, not background levels of radiation.

Changes in material fluxes

Medical X-rays; radioactive particles carried downwind from nuclear power plants and weapons plants; radioactive water leaks within the nuclear power plants and weapons plants; radioactive particles seeping into land and water from nuclear waste sites.

Changes in valued environmental components

Increased radiation in land, water, and air.

Exposure

Inhalation; ingestion through food supplies and water; absorption through the skin.

Consequences

Human mortality and morbidity. Damage to ecosystems.

14. Chemicals in the Workplace

Human activities

Use of chemicals (including pesticides) and biological pathogens in the workplace.

Changes in material fluxes

Increased levels of chemical contaminants in work environment.

Changes in valued environmental components

The safety of the occupational environment is reduced.

Exposure

Inhalation, ingestion, absorption through the skin.

Consequences

Human mortality and morbidity.

15. Accidental Chemical Releases

Human activities

Use, storage, and transport of chemicals.

Changes in material fluxes

Release of toxic chemicals to the environment. This usually entails a high-concentration release over a short period of time.

Changes in valued environmental components

Toxicity of atmosphere, land, water.

Exposure

Humans and ecosystems exposed to chemicals.

Consequences

Human mortality and morbidity. Ecosystem damage. Commercial losses due to shut-downs, cleanups, and reduction in quality of resources.

16. Food Contaminants

Human activities

Use of herbicides and pesticides in agriculture; disposal of industrial chemicals; radionuclides; inorganic compounds; hormones in meat such as chicken or beef.

Changes in material fluxes

Increase in herbicides, pesticides, and toxins in soil and on food; impure fish due to contamination by chemicals and heavy metals.

Changes in valued environmental components

Decrease in purity of food or increase in toxicity of food.

Exposure

Ingestion.

Consequences

Increased cancer and other health problems in humans. Possible damage to wildlife that consumes human foods.

Resource-depletion hazards

Land

17. Agricultural Land: Salinization, Alkalinization, Waterlogging

Human activities

Irrigation.

Changes in material fluxes

Excess irrigation water percolates down, raising the water-table. Capillary action pulls near-surface water up, and evaporation results in increased saline and alkaline salt deposition. In addition, if the water-table is raised high enough, it can waterlog crops by depriving their roots of needed air.

Changes in valued environmental components

Increased salinity, alkalinity, and waterlogging, leading to decreases in soil productivity.

Exposure

For humans, decreased food supplies. For crops, exposure to changed soil conditions.

Consequences

Decrease in productivity of land or complete loss of productive land, leading to crop losses. This may cause hunger, which in turn leads to increased human morbidity and mortality.

18. Agricultural Land: Soil Erosion

Human activities

Clearing of land (deforestation, burning, or harvesting). Cultivation of marginal lands. Overgrazing. Cultivation techniques such as furrowing and mechanization.

Changes in material fluxes

Degradation or loss of vegetative cover, which leads to increased soil erosion.

Changes in valued environmental components

Reduction in soil productivity, which leads to a reduction in the food supply (crop losses).

Exposure

For humans, decreased food supplies. For crops, exposure to changed soil conditions.

Consequences

Reduction in crops and livestock produced on land. This may cause hunger, which in turn leads to increased human morbidity and mortality.

19. Agricultural Land: Urbanization

Human activities

Urbanization.

Changes in material fluxes

Land is removed from agricultural production and put to use for human settlements, industry, and commercial purposes.

Changes in valued environmental components

Loss of agricultural land that leads to a reduction in the food supply.

Exposure

Human exposure to reduced food supply.

Consequences

Productive losses in agriculture. This may cause hunger, which in turn leads to increased human mortality and morbidity.

20. Groundwater

Human activities

Irrigation, drinking-water extraction, industrial use.

Changes in material fluxes

Groundwater is extracted at rates greater than recharge, leading to depletion.

Changes in valued environmental components

Decreased supplies of groundwater.

Consequences

Losses to agriculture, industry. Significant material and productivity effects.

Biological stocks

21. Wildlife

Human activities

Hunting, poaching (including subsistence hunting).

Change in material fluxes

Quantity of wildlife decreases.

Changes in valued environmental components

Wildlife populations decrease.

Consequences

Loss of species, reductions in human food supply, loss of recreation, secondary effects on the viability and population of other wildlife and vegetation.

22. Fish

Human activities

Fishing (including subsistence fishing).

Changes in material fluxes

Quantity of fish decreases.

Changes in valued environmental components

Fish populations drop below minimum sustainable levels.

Consequences

Loss of species, food supply, recreation.

23. Forests

Human activities

Agroforestry, firewood cutting, burning of forests for conversion to agricultural uses, forest fires.

Changes in material fluxes

Quantity and quality of forests decrease.

Changes in valued environmental components

Forest productivity declines.

Consequences

Loss of lumber, firewood, recreation; altered hydrological processes; soil erosion.

Natural hazards

24. Floods

Human activities

Although caused by natural fluctuations in weather patterns and seasonal weather patterns (e.g. monsoons, hurricanes, and other storm systems), human activities that alter the flow of water can influence flooding. The activities of importance include construction of dams and levees, wetlands conversion, modifications of coastline and coastal areas, and irrigation. In addition, human settlement patterns help determine the magnitude of losses due to flooding.

Changes in material fluxes

Increased water flow in rivers and lakes. Change in physical environment along shorelines.

Changes in valued environmental components

Increased or decreased flooding.

Exposure

For humans, ingestion of contaminated food and water. For humans, wildlife, land and physical property, contact with the force of fast-moving waters.

Consequences

Human morbidity due to ingestion of contaminated food and water. Human mortality due to increased morbidity and drowning. Welfare losses from damage to crops and agricultural land, to food and water supplies, and to physical infrastructure.

25. Droughts

Human activities

Although droughts are caused primarily by natural fluctuations in weather patterns, human activity such as deforestation can contribute to changes in the average level of rainfall (i.e. it can affect local climates), whereas a range of activities (annual husbandry, water management) can influence the susceptibility of land and crops to drought.

Changes in material fluxes

Decreased or more erratic rainfall.

Changes in valued environmental components

Decreased quantities of water in lakes, rivers, and reservoirs; decreased moisture in soil.

Exposure

Dehydration of vegetation, crops, livestock, wildlife, and humans. Drying and hardening of soil and increased susceptibility to the processes of desertification.

Consequences

Loss of crops and livestock populations. Hunger and thirst leading to increased human morbidity and mortality. Reduced function of hydro-electric facilities. Damage to agricultural and non-agricultural land.

26. Cyclones

"Cyclones" are defined generically to include hurricanes, typhoons, and tornadoes.

Human activities

Not applicable.

Changes in material fluxes

Increased rainfall, wind movements.

Changes in valued environmental components

Stability of atmospheric conditions.

Exposure

Exposure to rising water and high-velocity winds, insanitary conditions, contaminated drinking water, and loss of food supply.

Consequences

Loss of human life and property. Adverse effects on local environment.

27. Earthquakes

Human activities

Not applicable.

Changes in material fluxes

Change in terrestrial environment.

Changes in valued environmental components

Stability of land.

Exposure

Proximity to human structures undergoing damage (e.g. buildings and roads), secondary exposure to ensuing fires, insanitary conditions, and contamination of drinking water.

Consequences

Loss of human life and property; ecosystem effects.

28. Pest Epidemics

Human activities

Pest epidemics are due to changes in natural environment, or to an environment that is always hospitable to a large pest population. Can be influenced by agricultural practices, such as use of pesticides, irrigation, and cropping practices that are favourable to pests (e.g. monocultures).

Changes in material fluxes

Natural changes include changes in water and temperature. Anthropogenic changes include increases in standing water, increases in food sources for a given species, decreases in predators for a given species, increases in species resistance to pesticides.

Changes in valued environmental components

Improved pest breeding grounds and increased susceptibility of ecosystem to pest infestations. This leads to growth of the pest population and pest epidemics.

Exposure

Insect bites of humans and animals. Insects ingest or otherwise destroy crops.

Consequences

Loss of crops. Increased human mortality and morbidity from pest-carried disease or hunger due to loss of crops.

Appendix C: Tips on applying the method

The following is a list of suggestions that will aid in scoring:

1. The scoring process frequently encountered insufficient data – a problem for each of the four countries scored – and thus required much judgement and extrapolation on the part of the scorers. We suggest seeking as much expert knowledge as possible so that scoring time is not spent agonizing over problems that have "no available data."

2. Although we have made every effort to define the descriptors clearly, an inevitable tendency will prompt different users to develop different interpretations and rules for scoring. Therefore, if more than one person is involved in the scoring process it is important to discuss the descriptors and be certain that a common interpretation is being used for each and that they are being applied in a similar manner across all the various environmental problems. Such communication should occur before scoring begins and also throughout the entire scoring process, as interpretation of the descriptors and their application is an ongoing process. We suggest that, before scoring begins, the scorers sit down together for a practice scoring or a "dry run" for a pollution-based problem, a resource-based problem, and a natural hazard.

3. It is difficult, but important, to be consistent in interpreting and scoring descriptors across countries and across different environmental hazards. Differences in data availability render it tempting to fit the in-

terpretation of descriptors to the available data. During the scoring process, one must be aware of this problem and attempt to use the correct descriptor definition uniformly, regardless of the information available about a specific hazard or for a specific country.

4. When scoring, it is important to pay careful attention to the shift in logic that must be made when scoring pollution-based problems, resource-based problems, and natural hazards. This has particular importance for the material flux descriptors.

5. In many instances, scores will reflect qualitative judgements of the scorer. For example, for the "future consequence" descriptors, a scorer extrapolates from the descriptor *Change in Perturbation* available "trends data" out 25 years, and from this is able to discern if the problem will become more than half and less than twice as large as in the past. Before deciding on such a score, however, the scorer must take into account such factors as the political climate, and laws that may be enacted or have been enacted but have not yet had any effects. It is beneficial, especially in a team-scoring situation, to discuss these judgements with another member of your scoring team in order to obtain feedback and double-check your logic.

6. It is important to remember to annualize when scoring the current consequence descriptors for the natural hazards. The data that you will use will generally be for specific occurrences (e.g. deaths due to an earthquake in 1989). This information must be annualized if this is the only earthquake that has occurred for many years.

7. If no history of a particular hazard exists in a country but the hazard poses an immediate threat, you must score on the basis of the risk that you perceive. For example, since no major accidental chemical release has occurred in Kenya, historical data on which to base a score are non-existent. At the same time, a great potential for such a release exists in the light of increased industrialization. Thus, scores must take into account the risk that Kenya is facing.

8. When a country did not have a particular resource-based environmental problem, scorers faced difficulties in scoring the problem for specific descriptors. For example, India does not have a problem with its stock of fish: there is no net overfishing in the country. For descriptors such as *Persistence*, a score of 5 was given because it takes 1–10 years for a fish to reach maturity. Similarly, in the United States for stock of forests there is net reforestation, yet *Persistence* scored a 7 because it takes up to 100 years for a tree to reach maturity. One may be inclined to score a 1 for *Persistence* if a problem does not exist in a particular country. In using the definition of *Persistence*, however, the approach to scoring in these examples makes sense because they

reflect the "time period required to regenerate a specimen of the resource extracted."

9. The scale for the *Natural Ecosystem Impacts* descriptor provides very little resolution. Unfortunately, we were unable to design a descriptor with greater resolution that could be scored across the range of hazards.

Appendix D: Data from case studies; India, Kenya, the Netherlands, and the United States

Table 2.16A **India: Hazard taxonomy scores**

	1	2	3	4	5	6	7	8
Environmental hazards	Spatial Extent	Distur- bance to Environ- ment	Change in Pertur- bation	Anthro- pogenic Flux	Changes in Anthro- pogenic Flux	Per- sist- ence	Recur- rence	Popula- tion Exposed
1. Fresh Water: Biological Contamination	5	9	7	9	7	3	1	9
2. Fresh Water: Metals and Toxic Contamination	5	9	6	9	6	9	1	7
3. Fresh Water: Eutrophication	5	7	8	7	8	7	1	7
4. Fresh Water: Sedimentation	5	7	8	7	8	3	1	7
5. Ocean Water	5	9	8	9	8	9	1	7
6. Stratospheric Ozone Depletion	9	9	8	9	2	9	1	9
7. Thermal Radiation	9	5	7	1	6	9	1	9
8. Acidification	5	9	8	3	9	1	1	5
9. Ground-Level Ozone Formation	5	9	8	9	9	1	1	7
10. Toxic Air Pollutants	5	9	8	9	8	3	1	7
11. Indoor Air: Radon	1	9	5	9	5	3	1	1
12. Indoor Air: Non-radon	1	9	5	9	5	1	1	9
13. Radiation: Non-radon	9	5	6	3	6	9	1	9
14. Chemicals in the Workplace	1	9	6	9	6	3	1	5
15. Accidental Chemical Releases	5	9	5	9	5	7	5	9
16. Food Contamination	5	9	8	9	8	9	1	9
17. Agricultural Land: Salinization, Alkalinization	1	5	6	9	5	7	1	5
18. Agricultural Land: Soil Erosion	5	5	6	5	6	9	1	9
19. Agricultural Land: Urbanization	1	5	6	9	6	9	1	3
20. Groundwater	5	5	6	1	6	3	1	5
21. Wildlife	5	5	7	5	7	9	1	3
22. Fish	3	1	5	3	5	5	1	1
23. Forests	5	7	8	9	8	7	1	9
24. Floods	5	9	8	9	8	3	3	7
25. Droughts	5	7	5	1	5	5	5	7
26. Cyclones	5	9	5	1	5	1	3	9
27. Earthquakes	5	9	5	1	5	1	3	7
28. Pest Epidemics	5	9	5	1	5	3	5	7

9	10	11	12	13	14	15	16	17	18	19
Land Ex-posed	Delay	Current Mortality	Current Morbidity	Current Eco-system	Current Material Product	Re-covery Period	Future Health	Future Eco-system	Future Material Product	Trans-na-tional
9	1	7	8	5	3	3	9	9	5	1
7	1	6	7	5	7	9	9	5	5	1
7	1	5	6	5	2	7	9	9	5	1
7	3	1	1	5	6	7	1	9	9	5
7	1	3	5	5	4	9	9	9	9	1
9	5	1	1	1	1	9	9	9	9	9
9	7	1	1	1	1	9	9	9	9	9
7	1	4	5	1	4	3	9	9	9	1
7	1	1	6	1	1	5	9	9	9	1
7	1	4	5	5	4	7	9	9	9	1
1	7	4	4	1	1	7	5	1	1	1
9	1	5	6	1	4	7	5	1	5	1
9	7	4	4	1	1	7	9	1	1	1
7	1	4	5	1	1	7	5	1	1	1
7	1	3	4	1	3	7	5	5	5	1
9	3	6	7	1	1	9	9	1	1	1
7	1	5	7	5	7	9	5	1	5	1
9	1	6	8	5	8	9	5	9	5	1
3	1	5	7	5	1	9	5	5	1	1
7	1	1	1	5	5	5	9	9	5	1
3	1	1	1	9	5	9	1	9	5	1
1	1	1	1	1	1	1	1	1	1	1
7	1	3	5	5	6	7	9	9	9	1
5	1	4	5	5	6	9	9	9	9	1
7	3	4	6	5	6	5	5	5	5	1
9	1	4	5	1	6	5	5	5	9	1
7	1	3	3	1	5	5	5	1	5	1
7	1	5	6	5	7	7	5	5	5	5

Table 2.17A **Kenya: Hazard taxonomy scores**

	1	2	3	4	5	6	7	8
Environmental hazards	Spatial Extent	Distur-bance to Environ-ment	Change in Pertur-bation	Anthro-pogenic Flux	Changes in Anthro-pogenic Flux	Per-sist-ence	Recur-rence	Popula-tion Exposed
1. Fresh Water: Biological Contamination	5	9	5	9	5	3	1	9
2. Fresh Water: Metals and Toxic Contamination	5	9	6	9	6	9	1	5
3. Fresh Water: Eutrophication	5	7	8	7	8	7	1	3
4. Fresh Water: Sedimentation	5	9	6	9	6	3	1	7
5. Ocean Water	5	9	7	9	7	9	1	5
6. Stratospheric Ozone Depletion	9	9	8	9	2	9	1	9
7. Thermal Radiation	9	5	7	1	6	9	1	9
8. Acidification	5	5	6	3	6	1	1	5
9. Ground-Level Ozone Formation	5	9	6	9	6	1	1	7
10. Toxic Air Pollutants	5	9	6	9	6	3	1	7
11. Indoor Air: Radon	1	9	5	9	5	3	1	1
12. Indoor Air: Non-radon	1	9	5	9	5	1	1	9
13. Radiation: Non-radon	9	5	6	1	6	9	1	9
14. Chemicals in the Workplace	1	9	8	9	9	3	1	5
15. Accidental Chemical Releases	5	3	5	5	5	7	9	7
16. Food Contamination	5	9	8	9	8	9	1	9
17. Agricultural Land: Salinization, Alkalinization	1	3	5	3	5	7	1	1
18. Agricultural Land: Soil Erosion	5	5	6	5	6	9	1	9
19. Agricultural Land: Urbanization	1	5	6	9	6	9	1	1
20. Groundwater	5	3	6	5	8	3	1	5
21. Wildlife	5	7	6	7	6	9	1	7
22. Fish	5	7	9	7	9	5	1	5
23. Forests	3	7	7	7	7	7	1	9
24. Floods	5	9	6	5	6	3	7	5
25. Droughts	5	5	5	1	6	5	5	9
26. Cyclones	5	3	5	1	5	1	9	1
27. Earthquakes	5	9	5	1	5	1	7	9
28. Pest Epidemics	5	9	5	1	5	3	5	9

9	10	11	12	13	14	15	16	17	18	19
Land Ex-posed	Delay	Current Mortality	Current Morbidity	Current Eco-system	Current Material Product	Re-covery Period	Future Health	Future Eco-system	Future Material Product	Trans-national
9	1	6	9	1	1	3	5	5	1	5
3	1	2	3	1	1	9	5	5	1	1
3	1	1	1	5	1	7	9	9	1	1
9	3	1	1	5	6	7	1	5	9	1
5	1	2	4	5	4	9	5	5	5	1
9	5	1	1	1	1	9	9	9	9	9
9	7	1	1	1	1	9	9	9	9	9
5	3	1	2	1	2	3	9	1	9	1
7	1	1	6	1	1	5	9	9	9	1
7	1	3	4	5	3	7	5	5	5	1
1	7	4	4	1	1	7	5	1	1	1
9	1	6	7	1	4	7	5	1	5	1
9	7	4	4	1	1	7	9	1	1	1
7	1	4	6	1	1	7	9	1	1	1
7	1	1	2	5	3	7	5	5	5	1
7	3	6	7	1	1	7	9	1	1	1
1	1	1	1	1	1	1	5	1	5	1
9	1	6	8	5	7	9	5	9	5	1
5	1	1	1	5	1	9	1	9	1	1
7	1	1	1	1	1	5	1	9	5	1
7	1	2	3	9	5	9	5	5	5	5
7	3	2	3	5	5	5	9	9	9	5
9	1	4	6	5	6	7	9	9	9	1
3	1	3	4	1	1	9	5	1	1	1
9	3	5	7	5	7	9	5	5	5	1
1	1	1	1	1	1	1	1	1	1	1
9	1	2	3	1	4	5	5	1	5	1
9	1	5	6	5	8	7	5	5	5	5

Table 2.18A **The Netherlands: Hazard taxonomy scores**

	1	2	3	4	5	6	7	8
Environmental hazards	Spatial Extent	Distur-bance to Environ-ment	Change in Pertur-bation	Anthro-pogenic Flux	Changes in Anthro-pogenic Flux	Per-sist-ence	Recur-rence	Popula-tion Exposed
1. Fresh Water: Biological Contamination	5	9	1	9	1	3	1	5
2. Fresh Water: Metals and Toxic Contamination	5	9	1	9	1	9	1	9
3. Fresh Water: Eutrophication	5	9	6	9	6	7	1	9
4. Fresh Water: Sedimentation	5	7	5	7	5	3	1	5
5. Ocean Water	5	9	5	9	1	9	1	7
6. Stratospheric Ozone Depletion	9	9	8	9	2	9	1	9
7. Thermal Radiation	9	5	7	1	6	9	1	9
8. Acidification	5	9	2	9	3	1	1	9
9. Ground-Level Ozone Formation	5	9	5	9	6	1	1	9
10. Toxic Air Pollutants	5	9	5	9	5	3	1	9
11. Indoor Air: Radon	1	9	6	9	6	3	1	3
12. Indoor Air: Non-radon	1	9	4	9	4	1	1	9
13. Radiation: Non-radon	9	7	4	5	4	9	1	9
14. Chemicals in the Workplace	1	9	4	9	4	3	1	3
15. Accidental Chemical Releases	5	9	5	9	5	7	7	9
16. Food Contamination	5	9	7	9	6	9	1	9
17. Agricultural Land: Salinization, Alkalinization	1	5	5	9	5	7	1	1
18. Agricultural Land: Soil Erosion	5	5	5	3	5	9	1	1
19. Agricultural Land: Urbanization	1	5	6	9	6	9	1	3
20. Groundwater	5	5	6	5	8	3	1	7
21. Wildlife	5	7	6	7	6	9	1	3
22. Fish	5	5	7	5	7	5	1	1
23. Forests	3	7	4	1	4	7	1	1
24. Floods	5	7	4	1	4	3	7	7
25. Droughts	5	5	5	3	5	5	5	7
26. Cyclones	5	3	5	1	5	1	9	1
27. Earthquakes	5	3	5	1	5	1	9	1
28. Pest Epidemics	5	9	4	1	4	3	5	1

9	10	11	12	13	14	15	16	17	18	19
Land Exposed	Delay	Current Mortality	Current Morbidity	Current Eco-system	Current Material Product	Re-covery Period	Future Health	Future Eco-system	Future Material Product	Trans-national
7	1	2	4	5	4	3	5	5	5	5
9	1	5	5	5	4	9	5	5	5	5
9	1	4	5	5	2	7	5	5	5	5
7	3	1	1	5	2	5	1	5	5	5
9	1	5	6	5	4	9	5	5	5	5
9	5	1	1	1	1	9	9	9	9	9
9	7	1	1	1	1	9	9	9	9	9
9	3	3	4	9	5	7	1	5	1	5
9	1	1	7	5	7	5	5	5	5	5
9	1	5	6	5	6	7	5	5	5	5
3	7	5	5	1	1	7	5	1	1	1
9	1	4	5	1	1	7	5	1	1	1
9	7	5	5	1	1	9	5	1	1	9
7	1	4	5	1	1	7	5	1	1	1
9	1	2	4	5	4	7	5	5	5	5
9	3	4	5	1	1	7	5	1	1	5
1	1	1	1	1	1	3	1	1	1	5
1	1	1	1	1	5	9	1	1	5	1
5	1	1	1	5	1	9	1	5	1	1
9	1	1	1	5	6	7	1	5	5	1
9	1	1	1	9	2	9	1	5	5	1
3	3	1	1	5	1	5	1	5	1	5
1	1	1	1	1	1	7	1	1	1	1
7	1	1	1	1	3	3	1	1	5	1
7	3	1	1	5	6	5	1	5	5	1
1	1	1	1	1	1	1	1	1	1	1
1	1	1	1	1	1	1	1	1	1	1
7	1	1	1	5	2	7	1	5	5	1

Table 2.19A **United States: Hazard taxonomy scores**

	1	2	3	4	5	6	7	8
Environmental hazards	Spatial Extent	Distur-bance to Environ-ment	Change in Pertur-bation	Anthro-pogenic Flux	Changes in Anthro-pogenic Flux	Per-sist-ence	Recur-rence	Popula-tion Exposed
1. Fresh Water: Biological Contamination	5	9	2	3	2	3	1	5
2. Fresh Water: Metals and Toxic Contamination	5	9	5	9	5	9	1	7
3. Fresh Water: Eutrophication	5	5	5	5	5	7	1	3
4. Fresh Water: Sedimentation	5	5	6	5	6	3	1	3
5. Ocean Water	5	9	6	9	6	9	1	5
6. Stratospheric Ozone Depletion	9	9	8	9	2	9	1	9
7. Thermal Radiation	9	5	7	1	6	9	1	9
8. Acidification	5	9	4	9	4	1	1	9
9. Ground-Level Ozone Formation	5	9	4	9	4	1	1	7
10. Toxic Air Pollutants	5	9	3	9	3	3	1	9
11. Indoor Air: Radon	1	9	5	9	5	3	1	5
12. Indoor Air: Non-radon	1	9	6	9	6	1	1	9
13. Radiation: Non-radon	9	7	5	5	5	9	1	9
14. Chemicals in the Workplace	1	9	4	9	5	3	1	3
15. Accidental Chemical Releases	5	9	5	9	5	7	3	9
16. Food Contamination	5	9	5	9	5	9	1	9
17. Agricultural Land: Salinization, Alkalinization	1	7	6	9	7	7	1	1
18. Agricultural Land: Soil Erosion	5	5	6	5	4	9	1	3
19. Agricultural Land: Urbanization	1	5	6	9	6	9	1	1
20. Groundwater	5	5	6	5	6	3	1	5
21. Wildlife	5	9	6	5	4	9	1	3
22. Fish	5	7	7	7	7	5	1	3
23. Forests	3	7	4	3	4	7	1	1
24. Floods	5	9	5	1	5	3	3	7
25. Droughts	5	7	5	1	5	5	5	7
26. Cyclones	5	9	5	1	5	1	1	9
27. Earthquakes	5	9	5	1	5	i	3	7
28. Pest Epidemics	5	9	4	1	4	3	5	5

9	10	11	12	13	14	15	16	17	18	19
Land Ex-posed	Delay	Current Mortality	Current Morbidity	Current Eco-system	Current Material Product	Re-covery Period	Future Health	Future Eco-system	Future Material Product	Trans-na-tional
5	1	2	5	5	6	3	1	1	1	1
5	1	5	6	5	6	9	5	5	5	5
5	1	1	2	5	6	9	5	5	5	5
5	3	1	1	5	6	9	1	5	5	1
5	1	2	3	5	6	9	5	5	5	5
9	5	1	1	1	1	9	9	9	9	9
9	7	1	1	1	1	9	9	9	9	9
9	3	5	6	5	6	7	1	5	1	5
7	1	1	6	5	6	5	1	5	1	1
9	1	5	6	5	6	7	5	5	1	1
5	7	5	6	1	1	7	5	1	1	1
9	1	5	6	1	1	7	5	1	1	1
9	7	5	5	1	1	7	5	1	1	1
7	1	5	6	1	1	7	5	1	1	1
7	1	2	4	5	5	7	5	5	5	1
9	3	5	6	1	1	9	5	1	1	5
3	1	1	1	1	5	7	1	1	5	1
5	1	1	1	1	6	9	1	1	5	1
5	1	1	1	5	1	9	1	5	1	1
7	1	1	1	5	6	9	1	5	9	1
3	1	1	1	9	4	9	1	5	5	5
7	3	1	1	5	1	9	1	9	9	5
3	1	1	1	5	1	7	1	9	1	1
5	1	3	5	1	7	5	5	1	5	1
7	3	1	1	5	5	5	1	5	5	1
7	1	3	5	1	7	5	5	1	9	1
7	1	3	4	1	6	5	5	1	5	1
7	1	1	3	5	2	7	1	5	5	5

Appendix E: Principal-component analysis

A brief introduction to principal-component analysis[1]

Principal-component analysis (PCA) is a method for expressing multi-dimensional data as new variables that are mutually uncorrelated, linear combinations of the original variables. These new variables are called principal components (PCs). The specific aims of performing PCA on data are as follows:

1. *Reducing dimensionality.* If some variables are correlated, then the effective dimensionality may be decreased by expressing their information content in their PCs. If the first few PCs account for most of the variation in the original data, then the rest of the PCs may be neglected, reducing the effective dimensionality of the problem.

2. *Arranging variables into meaningful groups.* It may be possible to interpret or assign meaning to the set of variables contributing significantly to the more important PCs. The PCs may also be rotated to maximize the difference in loadings between variables. Rotation modifies the PCs without altering their underlying structure, to allow easier determination of the significant variables in each PC and better interpretation.

3. *Discarding redundant variables.* Some variables may be redundant and may therefore be discarded. PCA gives a method for systematically identifying variables with the least contribution to the overall variability in the data.

4. *Using the components in subsequent analysis*. PCA reduces the effective dimensionality of the data. For example, if the first two components account for a large proportion of the total variation, then the data may be plotted in two dimensions. This can be used for classification and cluster analysis.

Results

The results presented below are based on a PCA of the descriptors. After identifying the important PCs, they were rotated using the technique of varimax rotation in an effort to improve interpretation of the PCs. Data generated by all four case studies were used in this analysis.

Correlations

The correlation matrix of the descriptors is shown in Table 2.20A. Listed in Table 2.21A are all the pairs that have correlations greater than 0.5. Of particular note are the first two pairs that have correlations greater than 0.85. For these two pairs, in future analysis, one descriptor may suffice.

At least part of the explanation for the high correlation between *Current Morbidity* and *Current Mortality* is a lack of data, particularly on morbidity. Thus, morbidity was most often estimated based on mortality data and generally assumed to be one order of magnitude greater. Better data could determine whether this correlation describes the actual relationship between these two indicators of human health consequences, or if it is simply an artefact of our scoring inferences. Without better data, it is probably valid to drop the *Current Morbidity* descriptor.

In contrast, the correlation between *Population Exposed* and *Land Area Exposed* is likely to correspond to the actual characteristics of environmental hazards. Thus, from the viewpoint of reducing dimensionality, a single descriptor would suffice; however, from the viewpoint of characterizing hazards, there may be some argument for including both.

Principal components

The PCs are shown in Table 2.22A. The 19 PCs are presented in order of increasing importance from left to right. The first 19 numbers in each column are the component correlations (or factor loadings); the final number is the cumulative percentage of variance explained by the PCs. We see that the first six PCs account for nearly 80 per cent of the variance in the data. This is not a surprising result: it is quite usual for data sets of these dimensions to result in six factors that describe most of the variance.

Table 2.20A **Correlation matrix for descriptors**

Descriptor no.

Descriptor no.	1	2	3	4	5	6	7	8	9	10	11	12	13	14	15	16	17	18	19
1	1.00																		
2	-0.04	1.00																	
3	0.21	-0.07	1.00																
4	-0.34	0.45	0.16	1.00															
5	-0.13	-0.17	0.67	0.05	1.00														
6	0.30	-0.21	0.29	0.14	-0.03	1.00													
7	0.08	-0.17	-0.22	-0.47	-0.12	-0.28	1.00												
8	0.49	0.40	0.09	0.10	-0.06	0.03	-0.06	1.00											
9	0.41	0.36	0.06	0.13	-0.07	0.02	-0.15	0.85	1.00										
10	0.43	-0.06	0.18	-0.10	-0.09	0.24	-0.18	0.19	0.15	1.00									
11	-0.19	0.41	-0.08	0.29	0.05	-0.03	-0.16	0.44	0.35	0.02	1.00								
12	-0.19	0.50	-0.10	0.35	0.08	-0.16	-0.15	0.50	0.39	-0.14	0.88	1.00							
13	-0.01	0.00	-0.07	0.14	0.04	0.15	-0.06	-0.04	0.09	-0.33	-0.08	-0.01	1.00						
14	0.03	0.12	-0.16	-0.12	-0.01	-0.16	0.05	0.27	0.21	-0.37	0.15	0.25	0.49	1.00					
15	0.15	0.18	0.26	0.32	0.01	0.60	-0.35	0.20	0.28	0.21	0.15	0.05	0.26	0.08	1.00				
16	0.33	0.27	0.51	0.16	0.25	0.18	-0.23	0.51	0.38	0.29	0.36	0.39	-0.26	-0.12	0.16	1.00			
17	0.39	-0.03	0.46	0.10	0.23	0.23	-0.20	0.21	0.28	0.00	-0.21	-0.12	0.44	0.13	0.27	0.30	1.00		
18	0.42	0.06	0.43	-0.12	0.15	0.02	-0.07	0.33	0.34	0.03	-0.26	-0.17	0.11	0.34	0.14	0.36	0.60	1.00	
19	0.55	0.10	0.11	0.04	-0.33	0.33	-0.16	0.31	0.33	0.44	-0.16	-0.19	0.01	-0.14	0.27	0.20	0.33	0.30	1.00

Table 2.21A **Highly correlated descriptors**

Pairs of descriptors	Correlation coefficients
Current Morbidity (11) and *Current Mortality* (12)	0.88
Population Exposed (8) and *Land Exposed* (9)	0.86
Persistence (6) and *Recovery Period* (15)	0.60
Spatial Extent (1) and *Transnational* (19)	0.55
Change in Perturbation (3) and *Future Health Consequences* (16)	0.51
Population Exposed (8) and *Future Health Consequences* (16)	0.51
Disturbance to Environment (2) and *Current Morbidity* (12)	0.50
Population Exposed (8) and *Current Morbidity* (12)	0.50

The magnitude of the component correlations for a given PC indicates the relative contribution of the descriptor to that PC. It is difficult to pick the descriptors most relevant to each PC, as the component correlations decline rather steadily. Accordingly, for each PC there is not an obvious cut-off point, nor are there only a few descriptors with high component correlations, although all others have low correlations. Therefore, we next performed a varimax rotation in order to maximize the difference in loadings between variables.

Varimax rotation

The results of the varimax rotation are presented in Table 2.23A. As mentioned above, the first six components account for about 80 per cent of the variance. In addition, selecting more than the first six components did not modify the results of the rotation; therefore, we present only the rotation using the first six PCs. Although the varimax rotation improves the distinctiveness of the factors when compared with the PCA without rotation, the problem of cut-off points remains. The analysis has still not provided a strong set of descriptor groupings.

Table 2.24A lists the descriptors that have component loadings greater than 0.5. The first four PCs lend themselves to meaningful interpretation and are also quite similar to the simple aggregation schemes which we developed without the aid of PC analysis (in other words, to aggregation schemes that are intuitively meaningful). The last two PCs are difficult to interpret.

PC1 is almost identical to the aggregation **Pervasiveness in Space**. PC2 is identical to the aggregation **Current Human Health** (descriptors 11 and 12). As we have discussed above, the two descriptors composing PC2 are highly correlated.

PC3 is perhaps the most interesting result. The heaviest loading is on the descriptor *Annual Change in Perturbation Level*. Continuing in order

Table 2.22A Results of PCA for descriptors' component correlations (factor loadings)

Component correlation	PC 1	PC 2	PC 3	PC 4	PC 5	PC 6	PC 7	PC 8	PC 9	PC 10	PC 11	PC 12	PC 13	PC 14	PC 15	PC 16	PC 17	PC 18	PC 19
1	0.548	0.506	-0.480	0.049	0.065	-0.140	0.063	-0.060	0.019	-0.070	0.241	0.237	0.046	-0.160	0.047	-0.050	0.160	-0.020	0.033
2	0.397	-0.580	-0.090	-0.010	-0.120	0.462	0.157	0.278	-0.140	-0.240	0.226	0.113	-0.110	0.020	-0.100	0.052	-0.070	0.000	0.021
3	0.470	0.371	0.549	-0.130	0.442	0.005	0.107	0.098	-0.100	-0.030	-0.090	0.141	0.009	-0.030	-0.070	-0.230	-0.080	0.064	-0.040
4	0.299	-0.410	0.501	-0.200	-0.350	0.377	0.159	-0.090	-0.050	0.104	-0.130	0.074	0.317	-0.080	0.049	0.015	0.068	-0.030	0.045
5	0.113	0.067	0.658	0.065	0.594	-0.170	-0.050	-0.110	-0.270	-0.040	-0.030	0.150	-0.110	0.032	0.027	0.186	0.046	-0.050	0.041
6	0.360	0.387	0.256	-0.270	-0.450	-0.450	0.180	0.117	0.231	0.145	0.044	0.106	0.040	0.038	-0.160	0.112	-0.070	0.018	0.035
7	-0.400	0.075	-0.450	0.249	0.294	-0.200	0.594	0.102	-0.150	-0.120	-0.160	-0.090	0.108	0.002	-0.010	0.030	0.000	0.005	0.035
8	0.774	-0.280	-0.400	0.120	0.131	-0.090	0.034	-0.070	-0.110	0.203	0.000	0.072	0.089	-0.010	0.032	0.071	-0.100	-0.060	-0.180
9	0.745	-0.220	-0.330	0.177	0.002	-0.020	-0.040	-0.180	-0.260	0.327	0.018	-0.090	-0.060	0.105	-0.050	-0.060	-0.010	0.048	0.132
10	0.343	0.322	-0.280	-0.600	-0.010	-0.120	-0.300	-0.070	-0.230	-0.330	-0.030	-0.120	0.217	0.011	-0.060	0.047	-0.050	0.032	0.012
11	0.341	-0.790	0.008	-0.180	0.059	-0.330	-0.010	-0.070	0.097	-0.140	-0.110	-0.070	-0.080	-0.070	-0.130	-0.100	0.049	-0.150	0.019
12	0.361	-0.850	0.023	0.000	0.127	-0.170	0.025	-0.100	0.156	-0.100	-0.070	0.001	-0.040	-0.090	0.011	0.064	0.072	0.192	-0.030
13	0.095	0.039	0.293	0.681	-0.460	-0.130	0.072	-0.280	-0.070	-0.240	0.078	0.024	0.062	0.216	-0.040	-0.050	0.059	0.006	-0.060
14	0.134	-0.240	-0.060	0.798	-0.060	-0.210	-0.290	0.248	0.069	-0.100	-0.120	0.152	0.123	-0.050	0.083	-0.010	-0.110	-0.010	0.081
15	0.538	0.090	0.301	-0.080	-0.510	-0.260	0.047	0.350	-0.250	0.000	0.000	-0.200	-0.120	-0.050	0.160	-0.020	0.062	0.001	-0.030
16	0.707	-0.070	0.064	-0.260	0.437	0.010	0.100	0.068	0.320	-0.080	0.100	-0.090	0.071	0.227	0.173	-0.020	-0.010	-0.030	0.035
17	0.551	0.431	0.277	0.398	0.007	0.166	0.119	-0.290	0.125	-0.110	0.067	-0.240	-0.060	-0.210	0.011	0.044	-0.120	-0.020	0.025
18	0.528	0.400	-0.020	0.454	0.269	0.269	-0.140	0.281	0.113	0.056	-0.120	-0.150	0.057	0.026	-0.160	0.060	0.152	0.000	-0.030
19	0.516	0.411	-0.350	-0.160	-0.330	0.232	0.048	-0.110	0.072	-0.130	-0.360	0.170	-0.210	0.062	0.056	0.027	0.000	-0.010	0.009
	22.4	39.3	51.0	62.6	72.4	78.1	81.8	85.1	88.0	90.6	92.5	94.3	95.9	97.0	97.8	98.6	99.2	99.6	100.0

Principal components — Decreasing importance

Table 2.23A **Result of varimax rotation of first six principal components**

PC 6	PC 5	PC 4	PC 3	PC 2	PC 1	Descriptor no.
-0.514	0.401	0.142	0.207	0.109	0.699	1
0.727	-0.017	0.012	0.289	0.141	0.286	2
-0.326	-0.113	-0.230	0.690	0.177	-0.213	3
0.537	-0.610	-0.297	0.520	0.144	-0.264	4
-0.140	-0.093	-0.065	0.447	-0.123	-0.496	5
-0.551	-0.488	-0.310	0.441	-0.219	-0.024	6
-0.158	0.530	0.332	-0.575	-0.139	0.168	7
0.207	0.385	0.196	0.441	-0.210	0.752	8
0.188	0.247	0.238	0.454	-0.116	0.677	9
-0.339	0.205	-0.533	0.141	0.048	0.421	10
0.547	0.033	-0.173	0.293	-0.660	0.175	11
0.670	0.068	-0.008	0.318	-0.546	0.174	12
-0.095	-0.525	0.610	0.237	-0.097	-0.197	13
0.111	0.003	0.795	0.081	-0.293	0.123	14
-0.203	-0.566	-0.133	0.615	-0.192	0.032	15
0.071	0.244	-0.267	0.632	-0.026	0.323	16
-0.304	-0.201	0.336	0.614	0.347	0.060	17
-0.229	0.192	0.449	0.438	0.424	0.295	18
-0.256	0.041	-0.088	0.248	0.396	0.547	19
78.20	72.40	62.60	51.00	39.30	22.40	Percentage variance explained, cumulative

Principal components (after rotation) Decreasing importance ←

of decreasing importance, the next three descriptors include all but one of the consequence descriptors that describe the future (*Future Human Health Consequences, Future Ecosystem Consequences, Recovery Period*). The last two descriptors with loadings greater than 0.5 are two more material flux descriptors (*Anthropogenic Flux, Recurrence*). This suggests that an aggregation designated "Trends" may be worth investigating in addition to, or as a replacement for, the aggregation **Future Consequences**. In such an aggregation, the material flux descriptors could serve as a balance to the "business as usual" scenarios by including descriptors that characterize current trends in pollution releases and trends in the perturbation of natural environmental cycles, rather than only those based on assumption-laden future scenarios. We had considered a similar approach as one of our aggregation schemes but chose not to present it in the final report owing to lack of space, its relative complexity compared

Table 2.24A **Descriptors forming each rotated PC**[a]

PC 1	PC 2
8. *Population Exposed*	11. *Human Mortality (Current*
1. *Spatial Extent*	*Annual)*
9. *Land Area or Resource*	12. *Human Morbidity (Current*
Exposed	*Annual)*
19. *Transnational*	

PC 3	PC 4
3. *Annual Change in*	14. *Material and Productivity*
Perturbation Level	*Losses (Current Annual)*
16. *Future Human Health*	13. *"Natural" Ecosystem Impacts*
Consequences	*(Current Annual)*
15. *Recovery Period*	10. *Delay*
17. *Future Ecosystem*	
Consequences	
7. *Recurrence*	
4. *Anthropogenic Flux*	

PC 5	PC 6
4. *Anthropogenic Flux*	2. *Disturbance to the Environment*
15. *Recovery Period*	*(Magnitude of Perturbation)*
7. *Recurrence*	12. *Human Morbidity (Current*
13. *"Natural" Ecosystem*	*Annual)*
Impacts (Current Annual)	6. *Persistence*
	11. *Human Mortality (Current*
	Annual)
	4. *Anthropogenic Flux*
	1. *Spatial Extent*

a. Includes descriptors with component correlations > 0.5. Arrows indicate direction of increasing importance.

with the other aggregation schemes, and the similarity of the results to those we obtained from the aggregation **Future Consequences**.

PC4 suggests that current ecosystem and material losses are more likely in problems in which consequences have long delays.

Conclusions

Compared with the simple aggregation schemes that we first developed for data analysis, the PCA did not add significantly to our ability to interpret the data:

1. The PCA, even after varimax rotation, did not result in strongly differentiated factors.
2. The first several factors were similar to the aggregation schemes that we had previously developed.

3. Only the first four factors lent themselves to meaningful interpretation.

This is not to suggest that the analysis was completely without merit. First, we did identify two pairs of highly correlated variables, suggesting places where we could reduce our number of descriptors in the future. (However, a simple analysis of pair-wise correlations would give the same result.) Second, we were encouraged to consider using a "Trends" aggregation, which would be based on both current trends and expected future trends. Although this would be more complex than the aggregation scheme **Total Future Consequences** that we used in this analysis, its expanded definition may help overcome some of the difficulties of relying on "business as usual" scenarios.

Note

1. Most texts on multivariate statistics discuss this method. Two that we used are *Multivariate Analysis* (Mardia, Kent, and Bibby 1980) and *A Primer of Multivariate Statistics* (Harris 1975).

3

The risk transition and developing countries

Kirk R. Smith

Economic growth through technological development helps meet many basic human needs in developing countries. It is clear, for example, that increased availability of chemicals through importation and local manufacture has improved public health, food production, comfort, and labour efficiency. On the other hand, tragic events (e.g. the Bhopal accident, the Chernobyl disaster) and long-term problems (e.g. chronic excessive exposures to pesticides) have raised questions about the way chemicals and other technologies should be introduced in the development process. In addition, the global impacts of particular chemicals (ozone-damaging greenhouse gases) are commanding growing attention in countries throughout the development spectrum. In some cases, the concern suggests that the benefits of certain classes and uses of chemicals may not sufficiently counterbalance the attendant risks; on the other hand, it is plausible that excessive attention to safety can sometimes inhibit industrial development to the point where no overall benefit is achieved (Kates 1978; Whyte and Burton 1980).

Other classes of health risk follow similar patterns. For example, many poor countries are extremely vulnerable to natural disasters such as those caused by violent cyclones and earthquakes (Hewitt 1997; Mileti 1999; Munich Re 1999; United Nations 1999; USNRC 1987a). Increased attention to disaster forecasting and mitigation would have benefits, but at some investment of financial, managerial, and scientific resources with, perhaps significant, opportunity cost (Mitchell 1988, 1999).

Strategies for the avoidance and mitigation of risks related to techno-logical and other hazards thus need to be blended into development strategies. In this way, perhaps developing countries can enhance well-being without experiencing unnecessary episodes of human illness, eco-system degradation, and costly cleanups of the kind that have occurred in Europe, North America, and Japan, as well as with apparently increasing frequency in the developing world itself. Such improvement calls for appropriate risk-analysis methods, exploration of which is, as noted in chapter 1, a primary purpose of this volume.

Criteria for evaluating and managing environmental risks and the hazards of development tend to be fragmented according to the source of risk (e.g. industrial waste), the type of impact (e.g. cancer), the nature of the political resistance to change (e.g. transnational corporations), and the group to which the victims belong (e.g. urban or rural). An integrated framework is needed in order to evaluate the pros and cons of decreasing one kind of risk at the expense of increasing another.

Keeping systematic inventories, performing meaningful risk compari-sons, and undertaking effective control measures for the vast and growing array of technological and other hazards are daunting jobs, even with the relatively impressive resources available in developed countries. Recog-nition of the need to engage these tasks has increased, however, as evid-enced by plaintive and eloquent pleas for quantitative and standardized techniques to deal with the growing number and complexity of hazards (Ruckelshaus 1983). Thus, although burdened by some controversies and remaining inconsistencies (Koshland 1987), government agencies in the United States and elsewhere are adopting various risk-assessment proto-cols to provide quantitative frameworks for evaluating technological hazards and options for their control (Lave 1982; USDHHS 1986; USEPA 1984, 1986, 1987, 1990; US Nuclear Regulatory Commission 1984; USNRC 1983b). In developed countries, therefore, risk assessment seems to offer attractive advantages in achieving a rational management of hazards, despite being frequently exasperating owing to its sometimes insatiable appetite for data and intransigent propensity to hide opinion under fact (White 1980; Whittemore 1983).

It is natural to hope that these techniques be transferrable in a more or less intact state to the developing countries so that risk issues can be ad-dressed during economic development (Gumley and Inamdar 1987; Lave and Lave 1985). It is argued here, however, that the developing world offers a quite different set of challenges – ones that impose substantially different demands on risk assessments.

First, of course, the size of the economic, technical, administrative, educational, and political barriers to applying modern risk-assessment and management techniques would seem to be considerably greater

(Bowonder 1981; Covello and Frey 1990). The most obvious barriers are the lack of trained personnel and comprehensive databases, whether industrial, medical, or environmental. These gaps have elicited calls for the development of "national risk profiles" in developing countries (Kasperson and Kasperson 1987; Kasperson and Morrison 1982; Whyte and Burton 1980). In addition, however, these deficiencies imply the need for risk assessments that are as simple as possible, a characterization that can be achieved only through careful matching of the system boundaries and assessment design to the question at hand (Carpenter et al. 1990; Smith, Carpenter, and Faulstich 1988). The multidimensional treatment needed for such characterization is explored in depth in the preceding chapter.

More fundamental differences, however, limit the transferability of risk-assessment techniques from developed to developing countries. These stem from the entirely different patterns and dynamics of risk in the two types of regions, as demonstrated by Norberg-Bohm and colleagues in chapter 2. Although this is also true for other classes of risk, such as those to natural ecosystems, here I focus on health risks, and particularly on the means to assess both those that result from technologies, whether old or new, and those that are mediated through environmental pathways.

The health transition

The "risk transition" to be discussed here is related to two other transitions – the epidemiological and the demographic – that have served as frameworks to understand the processes of economic and social development. Together, as suggested by Kjellstrom and Rosenstock (1990), the three make up the health transition (fig. 3.1).

Demographic transition

This transition, first described before mid-century (Teitelbaum 1975), involves a shift from Stage 1 (traditional equilibrium), in which population sizes are stable (owing to high death rates as well as high birth rates), to Stage 2 (population explosion), during which death rates decline more rapidly than birth rates, and eventually to Stage 3 (modern equilibrium), when birth rates decline and catch up with death rates.

Although many countries have gone through such a transition, it is neither inevitable nor irreversible. Rather it is a pattern with many local variations which is influenced by many factors. The third stage, where low death rates are matched by low birth rates, is clearly the most desirable, but it has been slow in coming for many developing countries. There is no

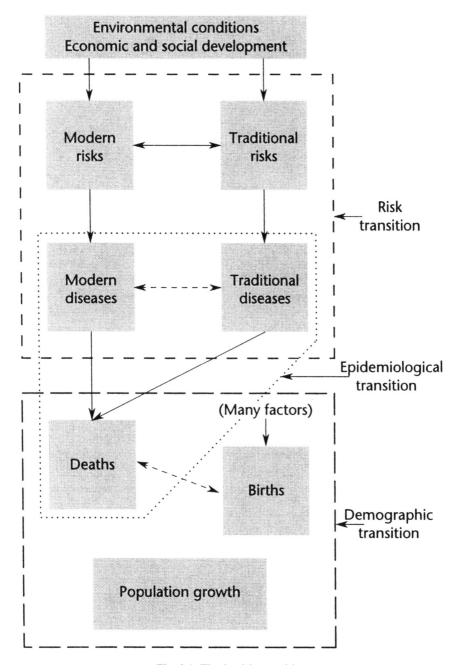

Fig. 3.1 **The health transition**

need to let it occur at its own speed, however, for it can be accelerated by proper management: death rates can be lowered more quickly by addressing specific risk factors (e.g. through vaccinations and mosquito control) and birth rates can also be lowered (e.g. by enhancing the availability of birth-control technologies and women's education). In turn, the two components of the demographic transition are linked, one weakly and one strongly, to the "epidemiological transition" (fig. 3.1).

Epidemiological transition

Nearly three decades ago, Abdel Omran (1971) proposed to demographers that information about patterns of disease and death could be more explicitly incorporated into demographic theory by adopting a framework he named "the epidemiologic transition."

Most developing countries are experiencing great reductions in some kinds of ill-health along with increases in others. The historically high "traditional" diseases associated with rural poverty trend downwards during economic development, although at different rates in different places and at different periods. Deaths due to natural hazards are a traditional disease in the sense that development usually leads to less health impact from these hazards, although to more economic damage over time (Burton, Kates, and White 1978, 1993). This overall downward trend in traditional disease is arguably the most important curve in human history.

The great reduction in traditional diseases that has generally accompanied development has usually led to a reduction in total ill-health, as can be seen in national and international settings and in both longitudinal and cross-sectional studies of trends in infant mortality, life expectancy, and other indicators of overall health (*Human Development Report* 1999, 2000; *State of the World's Children* 1999, 2000; *World Health Report* 1999, 2000). Figure 3.2, for example, shows the results of combining data from a number of countries and also the presence of a secular trend (i.e. the poor were healthier in 1970 than in 1940).

Paralleling the decline in traditional diseases has been a rise in the fraction of deaths due to modern diseases. These are principally the degenerative diseases, such as cancer, heart disease, and stroke, along with certain new types of accidents and occupational hazards. The curves of figure 3.3a, which show, for a cross-section of countries, the relative probabilities of dying from a modern compared with a traditional disease, illustrate that the transition occurred in most countries as life expectancy exceeded 50 years (Preston 1976). This corresponds today roughly to the average life expectancy of all "low-income" countries (except China and India) and to a 1987 income per capita of less than US$300 (*World Development Report* 1989).

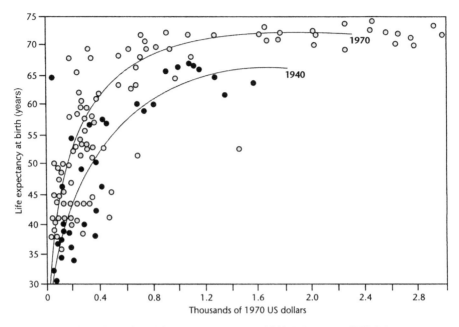

Fig. 3.2 **National income and health: 1940 (●) versus 1970 (○)**

Since everyone must die of something, overall changes are not shown in figure 3.3a. A more revealing index is life-years lost to each class of risks, as shown in figure 3.3b. Over all, life-years lost has generally decreased with development, which is consistent with figure 3.2. Loss due to modern diseases does not match that from traditional diseases until overall life expectancy reaches about 60 years. This corresponds to a 1987 income per capita of something over US$1,000, which is about equal to the average for "lower middle income" developing countries (*World Development Report* 1989).

Sweden is one of the few countries for which data are available to examine the trends of traditional and modern diseases over long periods. This is illustrated in figures 3.4a and 3.4b, which show trends for four disease categories – two traditional, one modern, and one mixed (Hofsten and Lundstrom 1976; Oden 1982). The traditional diseases (categorized in the original references as epidemic and infectious) show sharp declines during the late 1800s and early 1900s. In contrast, the rate of cancer mortality increased steadily in the twentieth century, partly owing to an ageing population structure.

It may seem mysterious that the rate of fatal accidents in Sweden seems not to have changed significantly for more than a hundred years. When

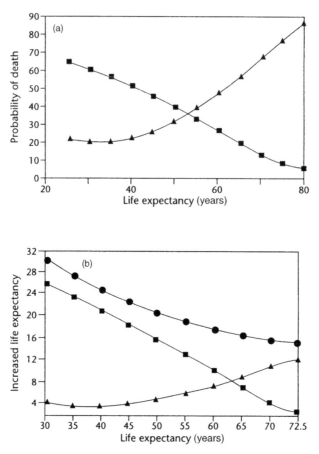

Fig. 3.3 **Trends in modern (▲) and traditional (■) diseases for a cross-section of countries: (a) relative probabilities of death; (b) life-years lost to different risks, and total risks (●)**

broken down by type of accident, however, the numbers show that the total rate masks a decline in traditional and a rise in modern accident rates. Many modern accidents involve motor vehicles, which seem to have substituted for the traditional accidents from drowning related to small boats (Oden 1982). The near-constant rate of total fatal accidents may be due to coincidence, or to the operation of "risk homoeostasis," a term used to describe the hypothesis that a sort of invisible hand operates within society to maintain some types of risk at nearly constant levels (Wilde 1982). Cross-national studies would seem to provide evidence of such a mechanism for they show that the life expectancy lost from violence (e.g. homicides, accidents, and suicides) varies relatively little

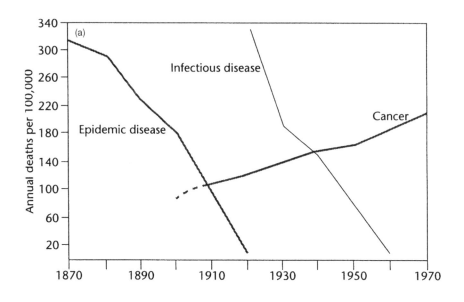

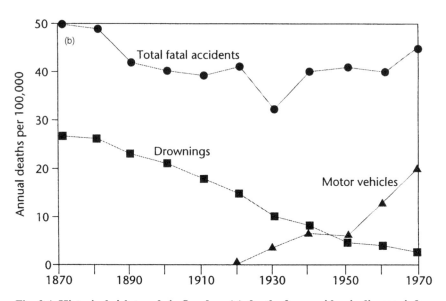

Fig. 3.4 **Historical risk trends in Sweden: (a) deaths from epidemic disease, infectious disease, and cancer, 1870–1970; (b) deaths from accidents (●), drowning (■), and motor vehicles (▲), 1870–1970**

across a range of countries with vastly different economic and overall health circumstances (Murray and Lopez 1996a, b, c; Preston 1976).

As shown in figure 3.4b, total ill-health has generally decreased with development because the reduction in traditional diseases more than counterweighs the rise of modern ones. In some cases, however, a stage may be reached where traditional diseases are all but squeezed out but modern diseases are still rising. There is evidence of this kind of upturn for infant mortality in the former Soviet Union (see for example, Dinkel 1985; *State of the World's Children* 1999). In recent years, the world has also witnessed too often the situation in which large groups of people, or even whole nations, have moved back up the traditional disease curve because of setbacks in development caused by wars, natural disasters, or human mismanagement (*World Development Report* 1987).

As shown in figure 3.4a for Sweden, three stages have typically characterized the epidemiological transition:

1. *The age of pestilence and famine*, with high and fluctuating overall rates of disease and a life expectancy of 20–40 years. With the assistance of international effort, most countries of the world have passed out of this stage.
2. *The age of infectious diseases and rural poverty*, during which poor sanitation and malnutrition keep life expectancies under 60 years. Most developing countries of the world are now within this stage.
3. *The age of degenerative and human-caused diseases*, often with long latency periods between the assault and the effect. Some developing countries and all developed countries are in this stage, which is characterized by life expectancies over 60 years (Olshansky and Ault 1986).

The risk transition

The shift of disease patterns from traditional to modern is due to changes in the underlying risk factors for the various forms of ill-health. Since one inevitably follows the other, it thus might be argued that no particular advantage accrues from differentiating between ill-health and risk. This might be so for some classes of disease, particularly those for which a fairly short delay separates changes in the risk factors and changes in ill-health. For these, monitoring disease rates can be expected to be a fair indicator of what is happening and, thus, a reasonably good guide for decision-making.

For other classes of disease, however, current patterns of ill-health are fairly poor indicators of actual risk. Consider the nearly archetypal case of cigarette smoking. Today's patterns of lung cancer reflect risk factors (smoking patterns) that existed 10–40 years earlier. Thus, it is possible to predict continued increases in ill-health from smoking for US women, who started smoking in large numbers some decades after men. Similarly, it is fair to say that, owing to tobacco use, "men in the formerly Socialist Economies of Europe and China may expect a higher risk of dying between the ages of 15 and 60 than they do today" (Murray and Lopez 1996b: 32). Indeed, China leads the world in tobacco deaths (*World Health Report* 1999: 66).

The distinction between risk and ill-health is also important for another category of hazards – pollutants that make their way slowly through the environment and eventually reach humans through food, air, or water. The release of the pollutant represents the creation of the risk (i.e. commitment to the health effect) even though, owing to environmental and physiological latencies, many years may elapse between the release and the disease.

In other situations, risk accrues (although it does not actually register) in health damage. Although most nuclear power plants apparently operate without creating any significant ill-health, all do create risk (i.e. even if a nearby community experiences no accidental radiation releases over the lifetime of the plant, it still has experienced risk). Thus, an important difference between ill-health and risk is the inclusion of a contribution from low-probability events in the latter but not necessarily in the former.

In a related way, modern risks also tend to differ from traditional risk in that many important outcomes do not have a unique cause. Unlike cholera, for example, which is caused by a single identifiable agent, a particular lung cancer cannot usually be confidently linked to a particular cause: it might have arisen from air pollution (human or natural), asbestos, or environmental tobacco smoke. We can only make probabilistic statements, at best.

This is true even when the impact is large. Consider again the archetypal case, smoking. The risk is huge – the World Health Organization (WHO) recently estimated the global death toll from smoking at three million deaths in 1995, or approximately 6 per cent of all deaths occurring that year (*World Health Report* 1997: 47). Moreover, "unless current smoking trends are reversed, that figure is expected to rise to 20 million deaths per year by the 2020s or early 2030s, with 70 per cent of those deaths occurring in the developing countries" (*World Health Report* 1997; see also Murray and Lopez 1996a, b, c; Peto et al. 1994). This risk is much higher than that of almost all major wars and plagues but, because it is delayed and non-specific, it is not perceivable by normal human senses. We require help from sophisticated scientific study and statistical interpretation to "see" it.

A more identifiably environmental example is lead pollution. Lead is now thought to have had disturbingly large impacts on children and adults throughout the world (Shy 1990; USNRC 1993). Because its impacts (lowered intelligence, low birthweight, stroke) are delayed and non-specific (even though large), they have been difficult to see.

Each category (long-latency, low-probability, high-consequence, and non-specific risks) not only encompasses local situations but also allows inclusion of two important global risks – damage to the ozone layer and global warming. These are among the *systemic* environmental changes described in chapter 1.

• Both the manufacturing and the use of chlorofluorocarbons (CFCs) represent an addition to global environmental risk, although it may be decades before they are released, react in the stratosphere, increase ultraviolet (UV) radiation, cause changes in body cells, and result in skin cancer deaths, which are non-specific.

• In a similar manner, releases of CFCs and other greenhouse gases represent a risk of global warming with attendant non-specific health impacts, despite a reasonable probability that the risk will never be expressed in actual health damage.

Global thermonuclear war is an extreme example of a never-yet-occurred but high-consequence hazard that has imposed substantial risk but few health effects, to date.

To summarize, the major reasons why it is valuable to focus on the risk transition separately from the epidemiological transition are that a focus on risk allows inclusion of the following:

1. Long latencies, either physiological, or environmental, or both, between the risk commitment and ill-health. These can operate at any scale, from the personal (smoking) to the global.

2. Low-probability, high-consequence events that may impose significant annual expected values of risk, even though they are very rare and may not even yet have been experienced. These can also operate at local through global scales.

3. Risks that can be described only statistically because they are non-specific and, in some cases, may never actually be subject to observation in any scientific way (e.g. low-level exposures to carcinogens).

Referring back to the epidemiological transition (fig. 3.3), it seems that non-specific, long-latency, and high-consequence events are more typical of modern than of traditional diseases. In other words, the traditional curves for risk and disease are much closer together than their modern equivalents. Part of the risk transition, therefore, would seem to be a decoupling of risk from ill-health, in time (owing to the latencies), in magnitude (owing to uncertainties in the probabilities), and in certainty (owing to non-specific unmeasurable impacts).

This raises a further rationale for focusing on risk rather than on ill-health – suitability of the risk for management. Waiting to count who dies of what, may be an adequate approach to understanding and controlling some traditional risks (e.g. dysentery) but is clearly inadequate for de-coupled risks (e.g. toxic metal poisoning, ozone depletion) for which the changes in risk factors greatly precede changes in ill-health. By the time actual changes in health occur, it is too late to do much about them (even if they can be measured).

For these hazards, we need to monitor further back along the causal chain – which runs (for example) from pollutant emissions, to environ-mental concentrations, exposures, doses, preclinical changes, and ill-health, before leading to death (Kates, Hohenemser, and Kasperson 1985; Smith 1988). Where along this chain it is best to measure depends on a range of factors, including not only the time constants involved but also the ease and cost of monitoring. Very generally, points further back on the chain tend to be easier to monitor than those (like dose) relatively closer to effects. On the other hand, they are more removed in causation, introducing uncertainties because of gaps in knowledge about, for exam-ple, how emissions translate into human exposures. The causal chain and, thus, risk monitoring, can also be extended back even further, to the basic decision steps involved in embarking on this or any technology at all, as compared (for example) to exploring other ways to meet human needs and wants (Kates, Hohenemser, and Kasperson 1985).

Although numerous implications attend the risk transition (Smith 1990a), here I discuss those related to environmental risk assessments of technological projects in developing countries, principally those that are designed to assist the development process.

Implications for risk assessment in developing countries

Omran (1971) offered five propositions in support of his epidemiological-transition framework, which, when rewritten, can do the same for the risk transition discussed here:

1. The theory of risk transition begins with the major premise that exist-ing mortality and morbidity patterns are fundamental but incomplete determinants of the ways in which risks are perceived and should be assessed and managed. To an increasing degree, risks are becoming decoupled from present-day patterns of ill-health.

2. During the risk transition, a long-term shift occurs in mortality and disease patterns in which traditional risks related to pandemics of in-fection and rural poverty are displaced at lower and lower levels of total ill-heath by modern risks of degenerative, human-caused, and global ill-effects.

3. Significant differences occur within, as well as between, countries in

the risk patterns of different population groups. Reporting average improvements, for example, masks the relatively greater progress made by women, children, and rural people when traditional risks decline.

4. For the currently developed European countries, social and economic factors largely determined the transition from one epidemiological age to the next in the nineteenth century whereas medical and public health interventions seem to have played the largest roles in the first two-thirds of the twentieth century. More recently, social and economic factors have again become dominant, through being determinants of technological choices that seem to enhance risk but not ill-health.

5. The pace of the risk transition has changed so that traditional risks persist whereas modern risks start earlier in the development process. This results in the "risk overlap," in which some populations are exposed to significant amounts of both types (but in a rapidly changing pattern). Even this overlap has its traditional and modern versions.

These propositions are not meant to imply that all countries at all points in history go through the same risk transition or that there is no hope of altering the pathway. Indeed, recent revelations from Eastern Europe would suggest that the societies there managed to reduce traditional risks and diseases but failed to control modern risks, even to the extent that modern disease rates are rising fast enough to lower overall health status (Feshbach 1995; *World Disasters Report* 1997, 1999, 2000; Nanda et al. 1993; Russell 1990). I wish to suggest only that the concept of a risk transition provides a useful framework for comparison and analysis of the changes in risk experienced by various countries and thereby offers a means to improve risk management. This changing risk pattern has many impacts on the demography, economy, and public health of developing countries. Here, however, the focus is on the implications for risk assessment. The basic premises are two: (1) effective response to environmental risk intimately depends on the ability to assess it; (2) substantially different risk-assessment methods are necessary in developing countries because the risk patterns depart so markedly from those of the countries in which risk-assessment methods have been devised.

The risk overlap and resulting interactions

At the point marked A in figure 3.5, modern risks have begun to rise but traditional risks are still significant, although rapidly dropping. An environmental example shows up when pesticide runoff starts to add to water pollution caused by poor sanitation. Another example is offered by those places in which urban air pollution from fossil-fuel combustion is rising

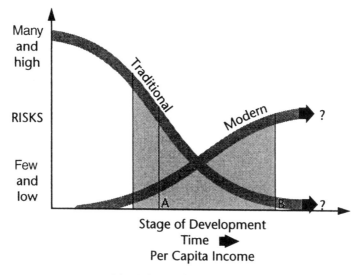

Fig. 3.5 **The risk transition**

while large village and urban air-pollution exposures from household combustion of traditional biofuels still exist. If these effects were merely additive, there might be no special need to consider interactions. Unfortunately, however, risk overlap can produce interactions with important implications for risk assessment by leading to effects that magnify or mask the separate impacts of modern and traditional risks. Borrowing, consolidating, and expanding from Greenberg's discussion of disease competition (Greenberg 1986), it is possible to identify six kinds of interaction:

1. *Risk genesis*, where risk overlap may lead to creation of an entirely different sort of risks, of which many examples exist (Farvar and Milton 1972), including the following:
   - mixing modern (motorized) and traditional (muscle-powered) vehicles leads to new kinds of accident risks and risk-management needs;
   - in areas in which crop residues are still used as cooking fuel, pesticide residues may lead to airborne chemical exposures that otherwise would not occur;
   - wide income disparity leads to intense garbage scavenging in dumps containing hazardous chemicals and other modern hazards;
   - in areas where drinking water supplies are still unprotected, use of nitrogen fertilizer can lead to methaemoglobinaemia in infants;
   - Use of modern hazardous materials before implementation of modern regulatory controls produces risks, such as the spread of

radioactive material from discarded hospital equipment that occurred in Goiânia, Brazil in 1987.

2. *Risk synergism*, where exposure to one agent causes immunity or sensitivity to other agents in the future. An example is the traditional in-
testinal diseases, owing to poor water quality, that increase sensitivity
to waterborne and airborne modern pollutants (Shakman 1980). In
some cases exposures to traditional risks may give some immunity or
resistance to modern risks.

3. *Risk mimicry*, where morbidity and mortality may be attributed to
traditional sources of risk but may be actually due to modern or synergistic risks. An example is attribution of lung-cancer deaths to acute
or chronic lung diseases. The deployment of the diagnostic tools
needed to determine modern diseases such as lung cancer may lag
substantially behind the development of the diseases themselves. In
addition, entirely different risk factors can lead to similar risk and ill-
health patterns.

4. *Risk competition*, where abnormally high or low risks of one disease
may actually be an indication of the decrease or increase in the risks of
an entirely different disease. Low rates of lung cancer in rural areas of
developing countries, for example, may not necessarily mean that the
risk factors for lung cancer are lower there than in places with high
rates; it may just mean that people are dying of other diseases with a
younger age pattern and that people are not living long enough to
develop cancer. Use of age-adjusted rates may compensate for only
part of this effect.

5. *Risk layering*, where the movement of people or their activities con-
centrates risk in one region and dilutes risk in another. This may be an
important effect in those many developing countries that are undergoing significant rural-to-urban migration, itself often triggered by environmental conditions, in which migrants are usually healthier than
the average person left back in the villages. Other types of movements,
of political and environmental refugees for instance, may have the
opposite effect. In addition, it is also possible for migration of people
or technologies simultaneously to lower (or raise) the risk levels at
both origin and destination areas. Also, the source of risks may be
difficult to pinpoint, because the place where the insult occurred may
not be where the disease first appears, especially for chronic diseases
or those with long latency periods.

6. *Risk transfer*, where efforts to control risk from traditional hazards
may enhance modern risks (or vice versa), as when pesticides are used
to control malaria. Such transfer can also occur from one type of
modern (or traditional) risk to another and often involves a shift in the
time, place, and population affected (Kates 1978; Whipple 1985;

Whyte and Burton 1980). The classic story of risk transfer is in Greek mythology, where one of Hercules' twelve labours was to clean in one day the great Augean stables, which had not been cleaned for 30 years. He diverted a river through the stables for this purpose, but, by doing so, merely transformed the hazard from that from solid waste to that of water pollution. Risk transfer and layering can too often result (whether by intention or inattention) in an increase in risk inequity, as, for example, in an increase in the disparity between the risk-lowering and risk-raising benefits of technologies or other activities.

It may also be useful to distinguish two general types of risk overlap – local and global. Listed above were cases of local risk overlap involving the interaction of traditional and modern risks on a local scale (as in a garbage dump with hazardous waste). These can be serious, but are amenable to various fixes. Dumping the hazardous materials somewhere else essentially eliminates overlap interactions (although not necessarily the individual, traditional, and modern risks).

Global risk overlaps, in contrast, will continue as long as great disparities exist anywhere on Earth among the risk patterns (and other characteristics) of different groups. The most important examples are ozone- and climate-threatening chemical releases. Although mostly created by modern activities in countries well along in the health transition, these pollutants affect the whole world. Indeed, they will be likely to have their biggest impacts among peoples still burdened by traditional risks, including those with high vulnerability to environmental stress (Blaikie et al. 1994; Tobin and Montz 1997; Tyler and Smith 1991; *World Disasters Report* 1997; *World Health Report* 1998: 123–125; 1999). Unlike local risk overlaps, the modern global versions cannot be avoided by releasing the pollution at another place. With these pollutants, for all practical purposes, emissions anywhere are emissions everywhere.

This approach is consistent with that of Mitchell (1989), who distinguishes between "local" and "universal" (e.g. global climate change) risks and points out that each requires different methods of risk assessment. Developing-country risk managers will have to assess the implications of these overlaps when plotting risk and development strategies (Smith 1990b). Those looking for areas of high vulnerability, as treated in chapters 5–7, may well find risk overlaps to be good indicators.

Net risk

Consider the different world views of the risk managers working in countries represented by points A and B in figure 3.5. The risk manager at point B sees an increasing number and, in some cases, intensification of modern risks from development. The traditional risks, largely under control, are of less concern. Thus, understandably, assessment efforts have focused on quantifying and managing modern risks.

At point A, however, important hazards of both traditional and modern sorts exist side by side, although trending in opposite directions. Introduction of technology and the resulting development (interpreted as a movement along the horizontal axis) will, therefore, have an impact on overall risk in developing countries that differs substantially from that in developed countries. On average, a country at point A will experience a significant degree of *risk lowering* (movement down the traditional-risk curve) from new technologies as well as *risk raising* (movement up the modern-risk curve).

Although the overall result of development may be general lowering of risk (as shown in fig. 3.5), each technological project will have its own pattern, which may not be risk-lowering in the aggregate. To understand the overall or net risk from any particular project, regulation, or activity, therefore, will require evaluation of both classes of risk. The developing-country risk manager will find that to adopt risk-assessment methods from developed countries, which focus solely on modern risks, will be misleading.

In principle, assessments on both sides of the risk transition should incorporate consideration of risk-lowering as well as risk-raising aspects of development. It has been pointed out, for example, that in spite of the considerable decrease in overall risk that has already taken place in the United States, a significant degree of risk lowering is still occurring (Whipple 1985). Indeed, focusing only on the modern-risk curve for each new project, as part of what might be called *microrisk* analysis, leads to the manifestly absurd *macrorisk* result that those communities with the most technology should be the least healthy, that Switzerland should be more unhealthy than Nepal.

A substantial body of literature relates health status to economic development (Brenner 1984; Murray and Lopez 1996a, b, c; *World Development Report* 1993). In general, a consistent and strong relationship seems to exist between national per capita income and various indicators of improved health status, including infant mortality and life expectancy at several ages (Preston 1975). These relationships can also be seen when other indicators of economic growth (e.g. per capita energy consumption) are used (Sagan 1987). Analysis indicates, however, that only about 20 per cent of the great secular shift in the relationship of economic welfare and health that occurred from 1940 to 1970 (fig. 3.2) seems to be directly attributable to economic factors: the bulk of the improvement would seem to be due to medical and public-health measures that were applied independently of income. Since development and dissemination of these measures were, at least to some degree, dependent on improving economic conditions in developed countries, however, it might be argued that a strong indirect connection between income and health existed (Preston 1976).

Further analysis of these data indicates that health status is associated with individual more than with national incomes. In other words, population groups of equal income levels will experience similar mortality risks, no matter what the national income of the countries in which they live (Preston 1980). In addition, the argument has been made that, in industrialized countries, ill-health has come to be associated with rapid economic changes and political instability more than with average income (Brenner 1980).

More recent analyses, however, seem to indicate weaker dependence after mid-century on factors other than income and education in enhancing life expectancy and decreasing childhood mortality. In general, health status seems now, more than previously, to be influenced by economic development in the developing countries of Asia and the Americas and to a lesser degree in Africa (Hill 1985a; *World Health Report* 1997). This seems to have occurred mostly through diminution of the impact of the exogenous variables, rather than through an increase in the effect of income.

These might be called "indirect" routes to risk lowering, to contrast them with "direct" routes (such as health care). Indirect routes to risk raising include, for example, the impact of urban stress, as contrasted with such direct routes as chemical pollution. Their being indirect, however, need not mean that they are small. A striking example of the effectiveness of indirect routes is provided by a recent study of all developing countries except China (Preston 1985): the estimated increase of life expectancy due to a general increase of 1 per cent in gross domestic product for these countries was estimated to be 2–3 weeks; this is to be compared with the impact of investing that much income on education, which was crudely estimated to be 5–20 years!

Studies of those countries that do significantly better (e.g. Sri Lanka and China) or worse (e.g. Libya and Senegal) than the average in converting income to better health show that it is possible, even at low incomes, to achieve better health through directly targeting educational and health programmes (Caldwell 1986). In doing so, the education of women has been particularly significant (Behm and Vallin 1980; PAI 1998; *State of the World's Children* 1999).

Sporadically, calls have emerged to quantify risk lowering as part of developed-country risk assessments (Bentkover, Covello, and Mumpower 1986; Keeney and von Winterfeldt 1986; Wildavsky 1988). In practice, however, little has been done, and none of the risk-assessment methods in common use seems to provide a framework from which to start. The previous chapter may, however, point to some promising directions. Whatever the resulting misrepresentation in developed countries, failing to conduct net-risk analysis in developing countries will produce much larger distortions owing to the more prominent role played by traditional risk.

Many studies have addressed the overall relationship of economic development with health status. Only a few, however, have conducted project-level analysis (Rodgers 1979), although some of the required techniques have been developed (Collins, Lundy, and Grahn 1983). It is at the project level, however, that individual decisions are made and policies implemented. To conduct such an analysis would necessitate understanding first the changes wrought by a project on income, employment, training/education, housing, and other relevant factors and then the impact of each of these on risk. Such analysis could focus on local impacts only through process analysis. To be sure that risk lowering is not due simply to transferring risk from one place to another, input–output techniques can be applied (Myers, Genter, and Werner 1984; Waterstone 1987; Wilson 1979).

Past failure to consider net risk provides intriguing hypotheses to explain two apparent inconsistencies in the way people react to risk. It has been confusing to some observers that the richest and healthiest communities in human history (those near point B in fig. 3.5) have become so preoccupied with risks, even those risks that seem to be extremely small when compared quantitatively with many others (Beck 1992; Wildavsky 1979a). The framework of the risk transition, however, provides another explanation to add to those offered previously (Clark 1980; Kates and Kasperson 1983). Although formal risk assessments have not done so, people intuitively have been conducting net-risk assessments all along. Thus, people at point B (fig. 3.5) recognize that they are obtaining much less risk lowering from new technologies and economic growth than they did in the past (Brenner 1980); consequently, the risk-raising aspects of technologies loom more threateningly.

This reality is undoubtedly further exacerbated by the decoupling of risk and ill-health that has occurred in modern societies. People may be intuitively adding risks (such as global warming, the future uncertain impacts of chemical pollution) that do not yet (and may never) appear as ill-health. More simply, it makes considerable sense to be more worried about rising hard-to-measure risks than falling easy-to-measure risks.

On the other side of the risk transition, intuitive net-risk assessments make more understandable the common observation in risk studies that "life is cheap in developing countries." This seems to be the inescapable conclusion of observations that developing-country people and governments have been willing to subject themselves to modern risks that could be avoided at relatively small marginal cost. What are, by modern risk standards, lax or incompletely enforced pollution and occupational-health regulations provide examples. Again, however, the risk-transition framework provides a partial explanation, one additional to standard economic calculations of "willingness to pay" (Robinson 1986). In contrast to conditions on the developed side of the risk transition, so much risk lowering often occurs as the result of a new technology that the marginal risk-lowering value of additional economic benefits exceeds the risk lowering those resources could bring if devoted to such direct controls on modern risks as pollution-abatement equipment. In other words, it may be less expensive to buy risk reduction through indirect pathways (such as increasing incomes) than by the direct pathway of modern-risk management.

An important component of developing-country risk assessment, therefore, should be net-risk analysis. This would acknowledge that the risk-lowering and the risk-raising aspects of projects require examination. To some readers, this may seem to be a thinly veiled rationalization for continued laxity in implementing health and safety standards in developing countries. This is not the case. Rather, it is a call to judge technologies on a net-risk basis so that those that fail to reduce net risk, or do so only weakly, can be rejected or modified to meet the risk-reduction goals of the society. The criteria remain the same as in past risk assessments that focused only on modern risk. In short, those projects that are associated with much risk raising will be penalized. It seems appropriate in many circumstances, however, also to penalize those projects that realize little risk lowering (Shrader-Frechette 1985). Risk is only one of the criteria by which projects are compared, but it is best done on a net-risk basis.

A significant difference may also exist in the amount of uncertainty associated with investments in programmes to reduce risk directly – say, by installing pollution-control equipment – compared with alternative actions that reduce risk indirectly through education and income. In many cases it may be advantageous to take the simpler and better-known route, even if the average expected risk reduction is forecast to be somewhat lower (Lumbers and Jowitt 1981). It is not clear, however, whether the environmental or the social route can be said to be generally more uncertain.

Risk distribution

Economic evaluations of projects, as noted in chapter 2, have sometimes considered distributional characteristics (equity) as well as overall benefits (efficiency). Some risk assessments, too, have tried to determine who wins and who loses, for example in regard to income (Schwab 1988) or age and sex (Sullivan and Weng 1987). The archetypal NIMBY (not in my back yard) phenomenon is an example of how some technologies that are acknowledged to be of overall societal benefit may not be pursued because of what is perceived as excess risk to some particular social group (Kasperson 1986; Lesbirel 1998; Shaw 1996).

Both theoretical and empirical evidence shows that national mortalities are functions of income dispersion as well as mean income and that greater income dispersion is associated with lower life expectancy (Rodgers 1979). Since income dispersion tends to be greater in developing countries, examination of distributional aspects will be more important in risk assessments in those countries (*World Development Report 1999*). This can also be seen in differences between points A and B in figure 3.5: at A, the distribution of wealth, income, health, education, and other indicators of well-being is generally less equitable; in addition, the slopes of the traditional and modern risk curves probably tend to be steeper (e.g. marginal risk alteration per unit change in income is higher; Behm and Vallin 1980). Consequently, a risk assessment that calculates risk, even net risk, based on nationwide means or medians may distort the picture. The populations of many developing countries will be spread along a rather large part of the risk curves, depending on their relative well-being.

A transnational project may also distribute risk in a non-uniform fashion between the developed- and developing-country participants. This may lead to serious problems, both actual and perceived (Ives 1985). Under some circumstances, however, it seems possible to minimize overall risk by carefully allocating hazards according to the comparative advantage of each participant (Pearson 1987).

Thus, in order to determine true net risk, a developing-country risk assessor may need first to separate the affected population into several groups based on income or other social and economic indicators. Combining the population-weighted net risk of each group would then reveal overall net risk. This would give a better picture of what actually occurs than, for example, using national averages for the marginal risk-lowering efficiency of income from the project. Net-risk lowering would be substantially less, for example, if the project benefits accrue to investors in the capital city rather than as training, housing, and wages for formerly unemployed and landless poor. In other words, economic growth among lower income groups will raise life expectancies more than the same growth among wealthier groups (Preston 1975).

Indeed, analogous to economic analysis, it might be useful to introduce the concept of "Pareto safety." A project meets the Pareto economic criterion if no group becomes worse off and at least some people are better off. A "Pareto safe" project would cause no increase in risk in any group and would result in some risk lowering in at least one group.

Dynamic risk patterns

Another implication of the rapidly changing patterns of risk in developing countries is that, to be realistic, risk assessments must take more cognizance of the dynamics within the economy and risk environment. For example, not only should a risk assessment consider risk overlap interactions, risk lowering, and the risk raising created by a technology, but it may be required to do so with parameters that change dramatically (i.e. are time or income dependent) over the life of the project. The lifetimes of many projects extend for several decades into the future, during which significant shifts may occur in a nation's position (fig. 3.5). Indeed, a country may well have moved from position A to B during such a period, thereby greatly changing the relationship among traditional and modern risks.

In general, the risk transition can serve as a framework for comparing development paths taken by different countries in different periods. Just as with the demographic transition, it may well be worth examining how and why developing countries follow or diverge from the risk-transition pathway taken previously by the now-developed countries and exploring what might be predicted and guided with regard to local, regional, or global risk patterns.

To be most useful, risk assessment in developing countries should be able to give guidance not only about specific actions, such as siting a particular facility or embarking on a new industry, but also on pathways of development that may incorporate trade-offs between economic growth and risk over many years. An example of this type of decision, albeit apparently made without systematic quantification, is that of one Asian country in which government policy reflected a conscious choice to use cheaper but high-sulphur fuels for a defined period in order to quicken economic growth. Once a trigger level of income had been achieved, however, the government phased in less-polluting fuels. More persistent or long-range pollutants, or those with irreversible effects, would complicate this type of assessment.

Another dynamic is the trend of some modern risks to become significant at earlier stages in the development process than in the past because of the impact of international and internal trade, communication, and foreign aid. Consequently, the period of risk overlap and need for net-risk assessment may be increasing.

Additional implications

As treated by Norberg-Bohm and colleagues in the previous chapter, a modified risk-transition framework should include the important nonhealth risks attendant on development. Risks to the sustainability of ecosystems, for example, may fit into such a pattern. Even political and economic risks may have interacting traditional and modern components. In addition, I have not delved into the implications of the risk transition for international interactions, such as technology transfer, among nations at different points in the transition. As with environmental health risks, it may be possible to derive testable hypotheses about these other risks as well.

Unfortunately, systematic, but *simplified*, protocols for health risk assessments to determine *net* and *dynamic* risk *distributions* and to address the interactions of *risk overlap* are not well developed, although the need for one or another of these characteristics has been recognized to some extent by a number of observers within (Bowonder and Kasperson 1988; Kim 1987; Pe Benito Claudio 1988) and outside (Covello and Mumpower 1986; Urquhart and Heilmann 1985) the developing world. Moreover, much international interest promotes risk assessment and management in developing countries (UNCTAD 1987; UNIDO 1987; USAID 1987; WHO 1984; 1985a; WHO and UNEP 1987; World Bank 1985).

Owing to the different risk patterns in developing countries, risk perception there is also likely to differ from that measured in developed countries. For example, in the context of recently high traditional risks, public perception may not yet have taken into account newly achieved large reductions because they have occurred so rapidly. Indeed, time lags have greater potential for distorting perception when conditions are changing rapidly, an effect that could mask the true impact of the recent introduction of modern risks. Important cultural (Thompson 1983) and political/institutional (Douglas 1985) factors also influence risk perception. Relatively few empirical comparisons have been attempted, however, in spite of the fairly voluminous literature in developed countries (Johnson and Covello 1987). What has been done seems to have focused on what are, by world standards, fairly similar cultural settings (Covello et al. 1988; Englander et al. 1986).

The risk transition has important implications for environmental-monitoring strategies (Smith and Lee 1993). These relate to the environmental pathway leading, as in the case of toxic chemicals, from inventory to emissions, environmental concentrations, human exposures, organ doses, and finally to health effects. Many traditional risks can be understood and controlled fairly well by monitoring oriented toward the health-effects end of this pathway (e.g. by monitoring causes of death). The further along in the risk transition, however, the less viable this approach, because each modern cause of death may have many risk factors that may reveal themselves only long after the original insults. There is a tendency during development to direct control measures to the beginning of the environmental pathway or even earlier in the causal chain of hazard that starts with human wants and needs (Hohenemser, Kasperson, and Kates 1985). This also places a burden on monitoring to supply knowledge about the linkages between the point of control and the point of concern (i.e. health effects).

The pattern shown in figure 3.5 can be seen as a circumscribed snapshot of long-term patterns in which risks rise and fall, first starting out as the modern risks of their time, but peaking and becoming traditional declining risks after some period during which a new set of modern risks may be rising. Ambient particulate air pollution provides an illustration of this phenomenon. Recent studies provide evidence that poor urban air quality is both a traditional and a modern hazard (Ezcurra and Mazari Hiriart 1996; Ezcurra et al. 1999; McCormick 1991; Smith 1995). Its contribution to risk is still increasing with economic development for the poorest countries but is decreasing in the cities of wealthier nations (WHO and UNEP 1987, 1988). The long-term risk pattern, therefore, would seem to involve, first, a rise with time and development, followed by a fall in what could be called a "long wave" of risk, analogous possibly to the long waves identified by Marchetti (1980) in energy and other technologies and by Kondratiev (1925) within economic development itself. To explain this phenomenon, Kjellstrom and Rosenstock (1990) have proposed the concept of an "intervention transition," in which the relative risks change as the result of intended or unintended risk-management actions.

Urbanization has been one of the most important influences in the transition from traditional to modern risks, but its perceived influence on risk is an excellent example of distortions that can develop by failing to consider risks in the transition framework. The tendency, for example, is to focus on the terrible modern risks imposed by the megalopolises of the developing world – the air and water pollution, the traffic jams, the garbage dumps, and unfettered pollution emissions of all sorts. Although it may seem contradictory, the (admittedly poor) evidence is that cities in general bring improved health, well-being, and environmental quality (Satterthwaite 1997). Their residents tend to be better off than they were before or than they would have been if they had stayed in the rural areas. Mexico City is a case in point (Ezcurra et al. 1999; Pezzoli 1998). In addition, although one might equally regard these megalopolises as ecological disasters, urbanization is one of the only historically proven ways to reduce pressure on forests and other endangered natural hinterlands.

It may be, however, that urban dwellers exert an equal or larger impact, as a form of concentration and agglomeration. This human driving force of global environmental change is examined in depth in Kasperson, Kasperson, and Turner (1995). African cities that rely on charcoal for cooking greatly increase the magnitude, intensity, and extent of damage to forests. In some nations, for example, the making of charcoal in inefficient kilns relies on clear-cutting of forests some 800 kilometres from the urban market (USAID 1988).

Risk overlap probably exhibits its strongest interactions in such cities. Perhaps the most striking examples emerge when villagers bring rural behaviour patterns with them when they move to the city. Use of biomass fuels for cooking and open ground for defecation, for example, are risky enough in a rural setting, but more so in the city. Thus, even if the net ecological or health risk for any megacity has never actually been negative (risk lowering), looking only at the risk-raising side can be misleading (Smith and Lee 1993).

In many countries, urbanization now seems to be taking on a different character. Called *Kotadesasi*, which is a combination of the Indonesian words meaning "town" and "village," this new process results in widespread increases of peri-urban areas, often linked to an urban centre. *Kotadesa* regions are characterized by rapid increases in the variety and intensity of workplaces and land uses intermixed within traditional farming areas; a high degree of population mobility; increased female participation in the workforce; and uncertain, inconsistent, and incomplete governance by urban authorities (Ginsburg, Koppel, and McGee 1990). Such characteristics create additional challenges for the risk manager because they describe an area in which traditional risk may linger even as modern risks are making their debuts earlier and with less control than elsewhere. This creates a special need to evaluate the risk interactions and net-risk implications of alternative actions.

On both sides of the post-traditional risk transition, efforts to achieve behaviour modification at the family and individual level are important components of potential risk control (Jeffery 1989). For reducing traditional risk, the incorporation of hygiene and nutrition principles into household practices can have a large positive impact. For reducing modern risk, changes in lifestyle related to exercise and diet are most effective. Behavioural changes are needed to decrease the health impact of drugs, including tobacco and alcohol, among many groups on both sides of the risk transition. The importance of education in stimulating such behavioural changes is apparent from analyses that show the marked sensitivity of health status to education, and particularly that of women. It may well be, for example, that the principal route by which new technologies reduce risk in countries at a certain level of development is through their impact on women's education.

Whatever the importance of this particular pathway, it is nevertheless clear that understanding the risks imposed by new technologies during economic development will entail incorporating new methods and system boundaries into existing risk-assessment practices (Edgerton et al. 1990). The risk-transition framework offers a way to set this process in motion.

4

Global risk, uncertainty, and ignorance

Silvio O. Funtowicz and Jerome R. Ravetz

Science evolves as it responds to its leading challenges as they change through history. The problems of global environmental risk, along with those of equity among peoples, present the greatest collective task now facing humanity. In response, new styles of scientific activity are already under development. Traditional oppositions, such as those among natural-science disciplines and between the so-called "hard" and "soft" sciences, are being overcome. The reductionist, analytical world view that divides systems into ever-smaller elements, studied by ever-more esoteric specialties, is giving way to a systemic, synthetic, and humanistic approach. The recognition of real natural systems as complex and dynamic entails moving to a science based on unpredictability, incomplete control, and plural legitimate perspectives.

We are now witnessing a growing awareness among all those concerned with global risks that no single cultural tradition, no matter how successful in the past, can supply all the solutions for the problems of the planet. Forms of knowing other than those fostered by modern Western civilization are also relevant for an exploratory problem-solving dialogue. Moreover, we should harbour no pretence of an Olympian detachment from the fate of our own species and that of our neighbours or, indeed, from the special problems of those rendered more vulnerable to environmental change owing to nationality, race, class, gender, or disability. Owing to the new recognition of linkage among regions and interdependency among peoples, as chapters 1 and 7 suggest, issues of equity be-

tween peoples and generations are not seen as "externalities" to decisions or to science; rather, with all their difficulties, they are appreciated
as central to the solutions of the problems of the global environment.
Closely connected with the emergence of global risk assessment is a
new methodology that reflects and helps to guide its development. In this,
uncertainty is not banished but managed, and values are not presupposed
but explicit. The model for scientific argument is not a formalized de-
duction but an interactive dialogue. The paradigmatic science is not one
in which explanations are unrelated to space, time, and process; the historical dimension, including human reflection on past and future change,
is becoming an integral part of a scientific characterization of Nature and
our place in it.

Our contribution to this new methodology focuses on two aspects of
the emergent science. One is the quality of scientific information, analysed in terms of both the different sorts of uncertainty in knowledge and
the intended functions of the information. The other aspect refers to
problem-solving strategies, analysed in terms of uncertainties in both
knowledge and ethics. Applied to policy issues, science cannot provide
certainty in policy recommendations, and one cannot ignore the conflicting values in any decision process, even in the problem-solving work itself.

For quality of information, the authors have developed a transparent
system of notations, or NUSAP, whereby the different sorts of uncertainty that affect scientific information can be expressed and communicated among traditional and extended peer communities alike. NUSAP is
the acronym for the five categories – *Numeral, Unit, Spread, Assessment*,
and *Pedigree* – that make up the scheme (Funtowicz and Ravetz 1990b).
This process depends on the principle that uncertainty cannot be banished from science, but that high-calibre information depends on effective
management of its uncertainties (Funtowicz and Ravetz 1990a).

We use the interaction of systems uncertainties and decision stakes to
provide guidance for the choice of appropriate problem-solving strategies. The heuristic tool is a set of graphical displays of three related
strategies, from the most narrowly defined to the most comprehensive.
Two of them are familiar from past experience of scientific or professional practice; the third, in which systems uncertainties or decision
stakes are high, corresponds to the practice of the emergent sciences in
dealing with global environmental risk (Funtowicz and Ravetz 1985;
O'Riordan and Rayner 1990). Here, the problems of quality assurance of
scientific information are particularly acute and require new conceptions
of scientific methodology.

In this new sort of science, the evaluation of scientific inputs to decision
making requires an "extended peer community" (Funtowicz and Ravetz
1991a). This extension of legitimacy to new participants in policy dia-

logues carries important implications for society and for science as well. With mutual respect among various perspectives and forms of knowing, the possibility exists for the development of a genuine and effective democratic element in the life of science. The challenges of global environmental risks can then become the successors of the earlier great "conquests," of disease and then of space, in providing symbolic meaning and a renewed sense of adventure for a new generation of recruits to science.

The reinvasion of the laboratory by Nature

The place of science in the industrialized world was well depicted by Bruno Latour (1988), when he imagined Pasteur as extending his laboratory to all the French countryside, thereby conquering nature for science and for himself. Nature itself no longer needs to be approached as wild and threatening, but through the methodology of science it can be tamed and rendered useful to humankind. The miracle of modern natural science is that the laboratory experience – the study of an isolated piece of Nature that is kept pure, stable, and reproducible – can be successfully extended to the understanding and control of Nature in the raw. Our technology and medicine together have made Nature predictable, and thereby have enabled human life to be more safe, comfortable, and pleasant than was ever before imagined.

The triumph of the scientific method in deploying the technically esoteric knowledge of its experts has led to its domination over all other ways of our knowing Nature, and over much else besides. Common-sense experience and the skills of making and living have lost their claim to reality, having been displaced by the theoretically constructed objects of scientific discourse, which are necessary for dealing with invisible agents such as microbes, atoms, genes, and quasars. Just as Galileo used his telescope to transform the familiar moon into an object of applied geometry (in which light and shade became peaks and valleys), the control of instruments and theories has become the power to define reality for specialists and for the public alike. Although formally democratic (since no formal barriers now stand in the way of training for that expertise), it is in fact a preserve of those who can engage in a prolonged and protected course of education.

The assumption that the world can be converted into an extended laboratory gives primacy to science as effective knowledge and to scientific experts as its legitimate interpreters. The rationality of public decision-making must appear to be scientific; hence, social and human scientists (including economists *par excellence*) have come to be seen as leading authorities. A universal assumption, however uncritical and superficial, has it that scientific expertise is the crucial component of decision-making, whether concerning Nature or society.

Now the very powers that science has created have led to a new relationship of science with the world. The extension of the laboratory has gone beyond the small-scale intervention typified by Pasteur's conquest of anthrax. We do not have merely the familiar gross disturbances of the natural environment resulting from modern industrial and agricultural practices. Rather closer to the heart of science, we have the new "experiments," occurring on a large or even regional scale, resulting from a destructive interference with Nature caused by technology. Classic examples are Hiroshima, Chernobyl, Bhopal, *Exxon Valdez*, and Kuwait. Each is artificial, in an important sense, and each provides data on the behaviour of natural and social systems. They differ from the classic experiments of science, however, in a number of crucial respects: once started, they cannot be turned off at will; further, the events are not isolated, pure, or repeatable. For their scientific study, we do not have the balance of controlled quantitative experimental data and mathematical theory that have been paradigmatic for the classic natural sciences. Instead, the data come from weakly analogous laboratory experiments, ad hoc field studies, anecdotal reports, and expert opinion, and the main theoretical tools are simulations and untestable computer models.

The supremacy of scientific experts is no longer so obvious in the case of the new type of experiments that science-based technology has produced. First, the experts as a class (including their managers) are associated with the causes of disasters, and they are not always successful in their attempts to ameliorate or remedy the effects of unexpected and unwanted events. The techniques applied in these cases, inherited from the successful experience of the laboratory-inspired scientific method, are inadequate in varying degrees. Those experts who use them uncritically and then publicly defend them as "scientific" risk are weakening the credibility and legitimacy of science.

Those new experiments provide evidence for the thesis that the traditional laboratory science must evolve in response to the challenges of global risks. The scientific methodology for coping with these novel risk problems cannot be the same as the one that helped to create them. So much of the success of traditional science lay in its power to abstract from uncertainty in knowledge and values; this is apparent in the dominant teaching tradition, creating a universe of unquestionable facts. Now scientific expertise has led us into policy dilemmas that it is incapable of resolving on its own. We have not merely lost control and even predictability; now we face radical uncertainty and even ignorance, as well as ethical uncertainties lying at the heart of scientific policy issues.

To understand the new tasks and methods of science, we can fruitfully invert Latour's metaphor and think of Nature as reinvading the laboratory, for now the risks we face are global in extent and complex in their structure. The fertile fictions of Nature provided by laboratory set-ups and computer models are now liable to be caricatures or simulacra (Baudrillard 1988). The laboratory has not advanced into the field; rather, the wilderness has penetrated back into the laboratory. We see this in many ways: for example, the human benefits created by science-based technology depend on an exploitation of the environment that could not be borne by the planet if all humanity were to share in them. Thus, the apparently beneficial applications of science create deep and urgent issues of equity that are increasingly important in international politics.

Global environmental issues are very different from the sorts of problems for which traditional scientific problem-solving has been successful. Theories of deterministic chaos of non-linear systems have provided insights into the uniqueness and instability of global environmental systems. Contrary to early expectations, these theories do not furnish new tools for knowledge and control on the model of classical physical science; rather, they open the way to a new conception of science in which knowledge and ignorance will always interact creatively. The social aspects of science are undergoing similar transformation as the exclusive expertise of the scientists is lost. Once outside the laboratory, scientists are citizens among others, contributing their special knowledge, which is different but not dominant, among the other sorts of knowledge in the policy dialogue. The essential complexity of global environmental issues ensures that science can be but one complementary approach among several, all of which are legitimate and necessary. When we realize that global risks are not only systemic but also cumulative (Turner et al. 1990b; see also chapter 1, this volume), our perspective on science shifts even further, for, in the assessment of the cumulative risks, our uncertainties and ignorance swamp our knowledge. Hence, scientific inputs to any policy process are worse than useless unless their uncertainties – including ethical uncertainties, burdens of proof, and principles of prudence and precaution – are effectively managed.

History has witnessed other episodes in which science has been transformed, when a particularly successful problem-solving activity has displaced older forms and also become the paradigmatic example of science. These transformations – identified with such great scientists as Galileo, Darwin, and Einstein – have mainly affected theoretical science, because (until quite recently) technology and medicine were not generally influenced in the short term by the results of scientific research: the challenges to science were largely in the realm of ideas. Now, as the powers of science have given rise to threats to the very survival of humanity, the response will be in the social practice of science as much as in its intellectual structures.

The aims of science will no longer be the traditional ones of achieving truth and the eventual conquest of Nature; rather, they will primarily reflect the need for a harmonious relationship between humanity and Nature. Also, the creative interaction of knowledge and ignorance will be a central element of the theoretical structures of the new sciences, and the admission of other forms of knowing will be inherent to their social practice.

The centrality of uncertainty and quality

Now that global environmental issues provide the most challenging problems for science, issues of uncertainty and quality are moving in from the periphery (one might say the shadows) of scientific methodology, to become the central, integrating concepts. Hitherto they have been kept at the margin of the understanding of science, for laypersons and scientists alike. Whereas science was previously understood as steadily advancing the certainty of our knowledge and control of the natural world, it is now seen as coping with many uncertainties in urgent technological and environmental decisions on a global scale. A new role for scientists will involve the management of these crucial uncertainties: therein lies the task of assuring the quality of the scientific information provided for policy decisions. Moreover, as one scholar recently observed "Scientific evidence has a long row to hoe to have distinctive impact on policy" (Skolnikoff 1999: 19).

The new global environmental issues share common features that distinguish them from traditional scientific problems. They are global in scale and long-term in their impact. Data on their effects, and even data for baselines of "undisturbed" systems, are radically inadequate. Since phenomena are novel, complex, and variable, they are not well understood. Science cannot always provide well-founded theories based on experiments for explanation and prediction; it can frequently achieve, at best, only mathematical models and computer simulations that are essentially untestable. On the basis of such uncertain inputs, decisions must be made, under conditions of some urgency. Therefore, science cannot proceed on the basis of factual predictions, but only on that of policy forecasts.

Computer models are the most widely used method for producing statements about the future based on data of the past and present. For many, computers still exude a magical quality, since they seemingly perform reasoning operations faultlessly and rapidly. But what comes out at the end of a program is not necessarily a scientific prediction; indeed, it may not even be a particularly good policy forecast. The numerical data used for inputs may not derive from experimental or field studies. The best numbers available, as in many studies of industrial risks, may simply be guesses collected from experts. Instead of theories that provide some deeper representation of the natural processes in question, there may simply be standard software packages applied with the best-fitting numerical parameters. And instead of experimental, field, or historical evidence, as is normally assumed for scientific theories, there may be only the comparison of calculated outputs with those produced by other (equally untestable) computer models.

In spite of the enormous effort and resources that have gone into developing and applying such methods, little concerted attempt has sought to determine whether they contribute significantly either to knowledge or to policy. In research related to risk and environmental policy, so crucial for our well-being, very little effort has focused on quality assurance of the sort that the traditional experimental sciences take for granted in their ordinary practice. Whereas computers could, in principle, be used to enhance human skill and creativity by doing all the routine work swiftly and effortlessly, they have instead tended to become substitutes for thought and scientific rigour (MacLane 1988).

It is clear that the dilemmas of computer modelling in research related to policy cannot be resolved at the technical level alone. No one really claims that computer models alone are adequate tools, and yet traditional science can provide nothing better. The critics basically judge such models by the standards of mathematical–experimental science, and of course in those terms they are nearly vacuous. Their defenders advocate them on the grounds that they are the best possible (Jasanoff and Wynne 1998; Keyfitz 1988); however, they do not appreciate how very different these new ecological sciences are in respect to their complex uncertainties, new criteria of quality, and sociopolitical involvements. The need is for exceptionally dedicated efforts for the management of uncertainty, the assurance of quality, and also the fostering of the skills necessary for these tasks. Such skills will not be easy to acquire within the old framework of assumptions about the methods, social functions, and qualified participants in the scientific enterprise.

Even the empirical data that serve as direct inputs to the policy process may be of doubtful quality. Their uncertainties are frequently unamenable to management by traditional statistical techniques. As J. C. Bailar puts it:

All the statistical algebra and all the statistical computations are of value only to the extent that they add to the process of inference. Often they do not aid in making sound inferences; indeed they may work the other way, and in my experience that is because the kinds of random variability we see in the big problems of the day tend to be small relative to other uncertainties. This is true, for example, for data on poverty or unemployment; international trade; agricultural production; and basic measures of human health and survival.

Closer to home, random variability – the stuff of p-values and confidence limits – is simply swamped by other kinds of uncertainties in assessing the health risks of chemicals exposures; or tracking the movement of an environmental contaminant, or predicting the effects of human activities on global temperature or the ozone layer. (Bailar 1988: 19–22)

Thus, in every respect, the scientific status of research on these policy-related problems is dubious at best. The tasks of uncertainty management and quality assurance, performed in traditional science by individual skill and communal practice, founder in confusion in this new arena. We must develop new methods developed for making our ignorance usable (Ravetz 1990), but the path lies in a radical departure from the total reliance on techniques – the exclusion of methodological, societal, or ethical considerations – that has hitherto characterized traditional science.

The NUSAP system provides an integrated approach to the problems of uncertainty, quality, and values (Funtowicz and Ravetz 1990a, b). In its terms, different sorts of uncertainty can be expressed and used to evaluate the quality of scientific information. We must distinguish among the technical, methodological, and epistemological levels of uncertainty, which correspond to inexactness, unreliability, and "border with ignorance," respectively (Funtowicz and Ravetz 1990b).

The management of uncertainty occurs at the technical level when standard routines are adequate. Such routines will usually be derived from statistics (which themselves are essentially symbolic manipulations), as supplemented by techniques and conventions developed for particular fields. The methodological level comes into play when more complex aspects of the information, such as values or reliability, are relevant. Then personal judgements depending on higher-level skills are required, and the practice in question is a professional consultancy, a "learned art" like medicine or engineering. Finally, the epistemological level is involved when irremediable uncertainty is at the core of the problem, as when modellers recognize "completeness uncertainties" that can vitiate the whole exercise, or when "ignorance of ignorance" (or "ignorance squared") is relevant to any possible solution of the problem. In NUSAP these levels of uncertainty are conveyed by the categories of spread, assessment, and pedigree, respectively (Funtowicz and Ravetz 1990b).

Quality assurance is as essential to science as it is to industry. Whereas, in traditional research science, quality assurance could be managed informally by a peer community, the new problems of global environmental risks urgently demand attention to the quality of the science. The research community has extensively analysed the inadequacy of traditional peer review for core science (Turney 1990), "mandated" science (Salter 1988), and "regulatory" science (Jasanoff 1990), so how then can the old methods and concepts manage the manifold uncertainties of the new ecological sciences? As we shall see, the evaluation of quality in this new context of science cannot be restricted to products; it must also include process and, in the last resort, persons as well. This "*p*-cubed" (products, process, persons) approach to quality assurance of science necessarily involves the participation of people other than the technically qualified researchers: indeed, all who have a stake in an issue constitute an "extended peer community" for an effective problem-solving strategy for addressing global environmental risks.

Problem-solving strategies

To characterize a global environmental-risk problem, we can think of it as one for which facts are uncertain, values in dispute, stakes high, and decisions urgent. A simple, linear methodology based on the example of a "pure" science of laboratory research will be unlikely to provide much guidance. On the other hand, the new problems do not render traditional science irrelevant; the task is to choose the appropriate kind of scientific problem-solving strategy for each particular issue.

Figure 4.1 shows three distinctive features of global environmental risks. First (and this is an innovation for scientific methodology), the interaction of the epistemic (knowledge) and axiological (values) aspects of

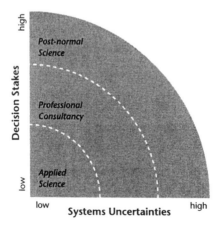

Fig. 4.1 **A diagram of problem-solving strategies**

scientific problems form the axes of a diagram, representing the intensity of uncertainty and of decision stakes, respectively. We notice that uncertainty and decision stakes are the opposites of attributes that had traditionally been thought to characterize science, namely certainty and value-neutrality. (This is the second innovative feature of our analysis.) Finally, each of the two dimensions themselves is displayed as comprising three discrete intervals. The resulting diagram shows three zones representing and characterizing three kinds of problem-solving strategies.

The term "systems uncertainties" conveys the principle that the problem concerns not the discovery of a particular fact but the comprehension or management of an inherently complex reality. By "decision stakes" we understand all the various costs, benefits, and value commitments that are involved in the issue through the interests of various stakeholders. It is not necessary to attempt now to make a detailed map of these as they arise in the technical and social aspects of dialogue on any particular policy issue. It is enough for the present conceptual analysis that it is possible in principle to identify which elements are the leading or dominant ones and then to characterize the total systems by focusing on them.

Applied science

The explanation of the diagram of problem-solving strategies starts with the most familiar strategy, which we call *applied science*. When both systems uncertainties and decision stakes are low, the systems uncertainties will be at the technical level and will be managed by standard routines and procedures that will include particular techniques to keep instruments operating reliably, as well as statistical tools and packages for the treatment of data. The decision stakes will be both simple and low; resources have been allocated to the research exercise because some particular straightforward external function exists for achieving results. The resulting information will be used in a larger enterprise, which is of no concern to the researcher on the job. We illustrate this in figure 4.2.

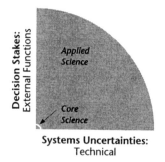

Fig. 4.2 **Applied science as a problem-solving strategy**

This diagram shows a concentration of traditional "pure," "basic," or "core" science around the intersection of the axes. By definition, no external interests are at stake in curiosity-driven research, so decision stakes are low. Also, the research is normally not undertaken unless confidence prevails that the uncertainties are low and that the problem can probably be solved. Thus ordinary "core" science, such as the applied science depicted in figure 4.2, is "normal" in the sense of being devoted to solving research puzzles that are assumed to have answers (Kuhn 1962). Clearly, highly innovative or revolutionary research, either pure or applied, does not lie within this category, since the systems uncertainties are inherently high and, for various reasons, the decision stakes are also high. Thus, Galileo's astronomical research involved the whole range of issues from astronomical techniques to religious orthodoxy; therefore, even though it was not directly applicable to industrial or environmental problems, it was definitely extreme in both its uncertainties and its decision stakes. The same could be said of Darwin's work *The Origin of Species*. In this respect a continuity links the classic "philosophy of nature" and the post-normal science that is now emerging.

We can usefully compare core science and applied science in relation to quality assurance. Where uncertainties and external decision stakes are both low, the normal processes of peer review of projects and refereeing of papers have worked well enough in spite of their known drawbacks (Jasanoff 1990; Turney 1990). When the results of the research exercise become important for some external function, however, the relevant peer community is extended beyond one particular research community to include users and managers of all sorts. The situation becomes rather more like that of manufacturers and consumers, bringing different agendas and different skills to the market. For an example of how criteria of quality can differ between producers and consumers, product safety is a relevant case. A rare accident may not be considered significant for the general performance of a device, especially if product-liability laws are lax, but it may be quite important for consumers, individually and as a class. In the case of applied science, a result validly produced under one set of conditions may be inappropriate when applied to others. Thus, if measurements of a toxicant are given as an average over time, space, or an exposed population, they may be adequate for general regulatory purposes but could ignore damaging peak concentrations or overlook susceptible groups.

It may happen that the results of applied science reflect not "public knowledge" (i.e. that freely available to all competent users) but rather "corporate know-how" (the property of the private business or state agency that sponsors the research exercise). Then the tasks of quality assurance may become controversial, involving conflicts over confidentiality, and decision stakes may be raised over non-scientific aspects. At this point, the problem-solving strategy is no longer applied science, for it may involve struggles over administrative and political power and principles of citizens' "right to know" (for example, concerning environmental hazards or technological risks). The relevant peer community extends beyond the direct producers, sponsors, and users of the research to include all with a stake in the product and process. This extension of the peer community may include investigative journalists, lawyers, and pressure groups. Thus, a problem that may appear totally straightforward scientifically may become one that transcends the boundaries of applied science, giving rise to one or another of the more complex problem-solving strategies that we discuss below.

Applied science has hitherto generally been accepted as the preferred problem-solving strategy for environmental and also social issues. In the case of heroic successes like that of Pasteur, scientific input has been assumed to be the dominant element in any decision process; consequently, scientific experts are the main authorities. We can appreciate the pervasiveness and the previously unquestioned status of this assumption by considering the recent re-evaluations of the sorts of policies for development and the environment to which it had led. Thus, the "Green Revolution," which presented itself as a research exercise solving the problems of peasant tropical agriculture on the basis of the applied science of temperate-zone agriculture, eventually drew effective criticism for its insensitivity to local conditions and for its adverse social and ecological consequences. As a result, in such cases the people with local knowledge, even though sometimes subliterate, came to be treated as legitimate participants in the decision process. In our terms, the peer community for quality assurance of policies and research (which originally seemed to be simple applied science) extends far beyond the traditional experts and their sponsors.

Professional consultancy

The diagram for a "professional consultancy" (fig. 4.3) has two zones, with applied science nested inside. This signifies that professional consultancy includes applied science but also that it deals with problems that require a different methodology for complete resolution. Uncertainty is not routinely manageable at the technical level, because more complex aspects of the problem – such as reliability of theories and information – are relevant. Then personal judgements, depending on higher-level skills, are required, and uncertainty is at the methodological level.

The decision stakes are also more complex. Traditionally, the professional task is performed for a client, whose purposes are to be served. These tasks cannot be reduced to clear, perfectly defined goals, for humans

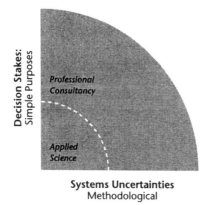

Fig. 4.3 **Professional consultancy as a problem-solving strategy**

are not machines or bureaucracies and are conscious of their own purposes. In the case of risks and environmental issues, the professionals may experience a tension between their traditional roles and new societal demands. The purposes relevant to the task are no longer the simple ones of clients: they are purposes in conflict, involving various human stakeholders and natural systems as well.

The relation between systems uncertainties and decision stakes is well illustrated by the task of incorporating costs into a decision. Exercises in applied science generally subsume these costs implicitly in standard statistical methods. Confidence limits and allowance for the two types of inference errors are normally employed at constant values, without reflection. However, in professional tasks, error costs may be high and may even endanger the continuation of a career. Hence they must be treated as risks, for which some calculation may be used but for which judgement will necessarily predominate. When in a forensic situation, the professional needs to take account of the burden of proof for a particular problem, and the values of a particular society become important (whose harm is most relevant?). Thus, the handling of a problem of environmental pollution may well differ, depending on whether a process is deemed safe until proven dangerous, or dangerous until proven safe. Such a simple opposition cannot encompass all practical situations, of course, and so professional tasks require an appreciation of the subtleties involved in establishing burden of proof.

Professional consultancy shares with applied science many features that distinguish both from core science. Both operate under constraints of time and resources, with problems defined by external interests. Their products generally do not lie in the domain of "public knowledge." Usually, professional tasks can be reduced to applied research exercises, as the routine work becomes standardized in technique and in the management of uncertainty. However, professional consultancy requires creativity, a readiness to grapple with new and unexpected situations, and a willingness to bear the responsibility for outcomes. Engineering straddles the border between these two, for most engineering work proceeds within organizations rather than for individual clients, and yet the problems cannot be completely reduced to a routine; "engineering judgement" is, therefore, a well-known aspect of this work.

As a problem-solving strategy, professional consultancy differs significantly from applied science. The outcomes of applied-science exercises, like those of core science, have the features of reproducibility and prediction. That is, any experiment should, in principle, be capable of being reproduced anywhere by any competent practitioner, since it operates on isolated, controlled natural systems; the results, therefore, amount to predictions of the behaviour of natural systems under similar conditions. By contrast, professional tasks deal with unique situations, however broadly similar they may be. The personal element thus becomes important. It is legitimate, therefore, to call for a second opinion, without impugning the competence of a doctor or other professional, or implying that either is simply wrong. After all, who would expect two architects to produce identical designs for a single brief? In the same way, it would be unrealistic to expect two safety engineers to produce the same model for a hazard analysis of a complex installation.

The public may become confused or disillusioned at the sight of experts disagreeing strongly on a problem apparently involving only applied science (and even the experts may be confused!). However, when it is appreciated that these issues involve professional consultancy, these disagreements should be seen as inevitable and healthy. Occasionally, however, the need exists for consensus among professional experts, as in the case of quantitative inputs for models of industrial or environmental hazards. "Expert judgements" are then invoked as a substitute for experimental or field data. When such a substitution is recognized as highly problematic, quality assurance is conducted at a higher level. The problem then becomes the assurance of the quality of the experts themselves! Such a process iterates without end, leading to what we might call the "(expertise)-nth problem." This is an indication that the professional tasks do not belong to applied science and cannot be solved as if they were.

This last phenomenon reminds us of the differences in quality assurance that emerge when we move from applied science to professional consultancy. We can envisage three components in the problem-solving task – the process, the product, and the person; this is the "p-cubed" approach to quality assurance mentioned above. In core science, the main focus in immediate quality assessment is on the process, for the product (the outcome of the research) is not usually reproduced by journal referees. Hence, the written reports of materials, instrumentation, and techniques are the objects of scrutiny by referees. This is why quality assessment requires expert peers and is necessarily a technically esoteric activity. Ideally, the persons (or their institutions) are not relevant to the quality assessment. In applied science, the focus of assessment extends to products and is done by users on whose behalf the research exercises are conducted. Quality assurance is then less esoteric, since the users have less need to understand the research process and there is, thus, an automatic extension of the community with a legitimate participation in evaluation.

We have already discussed the enhanced personal factor in professional consultancy and observed how the quality of the process reduces in the last resort to the quality of the persons performing the task. This contrasts with the case of core science with its consensus on process, and that of applied science with its external users and their pragmatic criteria for the products of the research exercise. In professional consultancy no simple, objective criteria or processes for quality assurance are possible: a "personal knowledge," in Polanyi's sense, is required in the choice and evaluation of experts (Polanyi 1958). The community of legitimate participants in evaluation becomes still broader and the technical aspects of the science are subordinate (although they can serve as evidence on the quality of particular experts). In addition, no personal knowledge (which can be as varied as people and their concerns) should predominate.

Post-normal science

We now consider the third sort of problem-solving strategy, in which systems uncertainties or decision stakes are high (fig. 4. 4). In addressing an issue in post-normal science, both professional consultancy and applied science can be part of the overall activity, since not all aspects of the issue will involve high uncertainty or conflicting values. The professional tasks or the applied research exercises, however, cannot dominate the decision-making process.

Post-normal issues may include a large scientific component in their description, sometimes even to the point of being capable of expression in scientific language. In this sense they are analogous to the "trans-science" problems first identified by Alvin Weinberg (1972b). But it seems best to distinguish the problems analysed here from that class of problems, for Weinberg imagined problems that differed from those of applied science only in scale or technical feasibility; they were scarcely different from those of professional consultancy as we have defined them. In the terms of our diagram, post-normal science occurs when uncertain-

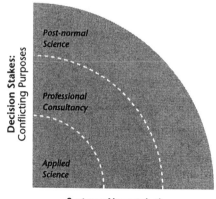

Systems Uncertainties:
Epistemological/Ethical

Fig. 4.4 **Post-normal science as a problem-solving strategy**

ties are either of the epistemological or the ethical kind, or when decision stakes reflect conflicting purposes among stakeholders (Funtowicz and Ravetz 1991b). We call it "post-normal" to indicate that the puzzle-solving exercises of normal science (in the Kuhnian sense), which were so successfully extended from the laboratory to the conquest of Nature, are inappropriate for the solution of global environmental problems.

It is noteworthy that in figures 4.2, 4.3, and 4.4, applied science appears three times and professional consultancy twice. Do these labels mean the same when they are part of a broader problem-solving strategy as when they are standing alone? In the sense of their routine practice, yes. However, when they are embedded in a broader problem-solving strategy, the whole activity requires reinterpretation: the problems are set and the solutions evaluated by the criteria of the broader communities. An analogy exists here with the evolution of scientific theories, as when (for example) Newton's mechanics was not so much refuted as included in, and reinterpreted by, Einstein's relativity.

The epistemological type of uncertainty has become familiar to experts, even in cases in which computer methods dominate problem-solving. Those experts were already accustomed to technical uncertainty, in the "errors" of data inputs, and to methodological uncertainty, in the response of the models to the inputs (as gauged, for example, by sensitivity analyses or comparison of models). But increasingly they are becoming aware of the insoluble questions of what, if anything, their models (the outputs of which are generally untestable) have to do with the real world outside. Thus, these experts discover in their own practice an extreme form of uncertainty, one that borders on ignorance. This sort of uncertainty cannot be reduced to the others and therefore cannot be treated by standard mathematical or computational techniques. This example allows us to appreciate how pervasive epistemological uncertainty is in all the scientific fields concerned with global environmental risks. Hitherto, such problems have been neglected because there seemed to be no systematic solution to them. But this is a form of ignorance of ignorance, a most dangerous state.

Examples of issues with combined high decision stakes and high systems uncertainties are familiar from the current crop of global environmental problems. Indeed, any of the problems of major technological hazards or large-scale pollution belong here. The paradigm case for a post-normal issue could be the design and siting, in Nevada USA, of a repository for long-lived nuclear wastes, to be secure for the next 10,000 years (Flynn et al. 1995). Such an undertaking would involve an enormous number of applied-research exercises as well as professional tasks of many sorts. The outcome of all this effort, however, has been effectively bypassed in the political struggle over the siting decision. It had previously been recognized that standard risk-assessment techniques were inadequate for the siting problem, and the original legislation of 1982 provided for equity in both procedure and substance (outcome). But in 1987 the US Congress scrapped essential elements of the Nuclear Waste Policy Act, with the result that science has dropped out of the debate (Parker et al. 1987), equity provisions have been jettisoned, and what Flynn and colleagues term "tunnel vision" prevails (Flynn et al. 1995, 1997).

Thus the process of siting a nuclear waste repository in Nevada became completely polarized. For the promoters, the overriding concern was the resolution of the increasingly urgent problem of disposal of nuclear wastes. Also at stake was their professional commitment to a scientific or engineering solution to all problems that arise in connection with civil nuclear power. For the opponents, the issue did not merely include the gross uncertainties in the performance of any human-made structure over such a time span: in addition was the keen sense of injustice that one group of people had this facility forced upon them. Furthermore, the dread of nuclear materials, however exaggerated it might be from a traditional scientific perspective, is a very real and powerful aspect of the policy issue. In this case, even the services of professional mediators were unavailing, given the absence of a basis for negotiation and compromise. Indeed, the more straightforward problem-solving strategies, relying more on objective and quantitative inputs, were ineffective for achieving a solution. The challenge of post-normal science is to prevent a recurrence of such struggles over all issues involving large uncertainties and high decision stakes.

Post-normal science has the paradoxical feature that its problem-solving activity inverts the traditional domination of "hard facts" over "soft values." Owing to the high level of uncertainty (approaching sheer ignorance in some cases) and the extreme decision stakes, we could even interchange the axes on our diagram, thereby making values (decision stakes) the horizontal, independent variable. Such an inversion characterizes the actions that will need to be taken in preparation for mitigating the effects of sea-level rise consequent on global climate change. The "causal chain" here starts with the various outputs of human activity, producing changes in the biosphere, and leading to changes in the climatic system, then changes in sea level (all these interacting in complex ways with varying delay times), and eventually effects on humans (Hohenemser, Kasperson, and Kates 1985). Out of this must come a set of forecasts that will be inputs to decision processes and that will result in policy recommendations that must be implemented on a broad scale. At stake may be much of the built environment and the settlement patterns of people (see chapters 11–13, this volume). Mass migrations from low-lying districts will be inevitable sooner or later, with consequent economic, social, and cultural upheaval.

Such far-reaching social policies will go forth on the basis of scientific information that is inherently uncertain to an extreme degree, all the more so because plans for mitigation require a long lead time to ward off the harm to engulfed lands and their peoples. A new form of legitimation crisis could emerge, for, if the authorities try to base their appeal on the traditional certainties of applied science or the skills of professional consultancy, they will surely fail. Public agreement and participation, deriving essentially from value commitments, will be decisive for the assessment of risks and the setting of policy. Thus, the traditional scientific inputs have become "soft" in the context of the "hard" value commitments that will determine the success of policies for mitigating the effects of a possible sea-level rise.

The traditional fact/value distinction has not merely been inverted: in post-normal science the two categories cannot be realistically separated, for the uncertainties go beyond the systems to include ethics as well. All global environmental risks involve new forms of inequity (as suggested by the Kaspersons and colleagues in chapter 1 and Kasperson, Kasperson, and Dow in chapter 7), which had previously been considered "externalities" to the real business of the scientific–technical enterprise. These risks involve the welfare of new stakeholders – such as future generations, other species, and the planetary environment as a whole (Kasperson and Dow 1991). The study of ethics has suddenly received a new stimulus, as these non-traditional subjects have come into focus. The problems attendant on extinctions of species, either singly or on a global scale, illuminate the intimate connection between uncertainties in knowledge and in ethics. It is impossible to produce a simple rationale for adjudicating between the rights of people who would benefit from some development and those of a species of animal or plant that would be harmed. The ethical uncertainties should not deter us, however, from searching for solutions, nor can decision makers overlook the political force of those humans who harbour a passionate concern for those who cannot plead or vote (see chapter 7).

All these complexities do not prevent the resolution of issues in post-normal science. It is important to view the diagram as dynamic rather than static, so that different aspects of the problem, located in different zones, interact and evolve. Issues with differing problem-solving strategies successively come to prominence, providing a means by which dialogue can eventually contribute to their resolution. For, as the debate moves from its initial confused phase, the result is a clarification of positions and a stimulation of new research. Although the definition of problems is never completely free of politics, an open debate ensures that such considerations are neither one-sided nor covert and, as applied research exercises eventually bring in new facts, professional tasks become more effective. A good example of this pattern of evolution is lead in gasoline (petrol), where, in spite of the absence of conclusive environmental or epidemiological information, a consensus eventually emerged that the hazards were not acceptable (Lovei 1998).

The dynamic of issue resolution in post-normal science involves the inclusion of an ever-growing set of legitimate participants in the process of quality assurance of the scientific inputs. As we have seen, the peer communities have already moved far beyond the traditional boundaries for applied science and professional consultancy. The manifold uncertainties in both products and processes have served to enhance the relative importance of persons in the post-normal dialogue. We have already noticed how the choice of experts, involving the quality assurance of their expertise, cannot be solved within the confines of professional consultancy. Hence, the establishment of the legitimacy and competence of participants will inevitably involve broader societal and cultural institutions and movements. For example, persons directly affected by an environmental problem will have a keener awareness of its symptoms, and a more pressing concern with the quality of reassurances, than those in any other role. They perform a function, analogous to that of professional colleagues in the peer-review or refereeing process in traditional science, which otherwise might not occur in these contexts.

On occasion, the legitimate work of peer communities extends even beyond the reactive tasks of quality assessment and policy debate. The relatively new field of "popular epidemiology" finds concerned citizens themselves doing the disciplined research that established institutions could, or perhaps should, have conducted (Brown 1987; Murdock, Krannich, and Leistritz 1999). In such cases they may encounter what may be called "institutional uncertainty," when they draw criticism either for lacking certified expertise or for being personally concerned about the problem. The creative conflict between popular and expert epidemiology not only leads to better control of environmental problems but also improves scientific knowledge. A classic case is "Lyme disease," for which local citizens first identified a pattern in vague symptoms that later became characterized as a previously unknown, but not uncommon, tick-borne disease (Barbour 1996).

The new paradigm of post-normal science, involving extended peer communities as essential participants, is visible in the case of AIDS. Here the research scientists operate in the full glare of publicity involving sufferers, care-givers, journalists, ethicists, activists, and self-help groups – as well as traditional institutions for funding, regulation, and commercial application of pharmaceuticals. The researchers' choices of problems and evaluations of solutions are equally subjected to critical scrutiny, and their priority disputes are similarly dragged into the public arena (Perrow and Guillén 1990).

Such cases are still the exception: extended peer communities generally operate in isolation on special issues in isolated localities, with no systematic means of financial support and with little training in their special skills. On many occasions, insufficient competence informs dialogue and communication with other stakeholders (Salter 1988). Recognition of their role is very variable: in the United States, with its populist traditions, "intervenors" in some decision processes receive support; elsewhere, they may be ignored or actively hindered. Within such extended peer communities the usual tensions will exist between those with simple NIMBY ("not in my back yard") demands and the outside activists with more far-reaching agendas, along with the inevitable divisions along lines of class, ethnicity, gender, occupation, and formal education (Hurley 1995; Inhaber 1998; O'Looney 1995). Nevertheless, all such confusion is inevitable and, indeed, healthy in an embryonic movement that is forcing a transition to a new era.

As in any deep transition, the present contains seeds of destruction as well as of renewal. Many participants in environmental struggles come to see scientists merely as hired guns, providing the data that "we" need and ignoring or concealing the rest, whereas others are impervious to arguments and evidence that contradicts their prejudged case. Are such participants legitimate members of an extended peer community? Even traditional science has always included such types, but an implicit ethical commitment to integrity has prevailed by which the community as a whole has maintained the quality of its work (Ravetz 1971). The maintenance of quality, without which all efforts to solve global environmental issues are doomed, is a major task for the methodology of the science of the future.

The extended peer communities are essential to the sort of science that can manage global environmental risks. We all know that global thinking must be complemented by local action – but "local" means "community" and, for that, the entire community must become peers in the shaping of the new sort of science. This is not to say that all jobs can be done by all people: post-normal science has a place for the technical work of applied science and also the judgemental skills of professional consultancy. The difference is that, whereas these components are still necessary, they are not sufficient in themselves. Seen in the context of post-normal issues, these special problem-solving strategies are ripe for enriched reinterpretation.

Our analysis rests on an awareness of uncertainty and ignorance, and this principle applies equally to our own arguments. We cannot predict the precise forms of post-normal science or which issues or institutions will be the major foci for the development of extended peer communities. We can be sure, however, that the examples we have discussed, as well as the many others that abound in the literature of protest on environmental issues, anticipate the form that the science of the future must take.

Conclusions

In every age, science takes shape around its leading problems and it evolves with them. The new environmental risks are global, not merely in their extent but also in their complexity, pervasiveness, and novelty. Up to now, with the dominance of applied science, the rationality of scientific research served as a model for the rationality of intellectual and social activity in general. However successful it has been, the growing recognition of global environmental risks shows that this ideal of rationality is no longer universally appropriate: diverse types of rationality will be needed in the exploration of approaches to global environmental risk that is the subject of this volume.

The activity of science now encompasses the management of irreducible uncertainties in knowledge and ethics and the recognition of different legitimate perspectives and ways of knowing. In this way, its practice is becoming more akin to the workings of a democratic society, characterized by extensive participation and toleration of diversity. As the political process comes to recognize our obligations to future generations, to other species (and, indeed, to the global environment), science will also expand the scope of its concerns. We are living in the midst of this rapid and deep transition, so we cannot predict its outcome. Nevertheless, we can help to create the conditions and the intellectual tools for managing the process of change for the best benefit of humanity and the global environment.

Part Two

Vulnerability

Editors' introduction

Risk, as we point out in chapter 1, is a joint product of environmental perturbations or (in the case of industry or technology) of the releases of energy, materials, or information on the one hand and the vulnerability or fragility of the receptor or affected system on the other. All the same, the history of risk and impact analysis, and other types of social-science analysis as well, has accorded much less attention to the second term than to the first. More than a quarter of a century after an intense debate over the need for greater prominence of vulnerability studies in hazard assessment (Hewitt 1984a), neither the scholars working in risk analysis nor their critics have fashioned a powerful theory of vulnerability. This is not to deny the increased attention to these issues and the significant progress, as the chapters in this section will attest. But this volume's call for a new global risk analysis, one in which vulnerability commands centre stage, suggests that an active search for theory and model development in vulnerability studies will need to be a high priority for risk studies over the next period.

Arguably, this development will nowhere be more important than for global environmental risks. Despite overall economic progress and enhanced societal coping resources throughout much of the world, the global toll from natural hazards is growing (Munich Re 1999; United Nations 1999; *World Disasters Report* 1999, 2000), as we detail in chapter 7. In addition, the growing loss of life is increasingly concentrated in developing countries, where vulnerability to events is greatest, long-term resilience weakest, and management institutions most underdeveloped. Growing inequalities in wealth form the social base upon which both slowly accumulating hazards and more extreme risk events are super-imposed. As linkages to an expanding global economy intensify in these developing countries, new patterns of vulnerability to, and buffering from, environmental changes emerge, as Smith underscores in chapter 3. As the traditional risks of natural disasters, poverty, and infectious dis-eases remain (or even, as in the case of cholera, reappear with new in-tensity), they are joined by the newer risks of advanced technologies, the facilities of multinational corporations, livelihood changes, or trans-boundary risks with sources in remote advanced industrial societies. To this newly configured portfolio of risks, third-world societies bring limited experience, scant expertise, and meagre economic resources (Lewis 1999; Twigg and Bhatt 1998).

It is scarcely surprising, then, that responding to global risks is not only a matter of systems of control and protection but intensely about fairness and equity. The World Commission on Environment and Development (WCED 1987) has argued, however simplistically, that the search for a sustainable world and the eradication of poverty are highly interrelated goals. The Commission also envisioned, perhaps prophetically, that effec-tive coping with global environmental risk would ultimately require a new global ethic. Nothing that has happened during the 1990s – at the Earth Summit, the Kyoto meetings, or the 1998 Conference in Buenos Aires, or the emerging international responses to global biodiversity loss – belies the Commission's insights. The various exercises in global modelling, and particularly the Polestar effort (Gallopin et al. 1997; Raskin et al. 1998), give credence to the powerful interaction that poverty and vulnerability more generally will almost certainly have with environmental degrada-tion in shaping future patterns of risk. It is also clear, however, that vul-nerability is far more complex than poverty, entailing entitlements and potential resource mobilizations that link households and extended social networks or leverages but also flows among institutions of different scales. Working Group 2 of the Intergovernmental Panel on Climate Change (IPCC) provided a highly useful service in giving prominence to the issue of vulnerability in impact assessment and in summarizing widely disparate relevant empirical data. But this very service underscores the need for more powerful models; for explanatory frameworks; and for approaches for identifying, characterizing, and measuring vulnerability. At the same time, what we do know speaks eloquently that managing global environmental risk is, at its heart, about choosing between alter-native world futures and exercising moral judgements as to which peoples (and which species) should be protected and on the basis of what moral claim.

The three chapters that follow incisively explore this nexus of risk, vulnerability, and equity. Liverman notes perceptively in chapter 5 that vulnerability issues are implicit in both the *drivers* and *impacts* of global environmental change and that both biophysical and political economic analyses are requisite to identify and assess vulnerabilities. In particular, continuing the observations above, political economic structures throughout the world are notoriously successful in passing on the burdens or prices of global environmental change to those least capable of responding to them. Indeed, as noted in chapter 8, peripheral and marginal regions often bear the brunt of environmental degradation. Liverman pays particular attention to the gender structuring of global risk and vulnerabilities. On the more optimistic side, she calls our attention to one area of improved responsiveness to one global risk – the Famine Early Warning System (FEWS) – and its counterpart Global Information and Early Warning System (GIEWS). At the same time, Liverman points out that even the movement to a more social model of famine is still well short of what is achievable through the use of social and economic indicators. Furthermore, clearly, future vulnerability monitoring will need to be attentive to the new risks involved in the transitions of concern to Smith in chapter 3.

Whereas Liverman focuses on socio-economic dimensions of vulnerability, Ezcurra and colleagues turn the lens on natural ecosystems. Loss of vegetation cover, biodiversity loss, overgrazing, and non-sustainable forest harvesting pose huge risks, they contend, for developing societies. Enquiring into the question of which natural environments are likely to be the most affected, they make a compelling case that tropical ecosystems are among the most vulnerable and that the cumulative types of global risk – population changes, soil erosion, tropical deforestation – pose the greatest threat. They illustrate these arguments by examining, in particular, the cases of coral reefs, mangrove forests and coastal marshes, and cloud forests. But, of course, it is not only the high proportion and richness of vulnerable species in tropical rain forests but of those in high-altitude tropical ecosystems, dry tropical forests, tropical savannas, and coastal ecosystems that are at risk. And Ezcurra and colleagues drive home the essential point about vulnerability to environmental degradation – that species loss will be concentrated in exactly those regions where human inhabitants are poor and the economic and social capacity to respond is most limited. Vulnerability is an interactive concept.

The editors and Kirstin Dow extend the treatment of vulnerability by examining in chapter 7 how it links to equity and fairness. Just as there are many sources of vulnerability, they argue, so are there many types or syndromes of vulnerability. Much of the global-change debate centres on managing what are somewhat incorrectly termed "north/south" conflicts; it also involves a host of subtle and cumulative equity problems, dealing with such matters as protection of future generations, intellectual property rights, protection of endangered species, and trade-offs in protecting the currently most vulnerable people versus protecting people as yet unborn. To facilitate more systematic and less "hit and run" approaches to equity assessment, they propose an analytic framework for more comprehensive analyses. This framework covers both distributional or outcome-oriented assessments and procedural equity investigations. The framework suggests that equity analyses can be very complex and that their effective engagement will require not only careful empirical analyses (to inform debates about the causes and nature of effects) but also selection of the moral principles that should guide societal responses and international interventions. The debate surrounding climate change and the host of issues addressing responsibility for the accumulated and growing environmental burden; the moral content of different types of greenhouse-gas emissions; major inequalities in capabilities to reduce emissions; competing principles (such as the precautionary and polluter-pays principles); and the opaque overlay of growing international equalities; suggest how difficult arriving at the new environmental ethic espoused in the Brundtland Report (WCED 1987) will be.

Taken together, the chapters highlight the importance of vulnerability in global environmental risk and also suggest the degree to which its sources, incidence, and modes of expression are heavily interactive in nature. It is also clear that risk, vulnerability, and equity are intimately intertwined. There is much that the risk analyst can contribute here, especially in the form of value-sensitive empirical analysis, to debates about moral obligations and alternative futures.

5

Vulnerability to global environmental change

Diana M. Liverman

The impacts of global change on environment and society are determined as much, if not more, by the characteristics of the ecosystems and people affected as by the magnitude of the change itself. Droughts of equal physical severity may have much less severe impacts on large, commercial, irrigated farms that can rely on insurance, good soils, and subsidized prices than on smaller, rain-fed, subsistence farms that lack institutional support. Irrigation, crop diversity, and flexible land-use and management strategies may buffer agricultural systems against climate variability. On the other hand, the systems may be more sensitive to change if economics, land tenure, and resource availability restrict options for land use, irrigation, and crop choice. Deforestation will have more severe effects in regions where slopes are steep, soils are fragile, or species are already being overhunted. Certain species, ecosystems, and cultures are disproportionately affected by desertification. Future generations, as well as the poor, sick, and powerless, are particularly sensitive to the widespread build-up of toxic substances in the environment.

The same environmental change will have different impacts in different places because some people and places are more *vulnerable* than others to environmental change. In this and subsequent chapters we argue (and illustrate) that understanding vulnerability is an important component of the risk assessment of global environmental change.

The definition of the term "vulnerable" originates in the verb "to wound," and the term has been used in a variety of ways to characterize the response of social and ecological systems to perturbations (Timmer-man 1981; see also chapter 14, this volume). Vulnerable people and eco-systems lack the strength to resist pests, diseases, and hunger; they may be unable to move away from danger; or they may not have access to the resources needed to provision or defend themselves. In a sense, they are more easily wounded and recover more slowly. Vulnerability is relative, it is almost always negative, and it generally describes the system that is affected rather than the agent of change (Dow 1992; Downing 1991a, b; Vogel 1998).

The physical and, to a lesser extent, biological sciences tend to domi-nate the debates, literature, and research funding on global change. The environmental-monitoring and climate-modelling communities are work-ing at the core of the effort, documenting the extent of biospheric change and projecting the climatic effects of changing atmospheric composition and land-surface conditions. The risks of global change are defined with technical estimates of the probability and environmental consequences of changes in the earth system. However, to understand the real risks of global environmental change for society (this volume argues throughout), research must move beyond basic environmental science and climate modelling to social-science analyses of causes and effects. A more so-phisticated and detailed analysis of the social *causes* of global change is required in order to project and manage the rates and locations of those human activities, such as industrialization and land-use transformation, that are driving environmental change. A more rigorous assessment of the potential social *impacts* of global change is needed to provide more circumspect forecasts and to guide policies for reducing change or miti-gating impacts.

Vulnerability to changes in the atmosphere, land, and biosphere very much determine the impacts of global change. But vulnerability also plays a role in causing global change, because those who are most vulnerable to environmental change are often those who are also vulnerable to other sorts of changes, such as war, inflation, and land expropriation. People who lose their income, land, or access to resources may be forced to cut down forests or overuse the land in order to survive. Thus, the vulnerable contribute to global change, especially through land-use transformations.

The analysis of vulnerability is an important complement to assess-ments of the nature and rate of environmental change. In the case of cli-mate change, the advantage of focusing on vulnerability is that it reduces dependence on the uncertain results of climate models, yet points to many potential ways to reduce the negative impacts of climate variation both today and in the case of a warmer future. Vulnerability studies, at regional and local levels, provide particular insights into the importance of demographic, political, economic, and technological changes in determining the impacts of climate on society in time and space. We can consider how such factors may change in the future and how they may determine the regional impacts of climate change (IPCC 1998). Developing countries, owing to their economic situation and great reliance on both rain-fed and irrigated agriculture, may well be relatively more vulnerable to a shift in climate (Agarwal and Narain 1999; Gleick 1989; Jodha 1989; Moltke and Rahman 1996).

Global change and changing vulnerability are part of a process of the mutual transformation of society and Nature at various scales. As industrial society is producing the greenhouse gases that warm the atmosphere, the world economy is changing, and often marginalizing, the conditions of subsistence agricultural production. Population growth is both adding to the output of greenhouse gases and increasing the stress that unequal land distribution places on the majority of the people who work small and inadequate plots. Adaptations to, and mitigation of, local and global environmental changes are constantly altering the basis of projections of the future.

One could, of course, argue that almost everything and virtually everyone is vulnerable to global environmental change. Many of the recent statements on global change do use "Spaceship Earth" analogies to emphasize that we are all in this together and that nobody will escape the impacts of the greenhouse effect, ozone depletion, or deforestation. Although such sentiments convey an appropriate urgency – and even promote international cooperation – they tend to evade the issues of differential susceptibility to global changes, the distribution of blame, and the possibility that some may benefit from environmental transformation. At the other extreme, some analyses suggest that the burden of global change will fall almost entirely on the poor people and the poor countries of the world, because political–economic structures always favour passing hardships on to the oppressed (Agarwal and Narain 1991, 1999; Blaikie et al. 1994; Seabrook 1993).

Although valid political reasons exist for promoting either one of these perspectives, neither is particularly helpful or profound in designing responses to changes in specific places. Since it seems unlikely that we shall marshall the resources needed to prevent all global environmental changes or to compensate everyone for their impacts, we need more precise estimates of who is vulnerable in order to decide where, when, and how most effectively to focus our responses.

Much of the popular concern about the risks of global environmental change originates in people's apprehension that they themselves may be vulnerable. Those who already live on the margins of existence fear that global change will destroy their basic life-support systems or that adjustments to global change will proceed at their expense. Those who have the resources to adapt to global environmental change fear that these resources – such as reservoirs, coastal defences, biotechnology, or advanced health care – may eventually be inadequate to cope with the projected changes. They also fear that impoverishment and suffering in the more vulnerable regions and social groups may result in migration, political instability, or international conflict that may indirectly affect them.

Many people may want to understand and reduce vulnerability to global change for purely humanitarian reasons. But it is important to recognize that some people and institutions will oppose detailed assessments that highlight inequality and conflicts within society, because the solutions to these problems will threaten their own interests. For example, some groups have opposed studies of drought and famine in Sudano-Sahelian countries or deforestation in Latin America because such assessments reveal exploitation, discrimination, and government failure (Franke and Chasin 1980; Maguire and Welsh-Brown 1986).

Some discussions of global environmental change have suggested that concern about the social causes and consequences of environmental degradation beyond one's own country or locality is in the national and individual self-interest (Mathews 1989; Myers 1989). Reinforcing the ability of other countries to cope with environmental problems is not only a way to develop strategic allies but also a means to ensure the stability of import and export trade. Linkages among environmental degradation, economic decline, and the debt crisis have served to demonstrate the vulnerability of global financial institutions to environmental change. Self-interest is at stake in protecting global commons, such as the atmosphere and oceans, and in preventing the destruction of biodiversity and other ecologically and economically important resources (Ostrom et al. 1999). International security concerns also arise from the politically destabilizing influence of environmental impoverishment and activism and from the possibility of thousands of "environmental refugees" moving across borders (Homer-Dixon 1999; Jacobson 1989a; Myers 1995).

Approaches to understanding vulnerability

Two contrasting frameworks for analysing vulnerability to environmental change can be termed the *biophysical* approach and the *political-economy* approach.

In the biophysical framework, the most vulnerable people are considered to be those living in the most precarious physical environments. Drought effects would be associated, for example, with low or variable rainfall and sandy soils. Biophysical criteria are used to delimit regions vulnerable to desertification despite the important role of human activities in the desertification process (Alexandratos 1995; Grainger 1982, 1990). Many basic regional geographies also begin by describing the constraints posed by the physical environment and its sensitivity to change. People are implicitly vulnerable, owing to their presence in these fragile environments.

Other examples of the biophysical approach are found in the risk literature. Natural and technological risks have been studied by documenting the human occupance of flood and earthquake zones (Mileti 1999; White 1974). The more people who occupy a biophysically vulnerable location, such as a flood plain or coast, the greater the total population at risk. Similarly, the more people who occupy regions near sources of toxic emissions or nuclear facilities, the greater the risks. Populations at risk live in areas of high-hazard magnitudes or frequencies, such as areas with severe rainfall deficits, flood plains, coasts subject to severe storms, and basins (such as the dramatic case of Mexico City) that trap air and water pollution (Blaikie et al. 1994; Ezcurra et al. 1999; Ezcurra and Mazari Hiriart 1996; Hewitt 1997).

Using this framework to assess the risk from global environmental change involves estimating changes in biophysical conditions. Projections of global transformations such as climate warming and vegetation loss indicate an extension of the area of fragile lands and an increase in the frequency and magnitude of natural hazards. The use of such criteria would include among the groups most vulnerable to global environmental change those living in areas likely to experience sea-level rise, increasing storminess, drier conditions, or heavier flooding. Those studies of global warming and sea-level rise that delimit populations at risk within topographic contours illustrate this approach (Barth and Titus 1984).

Assessments of the impacts of global climate change are frequently based on estimates of biophysical changes, particularly potential changes in agricultural yields and water resources. The direct approach traces the impact of a specific change in a physical input variable (such as temperature) on yields or biomass, and then, through a series of steps, to impacts on economy and society (Parry, Carter, and Konijn 1988). This type of assessment relies on (and is often limited to) physical models of the climate, water balance, and vegetation growth.

The demographic factor is important in many of the biophysical assessments of vulnerability that examine the capacity of the physical environment to support population. One example of this approach is a study by the Food and Agriculture Organization (FAO) of the United Nations on the food-production potential and population-carrying capacity of lands in the developing world (FAO 1984; Higgins et al. 1983). This study, on the basis of climate and soil constraints, estimates the agricultural production potential of lands and converts that production to cal-ories and hence to the number of people who can be fed in different agroecological zones. This process identifies certain regions and countries (e.g. the Sahel, Bangladesh, Haiti) as "critical zones" in which current demands for food already exceed the agricultural production potential – that is, the carrying capacity. The number of critical zones in the FAO study increases dramatically when population growth is included in the estimates of carrying capacity in the year 2000. Although the study does take into account different scenarios for population growth and technol-ogy, its underlying analytical framework rests on a biophysical determi-nation of vulnerability to food shortage (Hekstra and Liverman 1986). Such studies of carrying capacity treat humans as organisms (hence the classification as a "biophysical" framework) using resources and give little attention to the social organization or consumption levels of in-dividuals within the population.

The FAO study does not take global change into account in its sce-narios for the year 2000, although proposals have been made to use the same climate and soil database to assess the impacts of climate changes on carrying capacity. By changing climatic data in accordance with fore-casts of global warming, for example, it would be possible to estimate corresponding changes in agricultural potential, critical zones, and pop-ulations at risk. Climate-impact assessments, which combine estimates of changes in water availability or crop production with estimates of changes in population demand, use a similar framework.

The foregoing examples represent a dominant analytical framework in studies of the human impact of global change. The predominance of this framework guides and justifies the research focus on biophysical mon-itoring, modelling, and modification in global-change studies. It implies that, in order to understand and delimit vulnerability, we just need to know how and where the physical environment may change: physical indicators will provide adequate insight into the populations at risk. The framework also governs the selection of responses. If biophysical con-ditions define vulnerability, then modifying these conditions or moving people away from biophysically marginal areas can reduce vulnerability.

A contrasting different framework criticizes both biophysical and biophysical–demographic characterizations of vulnerability. The *political-economy* approach has become increasingly important in studies of cli-mate impact and environment and development. This perspective sees vulnerability as the creation of the political, social, and economic con-ditions of society rather than of the physical environment.

Susman, O'Keefe, and Wisner (1984), for example, define vulnerability to natural hazards as "the degree to which different classes in society are differentially at risk" (see also Blaikie et al. 1994). In hazards research, the political-economic framework shows how underdevelopment (flows of resources out of a region, land expropriations, exploitative labour conditions, political oppression, and other processes associated with colonialism and capitalism) have made people, especially the poor, more vulnerable to disaster and has forced them to degrade their environment (Blaikie et al. 1994). This mode of analysis characterizes many studies of the 1972 drought in the Sahel and other regions, studies that proposed a "structural" approach for diagnosing the impact of climate anomalies (Copans 1975; Garcia 1981). By analysing the historical evolution of social systems in various regions, such research sought to demonstrate how certain groups become so disadvantaged and exploited that they are unable to cope with drought or to struggle for the resources to overcome environmental stress.

Political economy can also serve to define vulnerability to other types of environmental change. Those most vulnerable to deforestation and species extinction are those most directly and exclusively dependent on forest products, such as the indigenous peoples of the tropical rain forests. Those most vulnerable to the accumulation of toxic substances in the environment are those who, because of their disadvantaged structural position in the economy, society, and housing market, are compelled to work and live in polluted conditions.

The biophysical and political economy perspectives provide two ways of analysing who may be vulnerable to global change. Of course, many scholars have tried to blend elements of both frameworks: for example, Parry (1985) notes that it is necessary to include economic and social vulnerability, as well as the sensitivity of the physical system, when studying the impacts of climate on agricultural systems. An increasingly popular integrative approach is *political ecology* (Blaikie and Brookfield 1987; Rocheleau, Thomas-Slayter, and Wangari 1996). Studies focus on the land manager and the physical, technological, economic, and political conditions that may constrain the use of land or the human response to environmental and social changes (Glantz 1994). The roles of the physical environment, political structure, and individual actions are all relevant to understanding the risks of global environmental change. In *Regions at Risk*, Kasperson, Kasperson, and Turner (1995) adopt an approach focused on trajectory analysis and what they term the "regional dynamics of change" (chapter 8, this volume; Kasperson, Kasperson, and Turner 1996, 1999).

To facilitate a contrast I have identified two frameworks. Clearly, both biophysical and political-economy perspectives identify important variables in analysing the risks of global change and can be brought together in an analysis. Separately, however, each tends to overlook the role of factors such as technology and culture in creating or mitigating vulnerability.

It is important to emphasize that the most vulnerable people may not be in the most vulnerable places – people can live in productive biophysical environments and be vulnerable because they are poor or landless, and people can live in fragile physical environments and live relatively well if they have money or technology as buffers. Often, the conditions are most critical where impoverished populations live in ecologically marginal environments (see Kasperson, Kasperson, and Turner 1995, 1996, 1999). Vulnerability has an important spatial component that may differ greatly from the spatial pattern of environmental change. Vulnerability is also sharply differentiated among different social groups living in the same place. If we can find ways to measure vulnerability, it should be possible to map such vulnerability both spatially and socially in different parts of the world.

Measuring vulnerability to global environmental change

There are many possible ways of describing, classifying, and measuring vulnerability to global environmental change using a range of physical and social indicators, as detailed in chapter 7. Table 5.1 is a preliminary attempt to develop a set of measures to assess the vulnerability of agricultural producers to climate change.

Biophysical vulnerability is defined in this case in terms of existing climate and soil conditions. For example, rain-fed farming regions of low and variable rainfall, high evaporation, and non-moisture-retaining soils (such as those of the semi-arid, shallow, sandy, eroded environments of the dry tropics) might well be especially vulnerable should global warming bring warmer and drier conditions. On the other hand, adequate and reliable rainfall, cooler temperatures, and moisture-holding soils (such as the deep, organic soils of the humid temperate zones) could buffer farmers against a shift to hotter temperatures and lower rainfall. Ezcurra and colleagues elaborate considerably on the above in the next chapter.

Technology can either increase or decrease vulnerability to climate change. A debate, for example, exists over the success of the new agricultural technologies associated with the "Green Revolution" in improving environmental and social conditions (Barkin and Suarez 1983; Hewitt de Alcántara 1976). On the one hand, irrigation, high-yielding seeds, fertilizer, pesticides, and machines were claimed to increase yields and decrease their variability (CIMMYT 1974; Wellhausen 1976). The expansion of irrigation, however, has also been associated with a loss of land

Table 5.1 **Some factors determining the vulnerability of agricultural producers to climate change**

Biophysical environmental conditions
- Climate that is already marginal or subject to extremes (e.g. low and variable rainfall, high evaporation, severe storms)
- Soils that do not hold water (shallow, sandy), are easily eroded (steep slopes, devegetated), or are of low fertility

Technological conditions
- Lack or loss of traditional knowledge and technologies for conserving soil and water
- Inadequate, insecure, or costly water resources such as wells and reservoirs for irrigation
- Limited range of seed adapted to a narrow range of biophysical or technical conditions
- Inadequate, expensive, or inappropriate fertilizer and pest-control techniques
- Lack of appropriate labour, machines, or energy at different points in the agricultural cycle
- Lack of roads and communication for marketing, access to inputs or information

Political and socio-economic conditions
- Poverty
- Small, poor-quality plots of land
- Insecure access to land (as owner or tenant) or oppressive sharecropping relationships
- Indebtedness
- Sexist, racist, or other discrimination in access to food, land, employment, or credit
- Market structures (such as artificially low food prices or high input costs) that make it difficult to recover costs or make a profit
* Lack of effective and fair public or private systems of disaster relief, insurance, health care, or agricultural extension
- Shortage of alternative employment or difficulty in moving to better locations

and production owing to salinization and waterlogging in regions such as India and the Middle East and with increased vulnerability to multi-year droughts in places in which agriculture has become dependent on shallow wells, small reservoirs, or declining water-tables. The new seeds have been shown to do well only in moist soils or irrigated environments, and to be vulnerable to diseases and weather extremes (Pearse 1980). A small range of improved varieties has replaced a diversity of traditional seeds adapted to a wide range of environments. In fact, some authors (e.g. Richards 1985; Wilken 1987) claim that the Green Revolution, together with modern development and education, has displaced traditional knowledge and technologies that were well adapted to marginal environments and varying climates in regions such as West Africa and Central America.

The critics of the Green Revolution also claim that it exacerbated social and economic inequality and thus intensified the political and economic vulnerability of some groups. The introduction of irrigation increased land prices and competition for land, pushing the poor or powerless off the land and into more biophysically marginal areas. Mechanization increased unemployment, and the purchase or import of expensive inputs (seeds, fertilizer, pesticides, tractors) fostered dependency and debt. In defence, proponents show that the Green Revolution increased food self-sufficiency, reduced famine, decreased food imports, and raised average incomes in parts of South America and Asia.

The starkest examples of how political and economic conditions create vulnerability to climate change appear in the literature on drought and famine in sub-Saharan Africa. Authors such as Franke and Chasin (1980) and Watts (1983) have shown how many African peasants and pastoralists have become vulnerable to drought. The restructuring of political boundaries and land tenure have reduced mobility and eliminated traditional rights to land and pasture. Integration into market economies and agricultural modernization have made it difficult to buy food, land, or inputs without going into debt. State policies of subsidizing urban food prices or export and cash crops have made the growing of basic food crops unprofitable. Foreign aid has sometimes brought inappropriate technologies, created dependency, or been a disincentive to local production.

In addition to, or underlying, the factors listed in table 5.1 are a number of individual characteristics that explain and differentiate vulnerability to climate change. For example, an individual's health status, age, and knowledge can increase vulnerability to drought, as shown in studies of famine (Currey and Hugo 1984; Downing 1991a, b; Downing and Bakker 2000; Downing, Watts, and Bohle 1995: 190–192).

Broader social structures and trends underlie many of the factors. Colonialism and economic dependency have created legacies of land tenure, export-oriented infrastructure and economy, and a loss of traditional environmental knowledge in regions such as West Africa. Population growth is an important driver of competition for (and fragmentation of) land, underemployment, and increased food demands. International and domestic conflict – in the Horn of Africa and Central America, for example – create poverty and push people into dense settlements and marginal areas. Long-term environmental changes, such as desertification, have reduced the availability and productivity of land. The analysis of these underlying causal or driving forces is vital for understanding and mitigating vulnerability to global environmental change.

One might develop parallel lists of factors for other types of global environmental changes at different scales. Individual vulnerability to the accumulation of toxic substances in the environment, for example, depends on location in relation to the sources and transport of toxics, as well as the degree of dependence on contaminated water and food. Bullard (1990) has argued that it is the poor, often minority Black, population of the United States that lives in the closest proximity to hazardous waste sites. The poorest countries are often encouraged to accept exported toxic waste in return for economic compensation (Moyers 1990). Children and pregnant women are especially vulnerable to heavy metals, such as lead and mercury, as are those people who are exposed to the synergistic and immune-suppressing effects of multiple contaminants in the home and workplace.

Some examples of vulnerability assessment

Although very few studies explicitly address the use of vulnerability analysis in assessing the risks of global environmental change, several case studies have examined vulnerability to existing environmental problems such as drought and resources depletion in different regions of the world. These studies illustrate the range of frameworks and indicators that might be used in vulnerability analysis.

One of the more explicit looks at the concept of vulnerability is a Clark University study that compares the Tigris–Euphrates Valley, the Sahel, and the US Great Plains to demonstrate shifting patterns of vulnerability to drought associated with changing technology, social conditions, and government policies (Bowden et al. 1981). In the case of the Tigris–Euphrates, increased vulnerability was associated with a lack of technological innovation and an inflexible sociopolitical organization. In the Sahel, population growth and export cropping caused diverse problems, although food aid reduced local vulnerability. The social consequences of drought in the Great Plains have changed in this century. Local vulnerability has declined in later years owing to insurance, economic adjustments, and soil and water conservation. Although technology and social organization have lessened local vulnerability, the regional and global impacts of drought in the Great Plains may have increased (Adams et al. 1990; Brooks and Emel 1995, 2000).

A study on environmental change in Latin America used physical criteria to delimit the "fragile" lands that are estimated to cover one-third of the continent of Latin America (Browder 1989). Fragility, which appears to be similar to biophysical vulnerability, is measured according to topography, climate, soils, and natural vegetation (Denevan 1989). For example, the 39 per cent of Latin America that is wet tropical forest is considered fragile and vulnerable to degradation owing to the poor quality of the soils – which lose their fertility rapidly if exposed to the elements – as well as the prevalence of disease and pests. The dry tropical lowlands (25 per cent of the area) are vulnerable to desertification because of their aridity, erodibility, and tendency to salinization. The highlands, such as the Andean region (9 per cent of the area), are fragile environments owing to their steep slopes and frost risks.

Denevan (1989) also shows, however, how population and technology can alter the fragility of these landscapes. Traditional technologies such as terracing, irrigation, and raised fields can reduce the vulnerability to erosion, drought, and frosts (Nahmad, Gonzalez, and Rees 1988). Population increases may well promote technological innovation, but they may also overtax the carrying capacities of fragile systems and result in degradation and collapse.

Gow (1989), in the same volume (Browder 1989), focuses on the social determinants of fragility, pointing out that people are often forced to overuse these environments because of their poverty, or because of inappropriate policies, technologies, and institutions. He suggests that inequitable land distribution, problems in credit, land titling, agricultural modernization, and tax incentives are some of the factors responsible for the degradation of the tropics.

Latin America also offers an important historical example of the vulnerability of ecosystems and peoples to environmental change. The arrival of Columbus in the Americas heralded an invasion by European organisms and diseases to which indigenous biota and people had little resistance. Crosby (1972) has shown how the Europeans inadvertently introduced diseases, such as the smallpox that decimated local populations, and exotic animals, such as the cats and rats that killed local birds and animals. The ecological vulnerability of long-isolated continents and islands, such as Australia and New Zealand, is also an important example.

Several of the case studies in Parry, Carter, and Konijn (1988) discuss vulnerability in specific regions of the world. The focus in most of the case studies is biophysical vulnerability through agroclimatic analyses of crop potential and crop yield with different climate scenarios. In section 6 of the Kenya case study, however, Akong'a and Downing (1988) discuss smallholder vulnerability and response to drought in terms of household food economy, ecological degradation, land tenure, agricultural technology, and institutional support. Other work (Downing 1991b, 1992; Downing, Gitu, and Kamau 1989; Downing, Watts, and Bohle 1995), has identified more specific measures of vulnerability: these are individual nutritional status; health and social position; household income; cultural preferences and demography; and regional food availability, physical geography, and institutional structure. In the Ecuador case study (Bravo et al. 1988), the vulnerability of indigenous farming systems is related to changing farming practices and labour strategies. The steep slopes of the

Andes have been terraced to improve microclimate, soil fertility, and water availability. In the India case study, the research team (Gadgil et al. 1988) discuss relationships among drought vulnerability, climate variability, agricultural technology, and government policies. Jodha (1989, 1991) has also discussed the general vulnerability of developing countries to drought in terms of their reliance on rain-fed subsistence agriculture and inadequate institutional support and infrastructure.

Several studies have analysed the differential vulnerability of women in developing countries to environmental change (Cutter 1995; Dankelman and Davidson 1988; Rodda 1991). Because in many cases women are responsible for agriculture, fuelwood, and water, they are disproportionately affected by drought, desertification, deforestation, and water pollution (see also Ali 1984; Rivers 1982). The argument is also well stated in *The State of India's Environment*:

Probably no other group is more affected by environmental destruction than poor village women. Every dawn brings with it a long march in search of fuel, fodder and water. It does not matter if the women are old, young or pregnant: crucial household needs have to be met day after weary day. As ecological conditions worsen, the long march becomes ever longer and more tiresome. Caught between poverty and environmental destruction, poor rural women in India could well be reaching the limits of physical endurance. (CSE 1985: 172).

The authors of the report are quick to point out, however, that the specific burdens of women are not attributable to ecological degradation:

They are located in the sexual division of labor, marked by a double work burden (at home and outside) and by the specific nature of the tasks that they do, and in the unequal distribution of resources like food within the household, which stems from women's inferior status in the household and from a lack of control over cash and productive resources like land. But given this situation, environmental destruction exacerbates women's already acute problems in a way very different from those of men. (CSE 1985: 177)

Another example of concrete attempts to assess vulnerability is the use of early-warning systems in famine response (Field 2000: 276–277; GIEWS 2000; Walker 1989). Such systems are generally designed to gather information on a set of indicators that can be used to anticipate the onset of famine, to target the people at risk, and to estimate food aid or other relief requirements. An early example of such a system shows up in the Indian Famine Codes developed to assist colonial administrators anticipate the onset of food shortages (Bhatia 1963; Brennan 1984).

Some contemporary early-warning systems, such as the FAO's Global Information and Early Warning System (GIEWS), focus only on the physical event in monitoring rainfall deficits and vegetation conditions. All of the current early-warning systems pay lip-service to the social explanations of famine vulnerability, backed by documents citing the importance of identifying vulnerable groups and downplaying the role of drought. Yet biophysical and crop-production indicators predominate in many warning systems because data are easier to collect, are less politically sensitive, and take advantage of existing weather networks and satellite capabilities.

In the United States, the Agency for International Development (USAID) has established the "Famine Early Warning System" (FEWS) to collect and synthesize existing physical and social information in order to identify regions and populations at risk from famine. Operational famine early-warning systems always entail compromises between using the best possible measures of vulnerability and using those that are easily, inexpensively, and rapidly available.

The indicators currently used in FEWS to assess vulnerability to famine include rainfall, crop conditions, remotely sensed vegetation indices, the nutritional status of populations, and the prices of food in markets. Increasingly, the literature on famine vulnerability emphasizes the value of social indicators. Both market prices and changes in population behaviour have been proposed as warnings of famine conditions. The FEWS system is designed to incorporate an understanding of vulnerability to famine. In the development of FEWS, it was accepted that a model that viewed famine narrowly as due to natural events was inadequate because it did not take account of the prevalent thinking on the social causes of famine.

Thus, FEWS shifted to a strong emphasis on the social model of famine, using the concept of vulnerability and attempting to measure a wide range of social indicators. Although the social model of famine, now used in FEWS, is linked to the structural and political economic analyses of famine vulnerability, the social indicators do not reflect the richness and complexity of the research literature. The emphasis in FEWS is on measures of individual health and nutrition, and, more recently, on market prices. The information is difficult to obtain, localized, and often unreliable in most Sahelian countries.

The primary use of the agroclimatological information in a famine-warning system is to provide early indications of the magnitude and location of crop-production declines and of stressed pastures, either of which might trigger livestock losses and human migrations. The aim is to make timely estimates of production and to use them to focus in on critical regions and vulnerable populations. In most countries, government agricultural statistics services produce a production estimate just before harvest and a final crop assessment several months later. In certain countries, however, the political sensitivity of production estimates results in considerable delays in the publication of reports and encourages debates about the credibility of the information. In many countries the focus of data collection is on cereals, even in areas in which roots or vegetables are important.

The 1990s have seen a major advance in agroclimatological monitoring in Africa in the form of satellite remote sensing. The great advantage of satellite imagery is that it provides comprehensive spatial coverage, filling the gap between meteorological stations and providing general indications of environmental and vegetation conditions in advance of ground observation.

Other, more social indicators of vulnerability considered in FEWS include poverty; landlessness; lack of entitlements to trees, water, land, and inputs; health, and nutrition. The onset of famine is monitored through monitoring the sales of jewellery and livestock, migration, the search for alternative work, the eating of famine foods, and the spread of diseases.

Conclusions

These examples illustrate some of the ways in which we might design vulnerability assessments for global environmental change. Although the examples primarily refer to drought and hunger, many of the key measures of vulnerability hold true for other global-change issues as well. Vulnerability to the accumulation of toxins, for example, may depend on geographical location and health or occupational status. Vulnerability to soil erosion will depend on farm practices, technology, and land tenure. Vulnerability assessments of natural ecosystems and human health to other global environmental change issues (e.g. ozone, loss of biodiversity), however, may depend more on ecological and medical factors (as discussed by Ezcurra and colleagues in the next chapter) than on those discussed in this chapter.

Such vulnerability assessments must be a key component of research on global environmental change. Focusing on vulnerability reminds us of the social context of environmental change, and it illustrates why we should disaggregate our analyses of global change by region and social group. In short, some people and countries will suffer and some will benefit (Glantz, Price, and Krenz 1990). Vulnerability analysis also expands the range of choices in responding to global change, as pointed out in chapter 1. It demonstrates, for example, that we can reduce impacts of global change not only by slowing the rate of climate change or ozone depletion but also by reducing the vulnerability of populations to these changes. Thus, it may permit more efficient and economic responses by highlighting the most vulnerable regions and peoples or those vulnerabilities that can be most easily reduced.

Many questions about vulnerability analysis are unresolved. Few empirical studies of vulnerability to environmental change exist outside the literature on famine and hazards. Little consensus exists on the appropriate definitions and indicators (although see Lonergan 1998). Yet we need to know how to deal with multiple vulnerabilities to multiple threats. We need to decide whether studying the existing pattern and degree of vulnerability to drought and other conditions is an accurate guide to future vulnerability or whether patterns of vulnerability are changing with rapid social, technological, and demographic transformations. And, above all, we need to communicate about vulnerability to other scientists, to policy makers, and to the public. Too many people still believe that the impacts of disasters and climate change are determined solely by the physical characteristics of events – that drought impacts are most severe where there is least rain – or want to believe that famine and soil erosion are uncontrollable acts of nature rather than socially created crises (Glickman, Golding, and Silverman 1992; Wijkman and Timberlake 1984). We can know very little about the social impacts of global change unless we work to understand, document, and communicate the nature of vulnerability.

6

Vulnerability to global environmental change in natural ecosystems and rural areas: A question of latitude?

Exequiel Ezcurra, Alfonso Valiente-Banuet, Oscar Flores-Villela, and Ella Vázquez-Domínguez

In July 1990, an intergovernmental group of experts, convened by the World Meteorological Organization and the United Nations Environment Programme, produced a document summarizing what was known at that time on the biophysical perspectives of global climatic change (GIESCC 1990; IPCC 1990a). One of their main conclusions was that it can be inferred with a high level of certainty that during the next century the average temperature of the earth will rise steadily at a rate of 0.3°C per decade if the emission of greenhouse gases continues at its present trend. According to this expert panel, the Intergovernmental Panel on Climate Change (IPCC), a sea-level rise of approximately 5 cm per decade will affect natural ecosystems in different ways and will significantly disrupt the water balance in many regions of the globe. The same panel recently revised its estimates to predict a temperature rise of 1.4–5.8°C and a sea-level rise of about 0.09–0.88 m by the year 2100 (IPCC 2001).

Large-scale climatic variation is an important element, but by no means the only one, of a series of large-scale disruptions in the global environment that are threatening the sustainability of the biosphere as a whole. Apart from the systemic effect of the emission of gases at a biospheric level, other phenomena are contributing significantly to so-called "global change." Among these, pollution of air and water resources, soil depletion and erosion, deforestation, overgrazing, and depletion of biological diversity are possibly the most important.

Global change is more, much more, than climate change or global warming. Most would agree that the global changes that will strike the hardest blows on people and other living things in the next 20 or 50 or 100 years will not be those of climate change. If the world is in environmental extremis today, it is more through the rapidly changing chemistry of the air and soils and water, and through the inexorable and wide-reaching pressures of urbanization and intensive agriculture and land use. (Eddy 1991: 3)

Arguing along a similar line, Turner and colleagues (1990b) defined two types of global environmental change: (a) "systemic" change, and (b) "cumulative" change (see also chapters 1 and 9). The first category includes driving forces that operate in a general, systemic manner throughout the globe and can directly affect regions that are removed from the original sources. Such is the case, for example, of chloro-fluorocarbons (CFCs), carbon dioxide (CO_2), and methane (CH_4) emissions. The second category includes driving forces that operate at a local level but can affect the environment of the whole planet by sheer quantitative accumulation at a very large scale. This is the case, for example, of deforestation, overgrazing, and the loss of biological diversity. Many environmental factors, of course, can operate in both a systemic and a cumulative manner, and the distinction between types of global change may blur easily. Nevertheless, the distinction is conceptually important, particularly when trying to understand the question of global environmental change in the less-developed nations (see also the discussions in chapters 9 and 10).

The problem of global environmental transformation has reached such a level of concern in the media that the Ecological Society of America (ESA) dedicated a significant part of its journal *Ecology* to a special report, *The Sustainable Biosphere Initiative* (Lubchenko et al. 1991), departing from a long tradition on publishing only original papers on basic ecological research. With this report, the ESA sought to alert ecologists to the need to focus ecological research on three basic priorities: (a) global environmental change, (b) biological diversity, and (c) sustainable ecological systems.

In developing countries, whose proportional contribution to greenhouse gases is still relatively small, the bioecological aspects of global environmental change are the main cause of concern (IPCC 1996b, 1997). The loss of vegetation cover and of biological diversity through deforestation, overgrazing, and non-sustainable harvesting in general, together with the uncontrolled growth of large cities, are possibly the main problems associated with global change. These problems are already strongly affecting the levels of equity and well-being enjoyed by inhabitants of the non-industrial nations. The transformation of tropical forests into cattle ranges is a good example of this type of process. In Central America,

40,000 km^2 of tropical forests were transformed into grassland between 1961 and 1978. During that period, exports of beef to the developed world increased by 65 per cent but the per capita meat consumption remained constant at around 10 kg/year; the per capita consumption in the United States was 56 kg/year for the same period (Myers 1981). In Brazil, the national cattle herd increased by 11 per cent between 1975 and 1981. During the same period the export of raw beef increased by 1780 per cent and the export of processed meat increased by 360 per cent. The reason why Brazil could manage such an increase was that, during the same period, the buying capacity in the country decreased from 47 kg of beef per minimum wage to 21 kg per minimum wage, and that the average per capita consumption decreased from 21 to 16 kg/y (*Informe Ganadero* 1983).

Among the bioecological consequences of global environmental change, one of the largest concerns is the loss of biological diversity in tropical countries. Towards the end of the twentieth century, the concept of biodiversity has become the symbol of the world's natural heritage that is being lost at an increasingly higher rate. Awareness of the irreversibility of species extinction is increasing, along with the awareness of the rapid and irreversible environmental transition that will transform the world that gave rise to our present cultures. The concern for the future consequences of these changes is possibly at the root of the sudden and growing interest that ensured that biodiversity was recognized in the Framework Convention on Biological Biodiversity adopted at the Earth Summit in 1992, in subsequent implementation efforts undertaken as part of Agenda 21, and in the United Nations Environment Programme's Global Biodiversity Assessment (UNEP 1995) intended to assist these implementation efforts. Toward the end of the 1990s, however, much of the content of the global biodiversity mandate remained undefined (Steinberg 1998). Humans, at all times of history, have betrayed a simultaneous need for change and apprehension of such a change. This contradiction is manifest in Western industrial civilization, which promoted a world view based on exploitation of natural resources and now shows a growing concern for the loss of biological diversity. This concern, however, is frequently not shared in the same terms by the less-developed nations, which face more immediate and urgent needs to satisfy the growing demands of their impoverished populations. Third-world decision makers frequently argue that the development of their nations has to take place at the expense of a certain amount of environmental damage and a certain level of social inequity, two variables that often go together. It is common wisdom among third-world politicians that a cost has to be paid, both socially and ecologically, if economic development is to become a reality.

It is difficult to imagine a scenario of human development that does not damage the environment in some way (see chapter 8) and that does not affect biological diversity in particular. In human history, development has usually gone hand in hand with environmental deterioration. It is becoming clear, however, that if postindustrial societies want to be able to control their destiny, they must be able to regulate their growth and consumption patterns. They should be capable of obtaining natural products without deteriorating the most important legacy of millions of years of biological evolution – diversity.

In this chapter we address the questions of which natural environments are likely to be more affected by global change, and which human groups depend primarily on these environments. Most of our data come from Mexico: this is because (a) most of our experience as researchers has occurred in Mexico; (b) we believe that Mexico, a country located in the transition between temperate and tropical regions, provides the ideal framework for a case study on the processes of global change; and (c) Mexico has been classified as one of the six most biologically rich countries in the world (i.e. it is classified as a megadiversity country; see Mittermeir 1988).

Biodiversity

Biological diversity is the result of the evolutionary process, manifested as the existence of a myriad of different life-forms. It is the result of differences at a genetic level; of differences in morphology, physiology, and ethology; and of differences in life-form and development, in demography and in life histories. Diversity can be analysed at all levels of the organization of life.

In a strict sense, diversity (a concept derived from systems theory) is simply a measure of the heterogeneity of a system. In biological systems, the concept refers to the amount and proportion of different biological elements. The measurement of biodiversity depends, among other things, on the scale at which the problem is defined. In a biogeographic context, biodiversity is measured by quantifying the heterogeneity of biogeographic zones (e.g. biotic provinces) that occur within a given region. Biogeographic diversity is simply the diversity of biomes and ecosystems that occur within a region. For many ecologists, this level of diversity is defined as γ-diversity (table 6.1).

At the ecosystem level, biodiversity has two well-defined expressions: these are point diversity, or α-diversity, and spatial heterogeneity, or β-diversity. The first, α-diversity, is a function of the quantity of species present in a given habitat, and it is the most commonly known and cited

Table 6.1 **Classification of the different hierarchical levels of biological diversity**

Biodiversity level	Scale of biological organization	Spatial segregation	Type of biodiversity
Geographic	Biome		γ
Multi-species	Community	Between habitats	β
		Within habitats	α
Genetic and demographic	Population		Variation and heterosis

component of biodiversity in tropical rain forests and coral reefs, to mention only two ecosystems with a very high species richness. The second, β-diversity, is a measure of the degree of environmental patchiness, or of the existence of a spatial mosaic of different habitats. This component of diversity is particularly important in traditional farming systems in the tropics, in multiple cropping systems and in multiple-use agroforestry systems, where the lower species-richness of human-managed systems is compensated by a human-made mosaic of crops and trees that generates spatial heterogeneity.

Finally, there is a genetic, or intraspecific, component to biological heterogeneity. Within a single species a large amount of variation can exist, maintained by the heterogeneity of different genetic alleles within the species (genotypic variations) and by the differences in the morphological characters that are coded by these alleles (phenotypic variations). Genetic diversity, also known as genetic variation, is a fundamental component of biodiversity: it is the building block of biological heterogeneity at all other levels. The crucial importance for humankind of genetic diversity is well known in the case of cultivated plants and domesticated animals, for which different agencies for decades have made a large effort to preserve the diversity of the original germplasms for future breeding programmes. Without genetic variation, the improvement of economically important species is not possible. It is, perhaps, less known that genetic variation is of crucial importance for the maintenance of natural populations, whose adaptation and survival are frequently conditioned to the maintenance of a minimum population density that can provide an adequate level of outbreeding and heterosis. Below this threshold, many populations become threatened by extinction, simply because they cannot adapt through natural selection to changes in their environment.

Biodiversity depends not only on species richness but also on the relative abundance of each species. Species are usually distributed in a hierarchy of abundances, from very abundant species to very rare ones. The greater the degree of dominance of some species over the rest, the lower the diversity of the community. This is well known, for example, in temperate pine forests where in some cases 90 per cent of the biomass of the community is contributed by only one or two species, and the remaining biomass is formed by a much larger group of low-abundance, rare species. To understand the problem of biodiversity, we must understand the problem of biological rarity. We define as "rare species" all those species that exist in such low numbers or restricted conditions that they may represent a conservation problem, and in many cases are threatened by extinction. Rabinowitz, Cairns, and Dillon (1986) found that, like biodiversity, the causes of biological rarity can be defined at three different levels – the biogeographic, the ecological, and the demographic.

Biogeographic rarity (or endemism) is found when species occur only in restricted ranges within a biogeographic region, although they may be locally abundant. These species, defined as *endemics*, may be endangered if the region in which they grow is itself threatened by environmental change on a large scale. Examples of this group are the freshwater organisms of the Basin of Mexico, most of which have disappeared from the area with the development of Mexico City and the drainage of the basin (Ezcurra 1990).

Habitat rarity (or stenoicity) occurs when a species is very specific with respect to the habitat in which it lives, although it may not be specific with respect to the biogeographic region in which it occurs. Narrow-habitat, or stenoic, species are associated with very specific environmental conditions, in contrast with the wide-habitat, or eurioic, species that can prosper in a wide range of environments. Examples of this type of rarity are the riparian (freshwater) organisms in desert oases (Ezcurra et al. 1988), gypsophilic plants (plants that grow on gypsum soils), coral reefs that prosper only at a certain water depth and with high water transparency, or mangroves that grow within a narrow tidal range.

Demographic rarity occurs when a species shows very low densities throughout its range, although it may be widely distributed and may show a wide habitat preference. Rarity, in this case, is a result of low population numbers, which necessarily imply high levels of inbreeding and low genetic variation.

Of course, some species may concentrate more than two levels of rarity and, in some cases, rare species may join the three characteristics – that is, being endemic to a small biogeographic area, being highly stenoic in their habitat preferences, and occurring at low population densities. One of the most remarkable examples of this type is the recently described *Lacandonia schismatica* (Lacandoniaceae, Triuridales), that has the dubious privilege of growing only in a one-hectare patch of the Lacandon forest (high endemism), associated with tropical peaty soils (an extremely rare habitat), in very low numbers and with practically no genetic variation (demographic rarity). It is, of course, one of the rarest plant species ever described (Martínez and Ramos 1989).

It is probably obvious, at this point, that the levels of rarity that most matter in relation to global change are the first two (endemism and stenoicity). Any process likely to change the environment at a global level is likely to produce significant changes in whole habitats and, in some cases, in entire biogeographic subdivisions. Clearly, any species closely linked to a given habitat or to a given biogeographic subdivision will be more vulnerable to large-scale environmental change. Some important questions may be posed in relation to the problems of biodiversity and rarity: is there a pattern of endemism and stenoicity at a planetary level that will allow us to predict where the most vulnerable ecosystems exist; is it possible to identify trends in endemism and vulnerability; how are these trends related to the productivity of ecosystems used by human populations? In the next sections we explore some of these questions.

Diversity patterns: The temperate–tropical gradient

One of the most often described (but also most poorly understood) patterns of biological diversity is the trend exhibited by most taxa to increase their species richness towards the tropics (Gentry 1982; Pianka 1966). Stevens (1989) has shown that, when the latitudinal extent of the geographical range of organisms is plotted against the mean latitude of their distribution, a positive correlation is found in which the geographical ranges of the species decrease towards the tropics (fig. 6.1). This pattern, called by Stevens (1989) "Rapoport's rule" (after Rapoport 1975), indicates in general terms that tropical species have extremely narrow ranges of distribution, whereas high-latitude organisms have comparably very large ranges. In terms of Rabinowitz's classification of rarity, Stevens' argument means that tropical species have a marked tendency towards high endemism and/or high stenoicity. That is, there is good evidence showing that most of the ecologically and biogeographically rare species are concentrated in the intertropical belt (fig. 6.2).

According to Stevens, the large latitudinal extent (i.e. the low level of endemism) of high-latitude organisms is a consequence of the selective advantage of individuals with wide climatic tolerances in regions with large differences between summer and winter temperatures. In contrast, for tropical organisms, selection for wide-tolerance characters is not expected or even may reduce the efficiency of exploitation of particular microhabitats (Stevens 1989). A similar argument was formulated by

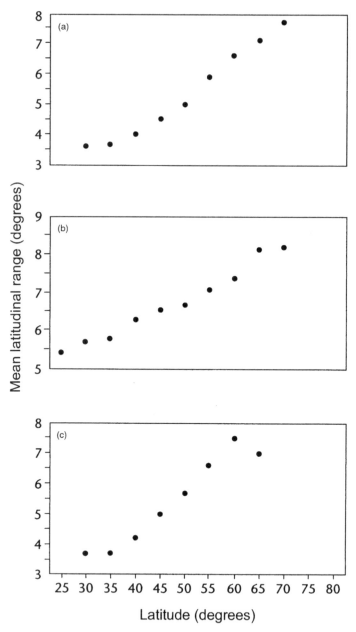

Fig. 6.1 **Mean latitudinal extents of (a) North American trees native to various latitudes, (b) North American marine molluscs with hard body-parts, and (c) North American herpetofauna (amphibians and reptiles). Source: modified from Stevens (1989)**

Fig. 6.2 **Micro-endemism in North American mammals, modified from Rapoport (1975). Note the concentration of endemic species (circles) and isolated subspecies (crosses) towards the tropics**

Janzen (1967), who argued that, in the tropics, terrestrial temperature regimens are more uniform than those in temperate regions and that this uniformity should affect the distribution of tropical plants and animals that are adapted to the temperatures normally encountered in their habitat. Janzen concluded that mountain ranges are more important elements of geographic isolation in the tropics, where organisms are not so flexible to temperature changes encountered during altitudinal migrations, and that this phenomenon partially explains why the ranges of species are narrower and endemism is higher in the tropics.

In conclusion, the levels of endemism and stenoicity are likely to be higher in tropical organisms than in temperate ones. By "tropical" we are alluding not only to the wet tropics – the most common, almost stereotypical, concept of tropical ecosystems; tropical areas also include other very important ecosystems, such as savannas, tropical arid scrubs, cloud forests, marshes, coastal mangrove forests, and high-altitude ecosystems (such as the Andean Páramo and the Peruvian Puna), to name just a few. Many of these systems show restricted distributional ranges and are closely associated with very narrow environmental conditions (such as altitudinal belts, flooding levels, or specific rainfall patterns). These ecological communities are particularly non-resilient to environmental change in general, and one might argue that they have been hitherto the areas most affected by the cumulative "proximate" forces of global environmental change (deforestation, overgrazing, desertification) and that they will very possibly be the most vulnerable ecosystems in the future.

A case study: Mexican columnar cacti

The pattern discussed by Janzen, Rapoport, and Stevens can be analysed in detail in certain taxa of particular interest. For this purpose, we have chosen Mexican columnar cacti. This group of plants (belonging to the subfamily Pachycereae within the Cactaceae; Gibson and Horak 1978), with 69 species in Mexico, is formed by succulent leafless plants with cylindrical stems and, in general, possesses a very simple architecture (Cody 1984). Like almost all cacti, they exhibit crassulacean acid metabolism (CAM), characterized by CO_2 fixation at night, which minimizes transpirational water loss (Nobel 1988). CAM has collateral problems, such as the high metabolic cost involved in maintaining a large amount of non-photosynthetic tissue. Cacti cannot thermoregulate by transpiration during the day and therefore need passive ways to dissipate heat and to avoid excessive radiation. Variations in body architecture, such as the vertical stems of columnar cacti, have been interpreted as adaptations to avoid excessive heat (Gates 1980). Columnar cacti are plants of great importance as a source of food for indigenous populations throughout

Mexico and the south-western United States (Felger and Nabhan 1976; Nabhan 1985). Many species of columnar cacti are cultivated for their fruit production and as hedges around houses and fields. A significant proportion of the fruit marketed in indigenous markets in southern Mexico comes from columnar cacti, from both cultivated and wild species.

The geographical patterns of species richness of these plants in Mexico, when the number of species is plotted in a map with a grid of 1° latitude and 1° longitude, indicate that most of the species have a narrow distribution range and are concentrated in the semi-arid, hot regions of central-southern Mexico (fig. 6.3a). In contrast, north-western Mexico – referred to as a zone of high diversity of succulents by Burgess and Shmida (1988) – has four times fewer species than the dry tropics. A regression model fitted to the species richness with the mean, minimum, and maximum temperatures and with the annual mean precipitation as independent variables shows that the species richness is limited primarily by low temperatures and either very high or very low rainfall. Therefore, areas with freezing events and low precipitation (e.g. the Mexican Plateau) contain few or no species. Comparing the number of species predicted by the model with the observed number of cacti and plotting both on the same grid permits an analysis of the species richness not explained by the model. Accordingly, it is possible to see on the map (fig. 6.3b) those areas with a higher number of species than predicted by present climatic conditions (i.e. the mountains of northern Oaxaca and the Balsas Basin). These areas are "hotspots," or biogeographically restricted areas of unusually high diversity and endemism. Thus rarity, in the case of columnar cacti, tends to concentrate in the dry tropics. Whereas the abundance of columnar cacti in other regions (e.g. the Baja California Peninsula) can be explained by the prevalent climatic conditions, the extremely high species richness in restricted tropical areas can be explained only in terms of historical events, or in the existence of a large topographic fragmentation in the area, or both (Janzen 1967). The theory of Pleistocene refugia is one of the most plausible and best explanations for this pattern. The present distribution of columnar cacti reflects the concentration of many species in refuge areas during the Wisconsin Glaciation, when the present deserts of northern Mexico and south-western United States were cooler, moister, and occupied mostly by grasslands.

Tropical endemism in other groups

Is the pattern observed in Mexican cacti maintained in other groups? The available information indicates that similar patterns exist and may be common in many other taxa. Several authors have pointed out the im-

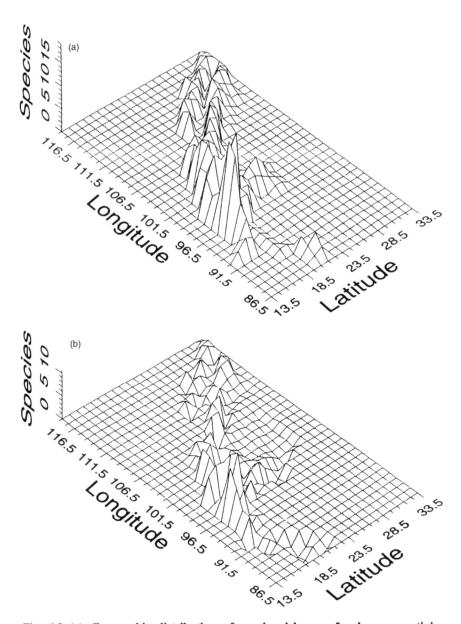

Fig. 6.3 **(a) Geographic distribution of species-richness of columnar cacti in Mexico (modified from Mourelle 1997). Note the higher diversity of the southern dry tropics. (b) Geographic distribution of the residuals of species diversity once the best-fit climatic model has been adjusted to the data. Positive residuals mean that present diversity is not explained by climatic variables alone. The high diversity in the south is much greater than would be expected in view of the present climate**

portance of biological refugia during Pleistocene glaciations. The refuge theory very robustly explains patterns of tropical biodiversity in restricted areas (Haffer 1982; Sears and Clisby 1955; Toledo 1982). Indeed, some of these areas still currently hold a significant number of endemic species (Rzedowski 1978). Areas such as the Balsas Basin and the southern part of the Tehuantepec Isthmus in Oaxaca have been proposed as refugia of xeric species of birds by Hubbard (1974) and of small mammals by Mares (1979).

One of the authors of this chapter recently conducted a detailed study of the biogeographic distribution of endemic reptiles and amphibians in Mexico (fig. 6.4 and Flores-Villela 1991). The results of this study showed that 25 per cent of all the endemic herpetofauna in Mexico has distribution areas of fewer than 2,500 km^2 and that most of these micro-endemics are distributed mainly in two biomes. One biome is formed by the tropical highlands of Central Mexico, mainly associated with cloud forests and pine-oak forests, and covers about 0.5 per cent of the country (Leopold 1950). The second biome with high endemism is the tropical deciduous forest of the Pacific slopes (about 8 per cent of the vegetational cover of Mexico; Flores-Villela and Gerez 1988). All areas of extremely high endemism for the herpetofauna are located in the highlands of central-southern Mexico, which are the most populated sections of the country and in which more than one-half of the total human population is concentrated. These areas of high endemism for amphibians and reptiles coincide with high-endemism regions for butterflies and birds (Escalante Pliego and Llorente 1985), for mammals (Ramírez-Pulido and Müdespacher 1987), and for birds (Escalante Pliego, Navarro, and Peterson 1993).

The results of these studies support the idea that non-resilient, narrowly distributed species concentrate in the tropics, although they do not necessarily coincide with the wet tropical forests. A similar situation occurs in South America, where it has been reported that a large proportion of the highly endemic species occurs in different altitudinal belts of the high-altitude ecosystems (see, for example, Vuilleumier 1986). It is also interesting to note that many of the modern temperate crops (e.g. maize, potato, tomato, and squash) have originated in these high-diversity areas, both in Mexico and in South America.

The idea presented in this chapter – that tropical systems may be highly vulnerable to global environmental change – seems to be in disagreement with the predictions of the general circulation models (GCMs) of the atmosphere, which forecast higher levels of climatic warming in the mid-latitudes (GIESCC 1990; IPCC 1990a, 1996a; Neilson et al. 1989). However, on the one hand, these regional predictions are of questionable reliability, as noted by the IPCC (1997) itself; on the other, our argument describing the existence of a general gradient of increasing vulnerability

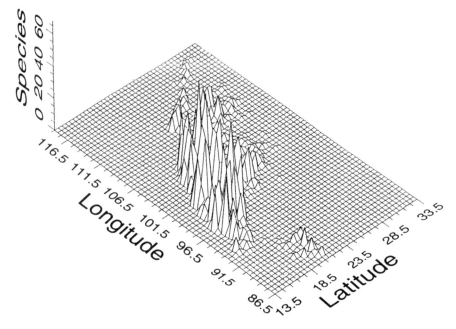

Fig. 6.4 **Geographic distribution of endemic amphibians and reptiles in Mexico (modified from Flores-Villela 1991)**

towards the tropics is based not solely on the predictions of global greenhouse warming but also on the other cumulative driving forces of global change (e.g. deforestation, erosion, and overgrazing).

Vulnerable habitats

Apart from the high-endemism tropical areas, which are (and will continue to be) extremely vulnerable to global environmental change, some natural areas are vulnerable because they occupy environments that are more likely to change or because the majority of the species growing in that habitat can be identified as extremely narrowly distributed. The best examples of this kind of vulnerable habitat are the ecosystems that are associated with very precise levels of a major environmental variable, such as tide or altitude. Obviously, changes in global temperature or sea level are likely to produce major disruptions in these systems. We next discuss three important ecosystems – coral reefs, mangrove forests, and cloud forests – that fall into this category, and analyse their vulnerability to global change. Although Neilson and colleagues (1989) have attempted a detailed identification of vulnerable ecosystems in the temperate part of North America, that work focuses exclusively on evaluating the possible impact of climate change and does not analyse other factors related to, or involved in, large-scale environmental change.

Coral reefs

Coral reefs, despite their confinement to tropical waters (considered as low-productivity regions), are among the environments with higher rates of photosynthetic carbon fixation, nitrogen fixation, and calcium deposition. They are also the aquatic environments with more species per unit area than any other marine ecosystem (Done, Ogden, and Wiebe 1995: 381; Goreau, Goreau, and Goreau 1979). The Mexican coral-reef systems cover approximately 1,000–1,500 km of coastlines, from the Caribbean coasts of the Yucatan Peninsula northwards into the Gulf of Mexico. Some reefs exist along the Pacific coast, although they are not as well developed as their Atlantic counterparts. The age of the present coral formations in Mexico is around 8,000 years (Wells 1988). The term "coral reef" does not refer to a single species but to an integrated community with a large variety of organisms. The hermatypic or "hard corals" (called reef-building species) and certain algae are directly responsible for the deposition of calcium carbonate on the reef. Other species – such as soft-bodied corals and sponges, sea urchins, burrowing worms, and some fishes – break down, rearrange, and cement this material into a solid basement layer on which coral growth continues. Through the subsequent deposition of new colonies, the reef is constructed (Robinson 1979).

The fundamental construction unit of the reef, however, is the coral animal itself (a polyp). The polyps of hermatypic corals contain symbiotic algae (zooxanthellae) that capture solar energy for the organism through photosynthesis. This symbiosis allows a tight chemical cycling between plant and animal in response to the nutrient deficiency, particularly of nitrogen, in tropical waters (Porter 1976). Corals have heterotrophic means of existence too, feeding primarily on plankton and absorbing nutrients from the seawater and sediments.

Numerous factors are important in controlling coral distribution and species diversity, including the environmental conditions required for growth and survival (Done, Ogden, and Wiebe 1995). Clear water and light, which decreases with increasing depth and with turbidity, are determinant for photosynthesis. Thus, coral reefs in the world are limited to the tropics, where more sunlight is available year round, and to shallow depths of less than 40 metres. Both tidal exposure and wave action, when extremely frequent or severe, decrease survival. Suspended sediments affect corals by reducing the quantity and quality of light and by settling directly on the coral, interfering with feeding, photosynthesis, or both, and reducing the amount of substrate suitable for coral growth (Houston 1985).

Dependable supplies of oxygen and planktonic food and the cleansing action of water currents are also important. Temperature requirements are related to the polyp's abilities to lay down calcium carbonate, so optimum water temperature is between 23 and 26°C, limiting the species to tropical latitudes. Temperature also influences the species distribution within the reef and the reproduction rates. Coral polyps present an optimum salinity range of 3.0–3.6 per cent (Gladfelter, Monahan, and Gladfelter 1978). All these factors have consequences for the colonization and succession patterns and are of great importance in the regulation of biodiversity of the reef ecosystem. Coral reefs not only are important for the extremely high biodiversity they sustain and for their tight relation with many other environments but also are economically important because they harbour fin fishes and many other organisms harvested by humans, and may form islands where people live.

Tropical marine organisms live closer than temperate organisms to their upper temperature limits, and also live closer to their lower oxygen limits. Not only are their metabolic rates higher but also dissolved oxygen concentrations are lower in tropical seas: seawater saturated with air contains less oxygen at 30°C than at 8°C (Johannes and Hatcher 1986). Thus, any environmental perturbation that lowers the oxygen concentration or increases water temperature is likely to exert a greater effect on tropical biota, and therefore on coral reefs. As we have seen, the reef community is adapted to a low-stress environment characterized by the absence of significant seasonal changes (clear but slightly turbulent waters, specific temperature ranges, oxygen concentrations that are relatirely high for tropical waters, etc.). The finely tuned adaptation of corals to their environment, however, makes the community remarkably sensitive to environmental changes (Newell 1972: 56). The specific impact of environmental changes on coral reefs is often more difficult to predict than that in temperate marine communities, because of the higher species diversity and trophic complexity and because of the generally smaller sizes of the populations of coral species and of their individual harvesting sites (Johannes and Hatcher 1986).

Coral dieback has been reported in recent times as a growing concern in many parts of the world. The main causes of coral degradation have been reported to be sewage discharge, oil pollution, siltation from clearings of upland vegetation, and dredging and mining (Goenaga 1991). These causes are frequently associated with the development of tropical coastal areas. Several of the most characteristic features of polluted corals are the proliferation of green algae that colonize the corals from their bases and eventually overgrow them, and the growth of a mucilaginous coating formed by filamentous algae that thrive on the excess nutrients brought by sewage or other effluents.

Additionally, a phenomenon associated with coral deterioration, termed "coral bleaching," attracted scientific attention during the 1980s because of its widespread occurrence in different seas of the world. Bleaching, or loss of pigmentation, occurs when photosymbiotic corals lose a major portion of their zooxanthellae, or when the concentration of photosynthetic pigments in the symbiotic algae declines drastically (Glynn 1991). As the algae provide up to 63 per cent of the coral's nutrients and facilitate calcification, bleaching greatly affects the corals, which often completely stop growing (Goreau 1991). All varieties of reef-building corals can be affected, ranging from near-surface to deep formations at a depth of more than 30 metres. Bleaching has been linked to a physiological reaction of the corals and their symbiotic algae to elevated sea temperatures, and the unusually high frequency of recurrent bleaching events in the 1980s has been linked with global warming (Glynn 1991; Goreau 1991). Repetition of this type of events can alter the food chain of the coral-reef ecosystem, making corals less able to compete with rapid-growth algae and ultimately converting reefs into algal habitats with unpredictable environmental and economic losses (Goreau 1991).

Changes in total productivity and in the global distribution of phytoplankton have been associated with glacial to interglacial fluctuations in the level of atmospheric carbon dioxide and in global temperature. The physical processes that govern the availability of light and nutrients for phytoplankton are also affected by climate change (Mix 1989). The above, together with the increase in temperature and the elevation of sea level (with the consequent increment of sedimentation, turbidity, etc.) that global warming could bring, would have catastrophic consequences for the coral reef environments. Almost three decades ago, Newell (1972: 65) concluded in an often-quoted review paper: "Any collapse of the present reef community will surely be followed by an eventual recovery. The oldest and most durable of the earth's ecosystems cannot easily be extirpated." This thought is possibly true in the light of naturally occurring environmental changes that occur over long periods of time and allow organisms to evolve, adapt, and survive to the next favourable interlude. The IPCC predictions for human-induced global warming and sea-level rise, however, indicate that rapid large-scale environmental change may occur in the coming decades (IPCC 1996a; 2001). This, in turn, may have devastating and irreversible consequences on the long-term survival of coral reefs (Done, Ogden, and Wiebe 1995).

Mangrove forests and coastal marshes

Mangrove forests grow on coastal tropical areas protected from wave erosion. Their habitat includes coastal lagoons, riverine deltas, protected beach fringes, and coastal marshes and swamps. In Mexico, mangrove forests occupy 14,000 km^2 (0.7 per cent of the national territory) and have enormous importance as breeding territories for many marine organisms that spend part of their life in the nutrient-rich coastal ecosystems. Mangrove litter-fall contributes significantly to the food chain of estuarine species (Camilieri and Ribi 1986; Snedaker 1989; Twiley, Lugo, and Patterson-Zucca 1986). The annual productivity of mangrove litter reaches values as high as 5,000–10,000 kg/ha, comparable to the productivity of many irrigated pastures in Mexico (López-Portillo and Ezcurra 1985).

One of the best-known and -described phenomena in mangrove forests is plant zonation, the distribution of different plant species along vegetational belts that usually parallel the coast. Mangrove zonation depends primarily on water depth and substrate salinity (Ball and Farquhar 1984a, b; Eleuterius and Eleuterius 1979; López-Portillo and Ezcurra 1989a, b; Rabinowitz 1975, 1978). In most tropical areas, as in the Gulf and Caribbean coasts of the Yucatan Peninsula, tidal variation is very low, around 20 cm. Thus the level of flooding inside coastal lagoons is extremely constant, and mangrove distribution is regulated mostly by the inundation of the soil. Tidal sorting is the main (but not the only) factor determining mangrove distribution: the average salinity of the substrate can also be of great importance. Mangroves offer excellent examples of ecosystems that have an importance that goes well beyond their natural boundaries (Mooney et al. 1995). The organic matter produced by these ecosystems is of crucial importance to the productivity of many estuarine and non-estuarine fisheries, including the open-sea shrimp fisheries and the estuarine oyster fisheries of both the Gulf of Mexico and the Pacific coast.

The average tide level is of enormous significance in defining mangrove distribution and abundance (fig. 6.5). On the one hand, tidal sorting defines the distribution of mangrove seedlings (Rabinowitz 1975, 1978): large seedlings colonize deeper waters, whereas shorter seedlings colonize shallower substrates. On the other hand, most mangrove species possess pneumatophores, special root adaptations that emerge from the soil and allow breathing of the root system that is anchored in the anaerobic swamp. Each mangrove species has a characteristic pneumatophore length that allows the trees to survive under a certain level of flooding. A difference of 5–10 cm in flooding depth can mean the replacement of one forest type by another, or even the dieback of the forest. Oil exploration in the Tabasco lowlands has shown this many times.

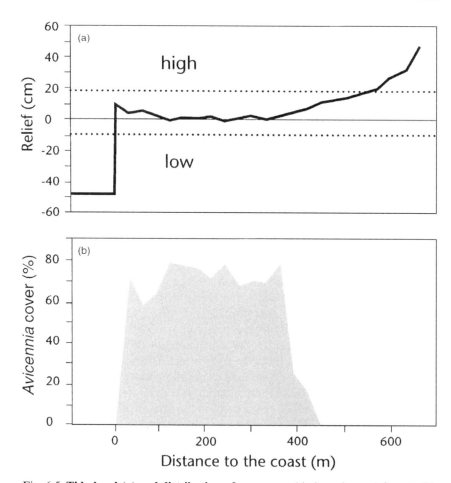

Fig. 6.5 **Tide level (a) and distribution of mangrove (*Avicennia* spp.) forests (b)**

Artificial closing or opening of the lagoon inlet (the canal connecting coastal lagoons to the sea) usually changes the concentration of salts and the tide levels within the lagoon in a sufficiently large amount to generate massive mangrove dieback. This very narrow adaptation of mangroves to the tidal level is a major concern with respect to the systemic components of global change. If global warming is going to generate a significant increase in sea level, then it is quite obvious that mangrove forests are likely to suffer significant damage.

Johannes and Hatcher (1986) have shown that important interactions occur among coral reefs, mangroves, and tropical seagrass communities. Barrier reefs function as self-repairing breakwaters, creating the low-energy conditions that favour mangrove development along tropical coastlines. Large quantities of calcareous sediments produced by reef organisms create the substrate in which mangroves and seagrasses grow. In turn, both mangrove and seagrass communities reduce the offshore transport of land sediments, thus helping to preserve the clear waters necessary for the successful development of the adjacent reef communities. All three community types exhibit considerable faunal overlap, especially of fishes. Reef fishes frequently forage on seagrass beds and migrate among the three communities to feed, seek shelter, or spawn. Juvenile reef fishes and lobsters use seagrass beds and mangroves as nursery areas. Obviously, anything that adversely affects one of these communities may ultimately affect the others.

Cloud forests

Just as mangroves are vulnerable to global change, owing to their narrow tolerance to tide fluctuations, other ecosystems are vulnerable because of their narrow and fragmented location along a certain altitudinal belt in mountainous reliefs. An excellent example of the latter is the cloud forest (Leopold 1950) or *bosque mesófilo de montaña* (Rzedowski 1978). In Mexico, this type of tropical forest is referred to as *selva nublada* by Beard (1946) and as Andean forest by Van der Hammen and Cleef (1986). This is distributed at an elevation of 600–3,200 metres, located in wet and cool mountain belts with annual mean precipitation more than 1,500 millimetres (occasionally near 5,000 millimetres), and with a mean annual temperature of 12–23°C. A general characteristic of this vegetation type is the high frequency of mists and fogs and, therefore, a high relative humidity. The present geographical distribution of cloud forests in Mexico is highly fragmented (fig. 6.6) and restricted to physiographically heterogeneous zones, some of which actually have been completely transformed to coffee plantations, mainly 1,000–1,500 metres in elevation, all over the country (Miranda and Sharp 1950; Rzedowski 1978). The degree of endemism in these forests is very high.

Several authors have pointed out that this ecosystem has had more extensive geographical ranges in the past than it has at present (González-Quintero 1968; Toledo 1982). During the Quaternary glacial events, strong vegetational displacements occurred in the Mexican tropics. In the cold glacial periods, the tropical forests at lower elevations were partially replaced by cloud forests, which formed thus a uniform corridor along the pediment of the different humid mountain ranges (Toledo 1982). Towards the end of the Wisconsin Period, some 14,000 years BP, the warming of the tropics by the present Holocene interglacial confined the cloud forests to their present disjuncted distribution. Cloud-forest species are mostly Pleistocene relics that have survived by climbing in altitude, a slow pro-

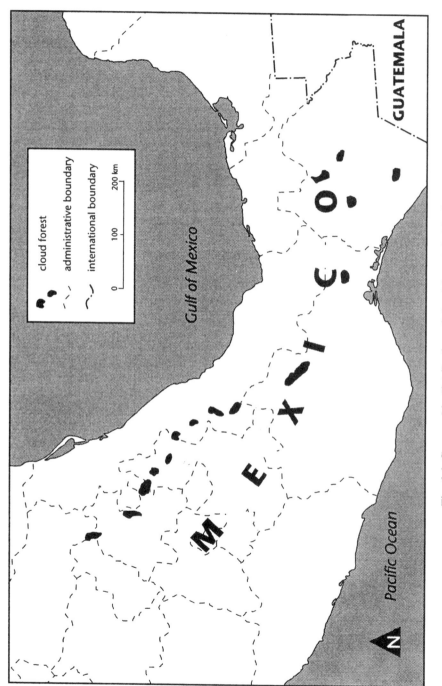

Fig. 6.6 **Geographic distribution of cloud forests in Mexico**

cess that took thousands of years (this, of course, does not imply that the evolutionary origin of the cloud-forest species was during the Plio-Pleistocene glacial periods, but rather that the glacial events shaped the present fragmented distribution of this vegetation type). Van Der Hammen and Cleef (1986) have similar vegetational displacements reported for the Andean forest in the Colombian Cordillera Oriental during the Pleistocene. Intermittent glacial and interglacial cycles were characterized by the domination of páramo high-altitude vegetation during the glacial maxima and its occupation of the area of the present Andean and sub-Andean forests. At the beginning of the Holocene, 14,000 BP, the Andean forest rose, occupying even higher elevations than at present, and then stabilized at its present altitudinal belt (Brown and Gibson 1983; Van Der Hammen and Cleef 1986).

Cloud forests are common all along the Andes from Argentina to Colombia, in the Brazilian Serra do Mar (the Atlantic coastal range), in the Mexican Eastern and Western Sierra Madre, and in the highlands of southern Mexico and Central America. They are very important agricultural areas, heavily used since the nineteenth century mostly for moist tropical plantations such as coffee, vanilla, citrus, and bananas, but also harbouring other important crops such as maize, beans, mangoes, avocados, and papayas (Luna-Vega 1984). The rich coffee-export regions of the Americas are located in cloud-forest areas. Owing to their high degree of endemism and isolation, the cloud forests are ecosystems of low resilience, and their occurrence is strongly linked to the present temperature regimen that defines current vegetation floors. The dominant species in the cloud forest are long-lived trees, with generation times of decades or centuries. Obviously, the possibility of the migration of cloud-forest species to a different altitudinal belt, in the case of significant temperature change at a planetary level, is ruled out by the slow dynamics of the system.

Hitherto, the main disturbance in cloud-forest ecosystems has been deforestation, which, apart from devastating the high biological diversity of the forest, increases the extremely high fragmentation of the system. Human-made patches have restricted the cloud-forest species to a few isolated areas (Luna-Vega 1984), dramatically increasing their probability of extinction. Additionally, the prospect of global temperature change brings forth a concern for the future of many altitude-belt, high-value tropical perennial crops – mainly those, such as coffee and vanilla, that have well-defined environmental conditions and which grow between 1,000 and 1,500 m. Plantations of relatively long-lived perennials are costly and cannot easily be replanted over short periods.

Hunter, Jacobson, and Webb (1988) have suggested the use of ecological corridors to protect highly fragmented areas such as cloud forests, which are rich in endemism. However, the capacity of the narrowly adapted cloud-forest species to change or migrate when confronted with large-scale environmental change remains to be seen. Current evidence suggests that such environmentally restricted ecosystems will suffer the most.

Agroecosystems and vulnerability

Modern agriculture is based on the cultivation of a few high-yield commercial varieties, but it needs an immense reserve of seeds of different origin from which to breed new crop varieties. Most of these commercial crops have a life span of 6–15 years, after which the agricultural pests and pathogens become so adapted to the varieties that they need to be changed. In open contrast, traditional agriculture maintains genetically diverse crops that are often in equilibrium with their pests and do not need periodic replacements. The areas of the Americas in which traditional agriculture survives are a major source of genetic diversity for future breeding programmes (Nabhan 1989). Modern techniques of genetic engineering notwithstanding, variation has to be obtained from the field.

The majority of the world crops originated in tropical areas that make up much of the less-developed countries, mostly those with nutrition deficiencies. Out of the first 1,000 major crops harvested in the United States, for example, only three (sunflowers, Jerusalem artichokes, and cranberries) are strictly native to the country (see table 6.2 and *New Scientist* 1981). The world's agricultural production depends fundamentally on wild or indigenous varieties, which are often collected by plant breeders and kept in large germplasm banks. The importance of these germplasm banks is so strategic that in 1977 the US Department of Agriculture decreed that all the seeds contained in the bank at Fort Collins were the property of the US Government, and that political considerations dictated at times the exclusion of some countries from the free exchange of seeds (*New Scientist* 1981).

Mexico, and particularly the area known as Mesoamerica, is recognized as one of the centres of the origin of agriculture and as one of the most important centres of domestication of plants in the world (Byers 1967; Caballero 1990; Mangelsdorf, MacNeish, and Galinat 1967). The south of Mexico (coinciding mostly with the high-endemism areas described above) was aptly described early this century by the great Russian geneticist N.I. Vavilov as "the furnace of creation" (quoted in Rzedowski 1978). Currently, 5,000–7,000 plant species in Mexico are known as sources of food, medicine, fibres, building materials, and other raw materials (Caballero 1990). For thousands of years, Mesoamerican cultures have developed an interaction with these plants that has increased the morphological and genetic intraspecific variability (Nabhan 1985, 1989). Among the contributions that the Mesoamerican cultures have made to crop diversity is the immense genetic variability of maize, bean, chillies, tomatoes, and squash, among many other species, which has been generated over millennia of plant–human interactions.

Plant variability has both a cultural determinant and a genetic component. In the case of Mexico, some 56 different ethnic groups that can be found in the country have increased significantly the variability of a great number of useful plants. In annual crops, such as maize, at least 25 races have been recognized, each with a great within-race variability. It is a well-known fact that indigenous groups cultivate and maintain different races according to environmental conditions (Caballero 1990). In some mountainous parts of Mesoamerica, neighbouring valleys can harbour different maize varieties, while at a regional level the diversity of crop varieties is very high. The same applies to bean crops, of which 427 varieties have been collected and identified in Mexico (Hernández-Xolocotzi 1985). A similar situation prevails in the Andean regions of South America: Camino, Recharte, and Bidegaray (1985) have shown, for example, that in a small district of the Cuyo province in Peru more than ten different races of potato are cultivated along the altitudinal gradient. Agricultural production is divided into seven narrow altitudinal belts of 200–400 metres each, and the varieties cultivated in each belt differ considerably.

We might well hypothesize that the effect of global climatic change upon annual crops could be of minor importance considering the enormous variability of these crops. At present, however, many researchers believe that the main problem related to agroecosystem management in Latin America is the general tendency towards genetic and ecological uniformity imposed by the development of modern agriculture (Caballero 1990; Toledo 1989). This trend implies the adoption of more technologically developed modes of production, including the use of genetically modified seeds. The cost of this transformation is the abandonment, and often the extinction, of natural varieties. Several authors have shown how modern commercial varieties, which are genetically uniform, generate high yields under optimum field conditions but can produce catastrophically low yields in less favourable environments. In contrast, local indigenous varieties have small distribution areas and are much more finely tuned to different environments, where they have been selected for centuries or millennia. Like many species of tropical ecosystems, local indigenous varieties are endemics of narrow distribution, but they have a great economic value in plant breeding. Many local varieties of different crops have already succumbed to the growing demand for market uniformity in rural production, dangerously narrowing the gene pool of many crops.

What will happen with these varieties in the scenario of global environmental change? It is difficult to say, but as a general rule a higher probability of extinction is expected on local crops with small distribution than in the widely dispersed, genetically uniform commercial crops.

The economic importance of native crops is still enormous (Barbier et al. 1995). Early this century, Vavilov recognized 12 "centres of diversity" for crops (table 6.2). According to Vavilov, different crops originated in different centres, each of which has contributed to modern agriculture with a large number of crops. The real meaning of Vavilov's model is that crop domestication has been historically associated with certain geographic areas in which sedentary agricultural groups have evolved (Juma 1989). Vavilov's work highlighted the uneven geographic distribution of the genetic material supporting the world's crops and the dependence of some nations on the supply of genetic resources in others (and usually tropical countries). This dependence has been analysed by Kloppenburg and Kleinman (1987; table 6.3), who calculated how much of the present food production in different parts of the world depends on germplasm (i.e. genetic material) that originated in other regions. According to this study, the West Central Asiatic and Latin American regions account for 65.5 per cent of the world's genetic resources for the major crops (Latin America has produced maize, potato, cassava, and sweet potato; West Central Asia has generated wheat and barley). Adding to this the contribution to the world's food production of germplasm from Africa, China, Indo-China, and the Hindustan region, it can be calculated that the world's poorest nations, as a group, account for more than 90 per cent of the world's genetic resources for food production (Juma 1989).

A question of rates and scale

May (1989) has calculated that, at the present rate of ecosystem destruction, approximately one-half of all species on Earth will meet extinction during the next century. It is estimated that biological evolution took 10–100 million years to produce this same number of species through the mechanisms of speciation. This makes the current rate of species extinction around a million times higher than the rate at which evolution is producing new species. In short, we are witnessing now one of the greatest biological catastrophes that has occurred since the beginning of life on Earth, and the rate at which this process is occurring completely excludes the possibility that massive extinctions generated by human activities can be compensated by the evolution of new species within a sensible time frame (Barbault et al. 1995). The World Conservation Monitoring Center estimated the minimum numbers of globally threatened animal and plant

Table 6.2 **Vavilov centres of crop diversity, with selected species**

Centre of diversity	Crops
Mexico–Guatemala	Amaranth, avocado, beans, corn, cacao, cashew, cotton, guava, papaya, red pepper, squash, sweet potato, tobacco, tomato, vanilla, tabasco pepper, guayule, luffa gourd, chayote, curuba, prickly pear, sisal hemp, sapote, mamey, agaves
Peru–Ecuador–Bolivia	Beans, cacao, corn, cotton, guava, papaya, red pepper, potato, quinine, quinoa, squash, tobacco, tomato, begonia, malanga, canna, cucumber, pumpkin, coca, tree tomato, marigold, peach palm, ground cherry
Southern Chile	Potato, Chilean strawberry, Chile tarweed
Brazil–Paraguay	Brazil nut, cacao, cashew, cassava, rubber, peanut, pineapple, passion fruit, maté, Surinam cherry, jaboticaba
North America	Blueberry, cranberry, Jerusalem artichoke, pecan, sunflower, black walnut, black raspberry, muscadine grape, wild gooseberry, wheatgrass
Ethiopia–Kenya–Somalia	Banana, barley, castor bean, coffee, flax, okra, onion, sesame, sorghum, pearl millet, bread wheat, garden cress, cowpea, mustard, date palm, watermelon, cantaloupe, yam, pigeon pea, Egyptian cotton, tree cotton, short-staple cotton, teff, veldtgrass, lupine, safflower, coriander, fennel, khat, spurge, indigo, vernonia
Central Asia	Almond, apple, apricot, basil, broad bean, cantaloupe, carrot, chickpea, cotton, coriander, flax, grape, hazelnut, hemp, lentil, mustard, onion, pea, pear, pistachio, radish, rye, safflower, sesame, spinach, turnip, wheat, garlic
Mediterranean	Asparagus, beet, broccoli, cabbage, cauliflower, celery, broad bean, artichoke, dill, fennel, mint, desert date, rape, crimson clover, lavender, chicory, hops, lettuce, oat, olive, parsnip, rhubarb, wheat, flax, carob, savory, thyme, rosemary, sage, hop
Indo-Burma	Amaranth, betel nut, betel pepper, chickpea, cotton, cowpea, cucumber, eggplant, hemp, jute, lemon, mango, millet, orange, black pepper, rice, sugar cane, taro, yam, balsam pear, Indian lettuce, Indian radish, tangerine, citron, lime, jambolan plum, jack fruit, coconut, bilimbi, safflower, leaf mustard, tree cotton, crotalaria, kenaf, sesbania, cardamom, indigo, madder, henna, Indian almond, senna, croton, cinnamon, Indian rubber, bamboo, palmyra palm

Table 6.2 (cont.)

Centre of diversity	Crops
Asia Minor	Alfalfa, almond, apricot, asine, barley, beet, cabbage, cantaloupe, cherry, clover, coriander, date palm, carrot, fig, flax, grape, leek, lentil, lupine, leaf mustard, oat, onion, opium, poppy, parsley, pea, pear, pistachio, pomegranate, purslane, rape, rose, rye, sesame, wheat
Siam–Malaya–Java	Banana, betel palm, breadfruit, coconut, ginger, grapefruit, sugar cane, yam, mung bean, sour orange, air potato, pokeweed, calamondin, jackfruit, durian, salacca palm, candlenut, curcuma, nutmeg, cardamom, clove
China	Adzuki bean, apricot, buckwheat, Chinese cabbage, cowpea, tea, sorghum, millet, oat, sweet orange, peach, radish, rhubarb, soybean, sugar cane, Chinese yam, bamboo, horseradish, cherry, eggplant, hawthorn, arrowhead, walnut, litchi, sesame, mulberry, ginseng, camphor, lacquer, hemp, fibre palm, aconite, ramie

Source: modified from Juma (1989: 17–18).

species in 1994 at 5,366 animals and 26,106 plants (Barbault et al. 1995: 234), and it is also clear that the current rate of extinction is the highest that it has been in at least 65 million years (Reid 1997: 17).

Many questions arise when analysing this problem. If species extinction has been common throughout the evolution of life on Earth (as a matter of fact, 99 per cent of the species that ever existed are now extinct), how worrying is it, really, that a certain number of species will meet extinction during the next century? Furthermore, if it has been clearly shown that complete flora and fauna migrated into refuge areas during other events of climatic change (e.g. the Pleistocene glaciations), will not the same flora and fauna migrate, or adapt to global environmental change, or both, during the coming century?

Gould (1991) has partially answered these questions. The mean duration of a species is of the order of 1–3 million years. After periods of massive extinction, the recovery of species diversity to its original levels by evolution and speciation has taken of the order of 10–100 million years, as evaluated by the fossil records. That is, biological diversity accumulates at a geological time-scale of millions of years, but it is being destroyed at human time-scale of decades. If global environmental change has a significant effect on life on Earth, evolution will have no chance whatsoever to compensate the loss with the development of new species on a comparable time-scale. If we were to represent the scale of

Table 6.3 **Regions of diversity**[a]

Regions of production	Japanese	Chinese	Chino-Australian	Indo-Hindi	West Central Asian	Mediter-ranean	African	Euro-Siberian	Latin American	North American	Total dependence
Chino-Japanese	37.2	0.0	0.0	0.0	16.4	2.3	3.1	0.3	40.7	0.0	62.8
Indo-Chinese	0.9	66.8	0.0	0.0	0.0	0.0	0.2	0.0	31.9	0.0	33.2
Australian	1.7	0.9	0.0	0.5	82.1	0.3	2.9	7.0	4.6	0.0	100.0
Hindustani	0.8	4.5	0.0	51.4	18.8	0.2	12.8	0.0	11.5	0.0	48.6
West Central Asiatic	4.9	3.2	0.0	3.0	69.2	0.7	1.2	0.8	17.0	0.0	30.8
Mediter-ranean	8.5	1.4	0.0	0.9	46.4	1.8	0.7	1.2	39.0	0.0	98.2
African	2.4	22.3	0.0	1.5	4.9	0.3	12.3	0.1	56.3	0.0	87.7
Euro-Siberian	0.4	0.1	0.0	0.1	51.7	2.6	0.4	9.2	35.5	0.0	90.8
Latin American	18.7	12.5	0.0	2.3	13.3	0.4	7.8	0.5	44.4	0.0	55.6
North American	15.8	0.4	0.0	0.4	36.1	0.5	3.6	2.8	40.3	0.0	100.0
World	12.9	7.5	0.0	5.7	30.0	1.4	4.0	2.9	35.6	0.0	

Sources: Juma (1989: 22); Kloppenburg and Kleinman (1987).

a. Global genetic resource interdependence in food production. The columns are Vavilov's regions of diversity, and the rows are the same regions evaluated by their present production. The values in the table indicate the percentages of the present food production in a region that depends on germplasm from different regions of diversity. For example, 18.7 per cent of the food production in Latin America depends on germplasm derived from the Chino-Japanese region. The bottom row (World) indicates the relative contribution of each region to the world's crop germplasm. The extreme right-hand column (Total dependence) indicates how much of the food production in each region depends on germplasm from other regions.

244

speciation and of extinction on a pair of orthogonal axes, the horizontal axis (representing speciation times) would be 10,000 km long, whereas the vertical axis (representing extinction rates) would be around 10 cm high, not much higher than a matchbox!

A similar argument, of course, can be put forth for the migration of floras. The high-endemism relict refuge areas discussed earlier indicate that, 14,000 years after the last glaciation, flora and fauna still retain a biogeographic "memory" of their Pleistocene refuges. Although we cannot see the process because the time-scale is so slow, the clumped distribution of biodiversity in terrestrial habitats (see figs 6.2, 6.3, and 6.4) indicates that the biota are a long way from biogeographic equilibrium (i.e. a slow migration is radiating from the refugia). The migration rate of groups of organisms is, indeed, very slow. In some long-lived perennial plants, migration may occur at a rate of a few hundred metres per generation, and a generation may span more than 100 years. There is, quite obviously, little chance for the refuge mechanism to operate in response to human-induced global environmental change if the latter produces significant transformations in less than a century; the time scales simply do not match.

Conserving the planet's biodiversity is, therefore, a problem of the highest priority. Humans need most urgently to learn how to make the growing needs of human populations compatible with the necessity to conserve the threatened species and habitats and how to use exploitable ecosystems in a sustained way. Otherwise, future generations will never understand how our legacy was culturally so rich and biologically so degraded.

Conclusions

We started this chapter with two important questions: (a) which natural environments are more likely to be affected by global change; (b) which human groups depend primarily on these environments? Throughout the chapter, we have analysed several points that we believe are still poorly understood in the context of global environmental change.

1. Uncertainty seems to prevail in the analysis of global environmental change. We still do not know with precision, despite the efforts of the IPCC, how much climatic change will occur and how such change will spread across the globe. On the other hand, the real influence of some cumulative variables – such as population change, soil erosion, and tropical deforestation – on the global environment needs more careful evaluation. These variables may be of crucial importance as driving forces of global environmental change.

2. As a general rule, tropical areas concentrate a higher biological diversity and a large number of rare, endemic species that are highly vulnerable to environmental variation (UNEP 1995). From the point of view of biological richness as a natural resource, the highest proportion of vulnerable species lies within the tropics, including species not only from tropical rain forests but also from high-altitude tropical ecosystems, dry tropical forests, scrubs, cloud forests, tropical savannas, tropical swamps and marshes, and coastal ecosystems (see the discussion by Jodha in chapter 9).

3. Because environmental conditions are much more constant in the tropics, many ecosystems as a whole have developed a narrow association with low-amplitude environmental factors. This is the case of the altitudinal belts of cloud forests, the tidal ranges of mangroves, and the association of coral reefs with very specific conditions of water temperature, transparency, and tidal height. Without considering the degree of biological diversity or of rarity and endemism in these ecosystems, it is clear that they can be, as a whole, highly vulnerable to environmental change.

4. The majority of world crops have originated in tropical and subtropical areas. Traditional crop varieties are often highly endemic and ecologically rare, and many of them still survive in vulnerable ecosystems such as the altitudinal belts of the Andes or the dry tropical plateaus of southern Mexico. Many areas in which traditional agriculture survives are major sources of genetic diversity for present and future crop-breeding programmes. Global environmental change could affect the capacity of the world's food system to develop new varieties for future cultivation.

5. In conclusion, good evidence indicates that tropical ecosystems and their inhabitants may suffer strong losses from global environmental change. This effect could be made even stronger, as Liverman shows in the preceding chapter, by the fact that many of these regions are poor and undeveloped. In consequence, their inhabitants have a lower economic and resource capacity to respond to, and to cope with, quickly shifting environmental conditions – an interaction between human and ecological vulnerability that concerns Kasperson, Kasperson, and Dow in chapter 7.

7

Vulnerability, equity, and global environmental change

Roger E. Kasperson, Jeanne X. Kasperson, and Kirstin Dow

Whether at national or global levels, environmental problems increasingly are fraught with questions of equity and fairness. In the United States, an environmental justice movement is in full flower and signs of a developing fairness debate are all around us. In 1985, the National Council of Churches established an Eco-Justice Working Group to address issues of environmental equity. In 1987, the United Council of Churches analysed the national pattern of hazardous waste facility sites in *Toxic Waste and Race in the United States* (UCC 1987; see also update in Goldman and Fitton 1994), arguing that such facilities were concentrated in poor communities with large minority populations. A 1990 "Conference on Race and the Incidence of Environmental Hazards" at the University of Michigan School of Natural Resources spawned an informal action group. This "Michigan Coalition" petitioned William Reilly, then director of the US Environmental Protection Agency (USEPA), to undertake a broad programme of environmental-equity initiatives with the following aims: to commission research geared towards understanding environmental risks faced by minority and low-income communities; to initiate projects to enhance risk communication targeted to minority and low-income population groups; to require, on a demonstration basis, that racial and socio-economic equity considerations be included in regulatory impact assessments; to include a racial and socio-economic dimension in geographic studies of environmental risk; to enhance the ability of minority academic institutions to participate in and contribute to the development of environmental equity; to appoint special assistants for environmental equity at decision-making levels; and to develop a policy statement on environmental equity (USEPA 1992: 7).

Reilly established within the USEPA an environmental equity workgroup composed entirely of agency staff. Since the agency had no explicit policy or programme directed at equity considerations and lacked relevant expertise to address such issues, its mandate "to assess the evidence that racial-minority and low-income communities bear a higher environmental risk burden than the general population, and consider what the EPA might do about any identified disparities" (USEPA 1992: 1; 2) posed substantial difficulties. Indeed, when one of the authors (Roger E. Kasperson) met with the USEPA workgroup to conduct a seminar on the concept of environmental equity in November 1990, it was apparent that the workgroup members had no clear conception of what equity was and, since the agency had not collected information concerning its programme in equity-relevant categories, the group was poorly positioned to conduct the mandated assessment. It is not surprising, then, that its report, *Environmental Equity: Reducing Risk for All Communities* (USEPA 1992) was anaemic and drew widespread criticism. None the less, President Clinton signed Executive Order Number 12898 in 1994 requiring each federal agency to adopt the principle of environmental justice in programmatic decisions. Seven years later, efforts are mired in definitional and measurement problems, with the prospect of huge intervention costs for minor gains in risk reduction; a bold far-reaching effort to address social inequities has yet to materialize.

Meanwhile, hazardous-waste management programmes, and particularly the siting of new facilities and the cleanup of defence-waste sites in the United States, are immersed in conflicts over a plethora of risk and value issues (Murdock, Krannich, and Leistritz 1999), but particularly over the fairness of distributive impacts of siting; the issue of "how clean is clean enough?"; the allocation of roles and power among risk managers and risk bearers; and debates over procedural equity in decision-making (USDOE 1998). A lack of sensitivity to these issues led the US Congress in 1987, in its "Screw Nevada" budget amendment, to destroy the architecture of equity underlying its radioactive-waste legislation adopted five years earlier, resulting in a "tunnel vision" approach to programme development and a federal–state stalemate in the radioactive-waste management programme (Flynn et al. 1997). Meanwhile, several recent works have documented the extensive conflict surrounding equity problems in siting facilities in various Asian countries (Lesbirel 1998; Shaw 1996).

At the global level, equity concerns in environmental policy are equally evident. The World Commission on Environment and Development (WCED 1987), in its influential report *Our Common Future*, accorded international equity a central place in any global strategy for sustainable development:

> ... sustainable development requires meeting the basic needs of all and extending to all the opportunity to fulfill their aspirations for a better life. A world in which poverty is endemic will always be prone to ecological and other catastrophes.... Meeting essential needs requires not only a new era of economic growth for nations in which the majority are poor but an assurance that those poor get their fair share of the resources required to sustain that growth. (WCED 1987: 8)

Distributional equity issues, especially as they involve development goals and the financing of global mitigation efforts, were central to the accords reached at the 1992 Earth Summit, and capacity-building and the amelioration of global poverty are major planks in Agenda 21 and Capacity 21 (UNDP 1996). Indeed, equity typically appears as an integral part of conceptualizations or indicators of sustainability (two of the ten sustainability goals of the US President's Council on Sustainable Development [1996] address equity principles).

The Kyoto Protocol of 1997 has accorded equity issues a central place in international efforts to reduce greenhouse-gas emissions. The Kyoto Convention formulated a set of equity-related criteria to guide the further development and implementation of the Protocol, namely

- intergenerational equity;
- equity among countries according to their responsibility in the historical generation of greenhouse-gas emissions and their response capacity;
- the consideration of the needs and special priorities of the developing countries and their vulnerability;
- the precautionary principle;
- the search for a better cost–benefit rate at world level; and
- the search for cooperation among interested parties. (Mazzucchelli and Burijson 1999: 5)

How such broad criteria will be developed into instrumental mechanisms and become the basis for concerted action will be daunting challenges in the coming years, especially without US backing.

Equity in the context of climate change will, Agarwal and Narain (1999) argue, be possible only if agreement is secured on two major principles – the principle of convergence and the principle of equitable entitlements. The principle of equitable entitlement seeks to bring responsibility and burden into accord. Defining responsibility for the accumulated burden of past and future emissions has been highly contentious, with the IPCC (1997) estimating that emissions of developing countries will equal those of industrialized countries by 2037, with the World Resources Institute (Hammond, Rodenburg, and Moomaw 1991) calculating that date in the latter part of the twenty-first century, and with others arguing that responsibility needs to be calculated on a per capita rather than a nation-state basis. Broad agreement does appear to exist that the atmosphere is a global common property essential to all human beings for life support and economic growth. What seems clear, but has yet to be widely accepted, is that nations that have contributed most to the problem and which have the greatest capability to reduce the risks should bear most of the financial burden. Supporting this criterion is the principle of convergence (or what we would term burden sharing) – that all nations need to participate in reducing their greenhouse-gas emissions for the overall global benefit.

An African perspective suggests why even well-intentioned equity arrangements may falter in implementation (Sokona, Humphreys, and Thomas 1999). Although Africa is highly vulnerable to potential future climate change, it has been in the throes of relentless environmental and economic crises for the past two decades. Climate change, with its major effects highly uncertain and well in the future, is not at or near the top of priority concerns. The more pressing needs involve improving the standard of living and environments of the continent's most disadvantaged and marginalized peoples. To achieve this goal, the development of technical, social, and institutional infrastructure will be critical, as will be the transfer of financial support and technology to enable such changes.

Kates (1986: 216) argued, over a decade ago, that society was entering what he called a period of "third-generation" ethics in regard to risk management. The *first generation* of risk management reflected the ethics of non-maleficence – do no harm to persons or nature. Such were the ethics of Earth Day 1970, which have been extensively institutionalized in corporate codes of social responsibility and the creeds of environmental organizations. *Second-generation ethics* attempted to weigh harms – to consider both benefits and the value of lost benefits, as well as risks. The ethical underpinnings of this approach, dominated by principles of utility, have been very much apparent in the United States in efforts to use cost–benefit analysis and comparative-risk analyses to establish management priorities and budget allocations. *Third-generation ethics*, according to Kates, are those of equity, fairness, and distributive justice. They not only are concerned with aggregate balancing of benefit and risk but also involve the distribution of benefits, harms, and relative protection and the fairness of process as well as outcomes. Since 1986, when Kates argued this, third-generation ethics of risk management has indeed assumed an increasingly central place in environmental and risk policy, most prominently in the US National Research Council's report *Understanding Risk* (Stern and Fineberg 1996), the American Chemical Society's symposium

(Cothern 1996) on ethics and values attendant on risk decisions, and the United Kingdom (UK) Interdepartmental Liaison Group on Risk Assessment (1998). Such ethical imperatives are being adopted as major goals of sustainable development strategies and programmes aimed at comparative-risk management and they will require the type of participatory and humanistic risk analysis espoused in this volume and elaborated in chapter 1.

What has been less appreciated than the increasing visibility of these equity and justice issues is the extent to which they intertwine with, and depend upon, considerations of human and ecological vulnerability. In this chapter we explore the intertwining of risk, vulnerability, and equity, arguing that the prospect for acceptable outcomes in terms of equity and social justice depends heavily upon explicit analysis and handling of *both* vulnerability and equity problems, that vulnerability is an increasingly important term in the risk equation for global environmental risk problems, and that the redressing of vulnerabilities is often a preferable but overlooked policy strategy for reducing risks and achieving greater equity.

Risk and vulnerability

Throughout this volume, risks refer generally to threats to humans and to the things they value. Environmental risks, it is well known, are the products not only of the magnitude of the environmental insult but also of the degree of vulnerability of the human or ecological receptor. Thus, global environmental risks may be considered to have the structure depicted in figure 7.1.

As this simple schematic suggests, environmental change and perturbations are both human induced and the result of natural variability. Environmental degradation has a reflexive and two-way relationship with society, while prevailing levels of socio-economic vulnerability and ecosystems fragility are also closely related, as in the spirals of land degradation to which Blaikie and Brookfield (1987) refer, or the interactions between human systems and environment often involved in poverty (Gallopin, Gurman, and Maletta 1989; Kates 2000; Kates and Haarmann 1992). These interactions produce the adverse consequences that represent human and ecological harm. Meanwhile, the design of management systems may accommodate interventions to control the evolution of risk, focusing alternatively "upstream" (as by altering human driving forces), "mid-stream" (as by mitigating environmental degradation or reducing vulnerabilities), or "downstream" (as by ameliorating damage that has already occurred). What is essential is to assess vulnerability as an integral part of the causal chain of risk and to appreciate that altering vul-

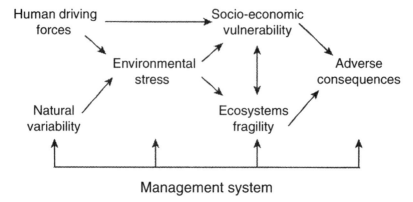

Fig. 7.1 **Highly simplified model of the structure of environmental risk**

nerability is one effective risk-management strategy, as noted in the Introduction and detailed in the two preceding chapters.

Fragility and vulnerability

The concept of *vulnerability*, defined here simply as "the capacity to be wounded by a given insult," is a central but still incompletely developed concept in analyses of environmental change. A variety of efforts across disciplines to address vulnerability has generated a pot-pourri of terms, differing approaches, and concepts such as *resistance* (the ability to absorb impact and maintain functioning), *resilience* (the ability to maintain a system and to recover after impact), and *exposure* (the presence of a threat to a group or region); for an extensive review, see Dow (1992) and Cutter (1996). These three dimensions account for most of the discussion that occurs under the general rubric of vulnerability. Table 7.1 overviews the range of factors that have been treated in empirical analyses of vulnerability.

Although the terms, concepts, and criteria used to address vulnerability differ, two basic concepts seem central: these are (1) the ability of a nature–society system, despite perturbation, to continue to function within a normal range; and (2) the ability to recover from perturbations that substantially disrupt the normal functioning of the system. These two concepts address internal characteristics of the affected system, or receptor, and are explicitly recognized in the approach of the second and third assessments of the Intergovernmental Panel on Climate Change (IPCC), which, addressing vulnerability to climate change, considers both sensitivity and adaptability (IPCC 1996b, 1997, 2001). External processes in

Table 7.1 **Factors contributing to vulnerability to global environmental change**

Environmental Conditions	**Demographics and Health**
• Existing biodiversity	• Health status
• Latitude	• Age
• Record of extreme meteorological events	• Population growth, density, pressure on resources
Technological Conditions	**Land Use and Ownership**
• Dependence on certain energy sources	• Land use/tenure
• Agricultural technology	• Changes in productive capacity of land
• Water-supply systems	• Common property systems
Social Relations	**Economy and Institutions**
• Class	• Income
• Gender	• Access to markets
• Race	• Level of government support
• Ethnicity	• Debt
	• Fluctuating prices

the economic, political, or biophysical system may, however, interact to alter the experience of resistance, resilience, or exposure.

Focusing on available data, Lonergan (1998) has sought a more inductive approach to develop indicators to characterize regions vulnerable to migration flows generated or affected by environmental degradation. His index of vulnerability draws upon 12 indicators for which data are generally (but not universally) available: these are food import dependency ratio, water scarcity, energy imports as a percentage of consumption, access to safe water, expenditures on defence versus health and education, indicator of human freedoms, urban population growth, child mortality, maternal mortality, income per capita, degree of democratization, and fertility rates. A cluster analysis is then performed on the individual indicators to derive a value score for each of the countries analysed. From this, Lonergan (1998: 27–30) has prepared vulnerability maps for the globe and for continents such as Africa. In another approach, Schmandt and Clarkson (1992) have categorized world regions vulnerable to climate change, interrelating diverse threats and multiple impacts. Meanwhile, German scientists at the Potsdam Institute for Climate Impact Research have developed a "syndromes" approach to global change and used it to pinpoint "hot spots" and "critical regions" by continent and worldwide (Schellnhuber et al. 1997).

The impacts of global environmental change will reflect the pattern of environmental change, the fragility of ecosystems, and the vulnerability of social groups. Elaborating somewhat on this, Bohle, Downing, and Watts (1994: 38) see three basic coordinates of vulnerability: the risk of exposure to crisis, stress, and shock; the risk of inadequate coping capacities; and the risk of severe consequences of (and slow and limited recovery from) risks and shocks. The magnitude of the impact will depend not only on the *extent* of the environmental change (e.g. decreased length of the growing season or increased number of hectares inundated by flooding or sea-level rise) but on the ongoing processes and existing factors contributing to the resilience or capacity to be harmed of a group or region (i.e. its *vulnerability*). The serious nature of the threats potentially posed by climate change exemplify well that vulnerability is a pressing concern. Dow (1992) emphasizes the importance of three broad categories of factors and processes in determining the vulnerability of affected populations. In the discussion that follows, we draw upon her distinctions among *ecosystem fragility*, *economic sensitivity*, and *social-structure sensitivity*.

Ecosystem fragility addresses the spatial shifts as well as the regional changes accompanying environmental change. Much of the research on the impacts of climate change has focused intensely on this ecosystem dimension. General circulation models (GCMs) predict different levels of warming across latitudes, although much can be done to improve climate forecasting generally, as suggested by the recent US National Research Council report *Making Climate Forecasts Matter* (Stern and Easterling 1999). In some areas, warming may increase the growing season, with beneficial impacts; in others it may require change and adaptation in cropping or farming practices (IPCC 1996b). Other analyses focus on the *rate* of climate change and the *ability* of species to keep pace with shifts in favourable habitat conditions. Predictions of sea-level rise note particularly the vulnerability of small island states and coastal regions and populations (IPCC 1997). Depending on the sensitivity of the ecosystem, relatively small environmental changes may have tremendous impacts because of potential rapid rates of change, the isolation and fragmentation of many ecosystems, and the existence of multiple environmental stresses (e.g. land-use changes, pollution) and limited adaptation options (Bolin and Warrick 1986). Marginality in ecosystems suggests that the greatest impacts may often arise outside the areas experiencing the greatest quantitative climate change, in areas on the margin of their successful range of operation (see chapters 5, 6, 8, and 10).

Sensitivity is often used to measure the magnitude of negative impacts of a particular environmental change. Change can have positive and negative impacts on human societies. Blaikie and Brookfield (1987) use the terms "sensitivity" and "resilience" to describe the quality of land systems. *Sensitivity*, in their usage, refers to "the degree to which a given land system undergoes changes due to natural forces, following human interference," whereas *resilience* refers to "the ability of land to repro-

Environmental resilience

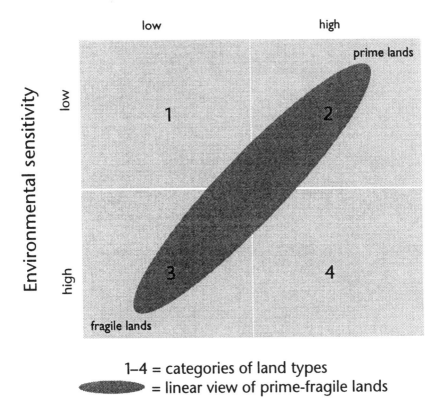

1–4 = categories of land types
 = linear view of prime-fragile lands

Fig. 7.2 **Types of ecosystem fragility. Source: after Turner and Benjamin (1994)**

duce its capability after interference, and the measure of the need for human artifice to that end" (Blaikie and Brookfield 1987: 10).

The concept of *fragility* combines two dimensions – the capacity to be wounded by a particular environmental perturbation (either nature- or human-induced), and the ability to maintain structure and essential functions and to recover. Turner and Benjamin (1994) view these definitions as indicating four distinct quadrants of fragility, each representing different risk situations and different "buffering" to environmental stresses (fig. 7.2). Some environments degrade rapidly but maintain themselves under stress and recover; others are sensitive to change and have low capabilities for maintenance and recovery; and so on.

Obviously, these different attributes of ecosystems are highly relevant to analyses of human-induced environmental stresses. Fragile environments degrade more readily under mismanagement and exact higher societal costs for risk management or for substitution in productive systems. More robust environments, by contrast, do not degrade so rapidly, and they respond to substitution and applications of technology more economically. It is noteworthy, however, that fragile environments are often those subjected to intense human stress and hence most subject to mismanagement and far-reaching damage.

The *vulnerability of the economic system* is a key issue in impacts. The economic marginality of an activity in relation to first- and higher-order impacts of change that cascade through the economic system becomes an important consideration. Economic marginality describes a situation in which the returns to an economic activity barely exceed the costs. A society or group may have resources sufficient to carry on through several extreme years, but prolonged stress may reduce those economic reserves and significantly increase vulnerability. The importance of assets and knowledge of the probability and magnitude of events are important to small-scale farmers' strategies for coping with climatic variability in drought-prone areas of Africa (Chambers 1989; Davies S. 1996; Krings 1993; Rosenzweig and Hillel 1998). Even poor people have household portfolios of investments and stores that are drawn down during times of environmental stress. The poorest households are typically compelled to dispose of larger assets earlier, since they have fewer options than do richer households. Meanwhile, as Vogel (1998) points out, a better grasp is needed of household livelihood strategies, particularly intra- and inter-household utilization and transfer of resources, and a better understanding of the "normal" conditions that shape the lives of the most vulnerable. This suggests the need to assess both baseline vulnerability and also how structural changes and events, together with potential countermeasures, interact to produce crises or disasters, as suggested in figure 7.3.

Social-structure sensitivity is broadly concerned with the response capabilities of particular classes or social groups. The bases of such sensitivity range from the impact of class and gender on access to societal resources to the nature of social institutions. Political-economy perspectives stress the importance of social divisions, often class, in accounting for differential access to resources and differences in the magnitude of impacts experienced. Indeed, a combination of class relations and empowerment forms one of the essential axes of explanation of those enlisting structural or political-economy interpretations of vulnerability (see, for example, Blaikie et al. 1994). Differences among states are another critical division in the international context of global change and response: those with adequate access to resources and information benefit from the global economy, whereas others struggle to survive. On the basis of such a perspective, Bohle, Downing, and Watts (1994: 42) see the groups vulnerable to global change as follows:

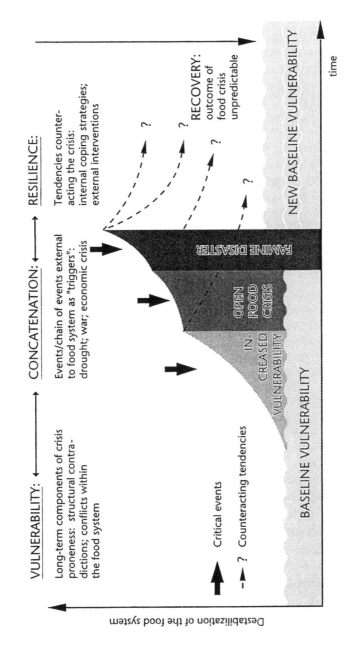

Fig. 7.3 **Schematic diagram of vulnerability to food crises. Source: after Bohle (1993: 23)**

- rural smallholder agriculturalists, with limited land and labour, who are resource-poor, and especially exposed if they are farming in marginal lands;
- pastoralists, often sensitive to drought and pests, and subject to dramatic collapse if entitlements fail in a political context of marginality;
- wage labourers, dependent on exchange entitlements with little or no direct agricultural production and who may be sensitive to market and political failures;
- urban poor, who often form a more visible vulnerable group with some power and access to the political economy;
- refugees, often nearly destitute, who remain dependent on the political economy for their subsistence; and
- destitute groups, who are no longer capable of productive activity, are unlikely to re-enter their former occupations, and are dependent on food aid from a variety of sources.

By contrast, Gallopin (1994) has advanced a conception of *socio-ecological impoverishment* in both the human and ecological subsystems in his more general systems approach to sustainable development. In his view, socio-ecological impoverishment involves the following:

- reduction in the availability or value of the resources necessary to satisfy human needs, desires, and aspirations;
- reduction in the capacity of the human subsystem to make adequate use of the resources it has available;
- reduction in the autonomy to use resources and to make decisions;
- reduction in the capacity to respond to internal and external changes; and
- reduction in the capacity for future improvement or maintenance.

These changes can be viewed as dimensions of vulnerability. The impoverishment can also have ecological roots, of course, including reduction in ecological productive capacity, reduction in the homoeostatic capacity of the ecological subsystem and in the ecological capacity to adjust to new internal and external changes, and reduction in the evolutionary capacity of the ecological subsystem (Gallopin 1994: 23–24).

The nature of social and economic institutions is yet another factor contributing to vulnerability. The infrastructure of social institutions and resource management pervasively shapes the breadth and depth of the response capability. Vulnerability can arise from the growth of technological specialization, homogeneity, and centralization (and the related loss of diversity, multiplicity, and redundancy). Whether institutions are highly adaptive or more narrowly prepared to respond to only a limited set of opportunities, the potential to benefit rests largely with the ability to identify and react to opportunities. Increasing the reliance on management systems to reduce the variability of an ecosystem may erode the overall resilience of the ecosystem and its associated social and economic systems (Holling 1986). In the context of global climate change and the potential shifts in water-resources availability, the ability of the legal and institutional system to adjust to changes in the resource base will obviously affect the magnitude of potential disruptions, a factor explicitly recognized at some length in the IPCC's assessments of climate change (IPCC 1996a, 2001) and explored in depth elsewhere by Frederick (1994) and Gleick (1998, 2000). The legal and institutional rigidity of natural-resource management across nation states is a clear constraint on adaptiveness, and the tendency of institutions to stabilize around a common procedural culture and to build up coherent world views or mindsets suggests that institutional change will often come at glacial pace and with limited capacity to learn from experience (O'Riordan et al. 1998: 395–397).

Trends in vulnerability

On the face of it, substantial reason exists to suspect that vulnerabilities will be major determinants of patterns of global environmental risk. The experience of natural disasters is informative in this respect. The annual number of disasters worldwide that result in 25 or more deaths has fluctuated from year to year but generally followed, until recently, an upward trend (Engi 1995; Glickman, Golding, and Silverman 1992; *World Disasters Report* 1999, 2000). Although improved reporting (and especially the inclusion of smaller events in more remote regions) may explain some portion of the trend, it is unlikely that the number of natural-hazard events (e.g. floods, hurricanes, earthquakes) has increased significantly. Rather, the explanation for the upward trend in disasters appears to lie in the substantial growth in world population and the increasing vulnerability of marginal groups. Disaster assessments (Glickman, Golding, and Silverman 1992) point in particular to population growth, increasing urbanization, land shortages, and economic hardships that force the poorest of the poor in less-developed countries to occupy more hazardous locations as poverty limits emergency preparedness and coping. It is not surprising, then, that the pattern of natural disasters shows a declining death toll in developed countries and a concentration of natural disasters in East Asia and the Pacific, Latin America, and southern Asia. This role of increased vulnerability as the major cause of worldwide natural disasters is in accord with assessments conducted by the US National Committee for the Decade for Natural Disaster Reduction (United Nations 1991), the analysis of Burton, Kates, and White (1993), and the International Federation of Red Cross and Red Crescent Societies (IFRC) (*World Disasters Report* 1999, 2000).

In an analysis of "great natural catastrophes" since 1960 – those disasters that overtax the ability of the region to help itself and make interregional or international assistance imperative – Munich Re (1999) finds that the number of such catastrophes worldwide has risen by a factor of three, economic losses (adjusted for inflation) by a factor of nine, and insured losses by a factor of fifteen or more. More and more of these catastrophes appear to be associated with a climate-related accumulation of extreme weather events. The Munich Re Geoscience Research Group see the major factors contributing to those most extreme disasters as (a) the concentration of people and economic value in an ever-growing number of larger and larger cities (with many in high-risk zones), (b) a greater susceptibility of modern industrial societies to catastrophes, (c) the worldwide accelerating environmental degradation, and (d) (in relation to insured losses) the growing amount of insurance for natural hazards being provided and bought. These trends present imposing challenges for global risk management and also for the international insurance industry.

The global patterns of famine and hunger also reflect the pervasive effects of human vulnerability. In 1990, an estimated 477 million people lived in countries where local crop production and import capacity failed to meet their usual levels of consumption and 828 million people suffered from chronic undernutrition in 1994–1996 (as defined by the Food and Agriculture Organization), up from 822 million in 1990–1992 (FAO 1998). In addition, more than 100 million children are estimated to have below-normal growth (Chen and Kates 1994). On the other hand, there is some good news: populations suffering from famine have been declining, from almost 800 million in 1957–1963 to about 100 million in famine-reporting countries in 1985–1991 (Chen and Kates 1994). These food-poor and impoverished households are primarily located in Africa south of the Sahara and in South Asia, and the number has almost certainly grown (Bread for the World Institute 1999; FAO 1998; *World Disasters Report* 1998). As an illustration, in mid-1997 the FAO (1997: 12) found 29 countries facing acute food shortages requiring exceptional and/or emergency food assistance, and more than half of these countries were in Africa. The recent increase in the number of hungry people in the world is directly connected with vulnerability – the lack of progress in reducing world poverty and the growing gap in income distribution in many parts of the world. In the 1985–1995 decade, the countries with 50 per cent or more of their population undernourished all had stagnant or worsening per capita income. Similarly, where more than 30 per cent of the people were chronically hungry, income growth had stagnated or declined (FAO 1998).

Famine studies have also made clear, as Sen (1981) powerfully demonstrated some two decades ago, that famines can occur without any loss in food availability – that famine incidence is primarily rooted in social and economic processes of growing intensity and not in a malevolent Nature. Similarly, the persistence of hunger in many countries is related not only to the lack of affluence but often to substantial inequalities in society (Drèze and Sen 1990a, b; Downs, Kerner, and Reyna 1991; Rangasami 1985; Sen 1990). Reviewing the literature, Watts (1991: 30) argues that famines can be understood only by analysing (1) long-term structural processes producing patterns of vulnerability, (2) contingent or proximate events producing reductions in food supply and changes in entitlements, and the locally specific social or molecular processes that give them a particular rhythm, motion, and timbre. Bohle (1993) has expressed these relations in schematic form (fig. 7.3) that delineates the interactions among continuing or "baseline" vulnerability, events, societal coping, and resilience. Bohle joined forces with Watts in 1993 to delineate a "space of vulnerability" that determines in their view the causal structure of hunger and famine (Watts and Bohle 1993).

Global urbanization is likely to be a major force in patterning vulnerability to food insecurity in the coming decades. By 2005, more than 50 per cent of the world's population will be urban and over the next 20 years over 90 per cent of the urban growth will occur in cities in the developing world. Food insecurity will increasingly be an urban problem worldwide, adding to the complex of problems that rapid urbanization will involve (FAO 1998).

In regard to climate change and world food scarcity, the IPCC (1996b) assessment of agriculture reached a number of important conclusions. Global agriculture production appears to be sustainable in the face of climate change, although the situation of countries in the low latitudes with low income is worrisome. Many of the countries of sub-Saharan Africa, dependent on isolated farming systems and with low income, are likely to be the most vulnerable to climate change. Other vulnerable regions worldwide are south of South-East Asia, where dependence on the food-production sector is high, and a group of Pacific island nations, where a combination of sea-level rise, salt-water intrusion into water supplies, and increased tropical storms pose substantial threats. In these areas, the capability for adaptation will be central, but current governmental policies and institutional constraints are likely to be major hindrances.

Vulnerability problems are also apparent in the growing global scarcity of high-quality water resources. The global water situation reveals a growing inequality and potential for social conflict (Falkenmark 1996). Over a billion people lack access to safe, clean potable water, 1.7 billion lack appropriate sanitation facilities, some waterborne diseases are still rising or are widely prevalent, groundwater overdrafts are rising, and irrigation development remains highly unequal (Gleick 1998). In his assess-

Table 7.2 **Countries with per capita annual renewable water supplies below 1,700 cubic metres per person per year (as of the mid-1990s)**

Country	Population (millions)	Per capita availability (m^3/person/year)
Kuwait	2.04	10
Malta	0.35	46
United Arab Emirates	1.59	94
Libya	4.55	132
Qatar	0.37	143
Saudi Arabia	14.13	170
Jordan	4.01	219
Singapore	2.72	221
Bahrain	0.52	223
Yemen Democratic Republic	11.7	350
Israel	4.6	467
Tunisia	8.18	504
Algeria	24.96	573
Oman	1.5	657
Burundi	5.47	658
Djibouti	0.41	732
Cape Verde	0.37	811
Rwanda	7.24	870
Morocco	25.06	1,197
Kenya	24.03	1,257
Belgium	9.85	1,269
Cyprus	0.7	1,286
South Africa	35.28	1,417
Poland	38.42	1,463
Korea, Republic of	42.79	1,542
Egypt	52.43	1,656
Haiti	6.51	1,690

Source: Gleick (1998: 113).

ment of national vulnerabilities to water-resources conflicts, Gleick (1998) lists countries in which annual per capita renewable water supplies were below 1,700 cubic metres in the mid-1990s (table 7.2). Falkenmark and Lindh (1993) describe 1,000 cubic metres per person per year as a level which, in the long term, allows a decent and realistic quality of life. Using the "basic water requirement" of 50 litres per person per day, 55 countries in 1990 with a total population of nearly a billion fell below this level (Gleick 1998: 44–45). Meanwhile, the World Health Organization estimates the high human toll from waterborne diseases suggested in table 7.3. The Stockholm Environment Institute assessment of future water-resources vulnerability suggests strong increases in such vulnerability between 1995 and 2025, concentrated in Africa, China, and the Middle East (Raskin et al. 1997).

Table 7.3 **Estimates of global morbidity and mortality of water-related disease (early 1990s)**

Disease	Morbidity (episodes/year or people infected)	Mortality (deaths/year)
Diarrhoeal diseases	1,000,000,000	3,300,000
Intestinal helminths	1,500,000,000 (people infected)	100,000
Schistosomiasis	200,000,000 (people infected)	200,000
Dracunculiasis	150,000 (in 1996)	–
Trachoma	150,000,000 (active cases)	–
Malaria	400,000,000	1,500,000
Dengue fever	1,750,000	20,000
Poliomyelitis	114,000	–
Trypanosomiasis	275,000	130,000
Bancroftian filariasis	72,800,000 (people infected)	–
Onchocerciasis	17,700,000 (people infected; 270,000 blind)	40,000 (mortality caused by blindness)

Source: WHO (1995, 1996a).

The argument implicit in these several examples is that vulnerability is an essential term, and sometimes the dominant one, in both the process (and perhaps "spirals") and outcomes of environmental degradation. If this is the case, then vulnerability is surely a central ingredient in environmental inequities. The next step is to explore how to incorporate vulnerability into equity assessments and public-policy interventions.

Incorporating vulnerability into equity assessments

As shown by the proceedings of the 1992 Earth Summit, as well as studies of environmental security (Homer-Dixon 1999), equity problems often dominate and drive social conflict over environmental risks, thereby shaping the broad patterns of societal response that emerge. It is not surprising, then, that equity problems are shadowy, poorly assessed, or not addressed at all by risk analysis (or other forms of assessment). This is readily apparent in the siting of hazardous facilities in advanced industrial societies in Asia, Europe, and North America, where conflicts over risk-equity issues have made facility siting one of the most difficult of public-policy problems (Shaw 1996). The failure to enact equitable siting programmes for radioactive-waste facilities, for example, has jeopardized the use of nuclear power as an energy source in a number of countries throughout the world.

But it is apparent, as suggested above, that third-generation ethical questions will be a central part of how environmental risks of various kinds will be constructed and negotiated during the twenty-first century. The "environmental justice" or environmental or economic "colonialism" debate over international responses to climate change and to the implementation of the General Agreement on Tariffs and Trade (GATT) will almost certainly become more endemic in future issues in global environmental change. Acrimonious debate has flourished, but also has been intentionally avoided, over the prospect of "winners and losers" owing to human-induced climate change and over the recognition that nations have differing responsibility for the costs of averting the impacts. Indeed, Glantz convened a 1990 workshop in Malta to explore the "winner–loser" issues and encountered significant controversy and opposition even to discussing the issues (Glantz, Price, and Krenz 1990), despite the fact that it was not the purpose of the meeting to identify individual countries that might benefit or suffer from climate change. Again, discussions have displayed a noteworthy lack of carefully documented and focused analysis on risk-equity issues or on the specific methodologies for assessment.

Here we are interested in the relations between vulnerability and equity. Many types of equity problems and considerations exist. Four of these – geographical equity, cumulative geographical equity, intergenerational equity, and social equity – involve "outcome" or "end state" considerations; another – procedural equity – focuses on the fairness of the decision-making processes. Figure 7.4 sets forth a structural model for a systematic approach to equity analysis, suggesting the major types of inequity to be addressed, their relation to principles of social justice, and how they link to policy and management (Kasperson and Dow 1991).

The primary concern in *geographical equity* is with the geographical patterns of benefits and eventual harms associated with a particular set of human activities. Thus, the global pattern of CO_2 and other greenhouse-gas emissions could, in principle, be compared with the pattern of harmful and beneficial impacts that would ultimately occur. In simple cases, such empirical patterns can be used as a basis for making inferences about the obligations and responsibility (if any) that gas emitters and other beneficiaries have for those harmed and the adequacy of legal structures and institutional mechanisms for meeting these responsibilities.

Assessing the geographical attributes of impacts associated with global environmental change is extraordinarily challenging, as the efforts of Working Group II of the second IPCC (1996b) process suggests. Only the broadest types of patterns resulting from climate warming may be estimated; effects may be expected in the geographical location of ecological systems and the mix of species. Forested and montane systems and coral reefs appear particularly vulnerable; water shortfalls will be exacerbated,

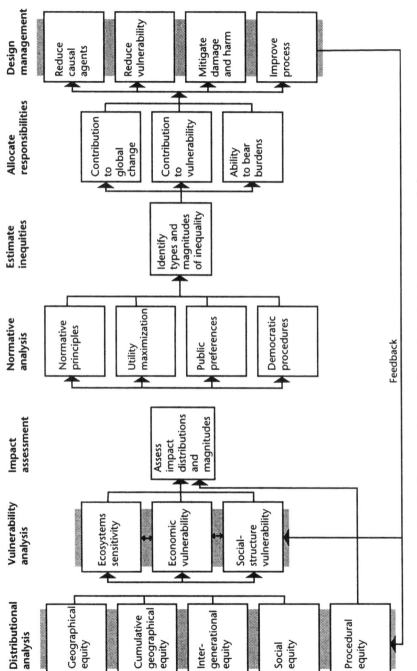

Fig. 7.4 **Framework for equity analysis of global environmental change**

especially in arid and semi-arid areas; flooding is likely to become a bigger problem in many temperate and humid regions; altered growing seasons are likely in rangelands; deserts may well become hotter but not significantly wetter; impacts on inland aquatic systems will probably be greatest in high latitudes; some coastal systems will bear especially high risk; and wide-ranging adverse health effects and significant loss of life are likely (IPPC 1996b, 2001). Even here the evidentiary base is murky for defining international obligations. Generally, widespread agreement exists that, because of the poor spatial resolution of global climate models, the prediction of impacts (and thus the extent of hazard) from global warming at scales smaller than continental regions is unreliable now and is likely to remain so for one or more decades. The second assessment of the IPCC, for example, concludes that "the reliability of regional-scale predictions is still low, and the degree to which climate variability may change is uncertain" (IPCC 1996a: 3). Whereas this does not obviate the need to act, the inability to discern who will be the winners and losers or who holds responsibility for human-induced climate changes is a major stumbling-block to the development of international societal responses. This is a primary reason, of course, why equity issues have often been "swept under the rug" in discussions of climate-change policy (Jodha and Maunder 1991).

What is known quite well from the analysis of past growth rates and CO_2 levels contributed by developed countries, however, is the geographical pattern of past and current *beneficiaries* from fossil-fuel burning. Gross predictions as to how these rates and levels may change over the next several decades under differing assumptions are also possible. So the past and current beneficiaries of human activities driving potential global warming are known at least grossly and they are strongly concentrated in developed countries. This information is highly relevant to allocating responsibilities and mitigation burdens. What can be said about future impacts is that substantial economic dislocation (depending on rates of change) is likely to attend the pattern of global environmental change. The loss of options for potentially affected regions can also be inferred, at least grossly. So, although it is impossible to determine the details of risk distribution, some general indications of future risk and their implications are known. How the baselines are set will be very important: should historical emissions be estimated from the start of the Industrial Revolution, or from 1950 when the post-war economic boom began? Should environmental sinks be included? Should emissions be calculated by country or on a per capita basis? The most compelling arguments, in the view of these authors, involve the need for spatially clear current beneficiaries to respond to spatially ambiguous but threatening environmental futures and the need to address the problems already created before future problems are negotiated. Furthermore, as Agarwal and Narain (1999) argue, it seems morally clear that the people of those nations who have contributed little to the current problem should not have their right to grow economically jeopardized, whereas those nations who have already contributed high amounts of emissions should bear most of the costs. Then, too (as China forcefully argued at the Fourth Conference to the Parties to the UN Framework Convention on Climate Change held in Buenos Aires in November, 1998), all emissions are not the same morally: "survival" emissions are not the same as "luxury" emissions (Jahnke 1999: 3).

Equity, however (it can be argued), cannot be satisfactorily assessed within a particular area, regime, or policy arena at risk. Climate warming cannot be divorced from other types of environmental change and past inequities accrued in the global interactions of peoples, nations, and economies. Indeed, the type of global environmental change (termed "cumulative" global change in the Introduction to this volume) with greatest impact on the planetary environment and developing societies is exactly that which is now receiving the least attention. We refer to this broader type as *cumulative geographical equity*, and it bears a direct relationship to our discussion of vulnerability. Inequities that correlate with other inequities suffered by disadvantaged societies or marginal groups are particularly pernicious because their effects are likely to be synergistic and not simply additive. Previous inequities are also almost certain to have increased, through gradual accumulation, the vulnerability of some nations and groups to global environmental change. The United Nations' *Human Development Report 1998* profiles in a telling way these inequalities in capability:

- private consumption has increased in industrial countries by 2.3 per cent per year in the last 25 years, whereas consumption in Africa has decreased by 20 per cent;
- the richest 20 per cent of the world's population has 86 per cent of private-consumption expenditures, whereas the poorest 20 per cent has 1.3 per cent; and
- the three richest individuals in the world hold assets larger than the combined wealth of the 48 poorest countries.

Thus, it is not surprising that developing countries object vehemently to admonitions from developed countries to reduce future fossil-fuel emissions for the well-being of the global environment (Agarwal and Narain 1999). Cumulative geographical inequities, and accumulated vulnerabilities, may be expected to form the core of many debates over global environmental policy. In addition, those inequities may be highly relevant to the sequencing of management intervention – particularly, which messes should be cleaned up first and by whom?

For *intergenerational equity*, some empirical realities are clearer. Widespread opinion exists, for example, that if present emission trends of greenhouse gases continue they will cause some degree of warming of the global climate during the twenty-first century. Although substantial disagreement exists as to the precise effects that such warming will have on ecosystems, water supplies, agriculture, and sea level, enough indications of substantial potential global harm exist to support international efforts to find prospective solutions, as indicated by the climate treaty (United Nations Convention on Climate Change) forged at the Earth Summit. Because two-thirds of the world's population lives in low-lying coastal areas, for example, sea-level rises of even 0.3–1.0 m would have major effects. In some areas, such as Bangladesh (where half the population lives at elevations below 5 m), the impact could be catastrophic (ch. 11). Although the impacts of global climate change and their precise distribution will remain opaque, assigning responsibility for the accumulated emissions of greenhouse gases is easier, for the causal mechanisms are at least generally (if imperfectly) understood.

Social equity provides another "cut" through the world population to contrast with equity analyses focused on states. It is essential to assess the distribution of impacts according to wealth, class, gender, or stages of development. Natural disasters, as already noted, continue to take a disproportionate and growing toll in developing countries. Many other hazards – such as water stresses and soil erosion – have remained "hidden" from the public view, eliciting neither global media attention nor institutional response (Kasperson and Kasperson 1991).

Finally, there is the *procedural equity* or fairness of the processes that have created global environmental problems and by which they may be resolved. This differs from the other equity analyses described above, each of which addresses the distributions of problem causation and projected outcomes. Here, the concern is with the adequacy and appropriateness of the economic and political structures processes leading to stratospheric ozone depletion, CO_2 emissions, and global deforestation. The issues are straightforward – unlike those in other regimes or policy areas wherein equity considerations have not entered into decision making so that self-interest has externalized damage, placing burdens on other places, social groups, and generations. The adverse impacts on the global environment have only recently gained recognition, so that the state or the corporation has typically been the locus of decision. International institutional mechanisms for incorporating equity, meanwhile, have generally been unavailable or undeveloped. Thus, the absence of procedural equity is one of the few unambiguous attributes of the problems attendant on global environmental change.

Each of these empirical analyses of risk or impact distributions provides a "face" of environmental change and a basis for formulating societal responses. The impact distributions depend, ultimately, not only on the distributional issues outlined thus far but also on the vulnerability and resiliency of the ecosystems, populations, and sociopolitical systems. The allocation of responsibility from a problem must also be decided. We see three primary bases for determining responsibility for mitigating or compensating for adverse impacts: these are (1) the extent of contribution to the environmental change or stress, (2) the extent of contribution to the creation of vulnerabilities among risk bearers, and (3) the ability to bear the burdens for reducing potential harm.

Various principles for responsibility and burden sharing have been proposed and reviewed by Gunnar Fermann (Fermann 1997). One prominent guiding principle for discussions of environmental equity is that "the one who causes the problem should pay to ameliorate it or clean it up." This is the "polluter pays" principle, which has become highly recognized, particularly in Europe, where it was adopted by the Organization for Economic Cooperation and Development as early as 1974. In this formulation, responsibility is equivalent to the contribution to generating an environmental stress or perturbation. Indeed, this perspective has provided the context of the contentious debate, sparked by the World Resources Institute analysis of national accountability, over greenhouse-gas emissions (Hammond, Rodenburg, and Moomaw 1991) and the spirited critique by the Centre for Science and Environment (Agarwal and Narain 1991, 1999). The *equal entitlements criterion*, a complement to the "polluter pays" principle, holds that all human beings are entitled to an equal share in the atmospheric commons (Grubb et al. 1992). Such a principle implies that the North would be required to reduce its emission, not only to the point of global-emissions stabilization but sufficiently to allow developing countries a margin for continuous economic growth (Fermann 1997: 182). The debate, alas, has focused narrowly on emissions and sinks as indicators of blame and responsibility, to the neglect of a broader equity analysis – although the discussion has been enlarged through the 1990s in the debate over implementation mechanisms, the so-called "flexibility" mechanisms of the Kyoto Convention.

There are at least two other perspectives. Those responsible for creating the vulnerabilities among people who will experience the burden of global environmental degradation should, it can be argued, bear some (or possibly much) of the responsibility for the harm that occurs. Because the harm that will occur is a joint product of *exposure* to environmental degradation and *vulnerability* to such change, this claim seems no less compelling than the first. However, in the absence of a social-science consensus on what vulnerability (much less its root causes) comprises, and because differing interpretations are embedded in conflicting ideologies, widespread support for these claims is likely to prove elusive. On the other hand, can anyone doubt that this equity claim underlies much of the debate over strategies for addressing global environmental change or realizing the equity goals embodied in notions of sustainable development?

Secondly, as explicitly recognized in figure 7.4, responsibility can be gauged by the ability to bear the burden of mitigation costs: the cost of emissions reduction, therefore, could well (and morally) be allocated disproportionately to countries on the basis of their ability to pay. This could help shape a compromise agreement between the North and South, although little political support currently exists in countries such as the United States.

Environmental-risk management strategies based upon third-generation ethics are essential within and among individual countries at the global level. The strategies will need to combine intervention into the often complicated and politically contentious system of human driving forces and political economies with a focus on reducing vulnerabilities, thereby enhancing vulnerable groups to be more flexible, more adaptive, and more resilient. The so-called "tie-in strategies" that pursue socially beneficial objectives simultaneously are a good example of a robust risk-equity approach. This approach recognizes that uncertainties may be difficult to narrow (in many areas they will, in fact, grow with further scientific work) and that increasing societal resilience and coping resources may be among the most effective means of reducing adverse global environmental impacts. Future generations, for example, can be provided with resources and skills that would increase their ability to adapt to future climate change. Similarly – whether the intent is to reduce worldwide famine and hunger, to head off the growing concentration of natural disasters in developing countries, or to develop more equitable siting strategies or environmental-protection programmes – interventions aimed at decreasing vulnerabilities and increasing resilience not only may often turn out to be highly cost-effective means of reducing specific risks but also may ameliorate longer-term, cumulative inequities and be socially meritorious for broader reasons. It is not surprising, then, that the G-77 countries, representing most of the developing world, hold the industrialized countries primarily responsible for global warming and have forcefully negotiated for increased financial and technological transfers to the developing world to offset the incremental costs there of adaptation and mitigation (Eleri 1997).

The selection of risk and equity management approaches and mechanisms suggested in the last column of figure 7.4 will be as contentious as choosing among the equity principles themselves. Indeed, it is likely that the equity issues will be fought out in the choice of implementing strategies and mechanisms. This is quite apparent in the struggle over the design and implementation of the Kyoto Protocol "flexibility mechanisms" – emissions trading, joint implementation, and the so-called Clean Development Mechanism (IGES 1999; Moomaw et al. 1999). Espoused by the developed industrialized countries as providing essential flexibility and efficiency in securing greenhouse-gas emissions reduction, these mechanisms are seen by some in developing countries as a means for the industrialized countries to buy cheap emissions from developing countries and to increase further future emissions. Indeed, it is argued that economic slowdowns in some countries (Russia) or restructuring in others (Germany) are already securing "emissions reduction" while a rich country such as the United States could buy its needed emissions credit for a rather paltry US$7 billion (Agarwal and Narain 1999: 18).

In another context, whatever the debate over its effects, GATT has explicitly incorporated equity considerations into its provisions. Developing countries, for example, are given more flexibility and longer transition periods for complying with major decisions reached by the World Trade Organization; they even enjoy certain exemptions from obligations and receive special technical assistance. Further, least-developed countries – the most vulnerable group – receive special attention in terms of rights and decreased obligations "consistent with their individual development, financial and trade needs, or their administrative and institutional capabilities" (Claussen and McNeilly 1998: 9). Similar equity accommodations may be expected in global environmental-risk management mechanisms and tools.

Agenda 21 specifically recognizes the interactions among mounting human-induced environmental change, ecological fragilities, and human vulnerabilities in addressing poverty:

The global effects of poverty contribute to a reduction in the quality of life throughout the planet. The struggle against poverty is the shared responsibility of all nations. Any efforts to protect the earth's environment must take the poverty of the world into full account. An environmental policy that focuses mainly on conservation and protection of resources without consideration of the livelihoods of those who depend on the resources is unlikely to succeed. Equally, any development policy which focuses mainly on increasing production without concern for the long-term potential of the resources on which such production is based will sooner or later run into problems. An effective and sustainable strategy to tackle the joint problems of poverty and environmental degradation should simultaneously focus on resources, production and people. (United Nations 1992: 33)

Unfortunately, neither Agenda 21 nor Capacity 21 sets out a broader conception of vulnerability, the needed research programme to support international Agenda 21 initiatives, or equitable prescriptions for addressing global vulnerability. The second and third assessments of climate change advance the science and politics of impact analysis by placing vulnerability and adaptation centre stage and by making an inventory of a wealth of relevant empirical data and results. However, social science research on vulnerability remains superficial and limited, owing to the weak conceptual and methodological foundations on which it rests. Exactly the same can be said of equity and environmental-justice debates: they are, as yet, at an embryonic stage and lacking a broad global consensus. For the new generation of risk and impact scholars, opportunity beckons.

Part Three

High-risk regions

Editors' introduction

Since the onset of scientific and public concern over global environmental risks in the mid-1980s, attention has focused prominently on the global scale of impact. Indeed, as noted earlier in this volume, *systemic* global environmental risks – such as ozone depletion and global warming – have commanded the lion's share of assessment work and the early international efforts to respond to environmental threats to the planet. At its core, however, global environmental risk is about *variability* – in human driving forces, in vulnerability to change, in the types and magnitude of impacts. Although the assessments of the Intergovernmental Panel on Climate Change (IPCC 1996a, b, c, 1997, 2001) surely advance our understanding of the science of climate change and the general state of knowledge concerning potential impacts, it is also clear that the regional and social patterns of impacts remain unknown at any level of precision. Yet it is the rate and pattern of accumulating stresses and their interaction with place-specific vulnerabilities that will drive the realities as to the eventual severity of these effects and the potential effectiveness of mitigation efforts and human adaptation.

The lesson from climate change is a more general one: risks do not register their effects in the abstract; they occur in particular regions and places, to particular peoples, and to specific ecosystems. Global environmental risks will not be the first insult or perturbation in the various regions and locales of the world; rather, they will be the latest in a series of pressure and stresses that will add to (and interact with) what has come before, what is ongoing, and what will come in the future. Risk is, in short, a byproduct of ongoing processes and developmental paths. Thus, whereas powerful analysis of global environmental risks must surely include an assessment of aggregate trends at the global scale and of what they mean for overall planetary systems and capabilities, just as surely it must be enriched by disaggregated analyses that take account of – indeed, build from – the specificity of particular environments, local cultures, and social institutions. Global environmental-risk analysis needs texture to be credible and to test the general propositions and policy inferences flowing from global-scale scenarios, models, and assessments. Regional and local assessments, in short, are indispensable for global risk assessment (Easterling 1997; IPCC 1996b, 2001).

This section also explores what it means for a region or place to be "impoverished," "endangered," "critical," or at "high risk." Much assessment approaches the subject of risk or endangerment from a narrow perspective. Accordingly, the extent of biophysical change or natural-resource depletion, or perhaps the degree of "pressure" emanating from environmental change, or the size of the ecological "footprint" (Holmberg et al. 1999; Rees and Wackernagel 1994), is often accepted as an adequate indicator of risk or endangerment. Recently, greater attention has also been directed to the potential overwhelming of environmental sinks as well as resource scarcity or resource depletion. Simplistic models of "carrying capacity" have attempted to gauge the implications of such biophysical changes, despite the evident inadequacy of the underlying assumptions that regions are closed systems and that response systems remain passive to growing threat. Assessments of ecological footprints are more sophisticated, estimating the biologically productive area necessary to support current consumption patterns, given prevailing technology and economic processes (Holmberg et al. 1999; Rees and Wackernagel 1994). History is replete with testimony to the extraordinary ingenuity and adaptability of the human species, as suggested by the long record of changing habitation of the Mediterranean Basin or the strides of economic growth, even in the face of long-overdue mitigation efforts, in such once-threatened regions as Lake Erie or the Baltic Sea. The global risk analysis that is needed is one that supplements global-scale assessments with detailed regional studies – but studies that recognize that regions are open systems, that risk is often exported to distant places or to unborn generations, and that humans have a huge capacity to shift economic production systems, to engage in technological innovation – to alter substantially nature–society relations, in short – so that pressures on particular environmental components, natural buffers, or resources ebb and flow with time and over space.

The analysis of high-risk regions also reflects a process dimension.

Earlier in this volume (chapter 4), Funtowicz and Ravetz argue persuasively that decision stakes and human values reside in particular cultures and that the requisite local knowledge, often indispensable in risk assessment, is rooted in particular human communities. As the scale of global risk analysis shifts to those regions and places where damage is likely to be concentrated, so the sources and conduct of risk analysis must follow suit. In short, those who will bear the risks need to be involved in creating and extending risk knowledge, in prioritizing the panoply of regional risk portfolios, and in fashioning regionally sensitive interventions and risk-management approaches.

The chapters that follow address these tasks in several different ways. The editors and colleague B. L. Turner II report on the findings of a broad-based study of nine regions at high environmental risk throughout the world (Kasperson, Kasperson, and Turner 1995). They devote particular attention to the trajectories of change within each of these regions, and particularly to whether these trends have eclipsed (or may soon eclipse) the capabilities of societies within the region to sustain their productive systems and general well-being through some combination of mitigation efforts and adaptation. The results of this comparative study are sobering as to the predicament of areas throughout the world at high environmental risk: in nearly all the regions the rate of environmental degradation is outpacing the rate of human responses. Indeed, the latter appear generically to be tardy, desultory, and ineffective, with a growing price entailed in the progressive loss of the most cost-effective interventions or adaptations associated with early responses. Finally, the research also speaks powerfully to the richness of the regional texture of human driving forces and localized vulnerabilities, questioning the standardized driving forces recognized in the IPAT (impact, population, affluence, technology) hypothesis of environmental impact or the structure that composes many of the global modelling initiatives or efforts at integrated assessment.

In the chapter that follows, N. S. Jodha examines a particular class of high-risk regions – mountain environments. Using the case of the Hindu Kush–Himalayan region, his approach is to detail the special qualities, or what he terms "specificities," that shape the nature of regional risks. Mountain regions are, he argues, areas in which inaccessibility, fragility, marginality, diversities, and special niches shape the structure of risk. In many areas, traditional resource-use systems are undergoing rapid intensification and pressures emanating from population increases, market demands, and state interventions. Jodha examines in detail how these changes affect regeneration and the ability of the system to withstand further stress. Although policy responses in recent history in this mountain region have all contained the seeds of further environmental damage and enlargement of the overall regional risk, Jodha identifies what he terms "dual-purpose" strategies aimed at enhancing the health and productivity of environmental resources.

Whereas Jodha treats the array of human-induced risks, Liverman in chapter 10 examines the particular threats associated with drought and climate change for vulnerable groups and regions in Mexico. Her account demonstrates that analyses centred upon physical climate and drought conditions in the arid north of Mexico provide a very incomplete picture of risk. Agricultural modernization and land reform are interacting with patterns of drought to produce a complex pattern of increasing and shrinking risk; furthermore, in many areas, traditional coping systems are being eroded by the effects of population growth, climate change, and (especially) political and economic transformations. Her account also reveals that the heterogeneity of landscapes can render even a regional assessment of vulnerability too general in scope. Accordingly, even this scale of assessments (in areas of great environmental and socio-economic variability) will require local studies to test regional generalizations. Cross-scale analyses need to become an integral part of spatially explicit risk assessment.

The three chapters that follow all treat potential sea-level rise as a source of regional risk. Taken together, they provide some telling illustrations of interactions within the risk–vulnerability–equity nexus and the nature of the challenges that lie in wait for the well-intentioned global efforts following in the wake of the 1992 Earth Summit. Broadus, in chapter 11, examines two of the highest-risk situations involved in sea-level rise – potential effects on the Nile and Bangladesh deltas. Ikeda, in chapter 13, inquires into the situation of the Sea of Japan, where large-scale human pressures are now emerging and where sea-level rise wields a large future potential to exacerbate those problems and add new ones, in a sea surrounded by very different levels of management capability (and reminiscent of the Baltic in this respect). O'Riordan, in chapter 12, examines early assessment and management responses around the North Sea, a region of affluence and high response capability.

Several particular insights emerge from a comparative review of these three chapters. First, all three authors see a very prominent role for adaptation, given the slow, accumulating nature of the problem. Even in the highly vulnerable Bangladesh and Nile deltas, where Broadus sees a potential economic impact that could be as high as 1–3 per cent of GNP, such a level of adverse effect will occur only if adaptation remains passive and feeble. Accordingly, it seems highly likely that mitigation measures, recombinations of productive factors, and technological adaptations would greatly reduce these potential losses. The countries around the Sea of Japan are early in their responses aimed at reducing environmental threats, but Ikeda sees a strong mitigative potential. O'Riordan, in his discussion in chapter 12 of responses in the North Sea, provides telling evidence of how rapid and effective assessment and management responses can be in regions with high stakes, high expertise, and strong motivation. The debate in the North Sea region is well past whether climate change and sea-level rise may occur; rather, it focuses upon how wealthy countries with a strong capability to protect themselves from this threat should best proceed, in order to minimize potential future losses without disrupting their economies in the process.

All this is telling support for the perceptive observation by Kates, that adaptation is crucial in assessing climate change and that it probably poses no more dangers than numerous other types of environmental change (Kates 1997: 31). Second, the chapters – with the enormous range of societal capability and vulnerability that they cover (from Bangladesh to the Netherlands) – indicate how inequitable global change is almost certain to be. The Netherlands have moved rapidly to consider needed future societal responses. Despite evident physical vulnerability, this small nation has already assessed the outline, scheduling, and magnitude of available responses and begun to commit the resources needed. In Egypt and Bangladesh, by contrast, the assessment process is finally under way but is only limping forward, and coping resources are certainly minimal.

This section carries conflicting messages and implicit warnings for the future. On the one hand, the chapters are encouraging in terms of identifying the types of study that will deepen our understanding of global environmental risks and inform the fashioning of potentially effective international responses under conditions of high uncertainty. Nevertheless, they also suggest that assessment traps exist by which broad-gauge studies may mislead, for important faces of risk experience appear only as analysis moves up and down scale and as risk is increasingly integrated into a holistic and synthetic picture of a particular region or locale.

8

Trajectories of threat: Assessing environmental criticality in nine regions

Jeanne X. Kasperson, Roger E. Kasperson, and B. L. Turner, II

The causes and consequences of human-induced environmental risk are not, as Norberg-Bohm and colleagues make clear in chapter 2, evenly distributed over the earth: they converge in certain regions and places where their impacts may threaten the long-term or even the short-term sustainability of human–environment relationships. Global environment risk, in short, is intrinsically geographical. International agencies, scholars, and the popular press have widely recognized this fact: the Food and Agriculture Organization (FAO 1998) of the United Nations, for example, identifies "acute food shortages" on the basis of "required and/or emergency food assistance;" Russian geographers over time have referred to "red data maps" that show the locations of "critical environmental situations" (Mather and Sdasyuk 1991: 159–175, 224); and publications such as the National Geographic Society's map of "environmentally endangered areas" have heightened public awareness (NGS 1989). Lonergan (1998) uses a broad array of indicators, as reviewed in chapter 1, to create an index of vulnerability, allowing him to map countries vulnerable to environmental stress.

Despite the currency of the terminology, most assessments of "threatened," "endangered," or "critical" places and regions lack well-developed frameworks for determining such categorizations and have certainly drawn very little on the risk literature. Their products, therefore, attest more to the idiosyncratic wisdom of the researcher(s) than to any well-founded conceptualization or definition of the problem. We suspect that the concepts in question will continue to prove important worldwide in the face of global environmental change and the search for sustainable development. Thus, we advocate the development of an analytic framework rooted in risk analysis that is capable of linking various needs but also of leading to soundly based empirical studies and comparative assessments of threats and criticality. Such an integrated framework and the studies following from it promise a more robust understanding of global environment risk and sustainability. To that end, the Project on Critical Environmental Zones (based at Clark University from 1988 to 1998) enlisted research teams from six continents to examine and compare nine regional candidates for criticality. Our book *Regions at Risk* (Kasperson, Kasperson, and Turner 1995) draws upon the nine case studies to explore and refine the concept of environmental criticality.

The concept of environmental criticality

Much recent work focuses only on adverse environmental change (degradation) and the ability of nature to repair the damage. Important as this focus may be, it is too narrow to capture the larger uses to which critical, threatened, or endangered places and regions have been (and will be) put. A wider lens admits a broader concept of criticality – one that recognizes those human–environment relationships that involve environmental changes, socio-economic losses, and the capability of society to cope with both, barring catastrophic or chaotic disjunctures in the environmental processes affected. This perspective reveals various levels of increasing criticality (table 8.1).

Critical, endangered, impoverished, and *sustainable* regions are areas characterized by these situations. To apply such a classification, however, it is necessary to define the characteristics in question, as well as their potential indicators. The complex issues inherent in developing our classification have been addressed elsewhere (Kasperson, Kasperson, and Turner 1995, 1996); a brief summary must suffice here.

Environmental degradation, in our usage, refers to the drawdown in the materials and flows that maintain the environment in human use. Such changes are complex, of course, and, in making them, the user or society at large may redefine what the base environmental condition is or should be. Our definition, therefore, is not an ecocentric one, and is very much linked to the human-use system(s) in operation in the environment. All the same, however, indicators of degradation – such as soil loss, declines in water quality, or atmospheric pollution – are environmental.

Human well-being refers to the wealth and health of occupants of the environment, focusing particularly on material standards of living and

Table 8.1 **A classification of regional environmental situations**

Environmental criticality: the extent and/or rate of environmental degradation preclude the continuation of current human-use systems or levels of human well-being, given feasible adaptations and societal capabilities to respond

Environmental endangerment: the trajectory of environmental degradation threatens in the near term (this and the next generation) to preclude the continuation of current human-use systems or levels of human well-being, given feasible adaptations and societal capabilities to respond

Environmental impoverishment: the trajectory of environmental degradation threatens in the medium to longer term (beyond this and the next generations) to preclude the continuation of current human-use systems or levels of well-being and to narrow significantly the range of possibilities for different future uses

Environmental sustainability: the environment can support the continuation of human-use systems, the level of human well-being, and the preservation of options for future generations over long time periods.

Source: Kasperson et al. (1995: 25).

quality-of-life indicators. Such indicators range from income (household or per capita) to access to high-quality water, and from nutrition to the longevity of life.

Finally, our conceptualization of environmental criticality involves an assessment of the capacity of the user or society to respond to, and to mitigate, environmental drawdowns that affect human well-being. Here we are concerned only with those responses that maintain the human occupation and use of the environment. The history of human responses of this kind is largely one of resource substitution, be it through clearing a new field in swidden (burning and clearing) cultivation (land substitution), or by importing energy and using technologies to make production and consumption more efficient, or by making interventions aimed at averting or mitigating the growing damage. Variable wealth and technological capacity affect the capacity of locales and regions to respond to the changes in question and (although we deal with the region) we recognize that this capacity may vary greatly among the different resource users and occupants of the region; depictions of regional risk, therefore, must be carefully disaggregated to reveal complex patterns of differential risk (and gain).

Regional dynamics and trajectories of change

To speak of various classes of growing regional criticality and the transition of regions from one risk level to another evokes the notions of dynamics and trajectories. "Regional dynamics of change," in our view, are conditions and processes that shape the changing nature of human–environment relationships of the area. The progression of these relationships on a distinctive path or course is, in our usage, a "trajectory of change."

The relentless globalization of economy and environment alike means increasingly that no region exists in isolation – all places are linked to one another, if only indirectly. Economic production systems in one region may degrade forests in another, and the capacity of a region to cope with environmental degradation often hinges on global markets or politics. Determination of a region's criticality, therefore, requires examinations of relationships at different scales and from varying vantage points, and over differing historical periods – not dissimilar to the "regional political ecology" espoused by Blaikie and Brookfield (1987) but without, as in that view, axiomatically elevating political economy to primacy or focusing unduly on the individual resource manager prior to careful empirical analysis.

Our analyses suggest that the situations of impoverishment, endangerment, and criticality, as well as the regional dynamics that cause them, take widely different forms and arise from different circumstances in diverse regional contexts, as the recent impact analyses of the Intergovernmental Panel on Climate Change (IPCC 2001) have made abundantly clear. In short, no simple evolutionary pattern or set of regional dynamics holds true across all regions. The relationship between growing environmental degradation and changes in the wealth and well-being of inhabitants varies markedly from region to region and assumes protean forms and expressions.

Linking critical regions to global environmental change and sustainability

Identification of critical regions is important in its own right. Moreover, it requires a breed of analysis that simultaneously provides many other insights. Thus, *Regions at Risk* seeks insights about the scope and nature of the human driving forces that are causing environmental change, as well as about the societal recognition and response to the emerging level of risk.

Each of our studies identified key human dynamics that gave rise to the trajectory of change in the region at hand and aimed to connect these dynamics and trajectories in an overall historical explanation. Given considerable temporal and spatial variability in both the human driving forces and the processes of environmental change, the interpretations of each study team accordingly varied in emphases. The inclusion of attention to interregional and global linkages helped to assess the growing dependency of many regions on world markets and shifting global demand as part of what we term the "regional dynamics of change." From our studies, it is clear that environmental degradation and emerging overall regional risk can be explained only with scrupulous attention to the restructuring of the regional economies over time and to how such changes interact with differential patterns of human vulnerability and well-being among subgroups or subregions.

Each study also addressed the nature of societal responses in the face of emerging trajectories of environmental risk, focusing especially on the differential societal responses at various scales or managerial loci. Such assessments can characterize the types, number, and effectiveness of responses undertaken at these various levels, as well as the societal options and constraints under which they operate. This analysis may put the individual environmental manager at "centre stage," as in the "farmer first" approach (Chambers 1989); alternatively, it may (as in much risk research) begin with state policies, international regimes, or multinational corporations. Risk analysts, it should be noted, often begin with the structure of the hazard or the consequences of environmental change and work backwards, or upstream, to identify and assess the driving forces of change (Hohenemser, Kasperson, and Kates 1985). Various analytical starting points are equally viable and valuable, as long as the analysis is sufficiently comprehensive to address both the complexity of factors shaping the trajectories of risk and the subsequent range of societal responses.

Selection of regions and comparative methodology

Regions at Risk addresses nine case studies of regions selected largely on the basis of the following criteria:

1. *The region must have been widely identified in the literature as one experiencing environmental crisis.* Candidates abound. Many areas around the world have been identified by someone (or some organization) as experiencing or nearing the linked decline in environmental condition and socio-economic well-being captured in our definition of criticality. However, this is not to say that, a priori, any region so identified in fact meets the conditions of criticality as we define them.

2. *The set of regions selected should represent a range of environmental and political-economic conditions.* Our framework focuses on human–environment relations, which vary enormously worldwide. The varia-

tion involves not only different environments but also widely differing human and ecological vulnerabilities to such changes and differing capacities to respond to growing risk. To capture this range, as well as to avoid any impression that critical environments are solely a third-world problem, we drew on a simple matrix of major life zones: economic structure (capitalist-dominated, noncapitalist-dominated), political structure (complex democracies, social democracies, centrally planned states), affluence (advanced industrial, industrial agrarian), and population density (from frontiers to megalopolises). We sought, in short, in our nine case studies to cover a substantial sample of this range (but not all combinations or permutations, obviously).

3. *The regions chosen should be ones in which ongoing research could provide a base upon which the case studies could draw and build.* Although relevant environmental research has been carried out in most candidate regions, we favoured regions in which individual (or teams of) researchers were already actively engaged in the kind of work that would facilitate our study.

These guidelines led us to select nine regions (fig. 8.1):

1. *Amazonia* – a region of frontier agricultural and pasture development that has experienced the loss of old-growth tropical rain forest, soil degradation, and other related environmental change, with major implications for global biodiversity, possible links to global climate change, and associated concerns about sustainability (Smith et al. 1995a, b).

2. *Eastern Sundaland* – another tropical forest frontier with the same environmental and development implications as those for Amazonia, but whose major source of risk is directly linked to international logging (Brookfield, Potter, and Byron 1995; Potter, Brookfield, and Byron 1995).

3. *Ukambani* – a semi-arid region of eastern Kenya where agropastoralists in dry forest and savanna lands have experienced colonization; widespread displacement and resettlement; male out-migration; and a transition to sedentary, permanent agriculture accompanied by deforestation, soil and water degradation, recurring drought, and continuing state intervention (Rocheleau, Benjamin, and Diang'a 1995).

4. *Nepal middle mountains* – a rugged high-altitude, high-energy region long containing substantial populations of smallholders who are currently shifting to commercial production under conditions of market development and rapid population growth, leading to major deforestation and soil loss and possibly to major downstream flooding and sedimentation (Jodha 1995; see also chapter 9, this volume).

5. *Ordos Plateau of China* – an arid region of Inner Mongolia, targeted for commercial development in a centrally planned economy and in-

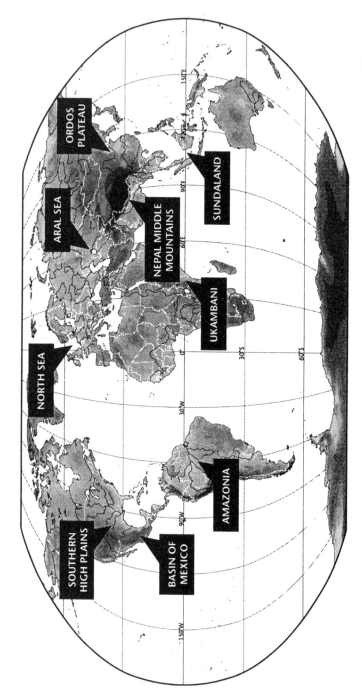

Fig. 8.1 **Map of case studies of environmentally threatened regions. Source: Kasperson, Kasperson, and Turner (1996: 6)**

vaded by large numbers of smallholder farmers, leading to massive soil erosion and sandification (Jiang 1999; Jiang et al. 1995).

6. *Aral Sea basin* – another arid region of a former centrally planned economy of the Central Asian republics in which one of the largest state-ordained irrigation schemes in the world, along with rapid population growth, has severely damaged the sea and the lands around it, with far-reaching adverse impacts on human well-being (Glazovsky 1995).

7. *Llano Estacado* – a semi-arid region of the US Southern High Plains where large-scale irrigated agriculture by wealthy farmers has drawn down the aquifer, increased soil erosion, and moved the region inexorably to higher (and possibly catastrophic) future risk over the very long term (Brooks and Emel 1995, 2000).

8. *Basin of Mexico* – one of the largest concentrations of people and of industrialization on earth, located within a geologically closed basin, with rapidly increasing contamination of water and air, stress on water supply, and dependence on surrounding regions (Aguilar et al. 1995; Ezcurra et al. 1999).

9. *North Sea* – in the mid-high latitudes, a common-pool resource, surrounded by one of the longest-industrialized and wealthiest areas in the world, which has become a sink for a mass-producing and consuming society (Argent and O'Riordan 1995; see also chapter 12, this volume).

Table 8.2 provides a comparative overview of these nine regions. Each study was undertaken by an interdisciplinary team of researchers following a common research protocol developed through the cooperation of the project study group at Clark University and the case-study team leaders. The protocol provides guidelines for defining the region, the time-scale of analysis, the major questions and topics to be addressed, the means of organizing each study, indicators to be used, and common forms of documentation. Each analysis of regional dynamics of change assessed human driving forces, historical trends and geographical patterns of environmental change, socio-economic vulnerabilities, impacts on human well-being, and major societal adaptations and responses. Case studies paid particular attention to the regional trajectories of change over time, addressing the various components involved with the regional dynamics of change. Throughout, we sought to achieve as much comparability as possible among the various case studies. This goal proved difficult, in part because so many of the data required new or reformulated information that could not be generated within the constraints of the study and available funds. None the less, the studies share a strong common focus, analytic approach, and key concepts that provide insights into the nature of emerging environmental threats in nine quite different regions spread throughout the world.

Table 8.2 **Characteristics of the study regions**

Region	Physical environment	Population	Forces for change	Threats
Amazonia	Tropical forest, riverine	Rapidly increasing; in-migration	Frontier extraction; development	Rain-forest destruction; non-sustainable uses
Eastern Sundaland	Tropical forest	Rapidly increasing; in-migration	Frontier extraction; development	Rain-forest destruction; non-sustainable uses
Ukambani	Semi-arid brushland	Rapidly increasing; seasonal migration	Marginalized land; inadequate access to technology	Desertification; marginal uses
Nepal middle mountains	Steeply sloped, rugged terrain with monsoons	Rapidly increasing, densely settled; seasonal migration	Marginalized land; inadequate access to technology	Deforestation; erosion; marginal uses
Ordos Plateau	Arid, sandstone plateau	Densely settled	Agricultural expansion; livestock intensification; command economy	Desertification
Aral Sea	Inland sea and exotic rivers in arid region	Rapidly growing	Massive irrigation; command economy	Death of sea; salinization; unsustainable irrigation system
Llano Estacado	Short-grass prairie	Stable growth; out-migration	Affluence; high-technology (intensive pump irrigation)	Desertification; mining aquifer; unsustainable agriculture
Basin of Mexico	Enclosed, highland basin; metropolitan	Rapidly growing; extremely dense settlement	Urban-industrial complex; affluence	Groundwater depletion; air quality; health problems
North Sea	Ocean; intensively used coast	Stable population; densely settled	Urban-industrial complex; international commons; affluence	Coastal and water pollution; depletion of ocean resources

Source: Kasperson, Kasperson, and Turner (1996: 7)

Regional trajectories of change: Towards criticality?

Although lacking thorough and comparable quantitative data, the nine case studies do provide information about, and insight into, broader human–environment trends threatening the human habitation and use of the regions. They permit estimation of general aggregate trends in environmental change, per capita well-being, and the sustainability of the current human–environment interactions for each region. These aggregate trends are depicted graphically in figure 8.2, where the horizontal axis is the time period (approximately 50 years ago to present) of the trend lines and the vertical axis is the amount of aggregate environmental change, the level of per capita well-being, and the judged long-term sustainability of human uses. The height at which the trend lines begin indicates the relative position of each at the time of the beginning of the study (some regions had already experienced significant change).

Two cautions are in order. First, we emphasize *aggregate* conditions because we are dealing with each region as a whole, not with specific individuals, subgroups, subregions, or ecosystems. It is important to note that these aggregate trend lines conceal marked internal variation that may be important in interpreting the human consequences of environmental change or in structuring the regional dynamics of change that determine the trajectory toward greater endangerment or criticality. Thus, a full analysis of the regional dynamics of change, as noted above, requires a series of disaggregated analyses to provide a richer portrait of trends among subgroups, subregions, or differing indicators of environmental change. Well-being per capita may be stable or rising in the aggregate, yet falling among certain social groups that constitute a significant proportion of the regional population. This impoverishment of particular social classes or groups may interact with existing vulnerabilities and ecosystem fragility to produce "spirals of land degradation" (Blaikie and Brookfield 1987; Kates and Haarmann 1992; Rosenzweig and Hillel 1998). Second, the specifics of each trend line are debatable, based as they are on our judgements concerning each case study and comparisons among them rather than on specific quantitative measures. We are relatively confident, however, in the overall shapes of the trajectories presented in figure 8.2, and herein lies their major significance: they chart the broad courses or paths of regional changes in environment, human well-being, and the long-term sustainability of human uses of the environment in the nine regions.

Human uses of a regional environment require certain inputs and offer certain economic returns. A rise in the cost of the human inputs or a decrease in the value of the economic returns will render the complex of human use unsustainable, especially if the deficit is not made up from

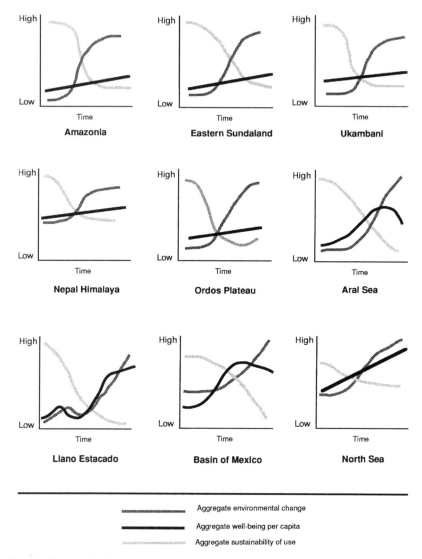

Fig. 8.2 **Regional changes in environment, human well-being, and sustainability of uses. Source: Turner et al. (1995a: 523)**

some other source. A region is in a critical situation, in our usage, if human-induced environmental degradation has made the current human occupation and economic production system unsustainable, either at present or in the near- to medium-term future, and if no alternative use is available to sustain the same level of wealth and well-being.

In this view, sustaining a region over time depends primarily on two factors: these are (1) *environmental recoverability* – the economic return available from the human uses of the environment, and (2) *the costs of mitigation and/or substitution* – the net costs involved in substituting alternative economic production systems and a changed natural-resource exploitation pattern and/or the costs to intervene to mitigate the environmental damage. Changing nature–society conditions can result from a decline in environmental recoverability owing to the drawdown of the environmental resource base, possibly exacerbated by a decline in the ability to sustain the current productive system for any of a number of reasons (such as a decline in the political power of the region, a decline in regional wealth, or changes in trade). The three states, or stages of change, that represent increasing environmental threat – impoverishment, endangerment, and criticality – are characterized by different trajectories and relationships in the two trend lines. Figure 8.3 suggests how

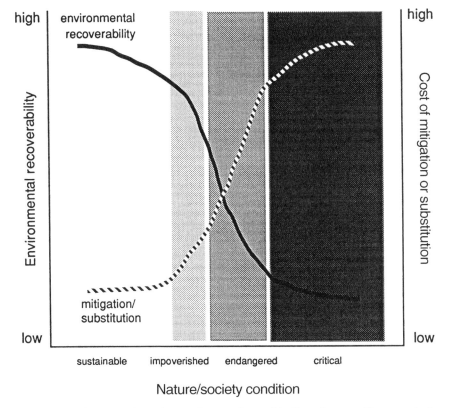

Fig. 8.3 **Regional trajectories and emerging criticality. Source: Turner et al. (1995a: 527)**

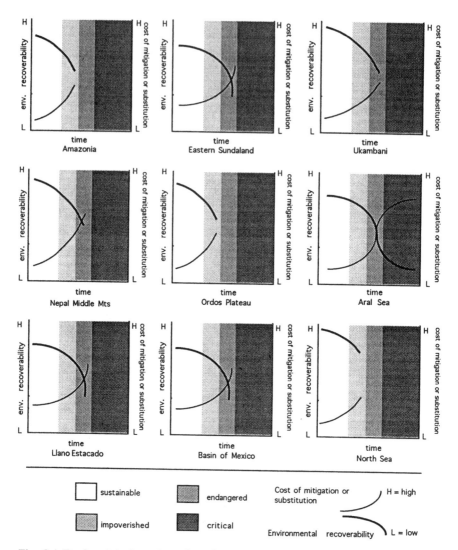

Fig. 8.4 **Regional trajectories of environmental recoverability and the costs of mitigation and/or substitution. Source: Turner et al. (1995a: 528)**

these trends relate to emerging environmental criticality in any given situation. Figure 8.4 depicts our judgements as to the trends in our nine regions and their trajectories toward greater endangerment and criticality.

It may be that all nine of the regions in this study (and perhaps many of the inhabited regions in the world), if recent past or current trends were to continue unabated, would eventually enter some stage of impoverishment or endangerment. In this sense, every region is one over which some level of long-term endangerment hovers. A mid-1990s assessment of 21 key environmental indicators in nine industrialized nations concluded that "overall, the quality of the environment in all of the countries studied has declined over the past twenty years" (Alperovitz et al. 1995: 2). Increasing fresh-water scarcity, as reviewed in chapter 7, is widely apparent over the globe (Gleick 1998, 2000). In assigning a regional case to one of our categories of impoverishment, endangerment, or criticality, we have judged the likelihood that recent trends will continue, whether they are amenable to alteration by feasible adjustments, and whether such human adjustments are possible in a time frame that would prevent the onset of severe consequences. Obviously, such judgements are debatable and we regard our analyses as heuristic and suggestive rather than conclusive. Ideally, such assessments should move from these subjective assessments to more quantitative expressions of diverse indicators, as Jiang (1999) has recently accomplished for several indicators and as Lonergan (1998) has accomplished for his 12 key indicators.

Human adjustments, often leading to partial recovery from the environmental damage and to total recovery or even improvement in human well-being, have been common throughout human history. Viewed over the short run, the trajectories of environmental and social change may appear to be inexorable; however, over the long run, regions typically display a fluctuating long-wave path in which environmental deterioration and social well-being are linked, only to be followed by a social rebound (Ayres and Simonis 1994; Berry 1991; Whitmore et al. 1990). Similar patterns are apparent in long waves of regional economic change (Marshall 1987).

Comparative findings

Human driving forces

Population growth was a major driving force in six cases (Amazonia, the Nepal middle mountains, the Ordos Plateau, the Aral Sea, the Basin of Mexico, and eastern Sundaland), all involving regions or regional systems of underdevelopment. Such growth, of course, was usually the product of other forces, largely those of political economy and structure, that stimulated in-migration or accounted for natural growth within the region.

Technological change was also important. Changes in technological capacity permitted new human uses of the regional environment and enhanced the overall level of technology available to users. Technological innovation supported the introduction of modern infrastructures and enlarged means of production, usually to frontiers or sparsely inhabited areas (Amazonia, eastern Sundaland, the Nepal middle mountains, the Ordos Plateau, and the southern High Plains of the United States). Damage initially followed from technologies unsuited to specific environments, in some cases producing initial impacts that were reduced over time by the adoption or development of new techniques (as in Amazonia). At least three broad types of interactions occurred: these were (1) poorly constructed, inefficient, and highly polluting technologies (the Ordos Plateau, the Aral Sea, the Basin of Mexico); (2) state-of-the-art technologies that exacerbated impacts, intended or not (the North Sea, the southern High Plains, the Basin of Mexico); and/or (3) technologies that abated problems through resource or economic substitution, such as water transfers or changes in fuel types (as in the Basin of Mexico).

Affluence was directly identified as a driving force only for the North Sea case (where high per capita consumption has had obvious impacts) and in the Basin of Mexico (where high per capita use of cars, for example, contributes to major pollution problems). It was implied for the southern High Plains, in the sense that local producers who are seeking to sustain and improve an already very high standard of living drive economic activities there. Increased per capita consumption, of course, is occurring in several other cases (especially in rural, developing areas), but its contribution as a driving force has been relatively small.

The opposite of affluence, or *poverty*, appears as an important driving force in four cases (Amazonia, the Nepal middle mountains, the Basin of Mexico, and Ukambani). In each of these cases, poverty drove inappropriate resource use that exacerbated resource consumption or depletion. In rural areas, this involved the intensification of land use without appropriate inputs (as in the Nepal middle mountains) and, in the case of Mexico, was a "push" stimulus for migration. Finally, if poverty is expanded to include country-scale "underdevelopment" or the imbalance of trade and debt, then all but two cases (the southern High Plains and the North Sea) witnessed the influence of poverty on state policies.

Two broad *political-economic forces* operated in all cases through state policy and the economy. The role of state policy was obvious and direct in centrally planned economies (the Aral Sea, the Ordos Plateau, the Basin of Mexico), but it was also important in poorly developed marginal areas and frontiers having little power *vis-à-vis* the national government (Amazonia, eastern Sundaland) or seeking state assistance to develop (the Nepal middle mountains). In the developed-world cases, the state operated through policy more directed at individual resource managers, such as subsidies for farming in the southern High Plains or the lack of regulation in the North Sea.

The economy operated as a driving force in several major ways.

First, the region provided a desired resource that would not necessarily have been exploited without external intervention (eastern Sundaland, Amazonia). Related to this was resource use predicated on external demands regulated through the market or government controls (the Aral Sea, the southern High Plains, the North Sea). Second, shifts from a precapitalist (traditional) towards a capitalist (market) economy heightened environmental change because of the transaction costs of changing the rules of resource allocation (the Nepal middle mountains). Finally, two regions – the North Sea and the Ogallala aquifer (southern High Plains) – suffered from inadequate or improper valuation of environmental impacts on a common-pool resource. (Air pollution in the Basin of Mexico may also reflect a common-pool situation.)

Beliefs and attitudes are usually difficult to identify as autonomous forces of change. Nevertheless, the case studies identified three forces that qualify for consideration here. The first is a "frontier" outlook, prevalent in Amazonia and eastern Sundaland (and once prevalent in the southern High Plains), that emphasizes rapid development of currently "underutilized" lands. The second is ethnic identity and associated religious beliefs that serve to increase natural population growth and hence demands on natural resources (as in the Aral Sea region) and to exacerbate conflicts over natural resources. Finally, several cases identified mass consumption and political corruption as important forces. In the cases involving mass consumption, such as the North Sea and Basin of Mexico, society has underestimated – perhaps denied – the connections between the level of consumption and its environmental consequences, whereas in the cases involving political corruption (eastern Sundaland, the Aral Sea), many layers of society have accepted behaviour that undermines the protection of nature and the prospects for long-term sustainability.

Our results stand in contrast to the first-generation global aggregate assessments (Meadows et al. 1972; USCEQ 1980; WCED 1987), as reviewed below in chapter 16. These works typically emphasize the triad of population, affluence, and technology (PAT) as the major driving forces of environmental degradation and even human well-being (or I, for impact). It takes the form of an equation, $I = PAT$, that has been both influential and controversial (for a history of the IPAT argument, see Holdren 1993; Meyer 1996: 23–24; Meyer and Turner 1992). The differences between these global assessments and our own are not necessarily irreconcilable, given that the spatial and temporal scales and the mode of analysis used affect the findings (Kasperson, Kasperson, and Turner 1996). At the global scale, the $I = PAT$ formula sums up the human pressures on natural resources in a closed system, but it says nothing about where within the system those pressures will be registered, or how they may be mediated. Regions are not closed systems, and, as our cases show, the export of demands on natural resources and environmental degradation to distant areas (or distant generations) is a common occurrence.

Indeed, driving forces external to a region are typically as important as (or more important than) internal forces. The external driving forces tend to be those of resource extraction or production for distant national and international markets and those causing out-migration from densely settled areas. In every one of our nine regional cases, resource extraction or production for external consumption was an important factor; only in the highly urbanized-industrialized regions (the North Sea and the Basin of Mexico) was consumption largely internal to the region a major force of change. Furthermore, in every case, state policy and institutions affecting resources and environment were key forces of change, and, with the exception of a few regions controlled totally from afar, external demands operated in the context of state policy and institutions, illustrating the confluence of external and internal forces of change. One institutional force that stood out was legal and institutional structures affecting common-pool resources, in which minimal or no rules of allocation or use operated.

The $I = PAT$ formulation may be employed to assert that poverty, rather than affluence, principally drives environmental degradation (Kates and Haarmann 1992; Leonard 1989). Our cases did repeatedly emphasize that the poor or marginal land users exacerbated environmental degradation because resource managers lacked access to capital and technology to use their resources efficiently and sustainably (e.g. Amazonia, the Nepal middle mountains, and the Basin of Mexico). Our cases of agglomerated environmental stressors (the North Sea and the Basin of Mexico, and perhaps the southern High Plains) did, however, illustrate the important role of affluence in increasing per capita consumption and accelerating associated drawdowns of nature, to which other studies have also recently added (OECD 1997a; Stern et al. 1997).

Our studies indicate a gap between the forcing functions depicted through specific case studies versus those emerging from statistical assessments conducted on continental and global scales (Meyer and Turner 1992). This gap may arise from data problems or biases on the part of practitioners whose favoured perspectives direct them either to IPAT analyses, often on the global scales, or to non-IPAT microlevel studies. Alternatively, the gap may arise as a result of inadequate attention to the details of linking local to global forces in a coherent and systematic manner. In any case, *Regions at Risk* clearly identifies this divergence in analysis, an issue that is increasingly making its way onto global-change and other research agendas (Turner et al. 1995).

Vulnerability

The nine regional studies explore human vulnerabilities to environmental change in less depth than they do the human driving forces of change. None the less, they do indicate several common features, mostly involving the social and economic dimensions of vulnerability.

Some socio-economic processes that threaten to lower regional wealth and well-being can at the same time spawn ecological recovery. In the US southern High Plains, for example, competition from other cotton producers in the world economy is one source of downward pressure on regional wealth. To the extent that it leads to land being taken out of production, it reduces soil erosion and the drawdown of aquifers. A past case of regional economic and demographic decline (resulting in reduced environmental stress) stemming from similar external competition is that of Amazonia, which grew wealthy from rubber exports but experienced a dramatic and long-lasting collapse in 1912 when the successful cultivation of rubber in South-East Asia undercut its profits (Smith et al. 1995a, b).

The nine regions also illustrate, interestingly, some obvious ways by which vulnerability has declined over time. Technological interventions have reduced the vulnerability of some regions to fluctuations in the physical environment. The southern High Plains is one part of the North American High Plains where widespread adoption of groundwater irrigation has insulated productive activities against the droughts that caused several major collapses of regional population and economy in earlier decades (Brooks and Emel 1995, 2000). Advanced pumping technology delivers water to a water-stressed Basin of Mexico, while it also removes pollution from the basin. (It should be noted, however, that these technological solutions bring with them even potentially greater environmental vulnerability – to the loss of irrigation water in the southern High Plains and the sustained growth of Mexico City in the basin.) A different form of reduced vulnerability shows up in the cases of the North Sea and southern High Plains: the increased wealth of the societies in which the regions are embedded has expanded societal capability to mitigate environmental degradation and its consequences. Finally, improved access to the world market expands the range of options for production systems and for interventions to ameliorate environmental damage – creating new uses and demands for regional natural resources once little valued, allowing natural resources to be imported as substitutes for ones locally exhausted, and making possible the adoption of innumerable technologies (developed elsewhere) to mitigate environmental problems or to increase the efficiency of natural resource uses.

In the nine regional cases, however (perhaps because they represent a sample weighted toward failure rather than success), processes that have increased vulnerability appear more prominently. The two major types of environmental change that increase human vulnerability are those that (a) immediately alter the physical characteristics of the environment, and (b) link the use of natural resources more closely with a highly variable, unreliable, or exploitative political economy.

Our regional case studies suggest that environmental degradation involves fragile or vulnerable settings such as the following: (1) extensive old-growth forests (Amazonia, eastern Sundaland); (2) arid lands or water sources in arid lands (the Ordos Plateau, the Aral Sea, the southern High Plains, Ukambani, and, partly, the Basin of Mexico); (3) high-energy environments (e.g. the Nepal middle mountains); (4) common-pool resources (the North Sea, the southern High Plains, the Aral Sea, and, partly, the Basin of Mexico). The Basin of Mexico dramatically illustrates the significance of the biophysical environment and regional ecology: here, an enclosed basin and thermal inversions exacerbate air pollution; aridity accentuates water-delivery problems; subsidence of an ancient lake bed due to groundwater depletion creates basic problems of land-surface stability; and steep slopes provide a high-energy source for erosion and other degradative processes. These environmental or natural-resource characteristics do not in themselves create environmental change or degradation, of course, because that is also a function of the economic production systems and institutions.

Such physical changes may increase the potential for environmental surprises and disruption, reduce the buffering and regenerative capacity of the land, or deplete natural resources so that the ability to cope with human-induced stresses is eroded. The degree of regional dependence on particular human uses of the environment is a key dimension of vulnerability in all of the cases. Long-term ecological vulnerability associated with the loss of genetic resources is specifically relevant in the cases of the North Sea and of eastern Sundaland.

Three aspects of changing regional environmental conditions suggest an increasing potential for higher or catastrophic losses and thus merit particular scrutiny, as follows:

1. *Vulnerability and overshoot.* In the Basin of Mexico, increasing demands are overwhelming finite sources of non-substitutable resources, such as clean air and water. The momentum of growth is outpacing the ability both to procure other supplies of necessary resources and to limit demand or mitigate damage. Meanwhile, the scale of the problems is escalating relentlessly, suggesting an "overshoot" situation, as assessed by Meadows and her colleagues (Meadows et al. 1972; Meadows, Meadows, and Randers 1992).

2. *Market conditions and overcapitalization.* Several regions confront

narrowing options and growing demands in dynamic global markets. The capital investments involved in the unsustainable production levels of cotton (the Aral Sea region, the southern High Plains) and timber (Amazonia, eastern Sundaland) products for the world market create livelihood systems dependent on those markets yet vulnerable to their fluctuations. Growing regional economic dependence on these environmental resources portends higher risks; transitions to other options and greater diversity, meanwhile, are increasingly painful. The difficulties faced by the fishing industry in the North Sea illustrate this case for a narrow portion of a regional population.

3. *Loss of options and safety nets.* For part of the Nepal middle mountains, environmental degradation has narrowed agricultural options and changes in the production system have eroded the capabilities of existing risk-management systems while failing to replace them with new systems. Environmental changes in Amazonia hold an increased potential for environmental surprises, and the extent to which the new productive systems will provide new buffers and sources of security is uncertain. Meanwhile, failures are likely to have far-reaching consequences, owing to the lack of entitlements and of back-up societal and livelihood options.

Societal response

Societal responses are central to the environmental futures of the regions because the trajectories of regional change, as noted above, depend heavily upon the capabilities of society to accommodate deteriorating situations and on the actions that are undertaken to mitigate damage and to alter the trajectory of deteriorating nature–society relationships. Such corrective interventions can occur at various levels and scales, ranging from individual resource managers to the state and to international governance regimes. Further, the interventions may be "downstream" (aimed at reducing exposure or ameliorating consequences), "midstream" (aimed at reducing environmental stresses or vulnerability to damage), or "upstream" (involving interventions into the more fundamental human driving forces that generate environmental stresses or shape social and economic vulnerabilities) (Kates, Hohenemser, and Kasperson 1985).

Conclusions and larger issues

We argue in this chapter, consistent with the discussion in this book in chapter 2 characterizing global environmental risk, that it is essential to move from current idiosyncratic treatments to a more systematic and comparative approach to understanding critical, endangered, impoverished, and sustainable human-environment situations within the world's regions. Our approach needs improvement and extension, of course, particularly in regard to still greater comparability of evidence, quantitative documentation of trends, and more detailed disaggregation of analysis. We believe, however, that our initial experiment shows much promise for the basic conceptualizations and analytic approach.

In summary, we find that from a human-environment perspective, few regions in our study have entered a stage of criticality, but even fewer suggest the prospect for sustainability. Most regions are in a state that is best described as impoverished or endangered. We cannot conclude that the existing regional trajectories of change necessarily imply that most regions will soon move into a state of criticality. The human adaptations and adjustments that will surely appear to combat environmental drawdowns may surprise us with their resilience. Intensive human occupation and use of regional environments rarely proceed without a concomitant drawdown of nature, but history is replete with cases of environmental recovery, vigorous human adjustments, or both, that make the altered environment compatible with new productive systems. The next stage of analysis should begin to identify those conditions that lead to criticality and those that halt at earlier stages of damage and revert to less threatening stages (Lee 1986). We conclude with several broad observations concerning societal responses to global environmental degradation experienced in diverse corners of the world.

Global environmental change through a regional lens

Discussions of global environmental change, as we have noted, abound with talk of long-term population growth, aggregate loss of global biodiversity, world energy use and depletion of non-renewable resources, water-resources depletion, and international summit conferences and institutions to combat these large-scale planetary changes. However, *environmental problems take on a different hue at the regional scale* (Easterling 1997). It is not, of course, that the issues and their causes – so apparent at the global scale – are not present in the world's regions; they are. Nevertheless, the change depicted in the smooth, averaged curves that portray global trends is placed in much sharper and variegated relief in the world's regions, and the mosaic and processes of change often appear very different from the averaged (or even dominant) global trends. The regional landscapes, economies, cultures, and variegated responses assert their distinctiveness in challenging broad global assumptions and interpretations.

The picture of environmental change in the nine regions treated in this analysis reflects the fact that these regions were selected because they were widely regarded as environmentally threatened. Thus, they may not depict the more general nature–society situation that prevails amidst the diversity of the world's regions. However, the trajectories of change in these threatened areas trumpet a warning – one that supplements recent surprises such as stratospheric ozone depletion or mounting CO_2 levels in the atmosphere at the global scale. In nearly all these regions, trajectories of change are proceeding to greater endangerment, sometimes rapidly so, while societal efforts to stabilize these trend lines and to avert further environmental deterioration are dawdling and, typically, ameliorating the damage only superficially rather than intercepting forcefully the basic human driving forces of adverse change.

This is not to say that disaster is imminent in most of these regions – although we judge the Aral Sea (Glazovsky 1995; Kotlyakov 1991; Micklin 1988; Rogers 1990, 1991) already to be an environmental catastrophe and the Basin of Mexico as rapidly nearing criticality (Aguilar et al. 1995; Ezcurra et al. 1999): it is to say that the trajectories of threat in most of these regions are rapidly outstripping societal responses; that these trends augur environmental impoverishment for future populations who will occupy these regions; and that the trajectories suggest accelerating long-term costs for substitution in economic production systems, adaptation, and remedial measures aimed at containing the damage. At some indeterminable point in the future, these regional trends will inevitably eclipse societal capabilities to respond.

A rich tapestry of human causation

At the regional level, the pattern of human causation is richly variegated. Rarely can a single, dominant, human driving force be discerned that explains the historical emergence of environmental degradation or that captures the complexity of change. Nor do the various grand theories – whether they arise from welfare economics, political economy, neo-Marxist thinking, global-dependency models, or development theory – provide satisfying broad interpretations, although all have elements to contribute. So what we term the *regional dynamics of change* – the interplay among the trends of environmental change, vulnerabilities and fragility, human driving forces, and societal responses – must be examined within their cultural, economic, and ecological contexts. The most satisfying interpretations invariably recognize the shifting complexes of driving forces and responses over time, tap diverse social science theory, and are firmly grounded in careful empirical work and field experience.

Increasing global integration

The mosaic of global change is increasingly the product of the growing linkages between regional productive systems and the global economy, between states and their policies and international financial institutions, between local aspirations and the value systems and lifestyles of advanced industrial societies, and between the exporters and importers of technological change. Global markets facilitate increasing agricultural specialization in the Aral Sea region, in the Nepal middle mountains, and in Ukambani in Kenya, whereas international markets for timber facilitate the cutting of forests in Amazonia and Borneo. But just as these global market opportunities, global trade agreements and financial institutions, and the activities of multinational corporations are key ingredients in the complex of driving forces, along with state policies of natural resource extraction and revenue production, they are also the levers for restructuring and the sources of more promising environmental futures.

Accordingly, regional trajectories of change and associated regional dynamics can be understood only in the context of changing extraregional linkages. Simultaneously, changes in global economy and trade policies, such as the General Agreement on Tariffs and Trade (GATT), dispense a major environmental fallout on the world's regions, as noted in chapter 7.

Needed focus

These issues raise basic questions about whether emerging international efforts to combat global environmental degradation have the needed focus and recognition of the "embeddedness" of causal factors in political economy, state policies, and social institutions to be effective. Our view is that a substantial regional tailoring of initiatives will be required. To this end, the regional structure well under way in the Global Change System for Analysis, Research, and Training Program (START 1999) and that sought in Capacity 21 (UNDP 1996) are promising, for only a strongly regionally based structure of programmes and development efforts is likely to succeed.

Beyond this, several other elements for needed focus in global initiatives, such as START, the Global Environmental Facility (GEF), and a changing World Bank, merit consideration:

1. The problems of *systemic global environmental change* (e.g. climate change, ozone depletion) so prominent in media attention and public concern in the North and at the 1992 Earth Summit are not, in our view, the dominant problems confronting the environmentally endangered regions of the world in the year 2000. Rather, *cumulative global*

environmental change (Turner et al. 1990b) involving land degradation, localized environmental contamination, and water depletion is the type of ongoing change that most imperils the people living in these endangered areas and is thus the priority for early interventions by international efforts aimed at averting global disasters and rapidly accumulating environmental threats.

2. Frontier areas that are marginal to state economies, or that are strongly dependent on the global economy, are particularly vulnerable to environmental change and typically powerless to intervene in human driving forces, the sources of which often lie outside the affected region. Our five regions that fall into this category are likely to prove particularly difficult to stabilize, as they often are either of low priority in the political units in which they are located or are expendable for other state objectives (or even candidates for sacrifice). In either case, outside intervention (as the cases of Amazonia and Borneo suggest) is likely to confront strong political resistance.

3. Areas of agglomerated stress, such as megalopolises, bring together in dramatic concentration many of the traditional driving forces of population, economic growth, and affluence with the complex of contributing factors in growing linkage to the global economy and political economic structures of the nation-state (Burton 1988). Typically, the rates of change are proceeding at an extraordinary pace that challenges capabilities of existing societal alerting systems, assessment, and societal response. The threats involved include both the ongoing depletion of regional and more distant natural resources to support these vast agglomerations of population and the overwhelming of local environmental sinks. Although urbanization and industrialization are often subsidized by the state, the appearance of metropolises of 20 million or more people, continuing to grow very rapidly and increasingly dependent on imported resources and technological change to combat ongoing environmental degradation, involves a relentless increase in the risk of eventual collapse from unanticipated futures (Satterthwaite 1997).

Discrepancies in rates of change

As noted above, the regional dynamics of change in the nine regions reveal a recurring disjuncture between the rapid rate of environmental change and the often glacial pace of societal response. Interestingly, the global scale reveals a much more mixed picture where societal responses to such change as stratospheric ozone depletion, global warming, and industrial accidents have often been quite prompt, if less than rigorously effective. Nevertheless, signals of environmental threat have been processed with considerable speed and responsive coping actions at the international scale and by particular advanced industrial societies. However, the trajectories of change in the nine regions provide considerable confirmation of the argument of overshoot put forth by Donella Meadows and her colleagues (Meadows et al. 1972; Meadows, Meadows, and Randers 1992). Only in Amazonia, the North Sea, the Ordos Plateau, parts of Sundaland, and perhaps Ukambani do responses appear to have some potential for stabilization or at least a significant "flattening" of the trajectory headed toward greater endangerment or criticality. But the dominant situation is one of divergence between the rates of environmental change and those of societal response, portending increasing environmental impoverishment of future populations and escalating costs of substitution in production systems and mitigative efforts aimed at containing the growing damage that must eventually occur. And the primary causes for disjuncture, our regional studies persuasively argue, lie less in inadequacies of scientific understanding than in those of sociopolitical goals and structures, institutions, and state policies.

Global inequity

The Earth Summit addressed with determined attention the broad issues associated with capacity-building in developing countries (United Nations 1992). Agenda 21 and Capacity 21 (UNDP 1996) call for national assessments of capabilities and specific initiatives aimed at strengthening such local and regional capabilities, and the Global Environment Facility (GEF) and the World Bank have important initiatives under way. Involved are not only scientific and financial resources but also the infrastructure of social institutions required to formulate and implement serious national programmes of sustainable development.

Our nine case studies underscore the importance of these initiatives. Among the various inequities that cause and structure the global pattern of environmental degradation and associated human harm, two stand out. First is the array of human driving forces in which inequalities in wealth, differential stages of development, patterns of debt, world poverty, and a global economy shape a continuing and devastating unequal worldwide distribution of environmental impoverishment, endangerment, and criticality. Second is the highly variable scientific and institutional capability to anticipate, assess, and respond to this degradation (Agarwal and Narain 1999; O'Riordan et al. 1998; South Commission 1990). Early and effective efforts to bring into consonance trajectories of economic development and long-term environmental protection are highly cost effective, as well as protective of generations yet to be born. The World Bank now calculates the "wealth of nations" by adding measures of "natural capital" (i.e. the endowment or stock of environmental assets such as soil, water, atmosphere, forests) and "human capital" (e.g. investments in education, nutrition, poverty alleviation, and population control) to the usual tallies of per capita income (World Bank 1995), and we recently have been treated to a calculus of the value of nature at the planetary level (Costanza et al. 1997).

But such efforts are necessarily the outcome of achieving minimal well-being, requisite institutional and scientific capabilities, and political will across the globe. The North Sea case, when placed in the context of the developing-country regions, suggests that the differential capability to respond to emerging environmental degradation is no less important than the levels of degradation found throughout the South, particularly among the poor who are both the victims and the unwitting agents of environmental degradation in many of the poorest societies on earth (Argent and O'Riordan 1995). Such differences will certainly work to widen existing inequalities and will require major international initiatives if differential societal response is not to exacerbate rather than ameliorate other global inequities.

9

Global change and environmental risks in mountain ecosystems

N. S. Jodha

Global environmental change is a major issue today. Owing to rather skewed perspectives of the work and debate on the subject (a result of overemphasis on the "systemic" type of changes as against "cumulative" changes – see chapter 1), however, the totality of the global environmental change – its processes, consequences, and possible remedial measures – are inadequately understood (table 9.1). In view of the greater certainty of issues involved and their regional disaggregation, a discussion focused on cumulative types of global change can prove very useful. Furthermore, to capture fully the cumulative type of changes, regions can be identified with ecosystems (e.g. mountain ecosystem) in the context of which nature–human interactions and their consequences can be more easily understood.

Such issues set the background for the discussion on environmental risks in mountain areas in the context of global environmental change. The emphasis is on understanding and handling environmental risks in mountains, which are affected both by specific features of mountain habitats and by the way in which imperatives of these features are ignored or considered by human interventions in mountain areas. The discussion draws heavily on earlier work related to the subject (Jodha 1990a, b, 1992b; Jodha, Banskota, and Partap 1992).

Table 9.1 **Indicators of the "skewed perspectives" on global environmental change**

Elements prominently stressed	Elements underemphasized
"Systemic" change	**"Cumulative" change**
Focus on biophysical variables and their interactive processes relating to the functions and operations of geosphere – biosphere systems of the earth.	Localized and widely replicated changes in different variables and process of resource use that (when accumulated) influence the global systems.
"Geocentric perspective"	**"Anthropocentric perspective"**
Focus on physical dimensions, typically in the natural science framework; concentration on geobiological variables and their complex interaction patterns, with little direct incorporation of human dimension of changes and change processes.	Primacy of nature – society interactions with focus on their importance to the society; potential mechanism for understanding and handling "cumulative changes", with some possibility of influencing impacts of "systemic changes."
Other associated aspects	**Other associated aspects**
Emphasis on long time horizon (decades/centuries), intergenerational issues; focus on terminal impacts involving selected variables (e.g. sea-level and temperature rise, shift of climatic zones) affecting fundamental equilibrium of world system and atmosphere; analytical methods and material used involve high degree of complexity and sophistication, information on several unknowns, limited transparency (for unlimited ones), multiple uncertainties, and conjectural nature of predictions.	Sensitivity to both intragenerational and intergenerational issues; analytical approaches simpler and oriented to integration of change processes in current problem-solving mode; predictions, action/advocacy focus on short or medium planning horizon, greater ease and possibility of associating causes of, consequences of, and responses to change; greater possibility of integrating geocentric and anthropocentric perspectives.
Advocacy and action	**Advocacy and action**
High "scare and noise" potential of issues covered (e.g. doomsday predictions); approaches to abate/adapt to changes: obstructed by uncertainty of change scenarios, induce higher discounting of the potential options, inject vagueness about gains and sacrifices, and create more panic and debate and concrete action.	Possibility of evolving options within the received (and modified) framework of handling current crisis situations in local contexts; greater scope for clearly associating cost and benefits; greater certainty of potential options and their easy acceptability to decision makers; possibility of dual-purpose options to handle current and future "impacts."

Source: Table adapted from Jodha (1992c: 209). For various issues and examples that could fit into this grouping of perspectives, see Chen, Boulding, and Schneider (1983); Clark (1985); Flavin (1989); Glantz (1988); Jiang (1999); Jodha (1989); Kasperson et al. (1990); Kasperson, Kasperson, and Turner (1995, 1996); Price (1990); Schmandt and Clarkson (1992); Turner et al. (1990b).

Introduction

Focus

This chapter focuses on four aspects of global environmental change:

1. Of the two types of global environment change, namely "systemic" change and "cumulative" change (see chapter 1), the latter forms the broad context for the present discussion. The former is covered only in a limited way, owing to the non-availability of sufficient information and the large uncertainties associated with the predicted changes.
2. Environmental risks are perceived as disrupting the basic biophysical processes and natural flows that determine the health, productivity, and stability of environmental resources and their interactions in a given ecosystem.
3. The chapter discusses environmental risk – the subject of the volume as a whole – in the context of mountain areas, where interactions among imperatives of specific mountain conditions (such as inaccessibility, fragility, and diversity on the one hand and varying degrees of human interference on the other) constitute the circumstances that influence the biophysical processes and natural flows.
4. The geographical context of the discussion is the Hindu Kush–Himalayan mountain region, although the chapter does refer to other mountain systems, such as the Andes, to a limited extent. Within the mountain regions, agriculture (including all land-based activities, such as cropping, livestock, forestry, and horticulture) is used as a focal point. This is due to both the predominance of this activity in mountains and to the recognition that the important environmental degradation/rehabilitation issues in most cases relate to agricultural-resource use.

Systemic and cumulative changes

For these two types of changes, we take the lead from Clark University conceptual work on the subject. Turner et al. (1990b) discuss two dimensions of global environmental change – "systemic" changes and "cumulative" changes – a distinction also set forth in chapters 1 and 8. Broadly speaking, a systemic change is one that, while taking place in one locale, can affect changes in systems elsewhere. The underlying activity need not be widespread or global in scale, but its potential impact is global in that it influences the operation and functioning of the whole system as manifested through the subsequent adjustments in the system. Emissions of CO_2 from limited activities that have impacts on the great geosphere–biosphere system of the Earth and cause global warming offer a prime example. The cumulative type of change refers to localized but widely replicated activities where changes in one place do not affect changes in other distant places. When accumulated, however, they may acquire sufficient scale and potential to influence the global situation in various ways. Widespread deforestation and extractive land-use practices and their potential impacts on the global environment serve as examples. Both types of changes are the products of nature–human interactions and are linked to each other in several ways.

This brief introduction to the important concepts may appear sketchy, but it should suffice for my purpose in this chapter. Emphasis on one or the other type of change, along with the relative focus on geocentric perspectives or anthropocentric perspectives (see chapter 8) in the work on global change will have very different implications. Table 9.1 briefly summarizes these, and several are developed in the chapter.

Environmental parameters

Environment, unless expressed in terms of its vectors (or contributing factors or resources), is a product of several interactive processes of different components of a system. The interactive processes between living and non-living things (such as soil, water, vegetation, and animals) generate products and services, which act both as inputs for the continuity and performance of the system and as its output. Hence, for practical purposes, it is often difficult to separate "environment as services and products" from "environment as manifested by the status of the resources." In other words, separating the conditions generated by the interactions among soil, water, and vegetation from the status of these resources themselves is difficult. However, the specific attributes of the latter and the way these attributes are manipulated in using the resources affect the overall dynamics and pattern of interactive processes of the resources. This brings in human intervention as one of the crucial components in the environmental matrix of an ecosystem.

The contributions of interactive processes to the stability and sustainability of the environment (both as inputs and products of an ecosystem) take place through the crucial biophysical functions and flows that are interrelated and often invisible. They could fit under categories of *regeneration, variability–flexibility, resilience, nature cycles*, or *energy and material flows*. These biophysical functions and flows (to be elaborated later) are the basic mechanisms through which the level and performance of environmental services (i.e. productivity), as well as the health and status of environmental resources of a system, are ultimately determined.

These scientifically well-recognized processes are often not readily visible but their operation is perceivable through an understanding of more easily visible, measurable, or verifiable circumstances.

In short, the environmental risks of a system, such as a mountain ecosystem, can be understood in terms of instability or destruction of (a) natural resources, (b) their productivity potential, and (c) the processes represented by the biophysical functions and flows stated above. The environmental risks can be characterized and identified with reference to a negative change in any of the three categories of variables. In the ultimate analysis, however, the extent and nature of environmental risk relate to disruptions in the biophysical functions and flows, which this chapter elaborates with reference to mountain areas. Owing to the complexity and direct invisibility of most of these functions and flows, however, the focus of the discussion will be on the more easily understood and visible circumstances that influence them.

Accordingly, I first describe the objective circumstances – that is, the specific conditions of mountains and their likely impacts on the aforementioned functions and flows in a relatively undisturbed situation. A discussion follows, of changes through human interventions, under relatively low pressure on resources as manifested by traditional resource-use systems in mountain areas. The next stage is characterized by increased resource-use intensification, following the high pressure on mountain resources generated by increased population, accentuated market demands, and state interventions. The processes in this stage represent some dimensions of the cumulative types of global environmental change mentioned above. The discussion reveals the degree of mismatch between the imperatives of mountain characteristics and certain attributes of resource-intensification strategies as the key sources of environmental risks in mountain areas. These risks are quite severe, even without considering systemic types of environmental changes. They tend to make mountain areas and people more vulnerable to the potential risks of systemic changes induced by global warming.

Biophysical functions and flows

As the environmental risks in mountains are discussed primarily with reference to the biophysical functions and flows stated above, a word of explanation is essential. Taking the lead from the ecological sciences (Conway 1985; Golley 1993; Krutilla 1979; Lowrance, Stinner, and House 1984; Monasterio, Sarmento, and Solbrig 1985; Shutain and Chunru 1988; USNRC 1986; Yaffee et al. 1996), we examine the stability of mountain environments in terms of the normal functioning of interrelated processes such as regeneration, the system's internal variability and flexibility, resilience, and natural flows. Regeneration – involving germination, growth, decay, decomposition, re-emergence, and growth – is the one important condition associated with the environmental health of a system. The process of regeneration and the ability of a system to withstand stress are facilitated by the internal variability of the system, in which input needs and output flows of different components (e.g. annual and perennial plants) are organically linked. The system's internal variability, involving organisms and mechanisms with temporally and spatially noncovariate input demand and output performance, offers to the system a degree of flexibility to adjust to different perturbations.

Somewhat related are the visible or invisible flows and cycles involving energy, moisture, and nutrients of different types and sources. Nature's pattern of energy and material flows, as well as their balancing in the context of a system, links the components from the geosphere and biosphere and helps to sustain the health and productivity of a system. The operation of the basic functions and flows, as mentioned earlier, is affected by the state of the natural resources. For instance, a system based on diversified vegetation would be more conducive to regeneration processes and the smooth functioning of Nature's flows. Similar cases may be the practice of zero tillage on fragile mountain slopes, systematic crop rotations, and indigenous agroforestry practices followed in mountain areas. Any practice disturbing the arrangements may disrupt the underlying biophysical functions and flows, thereby initiating environmental instability and risk.

The reasons for focusing on these biophysical–chemical processes and flows, rather than on the simple categories of resources, are several. In the overall context of environmental stability, sustainability, and productivity, it is the understanding of the dynamics of underlying processes and flows rather than the structure of environmental resources that can help to evolve strategies to minimize environmental risks in mountain regions. An understanding of the processes and their associated conditions can help to identify alternative resource structures, usage patterns, and their alterations to meet changing demands. For instance, in view of the unavoidable intensification of resource use in mountain areas to meet growing human demands, restoration of traditional resource-extensive management practices may not be feasible. By using the rationale behind traditional systems, however, it may be possible to create new resource-intensive systems that are compatible with new demands and conducive to resource conservation. For instance, balancing intensive and extensive land uses by putting some proportion of area under crops and retaining large parts under forest may not be possible.

For smooth operation of certain biophysical functions and flows, however, the key factor is the complementarity of annuals and perennials rather than the rigid proportions of specific land-use categories. This complementarity can be partly ensured by interplanting annuals and woody perennials, as under agroforestry systems. Similarly, reforestation using traditional species involving a felling cycle of, say, 100 years, may not be feasible today, but reforestation using early maturing trees (and especially those with multiple functions) can be promoted. Thus, in the microlevel context of a degraded watershed, interventions need to focus not on re-creating its past but on rehabilitation, using the rationale behind its past status. This, in turn, implies an emphasis on dynamic biophysical processes and flows, using the new understanding offered by modern science and technology blended with the rationale of traditional resource-management systems (Jodha 1991).

Another reason for emphasizing the basic biophysical functions and flows as the focal point for understanding environmental risks is that the environment (however defined) is an integrated product of several processes. Such processes cannot be properly addressed by focusing on individual environmental resources or their productivity: any approach addressed to them usually acquires a sectoral character and misses the required integrated focus. For instance, any strategy directed to the stability of the hydrological cycle in a mountain region, or to the vegetative regeneration of mountain slopes, will require the integrated use of multiple components affecting the biophysical processes and flows. Finally, for an integrated understanding of mountain characteristics, appropriate human interventions and their risk implications, the biophysical functions and flows (rather than environmental resources) offer a more useful and effective context.

Mountain specificities

Because the smooth operation or disruption of biophysical processes is largely a product of the specific attributes of a mountain system and the ways in which they are used as natural resources, it is essential to digress briefly into a discussion of relevant mountain characteristics and their risk imperatives.

The important conditions that separate mountain habitats from other areas are called here "mountain specificities," six (some of which might be shared by other areas) of which I consider here. The first four – namely inaccessibility, fragility, marginality, and diversity or heterogeneity – may be called first-order specificities. Natural suitabilities or "niches" (i.e. activities/products in which mountains have comparative advantages over the plains) and "human adaptation mechanisms" in mountain habitats are second-order specificities. The latter differ from the former in that they are responses or adaptations to first-order specificities; nevertheless, they are specific to mountain areas (Allan 1995; Jodha 1990a, 1995).

Before the major mountain specificities are described, it should be noted that not only are these characteristics interrelated in several ways but also, within the mountains, they show considerable variability: for instance, not all locations in mountain areas are equally inaccessible, fragile, or marginal, nor are human adaptations uniform in all mountain habitats. With full recognition of such realities, a brief description of the mountain specificities follows.

Inaccessibility

Owing to slope, altitude, overall terrain conditions, and periodic seasonal hazards (e.g. landslides, snow, and storms), inaccessibility is the best-known feature of mountain areas (Allan 1995; Hewitt 1995; Messerli and Ives 1997; Price 1995). Its concrete manifestations are isolation, distance, poor communication, and limited mobility. Besides being the dominant physical dimension, it has sociocultural and economic dimensions (Jodha 1990a, 1995). The implications of inaccessibility as objective circumstances influencing the operation of biophysical functions and flows can be stated as follows. First, it restricts the mobility and external linkage-related disturbance of the ecosystem. Second, the relatively closed character of mountain habitats imposes a number of compulsions for linking survival strategies to local availability of resources and their protection and regeneration. Third, meeting diversified human needs in a closed or isolated situation induces diversification in production and resource-use patterns, in both temporal and spatial contexts. Fourth, the limited scope for dependable external linkages and supplies induces the adjustment of demands to supplies (rather than the other way round) through various forms of demand rationing and periodic siphoning of pressure through transhumance and out-migration.

The coping strategies stated above are potentially more conducive to biophysical processes essential for environmental stability. A disregard of the foregoing imperatives, however, for any reason, can make mountain areas and mountain people vulnerable to serious environmental risks. For instance, increased internal pressure on resources owing to population growth within a relatively closed system may lead to overexploitation of resources and reduced diversification and flexibility of resource-use patterns. Similarly, the establishment of external linkages within the overall context of the general inaccessibility problem may accentuate selective resource extraction and external exchange at unequal terms.

Table 9.2 **Mountain specificities and their environmental stability/risk imperatives**

Mountain specificities and their implications[a]	Risk-reducing/stability-promoting biophysical processes and flows			
	Regeneration	Variability/ flexibility	Resilience	Energy/ material flows
Inaccessibility (isolation, limited mobility, limited external linkages, less disturbance to system, local resource-based diversification)	(+)[b]	(+)	(+)	(+)
Fragility (vulnerable to degradation through small disturbance and use intensity, slow recovery, limited and low productivity options)	(−)[b]	(−)	(−)	(−)
Marginality (limited, low potential, inferior options, vulnerable to shocks)	(−)	(−)	(−)	(−)
Diversity (basis for diversified, interlinked activities, organic integration of potential options)	(+)	(+)	(+)	(+)
Niches (products, resources, activities with comparative advantages for mountains, result of diversity and specific resource conditions)	(+)	(+)	(+)	(+)

Source: Adapted from Jodha (1995: 155–158).

a. **Human Adaptation Mechanisms** is another mountain specificity elaborated in the text. Its role in promoting environmental stability is sketched in table 9.4.

b. (+) and (−) indicate respectively more favourable and less favourable circumstances generated by mountain specificities for operation of biophysical processes and flows.

Such developments may generate circumstances that are less conducive to environmental stability of the mountain regions. Tables 9.2, 9.3, and 9.4 indicate some of these issues.

Table 9.3 **Mountain specificities, human adaptations, and implications for environmental stability/risk**

Mountain specificities and features of adaptations[a]	Implications (circumstances) potentially conducive to environmental stability			
	Regeneration	Variability	Resilience	Energy and material flows
Inaccessibility (isolation, closedness):				
• Local (diverse) resource-centred production systems		X		
• Local demand-driven, local capacity-based (low) resource extraction	X	X		X
• Limited external reliance/support, compelling rationing of demand and resource use; social sanctions	X	X	X	
Fragility and Marginality (limited and "inferior" options, high vulnerability to disturbance):				
• Land-extensive production systems (annual perennial linkages); sanctions against overuse (common-property resources), etc.	X	X		
• Resource upgrading (terracing, irrigation), collective sharing systems	X		X	X
Diversity (potential for multiple linked activities):				
• Diversified farming system, spatial–temporal linkages of land-based activities, food systems and other demands tuned to diverse supplies	X	X	X	X
Niches (options, possibilities with comparative advantage):				
• Local need- and capacity-based low and regulated extraction	X			X
• Diversified and linked activities	X	X		

a. **Human Adaptation Mechanisms** is another mountain specificity elaborated in the text. Its role in promoting environmental stability is sketched in table 9.4.

Fragility

Mountain areas – because of their altitude and steep slopes in association with geological, edaphic, and biotic factors that limit the farmer's capacity to withstand even a small degree of disturbance – are known for their

Table 9.4 **Interaction between resource-intensification factors and mountain specificities affecting environmental stability/risk in mountains**[a]

Factors causing resource-use intensification (human interventions)	Mountain specificities and implications				
	Inaccessibility (closedness, limited external linkages)	**Fragility and Marginality** (incompatibility with intensive use)	**Diversity** (high potential for diversification)	**Niches** (products, activities with comparative advantage)	**Adaptations** (activities, practices tuned to mountain conditions)
Population growth, per capita in-creased activities, increased animal numbers	Excess pressure on local resources with limited outlet (R, F, N)	Resource-use intensity beyond use capacity (R, F, S)	Pressure of food needs, reduced range of land-based activities (R, F, S, N)	Pressure of food needs, disregard or misuse of potential (R, S, F)	Disregard of resource-extensive, diversified practices (R, F, N, S)
Market forces, trade links, pressure of external demand	Integration with mainstream market situation, despite low physical accessibility (R, F)	Distant demand-induced overuse, backlash of cash cropping (R, S)	Narrow specialization, reduced diversification (R, F, S, N)	External demand-induced over-exploitation, marginalization (R, F)	Decline of environmentally sensitive local concerns, practices (R, S, F)
Public Interventions: a) Infrastructure for accessibility, integration, harnessing of niches, etc.	Reduced isolation, increased integration and level of activities (R, N)	Direct and side effects on fragile/ marginal resources, increased use level (R, S, N)	Increased use level, access-determined narrow specialization (R, F, S)	Over-exploitation of high-potential areas, products, disregard of side effects (R, N, F)	External contacts, loosening of traditional values, measures (R, S, F)

b) Technology with narrow focus on market signals, short-term needs, sectoral orientation, external origin/orientation	Application for improved mobility, integration (F, N)	Product maximization, indifference to resource limitations, inappropriateness (R, F, S)	Narrow specialization, focus on limited product attributes (R, F, S)	Commercial-extraction orientation, disregard of side effects (R, F, S)	Disregard of traditional wisdom, know-how (F, R, S)
c) Macro-economic policies – price, tax, trade, investment, extraction, development strategies	Disproportionate focus on accessibility, integration, disregarding side effects (F, N, R)	Focus on current production, disregard of resource limitations, long-term consequences (R, F, S)	Narrow specialization, through incentives support systems, disregarding organic linkages (R, F, N, S)	Focus on revenue generation, external demand, extraction disregarding the side effects (R, N, S)	Marginalization of traditional systems, increased dependency, subsidization (F, R)

a. Biophysical processes affected by intensive human interventions in mountains: R, regeneration; F, flexibility, variability; S, resilience; N, natural flows (energy and material flows).

fragility (DESFIL 1988). Vulnerability to irreversible damage, due to overuse or rapid changes, is manifested by limited options. Consequently, with disturbance, mountain resources and environment deteriorate rapidly and sometimes irreversibly (Eckholm 1975; Hewitt 1988, 1995; Johnson, Olson, and Manandhar 1982; Messerli and Ives 1997). Environmental risk related to fragility is commented on later in this chapter.

Marginality

A "marginal" entity (in any context) is one that counts the least in the context of a "mainstream" situation. This may apply to physical and biological resources or to people and their sustenance systems. The basic factors contributing to such a status of any area or community are remoteness and physical isolation, fragile and low-productivity resources, and several human-made handicaps that prevent one's participation in the mainstream patterns of activities (Blaikie and Brookfield 1987; Blaikie et al. 1994; Chambers 1987). Mountain regions, being marginal areas in most cases, share these attributes and suffer the consequences of such status in different ways (Jodha 1995). To this one may add that "marginality" implies a comparative context – that is, that a situation, an option, a resource, an area, or a community has marginal status in comparison to other entities of the same genre. Accordingly, an entity acquires marginal status when compared or linked with other entities. Some mountain areas, their production systems, and people's adaptation strategies, start off marginalized and, therefore, unequal, owing to their integration with the dominant, mainstream situations of the plains.

Although they are products of broadly different factors and processes, marginality and fragility share several common risk implications. Accordingly, unless resources are upgraded or strengthened, the existing use capacity and input-absorption capacity of fragile and marginal resources remain low. They are suited for less-intensive uses, with low productivity and low pay-off. These features, in turn, restrict the scope for diversification and flexibility and reduce the system's (physical and economic) abilities to absorb shocks, thereby making it more vulnerable to different sources of risk. The risks become a reality when increased pressure on resources or the side effects of external linkages and interventions push their usage level far beyond their use capabilities (Kasperson, Kasperson, and Turner 1995).

Diversity or heterogeneity

Mountain areas accommodate immense variations among and within ecozones, even within short distances. This extreme heterogeneity is a function of interactions of such factors as elevation, altitude, geological and edaphic conditions, steepness and orientation of slopes, wind and precipitation, mountain mass, and relief of terrain (Troll 1988). Biological adaptations (Dahlberg 1987) and socio-economic responses to these diversities also acquire a measure of heterogeneity of their own (Allan 1995; Price 1981; Rhoades 1997). The diversity or "heterogeneity" phenomenon applies to all mountain characteristics discussed here. From the viewpoint of environmental risk, the internal diversity of mountains is the most important factor in affecting the smooth operation of biophysical processes and flows and thereby ensuring environmental stability (Zimmerer 1995). This is both the basis for diversified, interlinked activities and a source of resilience for people's survival strategies in mountains. The imperatives of diversity, in terms of matching diversification in resource use and production practices, however, can be used as a basis for interventions in mountains only if the human demands are also diversified. These imperatives may be ignored with reduced diversity of demands on mountain resources, which may occur owing to increased population and consequent changes in resource uses focused on staple foods rather than diversified products. Similarly, market-induced narrow specializations can reduce diversification. Such changes can prove detrimental to the environmental stability associated with internal diversity of mountain ecosystems.

"Niche" or comparative advantage

Owing to their specific environmental and resource-related features, mountains provide a "niche" for specific activities or products or services. At the operational level, mountains may have comparative advantages over the plains in these respects. Examples include mountains serving as the habitat for specific medicinal plants, as a source of unique products (e.g. certain fruits or flowers) and as the best-known sources of hydropower production. In practice, however, niche or comparative advantages may remain dormant unless circumstances are created to harness them. Mountains, owing to their heterogeneity, have several specific niches, which are used by local communities in the course of their diversified activities (Brush 1988; Jodha 1995; Whiteman 1988).

A niche is a product of interactions of various biophysical (and even socio-economic) factors, the regulated use and protection of which are conducive to the environmental health of a region. Since a niche is a part of the diversity that characterizes mountain habitats, its implications for environmental risk are similar to those of "diversity." The role of niches in the circumstances affecting basic determinants of environmental stability (i.e. regeneration, resilience, energy, and material flows) is affected by the pace and pattern of extraction of the niche. Overexploitation of niches and disregard of the side effects of extraction methods can adversely affect the environmental situation. Evidence shows that both state interventions and market forces tend to contribute to overextraction of niches and thereby affect the environmental stability of mountains.

Human-adaptation mechanisms

Mountains, through their heterogeneity and diversity, offer complex constraints and opportunities. Mountain communities, through trial and error over generations, have evolved their own adaptation mechanisms (Guillet 1983; Jochim 1981; Jodha, Banskota, and Partap 1992). Accordingly, either mountain characteristics are modified (as through terracing and irrigation) to suit their needs, or the activities are designed to adjust the requirements to mountain conditions (as by zone-specific combinations of activities or crops). Adaptation mechanisms are reflected through formal and informal arrangements for resource management, diversified and interlinked activities to harness microniches of specific ecozones, and effective use of upland–lowland linkages (Allan 1986, 1995; Brush 1988; Whiteman 1988). Adaptation mechanisms have helped to ensure the sustainable use of mountain resources and stability of mountain environments in the past. With the changes related to population, market, and public interventions, however, some adaptation mechanisms are losing their feasibility and efficacy. An understanding of the rationales underlying such mechanisms can help in the search for options to reduce emerging environmental risks in mountains.

Mountain regions: Natural conditions and risk potential

Inaccessibility, by restricting mobility and limiting external linkages, helps to reduce disturbances and perturbations in basic biophysical processes. Similarly, diversity (internal heterogeneity) and specific niches characterizing mountains also help in regeneration and in supporting linkages among different animate and inanimate components of the system, and also facilitate intrasystem flows of energy, nutrients, and moisture. In contrast, fragility and marginality, indicating vulnerability to resource degradation and a slow pace of recovery and growth, offer limited scope for biophysical processes and flows.

Thus, under their natural state, mountains show a mix of favourable and less favourable circumstances that affect the operation of dynamic processes underlying the health and stability of mountain environments. Depending on which circumstances dominate a given area, it is possible to assess the threat of the risk to a mountain environment.

Furthermore, as a part of the natural withering, stabilization, and succession processes (especially in a young mountain system such as the Himalayas), circumstances and their environmental impacts do change (Allan 1995; Rhoades 1997; Thompson, Warburton, and Hatley 1986). However, the role of the natural processes is accentuated by human interventions (Allan 1995; Ives and Messerli 1989; Messerli and Ives 1997). Hence, the latter plays a crucial role in altering the circumstances (indicated by table 9.2) and their environmental impacts. In connection with human interventions, most of the mountain characteristics are interrelated, owing to their common cause or shared consequences. The environmental risk or stability implications of these interrelationships of mountain characteristics, in terms of impacts on biophysical processes and flows, become more clear in the context of increased intensification of resource use in mountains.

Human intervention: Low-intensity phase

The potential conditions of environmental risks or stability in mountains become a reality once the natural system is exposed to human-induced perturbations. Though often seen as difficult places for human habitation, mountains have historically been centres of flourishing civilizations and have sheltered religious and political refugees during different times in history. As a result, human interference with the natural ecosystems has also been widespread. The communities that lived and multiplied in mountains acquired a detailed understanding of the limitations and potentialities of mountain environments. Their survival and growth strategies exhibited great sensitivity to ecological and economic interdependencies. They adapted to mountain conditions well enough to minimize environmental risks (Guillet 1983; Hewitt 1995; Jodha 1995). The resource-management systems and practices they evolved generally proved conducive to the operation of biophysical processes and flows that help ensure environmental stability. These systems and processes worked reasonably well under the situation of low pressure on mountain resources. These resource management and production practices can be seen in the traditional measures often grouped under folk agronomy, ethno-ecology, ethno-engineering, collective sharing systems, and recycling practices. Instead of describing them individually, we may refer to their key implications *vis-à-vis* the environmental stability that existed through generation of changing circumstances. Table 9.5 lists some of the traditional practices in response to specific mountain conditions.

Most of the traditional practices were low-intensive resource uses, and were governed mainly by local needs and local capacities to extract resources. Accordingly, despite human habitation, inaccessibility neces-

Table 9.5 **People's traditional adaptation strategies in response to mountain specificities**

Adaptation measure	Mountain specificities[a]				
	I	F	M	D	N
Diversification and self-provisioning					
• Spatially, temporally linked activities	X	X		X	X
• Local resource-focused recycling, self-provisioning		X		X	X
• Scattered settlement patterns	X			X	
Folk agronomy					
• Annual–perennial plant complementarities (farming–forestry linkages, etc.)		X	X	X	
• Cultivars of varying attributes		X	X	X	
• Fallowing, rotations, top sequencing, intercropping	X	X	X		X
Ethno-engineering					
• Slope management (terracing, etc.)		X	X		
• Protective vegetation, contour farming		X	X		
• Traditional irrigation/drainage management			X	X	X
• Small-scale transport logistics (ropeways, trails, donkey tracks, etc.)	X				
Collective arrangements					
• Common property resources		X	X	X	
• Social regulations for use/protection of fragile resources		X			
• Community irrigation systems, etc.			X	X	X
• Crisis period-sharing systems	X	X	X		
Upland–lowland linkages					
• Petty trading in specialized mountain products (with high value, low weight, etc.)					X
• Periodic migration	X			X	
• Transhumance	X	X			X
• Externally planned extraction of mountain niches	X	X			

a. Table adapted from Jodha (1990a). The following letters stand for the respective mountain characteristics: I, inaccessibility; F, fragility; M, marginality; D, diversity; N, niche.

sitated diversification, local resource regeneration, and balanced resource use. Similarly, since local demand and local capacities to extract resources were small, the pressure on resources remained low. Local control kept pressure on resources deliberately low by various measures of demand-rationing, including social–cultural sanctions and periodic out-migration (Guillet 1983; Hewitt 1988). These measures contributed in different ways to generate circumstances conducive to the operation of biophysical processes and flows that helped to maintain environmental stability.

Fragility and marginality, the two features of mountain habitats that made mountain environments most vulnerable to degradation and prone to slow regeneration, were handled by a two-way adaptation process under the traditional systems. First, through land-extensive production practices and uses (e.g. pasture and forestry, instead of intensive annual cropping) and institutional regulations (e.g. through provision of common-property resources), use systems were adapted to the limitations of the resources. By implication, these measures helped in better regeneration, flexibility, improved energy and material flows. On the other hand, the use of ethno-engineering measures (e.g. terracing, water-harvesting, community irrigation systems) to upgrade fragile and marginal resources, contributed to greater environmental stability.

Diversification has been one of the most important features of traditional resource-use systems. At a macrolevel, diversification is reflected through balanced land use involving the provision of forest, pasture, and cultivable land. At a microlevel, it shows up in diversified cropping systems and other features of traditional farming systems. This sort of diversification, besides matching the imperatives of resource characteristics, also met the diversified needs of mountain communities in the relatively closed or isolated context of their habitats. More important, it helped in the processes contributing to regeneration, flexibility, and regulation of natural flows.

An important component of the overall diversification of activities was the focus of traditional systems on harnessing and protecting "niches," or areas and activities of comparative advantage for mountains. Since "niches" served as an important basis for upland–lowland linkages and a means of surplus generation and exchange, their protection and development were part of the survival and growth strategies of mountain communities. Given the crucial importance of protecting niches and the key role of local needs and local extraction capacities in determining the use level of mountain niches, overexploitation of resources did not usually occur. In short, the harnessing of niches did not disturb the dynamic processes and flows of nature.

All these features of traditional resource-use systems or farming systems are part of the human adaptation mechanisms that have evolved and been inherited by mountain people through the centuries. These adaptations involved various other practices – such as product recycling, flexible consumption patterns, transhumance and migration – that directly or indirectly facilitated regulation of pressure on resources and, by implication, proved conducive to the operation of biophysical processes for environmental stability.

As indicated earlier, however, most of the traditional adaptations to imperatives of mountain specificities evolved in the context of low demands on mountain resources. Low pressure, in turn, was the product of a smaller population- and subsistence-oriented, local resource-centred agriculture as the dominant activity of mountain areas. Trade and industry (cottage industry) were largely linked to local resources (niches) and agriculture. Furthermore, owing to limited external linkages, outside demand or market signals could not exert undue pressure on local resources. Besides low demands, the lack of means for large-scale extraction of mountain resources also prevented undue disturbance to the mountain environment through human intervention. However, the land-extensive, non-extractive features of traditional systems are incompatible with the resource-use intensification forced by rising demands on mountain resources (Jodha 1991, 1995).

Environmental risks: Resource-use-intensification phase

Intensification of resource use in mountains exposes the mountain environment to serious degradation, a process that manifests the cumulative type of global environmental change visible in developing countries (Turner et al. 1990b). Its more popularly understood or projected components are deforestation, overgrazing, extension of cropping to steep and fragile slopes, landslides and mudslides, periodic flash floods, soil erosion, disappearance of vital biophysical resources, and reduced resource productivity. Some of these have been documented as emerging indicators of unsustainability (Jodha 1990a, b, 1995).

These changes are reflected in biophysical processes and flows and are related to interactions among driving forces, resource intensification, and the imperatives of mountain specificities. The forces behind resource-use intensification are rapid population growth, market-induced demand, and resource-extractive public policies. The mechanisms (or immediate causes) include the creation of infrastructures for reducing inaccessibility and supporting extraction of mountain niches and the introduction of new technologies and macroeconomic policies to develop and integrate mountain areas with mainstream economies. Whatever their explicit or implicit goals or the nature of mechanisms to implement them, most of the public policies in mountain areas are insensitive to the imperatives of mountain specificities.

Table 9.4 summarizes some of these issues and their implications in terms of circumstances associated with environmental stability or risk in mountain regions. Accordingly, irrespective of the factors behind resource-use intensification, the invariable consequence is a disruption of circumstances conducive to biophysical processes and flows central to the stability and sustainability of mountain environments. Detailed discussion of these factors follows (pp. 329–335). At this stage, it is sufficient to indicate the consequences of these factors *vis-à-vis* the implications of mountain specificities.

The literature on changing resource-use patterns, productivity, and environmental deterioration, and their possible causes in the Hindu Kush–Himalayas and other mountain systems in developing countries, confirms the situation indicated above (Eckholm 1975; Hewitt 1995; Ives and Messerli 1989; Tapia 1992). Most of these changes can be analysed and interpreted both as manifestations of circumstances leading to disruption of biophysical processes and, in some cases, as consequences of such disruptions (table 9.6).

The emerging risk scenarios

Table 9.6 presents a broad picture of negative changes in mountain areas that can be interpreted as indicators of emerging environmental risk scenarios in the Hindu Kush–Himalayas. The table is based on macro-level data and observations as well as evidence from microlevel field studies in the selected hill areas of China, India, Nepal, and Pakistan (Banskota and Jodha 1992a; Jodha 1995). These changes may also be described as indicators of the unsustainability of the present pattern of resource use in mountain areas.

The negative changes may relate to (a) resource base (e.g. land degradation), (b) production flows (e.g. persistent decline in crop yields), and (c) resource management/usage systems (e.g. increased infeasibility of annual–perennial intercropping or specific crop rotation) (Jodha 1990a, 1995). More important, for operational and analytical purposes, it is possible to group the indicators of emerging environmental risks and vulnerabilities under three categories based on their actual or potential visibility.

Directly visible negative changes

Directly visible negative changes can include the increased extent of landslides or mudslides, drying up of traditional irrigation channels (kools), increased idle periods of grinding mills or sawmills operated through natural water flows, prolonged decline in the yields of mountain

Table 9.6 **Negative changes as indicators of emerging environmental risks in mountain areas**

Visibility of change	Changes related to:[a]		
	Resource base	Production flows	Resource use/management practices
Directly visible changes	Increased landslides and other forms of land degradation; abandoned terraces; per capita reductions in availability and fragmentation of land; changed botanical composition of forest/pasture Reduced water flows for irrigation, domestic uses, and grinding mills	Prolonged negative trend in yields of crops, livestock, etc.; increased input need per unit of production; increased time and distance involved in gathering of food, fodder, and fuel; reduced capacity and period of grinding/saw mills operated on water flow; lower per capita availability of agricultural products; etc.	Reduced extent of fallowing, crop rotation, intercropping, diversified resource-management practices; extension of plough to submarginal lands; replacement of social sanctions for resource use by legal measures; unbalanced and high intensity of input use, subsidization
Changes concealed by responses to changes	Substitution of cattle by sheep/goats; deep-rooted crops by shallow-rooted ones; and shift to non-local inputs Substitution of water flow by fossil fuel for grinding mills; manure by chemical fertilizers[b]	Increased seasonal migration; introduction of externally supported public distribution systems (food, inputs,[b] intensive cash cropping on limited areas[b]	Shifts in cropping pattern and composition of livestock; reduced diversity, increased specialization in monocropping; promotion of policies/programmes with successful record outside, without evaluation[b]

Development initiatives etc. – potentially negative changes[c]	New systems without linkages to other diversified activities and regenerative processes; generating excessive dependence on outside resources (fertilizer/pesticide-based technologies, subsidies), ignoring traditional adaptation experiences (new irrigation structure); programmes focused mainly on resource extraction	Agricultural measures directed to short-term quick results; primarily production- (as against resource-) centred approaches to development; service-centred activities (e.g. tourism) with negative side effects	Indifference of programme and policies to mountain specificities; focus on short-term gains; high centralization; excessive, crucial dependence on external advice, ignoring traditional wisdom; generating permanent dependence on subsidies

Source: adapted from Jodha (1990a; 1992b: 66; 1995: 146–147).

a. Most of the changes are interrelated and they could fit into more than one block.

b. Since a number of changes might occur for reasons other than environmental unsustainability/risk, a fuller understanding of the underlying circumstances of a change will be necessary.

c. Changes under this category differ from those in the two categories above in the sense that they are yet to take place, and their potential emergence could be understood by examining the involved resource-use practices in relation to specific mountain characteristics.

crops, reduced diversity of mountain agriculture, abandonment of traditionally productive hill terraces, and increased seasonal out-migration of the hill people.

Negative changes made invisible

People's adjustments to negative changes often tend to hide the latter. Adoption of shallow-rooted crops as substitutes for deep-rooted crops, following erosion of topsoil on mountain slopes; substitution of cattle by small ruminants, owing to permanent degradation or reduced carrying capacity of grazing lands; introduction of a public food-distribution system, owing to increased interseason hunger gaps (local food-production deficits); and small farmers leasing out their lands to concentrate on wage earning, illustrate this category of negative change.

Development initiatives with potentially negative consequences

Various measures are adopted for meeting present or perceived future shortages of products at current or increased levels of demand. Some of the measures (changes), despite enhancing productivity of mountain agriculture in the short run, may jeopardize the ability of the system to meet the increasing demands in the long run. Such risks are positively linked with the insensitivity of the interventions to specific mountain conditions and their imperatives for environmental stability. To illustrate, any farm technology that increases the crucial dependence of mountain agriculture on external inputs (e.g. fertilizer) and disrupts local regenerative practices may eventually increase environmental risks. Similarly, any measure that disregards the fragility of mountain slopes and ignores the linkages among diverse activities at different elevations in the same valley (e.g. farming–forestry linkages) and promotes monocropping may prove unsustainable. Likewise, any resource-extraction activity (e.g. hydropower projects) or service-centred activities (e.g. tourism) or welfare-oriented schemes (e.g. subsidies generating permanent external dependency of mountain people) that ignore side effects and long-term consequences may enlarge the environmental risks for mountainous areas and their inhabitants (Price 1995).

Table 9.6 summarizes some of the visible and less-visible negative trends relating to the resource base, productivity and management of mountain resources, largely in the context of agriculture. Evidence of resource degradation, productivity decline, and disruption of traditional resource-management systems from other fields such as mining and industry (Bandyopadhyay 1989), infrastructure development (Paranjpye 1988), and tourism (Kariel and Draper 1995; Price 1995; Singh 1989) could be presented in the same manner. In some cases, the changes listed in the table are causes, whereas in others they are consequences of disruptions of the biophysical processes and flows. Furthermore, ultimately, the circumstances that underlie the changes act as causes of disruptions of biophysical processes and flows and are associated with resource-use intensification in mountains.

Resource-use intensification processes: Increased human interventions

The causative factors or driving forces behind the process of resource-use intensification and consequent environmental risks in mountains resemble those observed in other ecosystems (Blaikie and Brookfield 1987). Broadly speaking, they include human (and animal) population growth, trade and market-induced pressure of demand, and public interventions with a general insensitivity to the imperatives of mountain specificities. This list has also recently been confirmed and extended by the comparative studies in Kasperson, Kasperson, and Turner (1995). The commentary on Table 9.4 has already alluded to the implications of these factors *vis-à-vis* mountain specificities and biophysical processes and flows. Now I describe the magnitude and role of these factors in environmental degradation induced by resource-use intensification.

Pressure on mountain resources

The first reason for environmental degradation in mountains is the sheer scale of demand on mountain resources *vis-à-vis* their carrying capacities and ability to regenerate. A series of forces accounts for the steadily mounting demand.

The population factor

One of the key factors to consider in the context of the scale of demand on mountain resources is the human population. The unprecedented growth in mountain populations has promoted a rapid increase in demand that threatens to thwart all efforts to achieve sustainable development. If current growth rates continue, most mountain areas in the Hindu Kush–Himalayas will easily have doubled their populations by the year 2030 (PRB 2000). This will further increase the pressure on natural resources, beyond their use capabilities, as reflected in the extension of cropping to steep slopes and the discontinuation of land-extensive practices (Sharma and Banskota 1992). During recent decades, population growth in some areas of the Hindu Kush–Himalayas has been unbearably high. Despite problems created by inaccessibility, marginality, and the inadequacy of facilities in the mountains, the "health revolution" has contributed to this growth. On the other hand, traditional pressure-management mechanisms – such as migration and the upgrading of resources through terracing, irrigation, and crop technologies – have failed to keep pace with the growth in population.

This failure carries both current and future economic and environmental consequences. Against a background of stagnant production systems, inadequate infrastructure development, and absence of alternative employment opportunities, people's sustenance strategies place a high premium on the over-supply of labour, making population increases inevitable in the mountains (Sharma and Banskota 1992). The qualitative changes in population characteristics (as reflected in increased individualism, factionalism, and commercial attitudes) also have had negative side effects in eroding traditional institutional mechanisms (e.g. provision of common-property resources and collective environmental security) in mountain areas (Jodha, Banskota, and Partap 1992).

Livestock

The increase in livestock numbers has also contributed to the rising demands on natural resources. In most mountain areas, the livestock population is equal to, if not greater than, the human population. The increase in livestock has been an important response mechanism of mountain farmers to deteriorating economic and environmental conditions, but it is clear that current growth rates are unsustainable in the context of widespread deforestation and overgrazing in the Hindu Kush–Himalayan region (Jodha 1990a; Sharma and Banskota 1992).

Market forces

Market-induced demands further accentuate the pressure on resources through rapid human and animal population growth. Governed initially by local revenue requirements and the desire to harness mountain "niches," resource extraction ultimately becomes a function of distant demands and market signals. The latter, being insensitive to local circumstances and indifferent to its side effects, accelerates the process of overextraction. Evidence concerning deforestation for commercial use, mining activities, and the environmental insensitivity of hydropower and irrigation schemes from various areas in the Hindu Kush–Himalayan region corroborate this (Banskota 1989; Banskota and Jodha 1992b; Jodha 1995; Rhoades 1997). At the microlevel, increased focus on cash cropping – especially horticulture and vegetable cropping in selected areas – has pushed staple food crops to more marginal, fragile slopes. Moreover, the "servicing" of horticulture development (e.g. through the provision of wooden boxes for fruits and support sticks for various vegetables) has a high environmental cost in terms of deforestation (Banskota and Jodha 1992a).

One important dimension of market-induced resource extraction relates to the terms of exchange between mountain regions and the plains/ urban areas that use the mountain products. As elaborated below, the factor and product prices are too low and fail to reflect their real worth. This induces overextraction of resources, with little or no concern for long-term sustainability and side effects.

The rapid resource-use intensification in the face of the massive growth in demand appears to be the immediate cause of several indicators of emerging environmental risks (table 9.6). The possible solutions lie in restraining and regulating the pressure of demand (or, rather, such driving forces as population growth) or in ensuring a higher use intensity of resources without degradation. The latter calls for high-productivity technologies with the potential for rapid resource regeneration and conservation, suited to mountain conditions. This, in turn, necessitates imparting the mountain perspective into research and development (R&D) policies (Jodha 1991; Messerli and Ives 1997; Rhoades 1992, 1997).

Macroeconomic policies

Macroeconomic policies are instrumental not only in influencing the pace and pattern of development but also in conditioning the nature of activities that influence environmental stability and sustainability of mountain resources (Banskota et al. 1990). In the Hindu Kush–Himalayas, many of the negative trends can be partly attributed to macrolevel economic policies. The missing mountain perspective is a significant gap in these policies, most of which are not designed for the mountain context but arise from the conventional practices or experiences in non-mountain areas (Banskota and Jodha 1992a; Jodha 1990b, 1995). This is so, whether one looks at investment priorities and resource allocation, factor/product pricing and other fiscal measures, infrastructure development and agricultural R&D, or choice of scale and technologies for various activities (Banskota et al. 1990; Jodha 1990b; Rhoades 1997; Sanwal 1989). Below is a brief treatment of some dimensions of macrolevel policies that seem to have adversely affected environment and hindered sustainable development in the mountains, for which evidence is available from different locations within the Hindu Kush–Himalayas.

Resource-extraction policies

Notwithstanding the recent focus on the welfare of mountain people and on the need for reducing interregional inequities, the focus of macroeconomic policies in the mountain areas has historically been directed towards the extraction of mountain resources, largely for use in the non-mountain hinterland (plains) or in urban areas within the mountain regions. The additional short-term consideration has been revenue maximization. A third dimension of the state's approach to mountain areas in the Hindu Kush–Himalayas has its origin in the geopolitics of the region. The patterns and goals of intervention depend largely on state perceptions of security. Meanwhile, the regeneration and sustainable use of resources and environmental stability in the mountains have seldom been major considerations in state policies (Banskota and Jodha 1992b; Rhoades 1997; Sanwal 1989). Both the mechanisms and procedures for resource extraction (e.g. classification of forests; system of contractors; and auction arrangements for harvesting of timber, development of irrigation and power potential without deferring to interests of local communities) are decided within this context. Similarly, product pricing and compensation mechanisms are guided by conventional yardsticks, rather than by the intrinsic worth of products and the sustainability implications of the pace and pattern of resource extraction. The phenomenal growth in demand for mountain resources, induced by distant market signals and with complete disregard for the "resource-use-intensification question" in fragile mountain ecosystems, can be attributed to the foregoing policies (Bandhyopadhyay 1989; Banskota and Jodha 1992b; Banskota et al. 1990; Jodha 1995; Paranjpye 1988).

Public sector investment: Allocative biases

In keeping with the "resource extraction" focus of development policies, the investment or resource-allocation patterns in mountain areas acquire certain specific features. Accordingly, most of the public-sector investment focuses on infrastructure development (e.g. roads) or on projects designed to harness mountain potentials (e.g. irrigation and hydropower). Unfortunately, in most cases, their gains in terms of helping mountain agriculture or supporting people's survival strategies are limited (Banskota and Jodha 1992a, b; ICIMOD 1990a–d). Owing to their scale and investment requirements, they leave few resources for ancillary activities that facilitate fuller use of the infrastructure and harnessing potential of diverse resources. Diversification and linkages of activities – the very preconditions (determined by mountain characteristics) for relevance and effectiveness of an intervention in these areas – are usually overlooked in investment allocations. Environmental auditing of investment decisions is, of course, a far cry from what is needed.

Besides the structure of investment, the low level of resource allocation to mountain areas also exacerbates the stagnation of mountain economies and the consequent degradation of natural resources, as poverty and environmental degradation are often closely linked (IPCC 1996b; Kates and Haarmann 1992; Rhoades 1997). The constraints imposed by inaccessibility, fragility, and diversity, raise the overhead and operational costs of development and service activities in mountains, both on a per unit and a per capita basis. Those very factors that cry for larger-scale investment in mountains are used for discounting investment opportunities in mountains by the conventional norms used in feasibility studies (Banskota and Jodha 1992b; Jodha 1990b). The consequence is persistent underinvestment in mountain areas, leading to stagnation, poverty, and environmental degradation.

A related aspect of public-sector investment is what may be called the *development culture* associated with public interventions in the mountains. The important features of public policies in mountain areas are centralized decision-making, perpetual subsidization of development activities, and substitution of traditional self-help and resource-protection devices by formal state interventions. Although initiated as a part of the extension of generalized public interventions in rural areas (in the mountains and elsewhere), these activities have had several negative side effects, including people's alienation from resources, resource degradation, increasing costs and subsidization of development activities, and a variety of inequities (Jodha, Banskota, and Partap 1992; Rhoades 1997; Sanwal 1989).

Technologies

Although science and technology have helped in resource-use intensification without undue environmental risks in different parts of the world, the mountain areas (and especially in the Hindu Kush–Himalayas) have witnessed no serious attempts to enlist such aid. On the contrary, applications of science and technology have gone forth with little concern for their side effects. Examples abound, such as the creation of massive power-transmission lines and networks of roads ignoring fragile rock alignments (Deoja and Thapa 1991); the extraction of minerals (Bandyopadhyay 1989); the generation of power through huge equipment and infrastructure, with little sensitivity as to their side effects (Paranjpye 1988); and the introduction of new cropping systems emphasizing monoculture and narrow specialization (Jodha 1991, 1995). Often, the primary goal here, as in other public interventions, is resource extraction and short-term gains. This entire approach to development and the use of scientific technologies for mountain areas merits vigorous challenge.

In most cases, technology, despite its irrelevance, is directly transferred to the mountains from the plains. In none of the countries of the Hindu Kush–Himalayan region does the existing R&D infrastructure match the requirements or proportionate importance of the mountain areas and their contribution to national economies. Even if some technology development is carried out in the mountains, the objectives and approach are seldom consistent with the imperatives of mountain specificities. Agricultural R&D, where work on new technologies (e.g. choice of cultivars, their attributes, and type of cropping systems) completely disregards the imperatives of mountain specificities such as diversity, fragility, and inaccessibility (Jodha 1991, 1995; Zimmerer 1995), is a prime example.

The features of public interventions that reveal an insensitivity to mountain specificities are corroborated by reviews of selected public policies and programmes in Himachal Pradesh, India (ICIMOD 1990a), West Sichuan and Xizang in China (ICIMOD 1990b), Nepal (ICIMOD 1990c), and Pakistan (ICIMOD 1990d). The site-specific case studies of farming systems in the same areas, which also covered the processes and impacts of development interventions, revealed several changes (table 9.6), that are less conducive to resource conservation and environmental stability (Jodha, Banskota, and Partap 1992). Conscious incorporation of a mountain perspective into public intervention can effect a reversal of these policies, although experience from the study areas belies this reasoning. At the same time, the experiences of a few "success stories" from the Hindu Kush–Himalayan region indicates some prospects of economic betterment without degradation of the resource base and environment. These cases include the integration of traditional and modern technologies for mountain agriculture; institutional innovations conducive to participatory development, with greater focus on stability and productivity of environmental resources; local, renewable, resource-centred cottage industries; and local niche-centred, integrated, area-development initiatives. The common factor in nearly half a dozen successful cases of rural transformation reviewed was their (conscious or unconscious) incorporation of mountain perspective into their programmes (Jodha, Banskota, and Partap 1992).

To sum up, the role of public policies in driving environmental risk in mountains can be stated as follows. Public policies and programmes are currently directed to the following: (1) integration of mountain areas with plains and urban areas through market and infrastructure; (2) extraction of mountain potential through technology and lopsided investment strategies; and (3) substitution of (a) traditional diversified resource-use systems by commercially and sectorally oriented, standardized arrangements, and (b) traditional self-help by subsidization, creating a permanent dependency of mountain areas and people on external help. All of these contain seeds of environmental instability in one form or another and enlarge the risk to the region and its people.

Linkages between cumulative and systemic changes

The preceding discussion has described various facets of the cumulative type of environmental change in the Hindu Kush–Himalayan mountain region. It highlights the central role of interactions between attributes of mountain resources and features of resource-use systems in creating environmental risks. The process of resource-use intensification, guided by several driving forces, has been described. With the unabated role of these forces, mountain areas and communities are liable to undergo greater environmental instability and its consequences. The level of instability and risks is already quite serious, even without the impacts of global systemic changes (e.g. global warming), which can further accentuate an already serious situation (IPCC 2001; Price and Haslett 1995).

Impacts of systemic changes

Details of potential systemic changes affecting the mountain areas are almost negligible compared with the information on cumulative changes. With full recognition of the limitation of the information on systemic changes (e.g. their conjectural nature and associated uncertainties of predicted change scenarios), however, a few possibilities may be examined. Accordingly, the potential changes in the Hindu Kush–Himalayas resulting from global warming, as summarized at a regional meeting in 1990 (Topping, Qureshi, and Sherer 1990), are still very relevant and include the following:

1. Owing to global warming, forests (the unmanaged ecosystems) may have both quantitative and qualitative changes. Some of the species may disappear; others may move spatially. This may accentuate the already known current negative trends relating to forest areas. The resulting reduced biodiversity may influence both biophysical func-

tions and flows governing environmental stability, thereby making the economy and survival strategies of people more vulnerable to risks.

2. The region may have higher rainfall (convective, high-intensity rains), which may cause increased runoff, flash floods, soil erosion, and mud- and landslides, and could influence overall farming systems. Impacts of such changes on the circumstances affecting basic biophysical functions and flows, on the one hand, and people's survival strategies, on the other, hardly need elaboration.

3. Increased warming would lead to increased snow-melting and consequent disturbance to hydrological cycles, seasonality of flows of water, and related impacts on land use and cropping intensities, disturbing the already threatened diversity and sustainability of mountain resource use. The environmental risks will, thus, be further accentuated.

4. To the potential changes one may add a few more possibilities. They include probable changes in the specific mountain conditions (such as fragility, diversity, or niches) and in their interrelationships; these changes may generate new constraints and opportunities, influencing the comparative advantages of mountains and their links with other regions, and perspectives of public interventions in mountain areas. At the microlevel, the agricultural systems covering all land-based activities may undergo several changes, including disturbance to well-adapted cultivars and management practices, product and income flows, and people's strategies for coping with risks (Jodha 1989, 1995). These changes, in turn, may influence resource-use patterns, with implications for environmental stability.

These potential effects, as well as their management implications, have also been explored by Price and Haslett (1995).

To sum up, the combined impact of all the above changes may result in increased compulsions or incentives (opportunities) for resource-use intensification, which may accelerate the already observed cumulative changes and their impacts on vital biophysical processes and flows.

Accentuation of cumulative change

A total view of the environmental risks in mountains due to global change can be obtained by a combined perspective of systemic and cumulative changes. Accordingly, table 9.7 presents some possibilities of current trends in resource degradation (cumulative changes) likely to be accentuated by impacts of global warming (e.g. systemic changes). The

Table 9.7 **Potential accentuation of cumulative environmental change under the impacts of systemic environmental changes**[a]

Current problem (cumulative type of change) likely to be accelerated by systemic change	Potential key manifestation of systemic change (impacts of global warming)		
	Vegetation changes: forest size, location, composition, growth cycle; biodiversity, interactive processes	*Increased convective rains*: floods, runoff, soil erosion, changes in growing season, hydrological cycle	*Warming-led snow melt*: increased water flows, soil erosion, changes in hydrology of mountains and flood plains
Deforestation, vegetation degradation, reduced diversity	X (R, F, N, S)[b]		X (R, N, F)
Soil erosion, landslides and mudslides, floods		X (N, F, S)	X (N, F)
Changes in land-use pattern, reduced diversity of farming systems, increased resource-use intensity and degradation	X (R, F, N)	X (S, N)	
Increased vulnerability of people's survival strategies to environmental instability due to resource degradation and disruption	X (R, F)	X (R, F, S)	X (R, S)

a. See table 9.6 for details of some of the negative changes indicating emerging environmental risks.
b. The capital letters indicate biophysical processes and flow likely to be affected: R, regeneration; F, flexibility, variability; N, resilience; S, energy and material flows.

impacts of the combined two types of changes on biophysical processes and nature's flows are indicated by the capital letters in table 9.7.

Potential impacts of global warming-induced changes in forests and vegetation may accentuate the current problems associated with deforestation, land-use intensification, overgrazing, and landslides (Hamilton 1995). The changes in flows of water due to warming-induced snow-melting or increased convective rains would also accentuate current problems. Thus the current crises reflected through cumulative changes in mountain areas in a way manifest the vulnerability of mountain habitats and the potential for negative impacts of systemic changes.

Socio-economic vulnerabilities

The discussion so far has focused on environmental change and risks largely in terms of disturbance to biophysical processes underlying the flow of environmental products and services used by human society. The disruption of the flows through cumulative and, to a limited extent, systemic types of environmental changes are apparent, as are impacts of these disruptions on the survival and growth strategies of mountain communities. These issues, however, require more systematic treatment. Hence, it is necessary, as chapters 5–7 emphasize, to examine the socio-economic–cultural vulnerabilities of mountain communities to environmental risks. A focus on environmental resources and their impacts on socio-economic variables (including opportunities and human decisions), rather than attention only to the basic biophysical processes determining the flows and stock of environmental services, must be part of this examination.

Socio-economic vulnerabilities, at the operational level, are apparent in the reduced range, viability, flexibility, dependability, and pay-offs of production and resource-use options to satisfy human needs. These problems may arise owing to the breakdown or infeasibility of diversified, resource-regenerative practices as well as to the degradation of the natural resource base. On the institutional side, slackening of resource-management/protection systems, reduced accessibility to resources, the reduced range and quality of options, and the marginalization of collective sharing systems all betray levels of socio-economic vulnerability. Some of these problems arise from disruptions in environmental and natural-resource situations; others cause such disruptions, as when socio-economic adjustments to environmental change create further changes in the environmental situation at secondary or tertiary levels.

One mechanism of change begins with increased scarcities owing to internal demands or external pressures on water, vegetation, and soil resources (as indicated by capital letters in table 9.8). It results in direct overextraction or promotion of adjustments that are more resource extractive. Each contributes in different ways to increased socio-economic vulnerabilities in terms of the reduced range and quality of options, particularly in time-tested resource-management systems. Table 9.8 indicates these possibilities, which relate mainly to the predominant activity (i.e. agriculture) of mountain communities. Such formulation, however, can

Table 9.8 **Environmental change and socio-economic impacts/vulnerabilities in mountain areas**

Environmental changes and underlying factors or responses to change	Socio-economic impacts/vulnerabilities[a]			
	Infeasibility of traditional production systems, regeneration, resilience	Reduced range/quality of options; control, access to resources	Increased external dependency, unequal exchange, subsidy, marginalization	Reduced collective sharing systems, low resilience, cultural breakdown
Physical degradation of land resources (W, S)[b]	X	X	X	X
Reduced variability, flexibility of production factors (V, W)	X	X	X^p	
Increased "ecological" subsidization through chemical, physical, biological inputs (V, W)			X^p	X^p
Vicious circle of resource degradation, overextraction-degradation (W, S)	X	X	X	
Niche, technology, market-induced overextraction, reduced resource availability/access (V, W, S)		X	X^p	X^p

Source: Adapted from Jodha (1995: 175).

a. Details presented in the table largely relate to agriculture dominated by stagnant production system but the items indicated by p apply to progressive agricultural areas as well.

b. The capital letters stand for worsening of the situation due to internal scarcities and external pressures with regard to the following resources likely to be affected by environmental degradation: W, water; V, vegetation; S, soils.

be present with respect to other activities, such as tourism, trade, and macroeconomic interventions.

Vulnerabilities to systemic change

The argument of linkage between cumulative and systemic changes can be easily extended to socio-economic vulnerabilities to environmental risks in mountain areas. The issues involved are apparent in several contexts.

First, to the extent that climate-change impacts are likely to be more severe for poor and marginal areas, mountain regions and communities currently experiencing environmental degradation and associated socio-economic disruptions may face even worse situations. These will occur as a result not only of erosion of their capacities to withstand the future crises but also of aggravation of resource extraction, imbalances, and scarcities (Price and Haslett 1995).

Second, systemic environmental changes are likely to aggravate the impacts of the off-site factors, such as external demand-induced resource extraction and marginalization of mountain areas and communities. In particular, one may think of increased external pressure on water, space, and vegetation (including biodiversity). The scale of applications and the specific technologies to be used in this process of resource extraction to meet the needs of off-site communities may further disrupt the survival and growth strategies of mountain communities.

Possible response: Dual-purpose strategies

If our understanding of linkages between cumulative and systemic changes is correct, the countermeasures for these socio-economic vulnerabilities can begin with strengthening of the ability of mountain communities to withstand the problems created by current resource degradation and environmental changes. This, in turn, calls for steps that enhance resource-use intensification with resource regeneration and conservation. In other words, it is necessary to achieve environmental stability and productivity in the current context to enable people to withstand impacts of systemic changes in the future. Accordingly, technological and institutional steps directed to enhancing the health and productivity of environmental resources (such as land, water, and vegetation) in the current context will also help their long-term sustainability. By implication, such steps will mitigate impacts of the cumulative type of changes and will also help to limit the effects of systemic changes. This is the essence of "dual strategies" directed to systemic and cumulative changes as well as "regional" and "global" environmental changes simultaneously (Jodha 1992c). The scientific and institutional prescriptions against global warming and its impacts (IPCC 1990b, 1996b) would have a greater chance of success if they were integrated into dual-purpose strategies. Such integration would reduce the role of uncertainty of modelled scenarios in obstructing the evolution of workable and readily acceptable strategies against systemic changes.

Some elaboration may be useful. One of the principal reasons for current societal inaction is the degree of uncertainty associated with modelled change scenarios, which in turn causes other problems (such as varying perceptions of different nations on the potential impacts of environmental change, the sharing of cost, and gains of strategies). In the broader context of uncertainty-induced inaction against environmental change and associated risks, an understanding of potential linkages between cumulative changes and systemic changes can offer certain useful leads (Jodha 1992c, 1995). Since the severity of impacts of cumulative changes is likely to be enlarged through systemic changes, and vice versa, measures against any one of them would help reduce the environmental risks. Since the cumulative changes are more certain and already exist as a reality, measures against them are less likely to be obstructed by the uncertainty phenomenon. Similarly, since the spatial context of cumulative changes is more concrete (e.g. deforestation in the Himalayas), the response measures against them (unlike systemic changes) would not be constrained by any lack of regional disaggregation of the problem. Finally, since cumulative changes are a part of the current problems in developing countries, any measures against them would need neither intensive lobbying and consensus building (as tried for global warming) nor any diversion of resources away from the current problems of poverty and underdevelopment.

If designed as dual-purpose strategies, the benefits of such measures (e.g. management of forests, land, and water resources) may strengthen people's capacities to withstand environmental risks associated with both cumulative and systemic changes (Jodha 1992c). Accordingly, a search for dual-purpose strategies should be the focal point of approaches to environment management in mountain areas.

The dual-purpose strategies against environmental risks cannot be confined to supply-side issues of the problem. In other words, no measures designed to regenerate, recycle, and diversify resources (for environmental stability) will help in the long run, without corresponding management of pressure on resources and associated extraction technologies.

Ultimately, the emerging indicators of cumulative and systemic changes are consequences of mismanagement or the free play of basic human driving forces, such as competitive and inegalitarian systems of resource exploitation, unequal systems of exchange, and population growth. In the face of these forces, no breakthroughs in resource regeneration and productivity gains may help in the long run. What has been stated above applies to the world economy in general, but applies even more so in fragile resource zones such as mountains, where the demand-induced resource extraction reaches its limits too soon. Hence, the solution to environmental risks in mountains is closely linked to the management of internal and external pressures on mountain resources (Messerli and Ives 1997; Rhoades 1997). A few important operational clues in this regard could be provided by better understanding of upland–lowland linkages, and the side effects involved in integrating mountain economies with mainstream economies through market, infrastructure, and administrative processes (Jodha 1995; Jodha, Banskota, and Partap 1992).

10

Vulnerability to drought and climate change in Mexico

Diana M. Liverman

In the southern Mexican state of Chiapas, thousands of hectares of tropical forest have been destroyed in the last hundred years. Timber companies, followed by ranchers and coffee producers, converted biologically diverse, soil-protecting forests into barren and desiccated landscapes (O'Brien 1997). Additional destruction of forests has accompanied the search for petroleum resources in the region and an influx of impoverished refugees from Central America and migrants from other parts of Mexico seeking farmland (Gonzalez-Pacheco 1983).

In Mexico City, contamination of air and water has already reached fatal or disabling levels for many residents of the metropolis, with increases in respiratory and cardiac diseases, gastrointestinal problems, cancer, poisoning and other toxic exposures (Aguilar et al. 1995; Ezcurra et al. 1999; Ezcurra and Mazari Hiriart 1996; Ortiz 1987; Puente and Legoretta 1988). Pollution has become a major political issue, and has prejudiced the international business and tourist reputation of the city.

In northern Mexico, 1989 was a year of severe drought and heat stress. As in the United States in 1988, people connected this drought with the predictions of global warming associated with the greenhouse effect and linked it to forecasts of the lowest agricultural production in nine years and a corresponding increase in malnutrition and expensive food imports.

These are localized examples of the types of global environmental changes that may face the world in the coming decades. Environmental changes in Mexico are typical of many countries and have similar social

343

causes and consequences. Who will be most vulnerable to these changes? In forest regions such as those of Chiapas, the most direct and severe impacts of deforestation probably fall on the indigenous peoples of the region, such as the Lacandon and Maya (Lobato 1980; O'Brien 1997). Forest clearance decimates the animal, bird, fish, and plant populations that provide the food and fibre base of traditional hunting-and-gathering livelihoods (Szekely and Restrepo 1988). Environmental changes also imperil the health and nutrition of the poorer refugees and migrants, who are trying to survive on small, infertile plots of deforested land. Larger farmers and ranchers, as well as timber companies, are experiencing diminishing returns to their unsustainable use of the Chiapan forests. Local and national governments are clearly concerned about political unrest stemming from the social and ecological impoverishment of the region (Aguayo 1987). The loss of biological diversity and vegetation cover in tropical forest regions, such as Chiapas, may also jeopardize the food and health security of the world as a whole and alter global nutrient and hydrological cycles (Myers 1984; Rosenzweig and Hillel 1998).

Although both rich and poor in Mexico City are exposed to ambient air and water pollution, the former can afford air conditioning, bottled water, rural retreats, good nutrition, and health care (Pezzoli 1998). Certain occupations – such as factory worker or driver – entail greater exposure than others, and pregnant women, children, and the infirm are often more vulnerable to pollution (Puente and Legoretta 1988). In addition to these individual health risks, air and water pollution incur economic costs to individuals, corporations, and governments in terms of prevention, cleanup, labour absenteeism, and material damage (Vizcaino 1975).

In the 1989 drought, newspapers reported many communities in northern Mexico without water, and children dying from dehydration during the peak of the heatwave. Important urban-industrial centres, such as Chihuahua and Monterrey, rationed their water supplies. The food supplies and livelihood of many small and cooperative agricultural producers of rain-fed crops were in jeopardy, and conflict over irrigation rights was intense.

Mexico is a useful case to consider the arguments developed in chapters 5–6, specifically that understanding and responding to the regional impacts of environmental change requires not only the best possible estimates of how the environment will change but also a detailed analysis of the vulnerability of society to such changes. The impacts of a climate change on environment and society are determined as much, if not more, by the characteristics of the regions and people affected as by the nature of the climate change itself (IPCC 1997, 1998). For example, rainfall declines of equal physical severity may have much less severe impacts on large commercial irrigated farms with insurance, good soils, and subsidized prices than on smaller, rain-fed subsistence farms. This differential ecological, economic, and technical vulnerability to climate variation and change is an important key to understanding the regional impacts of global warming, and parallels many of the arguments developed by Jodha in chapter 9.

Mexico is an important place to study the social causes and consequences of environmental change, in both the international and North American context (Liverman 1992). Mexico is a significant producer of greenhouse gases, ranking as high as ninth in the world according to some researchers. Mexico is important, ecologically, as a centre of crop genetic and biological diversity; economically, as an increasingly important importer of food; and politically, as a key actor among non-aligned nations on environmental issues (OECD 1998). These regional and international roles are played in the context of rapid growth in industrial development, urbanization, population, and consumption, as well as a volatile economic and political situation. In terms of global climate change, Mexico is a critical case study because of the sensitivity to any climate change that brings drier conditions or greater extremes. Much of Mexico has a seasonally dry climate, and droughts are frequent and severe. Only 20 per cent of Mexican cropland is irrigated, and even this is vulnerable to reservoir and well depletion in very dry years. Reliable water supplies are critical to cities, industry, and the production of hydroelectric power.

For the United States, Mexico is of tremendous economic, political, demographic and ecological significance. Mexico has long been an extremely important trading partner of the United States, and the enactment of the North American Free Trade Agreement (NAFTA) has tightened the ties. Environmental degradation in Mexico affects US imports of winter fruits and vegetables, the ability of Mexican consumers to afford US exports, and the ability of the Mexican government to service debts to US banks. The political stability and legitimacy of the Mexican government depends partly on its ability to respond to environmental and subsistence crises. The decision to migrate to the United States has been related to environmental crises in Mexico. Many birds and sea mammals, valued by US environmentalists, winter or breed in Mexico. A shared border means transboundary flows and conflicts over pollution and water.

Mexico is a challenging and exciting country in which to study vulnerability and climate change. Current climate models indicate significant potential increases in temperature and changes in precipitation in Mexico (Liverman and O'Brien 1991). It has also been suggested that global warming will be accompanied by an increase in sea level of up to one metre. For Mexico, continually striving to support a growing population with an agricultural system that relies on relatively low and variable rainfall, any warmer, drier conditions could bring nutritional and economic disaster. More than one-third of Mexico's rapidly growing population works in agriculture, a sector whose prosperity is critical to the nation's debt-burdened economy. Although only one-fifth of Mexico's cropland is irrigated, this area accounts for half the value of the country's agricultural production, including many export crops. Many irrigation districts rely on small reservoirs, and wells deplete rapidly in dry years. The remaining rain-fed cropland supports many subsistence farmers and provides much of the domestic food supply. Frequent droughts already reduce harvests and increase hunger and poverty in much of Mexico (Liverman 1990b).

A significant geographic mismatch exists between water and population in Mexico (Enge and Whiteford 1989): 7 per cent of the land, lying in the extreme south-east of the country, receives 40 per cent of the rainfall; only 12 per cent of the nation's water is on the central plateau that includes 60 per cent of the population and 51 per cent of the cropland. Any reduction in water availability would also create problems for cities, industries, and hydroelectricity production. At present, Mexico generates about one-fifth of its electricity from hydroelectric schemes and has developed less than 25 per cent of its potential. Water-supply infrastructure has been unable to keep up with rapid urban and industrial development. These other sectors compete with the agricultural sector, which consumes more than 80 per cent of total water supplies.

How might global warming affect Mexico? Higher temperatures could increase air pollution and heat stress, and sea-level rise could threaten agriculture, tourism, industry, and natural ecosystems along Mexico's extensive coastlines. These impacts would be likely to create economic, political, and demographic pressures that could aggravate tensions with neighbouring countries, particularly the United States (Gleick 1988, 1998; IPCC 1997: 12; Liverman 1990a).

Of course, many factors other than climate could affect Mexico's future. Economic crisis, social inequality, political instability, and population growth may well continue to reduce living standards over the next 25 years (Bailey and Cohen 1987; Castaneda 1986; Pastor and Castaneda 1988). Global warming is likely to accompany a doubling of the Mexican population to 160 million by 2025, with a corresponding increase in pressure on natural resources. Although changes in agricultural technology and land reform to some extent allowed agriculture to keep up with growing populations in the first part of the twentieth century, the distribution of economic and nutritional benefits of growth in agricultural production has been unequal among regions and social groups (Sanderson 1986). Further expansion of technology and land is unlikely in the next decade because of economic problems and land-resource constraints.

In any case, it is not clear that this type of change has reduced vulnerability to climate in the past or that it will do so in the future.

We should not – and need not – wait for improvements in the climate models before developing methods for assessing the regional impacts of global warming; we can acquire a general sense of possible changes in climate by examining a range of model results and sensitivities. More important, we can approach the study of the impacts of global warming from a different perspective by looking at vulnerability to climate change (Downing et al. 1999). The impacts of global warming will depend much less on the exact amount of the temperature or moisture change than on the characteristics of the regions and people that experience the changes.

In Mexico, we gain considerable insight into the possible regional impacts of global warming by analysing the vulnerability of agriculture and other resources to current variations in temperature and precipitation (Appendini and Liverman 1996). The following discussion of vulnerability focuses particularly on the factors that influence the impacts of warm, dry weather periods – droughts – in Mexico.

Historical context of climate vulnerability in Mexico

Palaeoclimatic and instrumental records suggest that significant shifts in climate have occurred in Mexico with long periods of drought (Metcalfe 1987). Such changes have been implicated, together with changes in social organization, in both the rise and collapse of powerful pre-Hispanic civilizations, such as those in the Yucatan and in the valley of Mexico. Historians have also blamed climate for famine and social unrest in the colonial period. Florescano (1969), for example, has linked variations in the price of maize in the sixteenth and seventeenth centuries with droughts and other climatic events. He claims that a severe drought preceded the majority of price rises, but he also notes the role of speculation and economic arrangements in triggering price rises and associated famines. Florescano suggests that the economic and land-tenure relations imposed by the Spanish crown created a tremendous vulnerability to drought among the poorer and indigenous *campesino* populations. The colonial political economy allowed the larger landholders and merchants to manipulate the price of staples in drought years to the disadvantage of poor consumers and small producers. According to Florescano, post-revolutionary Mexico betrays differential vulnerabilities to drought, inherited from the colonial land-tenure systems: "The most disastrous effects of drought, as in earlier times, are concentrated in the rainfed agriculture practised by the poorest *ejidatarios* and *campesinos*, lacking credit, irrigation, fertilizers, and improved seeds" (Florescano 1980: 17).

In this century, a steady and unvarying expansion of Mexican agricultural production has been necessary to meet the demands of a rapidly growing population and agricultural export market (Wellhausen 1976). In the drive to modernize and expand production, the Mexican agricultural system has incorporated some techniques (such as irrigation, improved seed varieties, and chemical inputs) that may reduce hazard losses (Venezian and Gamble 1969; Lamartine Yates 1981). Very little research addresses the role of the Green Revolution in reducing or increasing vulnerability to climatic variation. Michaels (1979) and Hazell (1984, 1991) have suggested, for example, that improved wheat varieties in Mexico and other countries may be more sensitive to drought and climate variability than traditional seeds. Some believe that the large regions of single-variety, high-yielding crops established through the Green Revolution are much more vulnerable to pests and diseases than traditional mixed cropping (Pearse 1980). Others suggest that these new techniques have replaced some traditional hazard-prevention strategies, such as mixed cropping and microclimate modification (Wilken 1987), and have allowed agriculture to expand into high-risk areas, such as deserts, mountains, coastal regions, and the disease-susceptible humid tropics.

Another question in Mexican agriculture concerns land reform, particularly the performance of the *ejido* sector. Although some authors (Wellhausen 1976; Whetten 1948), claim that the *ejidos* are inefficient and unproductive, others (Dovring 1970; Mueller 1970; Nguyen 1979) suggest that, in terms of input use, they are relatively efficient and produce yields equal to the private sector. The problems of the *ejido* sector have been explained in terms of its lack of political power, difficulties in obtaining access to credit and inputs, bad management, and poor land-resource endowment (Coll de Hurtado 1982; Hewitt de Alcántara 1976).

Agricultural vulnerability to drought in Mexican states

What might changes in agricultural technology and land tenure mean for vulnerability to climate in Mexico? It is possible to examine differential vulnerability to drought and other natural hazards in Mexico in terms of physical geography, access to technology, and land tenure. Standard weather and agricultural data have been used to examine the relationship between physical drought severity, reported drought losses, yields, land tenure, irrigation, and other inputs.

In the states of Sonora and Puebla, as reported in the 1970 agricultural census, the physical pattern of drought, expressed in terms of rainfall deficits and evapotranspiration, did not correlate strongly with the pattern of drought loss. In both states, some *municipios* (local administrative districts) with high drought losses were in areas of low rainfall, and some with low losses were in places with severe rainfall deficits.

Thus, vulnerability to drought loss was not directly linked to physical climate conditions. In general, lower drought losses seemed to be associated with the use of irrigation, high-yielding varieties, and fertilizer, suggesting that these technologies may reduce the impact of drought. In some *municipios* with reportedly high drought losses, however, irrigation did not seem to buffer the agricultural system against rainfall deficits. The *ejido* sector in both states was, on average, reporting drought losses twice those of the large private landholdings.

At the national level, the decadal census data from 1930 to 1970 indicate that, on average, more than 90 per cent of all hazard losses in Mexican agriculture are from drought. The area of total hazard losses increased in Mexico from 1930 to 1970, partly reflecting a change in the total crop area, which increased from 7.8 million hectares in 1930 to 11.8 million hectares in 1960; however, relative (percentage) losses also increased, indicating that some of the land expansion may have included more hazard-prone land. Moreover, both absolute and percentage hazard losses increased dramatically in 1970, when cultivated land area dropped to 10.2 million hectares. The meteorological record shows no indication of increasing severity of weather events in the later census years, especially from 1960 to 1970. The increase may, therefore, support the more general hypothesis set forth in chapter 7, that hazard losses have been increasing, irrespective of the incidence of events, owing to increases in vulnerability to natural disasters.

High drought losses tend to occur in the arid northern region of Mexico, where precipitation is low and highly variable. Parts of northern Mexico lost to drought, in the summer of 1970, more than 20 per cent of the area planted. States with high levels of irrigation, such as Sonora and Sinaloa, do have much lower drought losses than many of the states with smaller proportions of irrigated land. At the same time, Aguascalientes and San Luis Potosi, with about 20 per cent of their land irrigated, reported drought losses of 70 per cent and 50 per cent, respectively, indicating that irrigation may not always buffer Mexican states against drought.

In every census year the average losses on private croplands are lower than those on the *ejidos*. In 1970, losses were almost double the average on *ejido* lands. Explanations of this increased vulnerability may be that more biophysically marginal land was given to *ejidos* in the land reform and that *ejidos* are socially vulnerable because they cannot acquire access to irrigation, credit, improved seeds, or other resources.

Agricultural modernization and land reform clearly interact with drought and climate variations. The expansion of irrigation, the adoption of new seed varieties, the use of chemical fertilizers, and the modification of land-tenure systems will often reduce, but sometimes increase, vulnerability. Clearly, we must consider such factors in assessing the possible impacts of global warming in Mexico and elsewhere. In general, the 80 per cent of Mexican cropland that is rain-fed and the 45 per cent that is in *ejidos* may be the most severely affected by global warming. If global warming brings more intense droughts, however, as the IPCC (1997, 2001) suggests may occur, commercial agriculture that depends on irrigation and improved varieties may also suffer.

Vulnerability in local communities in Mexico

The tremendous heterogeneity of the Mexican landscape makes even regional vulnerability assessments quite difficult. Topography, soils, climate, and culture can all vary across short distances, especially in the central highlands. All but the most local studies may conceal the actual temporal, spatial, and social variations in vulnerability to climate.

Climate vulnerability at the local level may show up in studies of the use of land and water in Mexican communities. Geographers and anthropologists have conducted many studies in rural Mexico (DeWalt 1979; Enge and Whiteford 1989; Hewitt de Alcántara 1976; Johnson 1977; Kirkby 1973; Lees 1976; Sheridan 1988). What do these studies tell us about the nature of, and changing vulnerability to, climate change and variation in Mexico? The low, seasonal, and highly variable rainfall presents challenges to water management and social organization in Mexican agricultural communities.

Mexican farmers have developed a variety of technical and social means to use these micro-environments and to cope with a varying environment, but it is clear that population growth, environmental change, and (most important) economic and political transformations are reducing the effectiveness of these traditional mechanisms and increasing vulnerability to drought. The more recent adaptations to drought include shifts into wage labour and out-migration.

Many communities exhibit wide variations in vulnerability to climate variations because of inequities in access to land and water. Reliable agriculture usually requires irrigation, and irrigation systems have created complex social organizations and political institutions for managing them. Irrigation is a source of wealth and power in rural Mexico because it buffers people against variations in the climate. In some communities, a few large landowners control the best and most easily watered lands.

These farmers are increasingly growing cash and forage crops that do not contribute to local food self-sufficiency. Many farmers have plots of land that are too small to support a family and that have insecure access to irrigation water. Inequality is felt hardest when rains are limited or water resources become depleted. Further irrigation development may benefit some, but this has been demonstrated in the past to have increased the vulnerability of others and to have fostered inter- and intracommunity conflict. The studies also demonstrate how irrigation is exhausting water resources and leading to the salinization of fertile land. In some cases, drought, by impoverishing some, has created opportunities for others to obtain their land. The studies show how differences in land quality have given rise to competition for land in many communities. Historically, these land conflicts have been resolved by force or by government policy, which some claim is biased towards the more powerful groups.

In many cases, the availability of fertilizer, hybrid seeds, and machines has permitted some farmers to take advantage of irrigation and to stabilize their yields. However, some communities have found that fertilizer applications in dry years harm the crops and incur costs that cannot be recouped. New varieties of seed are costly and are sometimes too tightly attuned to average conditions, failing more easily than traditional open-pollinated varieties in extreme years. Irrigation has depleted groundwater and caused salinization of soils. Economic crisis means that many farmers can no longer afford the inputs that they were persuaded to use in more affluent times. Yet education and development have meant the loss of many traditional techniques and insights for reducing the impacts of climate.

Market integration, with the production of crops for sale and export, has benefited some people and communities: some of the cash crops are more drought resistant than traditional food crops, and receive higher prices; however, forage and fibre crops grown for sale cannot be easily converted to subsistence food sources for local people when markets or rains fail.

Finally, population growth and redistribution have clearly increased vulnerability to drought in some communities by fostering subdivision of the land into unsustainable parcels, and exacerbating pressure on land and water resources. In some cases this population pressure results from high local birth rates, but in other cases it originates in land concentration and agricultural modernization elsewhere which, in turn, stimulate migration.

These community studies provide useful insights into the distribution and dynamics of climate vulnerability in Mexico. They illustrate how local politics, local knowledge, land tenure, water rights, government subsidies, agricultural technology, and micro-environments define the risks of producing different crops under varying environmental conditions. They also provide insights into how global change might be locally experienced and adapted to in Mexico. If global warming means more frequent droughts in Mexico, then conflict over land and water, land concentration, and rural migration may increase.

In any community, physical heterogeneity and social inequality will mean that the impacts of global warming will be experienced differently in each micro-environment and by each individual. The pace of economic, political, technological, and demographic changes in many of these changed communities is rapid, and it is difficult to envisage how these will interact with climate change. However, it is clear that some Mexicans in many communities are already highly vulnerable to environmental and economic change, and that climatic change is likely to affect them adversely and prematurely. Population growth, market integration, technological change, and political development will change vulnerability to climate in Mexico's future, as they have in the past. Many researchers have advocated reductions in poverty, more efficient use of water, redistribution of land and resources, and the rediscovery of traditional agriculture in the communities that they studied. Such moves towards more equitable and sustainable agriculture in Mexico might also go a long way toward reducing the impacts of global environmental change.

11

Sea-level rise and the Bangladesh and Nile deltas

James M. Broadus[†]

The Bangladesh and Nile deltas are often cited as among the most vulnerable regions to sea-level rise driven by greenhouse warming (IPCC 1990a, b; Jacobson 1989b; Milliman, Broadus, and Gable 1989; Titus 1986). The geographic exposure of these two deltas, high population densities, poverty, and limited ability to defend themselves against natural disaster have placed them as archetypes among the world's "worst cases" of potential global-change impacts. Closer examination of these two cases, however, makes clear not only the importance of their vulnerability but also the huge uncertainties that remain about the impacts and the complex of options for response. As such, it emphasizes the critical role of uncertainties in global environmental-risk analysis, as argued in various chapters of this volume, beginning with chapter 1.

The presumption is that global warming, driven by the build-up of greenhouse gases from human activities almost entirely outside Egypt and Bangladesh and largely in industrialized countries, will increase the volume of water in the oceans (through a combination of thermal expansion and melting ice) and lead to sea-level rises throughout the world (IPCC 1990b; USNRC 1990). The impacts of sea-level rise include inundation of low coastal lands, shoreline erosion, loss or relocation of wetlands and beaches, increased exposure to storm surge and flooding, and increased salinity of rivers and aquifers (Bird 1986; IPCC 1990b). These impacts depend on the magnitude and pace of local relative sea-level change, of which change in global sea level is only one component (the others being various forms of local land subsidence or uplift). We still cannot distinguish local from global sea-level change empirically (Barnett 1984; Solow 1987; USNRC 1990).

The Intergovernmental Panel on Climate Change (IPCC) estimated that global sea level could rise by some 30 cm by 2050 (IPCC 1990a, d). It is useful, therefore, to attempt a rough estimation of the potential impact of future sea-level rise in these two especially vulnerable regions and to conjecture about possible responses and implications for policy.

Assessing future impacts

Economic impact assessments are a form of cost–benefit analysis. Generically, cost–benefit analysis is simply the counting up of estimated benefits and costs associated with some event or action to compare them with each other or with those estimated for some alternative (Smith 1986). Typically the estimated benefits and costs accrue from different sources and at different levels over time, and it is important to try to include them all and not to double count. A premise of cost–benefit analysis is that economic decision makers seek to act rationally, and that they act in such a way as to do the best they can with the information and other resources available. The goal in dealing with prospective actions thus is to identify the one with the greatest net benefits (or least net cost).

For potential events such as accelerated relative local sea-level rise, the goal may be to estimate the expected cost of the event in the absence of mitigation. This provides a basis against which to compare the costs of investment in mitigation. Of course, where information is sparse and uncertainty about future events is great, as Funtowicz and Ravetz point out at some length in chapter 4, the best we can hope for are rough approximations. Even those may be useful, however, to clarify relative magnitudes, to organize thinking about possible outcomes, and to make both ignorance and uncertainty relevant to policy.

Most attempts to assess the economic impact of projected relative sea-level rise use a kind of "colouring book" approach. First, a scenario is selected that describes a hypothetical rise, often within a specified time period. The scenarios are usually selected from a range of projections (Hoffman, Wells, and Titus 1986; IPCC 1990a, b; Yohe 1990) and may be adapted to account for local land subsidence (Barth and Titus 1984; IPCC 1991; Milliman, Broadus, and Gable 1989). The area subject to inundation from such a rise is then identified with topographic information and "coloured in" on a map. Expected effects other than inundation (such as saline intrusion, storm exposure, and structural damage) may also be identified to complete the scientific description of the physical "dose–response" function (IPCC 1990a, d; IPCC 1991). Cost–benefit techniques are then applied to estimate the potential loss in economic terms.

That is, in general, the approach (Broadus et al. 1986) employed here, and its focus on economic analysis complements other chapters in this volume. First, geological information is used to describe hypothetical scenarios for physical transgression of sea level beyond existing coastlines over the next 50–60 years. A scenario of a 1 m rise by 2050 is examined here. This is much higher than the IPCC "best guess" of 0.3 m but is consistent with the IPCC "boundary condition" of 1 m (IPCC 1991). It may be thought of as a conservative upper boundary representing some combination of global sea-level rise and local subsidence. The scenario is then used to describe the geographic areas within Bangladesh and Egypt that would be affected by landward transgression. Next, an examination of demographic and economic information helps to portray, within each nation, the scale of economic activities that currently originate within these potentially affected areas. It is most important to note that we are estimating only current levels of economic activity within areas that potentially will be affected. Using strong but reasonable parametric assumptions, this can be extended to an even cruder estimate of potential economic loss.

Bangladesh

This densely populated nation of an estimated 118 million people covers an area of 144,000 km^2 (fig. 11.1). Eighty per cent of this area is made up of the complex Bengal delta system created by the Ganges, Brahmaputra, and Meghna Rivers (Ahmad 1976). Together, the Ganges and the Brahmaputra currently deliver approximately 1.6 billion tons of sediment annually to the face of the delta (Milliman and Meade 1983). This sediment replenishment appears just to offset the natural compaction and subsidence of the delta, keeping its size relatively stable (Alam 1996; Milliman 1987). There is some evidence, however, that the delta may be (or may reasonably be expected to be) in net accreting as a result of increased upstream erosion (Alam 1996; Brammer 1993). The delta complex is, in any event, an extremely complicated, dynamic, and continuously changing physical setting. The uncertainty about whether its total land area is now roughly stable or in net accreting only reinforces the extent of the uncertainties about current conditions, not to mention those about future changes.

The country's population is widely distributed, with heavy concentra-

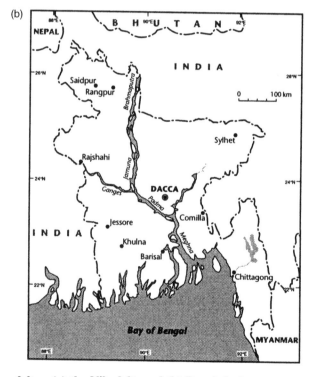

Fig. 11.1 **Two deltas: (a) the Nile delta and (b) Bangladesh**

tions in the major city of Dhaka (population over 4 million), in the south-western city of Khulna (population over 800,000), and in the eastern port city of Chittagong (population about 1.5 million). Most of the remaining population is rurally dispersed and dependent on subsistence agriculture. This is a highly traditional, largely rural society, strongly bound to customary relationships. "Poverty, unemployment, malnutrition, illiteracy, rapid population growth, and frequent national disasters continue to remain the endemic problems of Bangladesh" (Mahtab 1989: 2). Ownership of the principal productive asset – land – is notoriously maldistributed, and a chronically large landless class dominates the subsistence sector. Water routinely inundates some 30 per cent of the country's area during the annual flood season (Mahtab 1989), so coping with floods has become a customary part of rural life. Much of the population lives at the very edge of subsistence. Per capita gross domestic product (GDP) is approximately US$180, compared with that of Egypt which is US$700 and that of the United States which is more than US$21,000. Population increase is in the range of 2.5–3.0 per cent per year.

For Bangladesh, the 1 m scenario can be thought of as 10–50 cm global sea-level rise, consistent with the IPCC estimate, and local subsidence of 50–70 cm by 2050. Milliman, Broadus, and Gable (1989) assume local subsidence of 22–65 cm over this time period in response to natural background (0.6–20.0 mm/year; Mahtab 1989) and reduced sediment replenishment because of river damming. Alam (1996) infers a Holocene subsidence rate of 2.5 cm/year, which, if renewed, would imply an even greater rate of local subsidence than that assumed by Milliman, Broadus, and Gable (1989). Again, the degree of uncertainty about the problem is reinforced.

About 7 per cent of the country's habitable land lies within the area covered by a 1 m rise. Even this estimate is quite uncertain and differs from some other reports. It is smaller than reported earlier by Broadus et al. (1986) because it excludes large aquatic areas that were included in the earlier estimate, and it is based on more precise computer digitization. Mahtab (1989), however, estimated that about 15.8 per cent of the total area of Bangladesh would be inundated by a 1 m rise, which is closer to the earlier, unrevised report by Broadus et al. (1986). The Bangladesh Centre for Advanced Studies (BCAS), meanwhile, has reported that an estimated 17.5 per cent of the total land area would be inundated in the case of a 1 m rise (BCAS 1991a). The estimated extent of the affected area does appear to reach further inland in Mahtab's "colouring book" than in the estimate used here, and further inland still in the BCAS (1991a) portrayal. This underscores substantial uncertainty even about the location of the coastal 1 m contour line. Our estimate was provided from contour maps by geologist J. Milliman (Milliman 1987). It is not clear how Mahtab or BCAS obtained their contour lines or how they treated aquatic areas in deriving their percentage estimates.

The spatial distribution of population, if superimposed on the projected 1 m transgression of sea level, would mean that approximately 5 per cent of the nation's present population would live within the area affected by a 1 m rise. The densely populated area around the southwestern city of Khulna, where population density exceeds 2,900 persons per square kilometre, is situated just beyond the 1 m transgression line. Again, other estimates vary: Mahtab (1989) appears to estimate that about 9 per cent of the population is within the 1 m area, and BCAS (1991a) reports its estimate that "over 11 per cent" of the population "will be displaced" by a 1 m rise. The BCAS estimate seems to assume inundation of Khulna, but Mahtab does not.

Exposure of the population to storm surge is an extremely grave consideration in Bangladesh. Severe cyclonic storms attack the country at a frequency of approximately 1.5/year, with storm surge reaching as far as 160 km up the river courses in recent experience (Bird and Schwartz 1985). Of course, some of the most severe monsoon floods in Bangladesh are not caused by sea-level rise or storm surge at all, although some Western observers insist on attributing such disastrous floods as those in August and September 1988 to a combination of sea-level rise and Himalayan deforestation (Eckholm 1976; Jacobson 1989b; Myers 1986). In fact, there was no involvement of sea-level rise: most of that flooding occurred inland from the coast, and persuasive doubts have arisen about the connection with Himalayan land-use practices (Ives 1989). Nevertheless, the worst events do seem to be those at the coast caused by cyclonic storm surges, and these could worsen if future sea-level rise occurs.

The tragic cyclone of 29–30 April 1991 swept away entire island communities with its 7 m storm surge and winds of 230 km/hour (Martin 1991). Some 200,000 people may have drowned, and 10 million more were left homeless (*Economist* 1991; Nickerson 1991). The greatest casualties were among land-starved subsistence settlers on the transient "new islands" of river silt that congeal and disappear in the dynamic coastal delta. Their plight illustrates the region's strong linkages among poverty, landlessness, and exposure to natural disaster. The storm was well announced and had been tracked for four days before striking; many of the widely scattered rural poor never heard the broadcast warnings, however, and many of those who did would not risk abandoning their tenuous landholds. Tens of thousands were saved in about 300 elevated concrete cyclone shelters, but more than double that number of shelters would have been needed to accommodate the million most exposed coastal residents (*Economist* 1991).

The larger number of shelters, and enhanced disaster preparedness generally, were set as high national goals when Bangladesh won its independence from Pakistan in 1971. In fact, that independence movement was fuelled by outrage at Pakistan's indifferent response to the terrible November 1970 cyclone that may have killed as many as 500,000 people. Despite tens of billions of dollars in foreign assistance since then, however, the government has never been able fully to accomplish these goals. Indeed, the political priority on land grants sometimes competes with storm protection, as major land-reclamation projects capture river sediments to lure even more settlers to exposed sites (Brammer 1993; Koch 1988). In the wake of the 1991 storm, only rapid mobilization of a US military relief effort averted widespread starvation and disease, perhaps foreshadowing a problematic new international role for the military forces of the advanced powers (Weiss and Campbell 1991).

With increasing population densities in the future, through the combined effects of natural population increase and possible loss of land to coastal transgression, the exposure to severe storm surges will almost certainly increase. In that context, a noteworthy feature of the Bengal Delta is the 6,000 km^2 mangrove and nepa palm Sundarban Forest Reserve in Khulna District on the south-western coast. A maze of forested waterways, this reserve contains no permanent settlement. Immediately adjacent to it, however, are heavily settled areas, including the densely populated environs of Khulna. It is reasonable to asume that the Sundarban Forest provides vital protection for these settled areas by acting as a buffer against the force of storm surges. Loss of this buffer could increase the threat of storm surges, and the forest does appear vulnerable to a 1 m rise in relative sea level.

Agricultural production comprises nearly one-half of the nation's GDP, and it is estimated that over 85 per cent of the nation's population depends on agriculture for its livelihood. Net cropped/agricultural areas are by far the major use of land in Bangladesh. Judging from the area of distribution of rice and jute production, over 5 per cent of the nation's agricultural output (or about 7 per cent of the nation's crops) appears to originate in the area seaward of the 1 m contour. A further consideration for agricultural productivity is the intrusion of salt water into the nation's fresh groundwater resources. Current estimates in the National Water Plan suggest that salt-water intrusion may now extend seasonally more than 150 km inland in the western districts and along the Meghna River. If future damming or diversion of the Bengal rivers effectively cut off new sediments from maintaining the delta, and if global rise in sea level over the next century is severe, the intrusion of salt water could well reach beyond Dhaka (Mahtab 1989).

Egypt

Of Egypt's area of 1 million km^2, only about 35,000 km^2, or 3.5 per cent, is cultivated and settled. This results in an estimated population density of about 1,500 people for every square kilometre of arable land in the country. The population is densely clustered about the banks of the Nile River and throughout the river's delta eastward from the port city of Alexandria to the northern entrance of the Suez Canal. Alexandria, exposed since antiquity to the forces of the sea and historically reached by a causeway extending to its walled enclosure, today contains nearly 3.5 million inhabitants and 40 per cent of the nation's industry (El Sayed 1991). Port Said, at the eastern extremity of the delta, is the home of nearly half a million residents. Much of the delta between is densely inhabited and devoted to intensive, multicrop agriculture, and to urban and industrial uses.

The 1 m rise in relative sea-level scenario corresponds approximately to a 10–40 cm global sea-level rise, and severe local subsidence of 60–70 cm by 2050. El Sayed (1996: 222) reported archaeological evidence that Alexandria's site has experienced continuous subsidence over the past 3,000 years at an average rate of 1.2 mm/year; Neev, Bakler, and Emery (1987) show that the region immediately east of the Nile Delta has been subject to rapid tectonic movements; and Stanley (1988) indicates recent subsidence in the north-eastern delta of 3.5–5.0 mm/year (Hamilton 1990; Harney 1992). As in Bangladesh, projecting future rates of subsidence for the northern Nile Delta is extremely uncertain, but the range selected seems reasonable as an outer bound. The major existing natural defences against transgression by sea-level rise are the series of sand dunes along the delta's coast and the increasingly brackish lakes that lie behind them. These lakes are a major source of the nation's annual fish catch of approximately 100,000 tons, 80 per cent of which is freshwater fish. The area affected by the 1 m relative sea-level rise represents approximately 12–15 per cent of the nation's arable land and (based on the assumption that population is evenly distributed within each governorate) contains approximately 14 per cent of the nation's estimated 55 million population, or almost 7.7 million people.

Assuming that agricultural output in the delta is distributed as arable area, and assuming that all other sectors of economic activity are distributed as population, approximately 14 per cent of Egypt's current GDP appears to originate in the area likely to be affected by a 1 m increase in relative sea level. If, as in historical times, Alexandria's 3–5 m elevation were to spare it from an encroaching sea, then the 1 m figures should be adjusted downward (El Sayed 1991). Specifically, less than 7 per cent of the nation's current population is then in the potentially affected area, and closer to 6 per cent of current GDP. El Sayed (1991, 1996) argued that, in part because of Alexandria's elevation, in part because of the shielding effect of the coastal dunes, and in part because settlement and other economic activities are not evenly distributed along the northern Nile Delta, the impact of sea-level rise will be minimal. As the most likely exceptions, he cited the harbours of Port Said, Damietta, and the entrance of the Suez Canal, along with the recreational beaches near Alexandria.

An extension to crude loss estimates

Although techniques have been developed and employed elsewhere to measure the present monetary value of potential net economic losses associated with physical scenarios for relative sea-level rise (Barth and Titus 1984; Yohe 1990), the approach to impact assessment reported above is not intended to estimate such future damages. It is, rather, an entirely mechanical approach intended only to portray the current scale of economic activities and population in potentially affected areas, as a first approximation of the economic stakes involved.

By imposing an additional sequence of strong assumptions, however, it is possible to extend the analysis to generate even coarser estimates of potential economic losses associated with the inundation scenarios. Obviously crude, and at best approximate, this value provides another, more summary, estimate of the potential scale of economic impact, expressed in terms of net present value. It is useful primarily for early reconnaissance of the potential problem and should be read with great caution.

The method used for the extension can be described simply. It begins with current gross domestic product (GDPo), which is known. Assume a time horizon (T) with a constant economic growth rate (r), for cumulative GDP over the full time horizon without any sea-level effects:

$$\text{GDPo} \sum_{t=0}^{T} (1+r)^t.$$

Weighting this cumulative value by agriculture's land-rent share of GDP gives cumulative agricultural land rents:

$$\infty \text{GDPo} \sum_{t=0}^{T} (1+r)^t.$$

The last parcel lost to inundation over the time horizon will produce its full share of output right up until the last year of the time horizon. The annual rent value of the first parcel lost, on the other hand, accrues as an additional economic loss every year over the time horizon. The actual rate of land loss would probably be non-linear, but the simplest case is a rent loss that is a linear function of time (t), giving a cumulative loss of land rent over the time horizon:

$$L = \infty \text{GDPo} \sum_{t=0}^{T} \partial(t)(1 + r)^{t}.$$

As all of this rent loss from relative sea-level rise would occur in the future (much of it over many years from now), it is helpful to express the full stream of losses in discounted present-value terms. The social rate of discount is the factor by which future costs and benefits are reduced to make them comparable with present costs and benefits (JEEM 1990; Lind et al. 1982). With social rate of discount ∂, the net present value of the cumulative loss (NPVL) of agricultural land rent becomes:

$$\text{NPVL} = \infty \text{GDPo} \sum_{t=0}^{T} \partial(1 + r)^{t}(1 + a)^{-t}.$$

With the unlikely (but arguably realistic and clearly convenient) assumption that the economic growth of the economy just equals the social discount rate, $r = a$, then the "back of the envelope" expression for estimated loss simplifies to:

$$\text{NPVL} = \infty \text{GDPo} \sum_{t=0}^{T} \partial(t).$$

Empirically, we know that agriculture accounts for about one-half of GDP in Bangladesh and about 15 per cent in Egypt, of which we assume about 50 per cent is rent earned by the productive qualities of the land itself (with the other one-half accounted for by returns to labour, capital, and risk-bearing). Hence, for our purposes, rent is set at 0.25 ($= 0.5 \times 0.5$) for Bangladesh and 0.075 ($= 0.15 \times 0.5$) for Egypt. Examination of reported land-use practice in Bangladesh suggests that the overwhelming bulk of land in the affected area that is not uncultivable, waste, or forest reserve is cropped; it is, therefore, reasonable to treat all rent-bearing land as agricultural. For Egypt, however, as much as one-half of the rent-bearing land in the delta may be non-agricultural, so is divided by 0.5

(= 0.15) to account for the loss of non-agricultural land rents there. Subject to several strong caveats and assumptions, we might thus expect that the present discounted value of the economic loss in Bangladesh and Egypt by 2050 of an unchecked 1 m rise in relative sea level would be of the order of 1–2 per cent of current GDP in Bangladesh and 0.9–2.0 per cent of current GDP in Egypt.

Qualitative qualifications

There are, of course, several good reasons to approach these crude estimates with a strong sense of scepticism. Foremost among these is the huge uncertainty that sea-level changes of such magnitude will even occur. Even if they do, they will stimulate human responses, including national allocation of resources according to observed conditions. In view of the awesome loss of life to storm surges in Bangladesh, of course, any assumption that responses will be "optimal" must be seriously examined. Still, the bulk of potential costs would occur decades in the future, a time horizon over which even quite large economic values are apt to reduce to relatively trivial sums in present discounted value. (Again, the potential for loss of life, together with likely intergenerational effects, raises serious issues about the interpretation of discounting to present "value" [but see Cropper and Portney 1990]). Also, the accelerating pace of technological change promises to change the picture entirely over the time frame in question. Uncertainty, in short, abounds.

The method for extending current economic data to projections of potential future losses, reported here, not only abstracts from some key mitigating factors but also omits some important sources of potential loss. These compound substantially the other uncertainties. Again, it is assumed that human response is largely passive in the face of rising local sea levels. The method results in a certainty-equivalent value confined to land inundation and, aside from obvious sources of imprecision, underestimates lost capital structures; impacts on subsistence farmers; increased exposure to storm damage and interior flooding; and such secondary effects as saline intrusion, crowding, and factor reallocation costs. Even so, it is probably an overestimation, as it is not likelihood weighted, assumes a linear rate of land loss, and (most important) ignores cost reductions arising from human responses.

Confining estimated losses to lost land rents obviously misses immobile physical capital (such as houses, wharves, factories, and bridges) that might also be lost to inundation. Where data on property values are available, and where market institutions are not severely distorted, the value of capital services will be reflected in reported property values.

However, when only the estimated value of land rents is included, as above, then the omission of physical structures can be important. It should be recognized, even so, that only the *remaining* lifetime value of lost capital should be counted, not the original or replacement value, so that the abandonment of largely depreciated capital need not weigh very heavily.

Limiting the estimate of losses to lost land rents also underestimates the true potential impact on subsistence farmers. Bangladeshis who participate as sharecroppers in commercial production may share in a part of the rents lost to inundation, but since that loss is very likely to comprise most of what they possess, it hurts immeasurably more than the financial loss borne by the landowner. For independent subsistence farmers, the effect is much the same: lost financial value has little meaning when what is taken away is the means to produce food for survival. One of the downsides of land reform, in fact, is that the newly entitled landholders have their entire holdings at risk, with no means to reduce their risk through diversification or insurance.

For some settings, the omission of added loss of life and damages from increased future exposure to storms may be the largest source of underestimation in net loss. Whereas coastal storms may not be a major concern in Egypt, exposure of the coastal population to storm surge is profound in Bangladesh. Recall, however, that it is only the *additional* casualties and damages attributable to sea-level rise that should be considered in this regard.

Another source of economic loss not captured in the estimate above is saline intrusion into freshwater resources. This is already a serious problem for both Bangladesh (Mahtab 1989) and Egypt (El Sayed 1991, 1996), and can be expected to spread further inland with relative local sea-level rise.

Both Bangladesh and the Nile Delta are densely populated. Land loss from inundation and population migration away from threatened areas can only tend to increase population density and impose the additional costs in social services and management so often associated with crowding. The aggregate cost of relocation is itself a "passive response" cost not captured in the estimates reported here.

Clearly, even if the assumption of passive adaptation is accepted, some adjustment costs must be expected. How quickly and how completely economic adaptation will occur, clearly depends on the suddenness of the initial loss, on the organization of the economy, and on the availability of other opportunities for productive employment of the displaced resources. Some readjustment costs are inevitable, however, including some penalty for moving resources into their "next best" productive employment.

All these additional sources of cost would be included in a refined, data-intensive impact assessment, and all would tend to increase the estimate of net loss. Three other qualitative refinements, however, including the benefits side of human responses, would tend to have the opposite effect. Two of these are discussed briefly here; the third, the beneficial effects of human responses, is important enough to be treated in a separate section that follows.

Probability-weighted expected values

The method above assumes that the hypothetical scenarios are realized (certainty equivalence). If, in fact, we believe that there is only a 10 per cent chance of the "worst case" in Bangladesh by 2050, for example, then the actual expected net present value of cumulative loss (NPVL) should be only 10 per cent of the reported US$360 million estimate, or US$36 million; however, that implicitly assumes that there is a 90 per cent chance of no sea-level rise at all. What is needed, instead, is a probability distribution for each scenario, so that the expected loss associated with each can be weighted by its estimated likelihood of occurrence. A probability-weighted loss function has been assumed for application to a US site (Solow, Ratick, and Srivastava 1994), but no such estimate has been attempted for either Bangladesh or Egypt. Accurate estimation of the probability density function, perhaps based on elicitation of subjective probabilities from experts, is a difficulty.

Non-linear inundation rate

If inundation in a scenario approaches its ultimate percentage coverage of total land area at a logarithmic or exponential rate over the time horizon, rather than linearly as assumed in the NPVL example, then the present value loss would be smaller, as a larger proportion of the eventual land loss would occur later in the period, preserving the output of most land for longer periods. Such a non-linear rate of rise could generate a loss less than one-half the size of the linear case, and confirms the arguments made throughout this volume of the need to examine in detail non-linear and possibly chaotic environmental changes.

Human responses

Impacts will depend on human responses (Bird 1986; Bruun 1986; IPCC 1990b, c; Schelling 1983). Unless the rate of inundation is very gradual, it is ridiculous to imagine that people would passively accept such losses;

it is more likely that they would respond by taking action to cut the losses. Economists stress that people have a strong talent for adjusting to change and that their adjustments are most effective when crafted little by little to incremental changes spread over time (Schelling 1983). Of course, the ability of individuals to make such adjustments effectively varies in accordance with their social organization, enfranchisement, resource endowment, health, education, and other capabilities, as well as the efficacy of collective responses. Economists also point out the value of collective actions, especially where free exchange fails to account for effects that are spread across large social groupings. This highlights the value of cooperative planning and the responsibility of governments and other collective institutions to anticipate effective responses (USNRC 1987b).

The reasonable assumption of rational decision-making leads to a fairly strong expectation about human responses: this is that responses will be selected to reduce the expected costs and maximize the expected benefits of predicted change. Responses to imminent or observed rising sea level include the following: *mitigation* (reduce or reverse the causes); *defence* (dykes, sea walls, floodgates, etc.); *abandonment and retreat* (building setbacks, or simply using structures until the floods and then leaving); and *adaptation and development of alternative uses* (raising houses on stilts, reoccupying inundated areas with wharf towns or houseboats, converting rice fields to shrimp ponds, etc.). In this regard, it is worth emphasizing that people are good at incremental adaptation, risk management, and technological advance. Bangladeshis' behaviour in repeatedly remaining in and returning to hazardous areas need not be seen as contradicting this observation, for their range of options is severely constrained by resource limitations, and their decisions probably do reflect calculated risks taken to maximize their overall expected well-being.

Of course, mistakes may be made; nevertheless, in general, human responses can be expected to lower the net cost of sea-level rise. If they did not, they would probably not be undertaken. Nor is it necessary that responses be confined to the areas most affected by the physical changes: if the nation loses its low-lying coastal rice production to sea-level rise, for example, this will alter the allocation of resources in the economy in ways that extend beyond the loss of rice production. The resources formerly applied to rice (including materials, mobile capital, and labour) will tend to be redirected into other activities that eventually will help replace at least part of the economic benefits lost with the rice fields.

In Bangladesh it would be difficult to replace croplands lost to sea-level transgression because the countryside is already so extensively cultivated. Of the 24.5 million acres in the nation estimated to be cultivable, more than 90 per cent are already in cultivation (Zaman 1983). Fallow areas constitute a negligible percentage of land utilization. Further prospects for increasing agricultural usage in northern districts (the areas least likely to be affected by salt intrusion) are also relatively limited. Increasing the intensity of agricultural land utilization through multiple-cropping strategies may show some promise as a response to loss of agricultural lands. Of the nation's net cropped area, more than one-half is currently used for only one crop per year, whereas 39 per cent is used for two crops annually, and about 7 per cent is fully utilized in the production of three crops. Much of the land in the delta least likely to be affected by coastal transgression, however, is in relatively high topographic areas produced by old delta sediments of distinctly poor fertility.

Bangladesh annually produces about 700,000 tons of fish, 80 per cent of which are fresh-water species. Fish products are the nation's fourth leading export commodity after jute products, jute, and leather, exceeding even exports of tea. It seems likely that fishing centres are subject to considerable relocation, and that the fishing industry (in part owing to its largely artisanal nature) may be in a position to respond with great flexibility over the coming century to changing configurations and conditions in the distribution of the nation's aquatic resources. Ironically, flood-protection measures in the near term may themselves be having a harmful effect on inland fisheries: seasonally inundated flood plains have an important role in the life cycles of important inland fish and prawns, but the damming and regulation of water flows is disrupting this link and diminishing these artisanal fisheries (BCAS 1991b). This is another example of engineering interventions leading to harmful side effects that may well exceed the present value of future damages from sea-level rise, other examples being the tendency of flood-control dykes to impede receding floodwater by holding waters on the flooded lands (Brammer 1993) and the potential effects of upstream damming projects on subsidence and erosion in the delta by intercepting renourishing sediments (Milliman, Broadus, and Gable 1989).

Many large industries in Bangladesh are associated with jute products, and, as is the case with jute production, only a relatively minor proportion is found within the area affected by a 1 m relative sea-level rise. In general, the relative sparsity of major industrial and urban development in the potentially affected area could be a decided advantage in responding to sea-level rise. Whereas the nation may be strapped for resources to mount a costly strategy of structural defence, it has avoided also a commitment into a durable infrastructure in the region that demands defending.

Because the timing, magnitude, and physical effects of sea-level rise are still so uncertain, potential near-term responses are the more likely type of risk management. Such response will put a premium on research efforts that increase information and on institutional arrangements that permit diversification and pooling of risks. In the near term, responses to the increased risk of sea-level rise might thus include, in some form, institutional adaptations such as simple insurance arrangements, shared markets, contracts and futures markets, mergers, and governmental risk-sharing schemes. High risk is nothing new to the Bengal Delta, and existing practices in property management and redistribution of risk can be ascribed to this (Brammer 1993). For example, part of the explanation for the highly skewed distribution of land tenure, combined with a share-cropping system of production, may be that small, poor freeholders are less equipped to bear the risks of natural calamity than are larger, diversified landlords. However inequitable, the system may provide the sharecroppers with a means to share in production from the land without taking on the full weight of the risks. As physical changes call attention to the increased risk from sea-level rise, medium-term responses are also likely to include increased investments in embankments, refuge mounds, and perhaps other defensive engineering measures (Brammer 1993).

Economic considerations in human responses

Understanding human responses and the calculations on which they are based is important, not only for descriptive and predictive purposes but also for prescriptive use. To anticipate and guide human responses, several economic considerations warrant special notice. Five are discussed briefly here.

Advantages of incrementalism

It has already been noted that people have a talent for adjusting to change, and that their adjustments are often most effective when made incrementally to observed gradual changes in conditions (Schelling 1983). Moreover, decision makers closest to the facts of each choice often are likely to devise the most appropriate incremental responses (McFadden 1984). In a prescriptive sense, this view tends to be non-interventionist and favours letting things sort themselves out as the facts emerge.

Responding to uncertainty

High uncertainty, as in this case, puts a premium on research efforts that increase information and on institutional arrangements that permit diversification and pooling of risks. Economic theory supports several "rules of thumb" for dealing with uncertainty that are also as time honoured as is simple common sense:

1. *Diversification* (risk pooling). Simply put, don't put all your eggs in one basket. Total risk can be reduced by maintaining a "portfolio" of alternatives.
2. *Playing it safe.* Conservative choices that err on the side of safety are reasonable in the face of uncertainty. In effect, this implies assigning a greater penalty to underestimates of future hazards than to overestimates.
3. *Insurance* (risk spreading). By allowing a large number of individuals to share their risks, the total exposure of each can be reduced. A number of institutional arrangements, including governmental programmes, can achieve this.
4. *Procrastination* (playing for time). If waiting is not too expensive or risky, then uncertainty may be reduced by waiting to see how things develop. Also, as time passes, technology tends to improve.
5. *Getting more information.* Probably the most direct way to reduce uncertainty, and often the most cost effective, is simply to invest in information. Conduct research, study, and learn.
6. *Reducing fixed commitments.* Uncertainty puts a premium on flexibility, and flexibility is enhanced by a minimum of fixed commitments.

Capital mobility, durability, and the retrofit problem

Economic impacts are sensitive to the amount and values of *durable fixed capital.* For physical capital in exposed areas, some can be moved, and some will have worn out (or nearly worn out) before inundation. Therefore, it is really the value of long-lived or durable fixed capital (such as power plants, industrial factories, waste-treatment facilities, highways, and port infrastructure) that must be reckoned into estimates of future losses. The more the useful lifetime of such installations extends into the period of expected inundation, the greater the potential impact. At the same time, extra front-end investments made in anticipation of uncertain sea-level rise can reduce the lifetime cost of such facilities if the rise does occur (Gibbs 1984, 1986; IPCC 1990c). Anticipating the rise can thus afford great savings compared with the cost of retrofit or investments in defensive reaction (Wilcoxen 1986); if the rise does not occur, or is more gradual than anticipated, however, the extra investment will have been wasted. The choice depends on the likelihood of the rise and on the ratio of anticipatory investment to retrofit cost or abandonment. In practice, it is well to remember that the future cost of retrofit may be overestimated unless the advantages of waiting to see (better information and technological advances) are taken into account.

Discounting future values

It makes good economic sense to apply some discount rate to future values, because people do tend to value a current payment or benefit more greatly than a nominally equal payment at a future date. That is smart, because to do otherwise would be to ignore the additional earnings that could be gained from the current payment (say, through investing it for compounded interest payments) in the time before the future payment comes through. Selecting the appropriate discount rate to apply in practice, however, is quite difficult (JEEM 1990; Lind et al. 1982). It involves strong judgements about the preferences of society and encounters serious ethical complications when intergenerational effects are at stake. It may also be confounded, as noted above, when potential future loss of life is at stake (Cropper and Portney 1990). Higher discount rates depress the estimated cost of future sea-level impacts; lower discount rates make them appear larger. Relatively small changes in discount rate can exert a remarkable influence on the present value of future impacts.

Common property, externalities, transboundary effects, and information

Collective intervention may be essential when common property is involved and when economic "externalities" lead private decision makers to ignore the effects of their choices on others. Effects that cross boundaries, whether property lines or national frontiers, are obvious examples.

Recall that any significant reduction in the massive delivery of the sediments to the delta could disrupt its balance and expose it to net erosion and subsidence, thereby increasing the effects of sea-level rise. Careful attention to this possibility seems warranted in the planning and design of upstream water-management projects, such as dams and barrages.

Ensuring such careful attention, however, is greatly complicated by a sensitive international facet of the problem. Bangladesh shares the Ganges Basin with two of its neighbouring states, India and Nepal, and the Brahmaputra Basin with India, China, and Bhutan. Of the total drainage area of the Ganges–Brahmaputra–Meghna river system, only 7.5 per cent lies within Bangladesh (Zaman 1983). A serious dispute between Bangladesh and India concerns India's construction and use of the barrage on the Ganges at Farakka. This dispute led in 1972 to the creation of the Indo-Bangladesh Joint Rivers Commission and in 1977 to the Ganges Water Agreement. None the less, tensions continue over this issue, and it seems very likely that future difficulties will surface in allocation and management of the aquatic and sediment resources of this complex international river system.

Summary

The Ganges–Brahmaputra–Meghna Delta of Bangladesh and the Nile Delta of Egypt are among the areas of the world most vulnerable to the impacts of potential future sea-level rise. Closer examination of their vulnerabilities to future impact reveals an interrelated complex of extreme uncertainties about the problem, ranging from the relationship between greenhouse-gas emissions and actual future sea levels to the nature and effects of adaptive responses. The case of a 1 m rise in relative sea level by 2050 – which is much higher than current "best" estimates of future sea level but which accommodates some local subsidence and provides a conservative outer limit for use in the crude assessment of potential impacts over the next 60 years – was examined. In Egypt, about 12–15 per cent of the nation's arable area falls within the region affected in the 1 m scenario. Approximately 7 per cent of Bangladesh's land area is within the area affected in the 1 m case. In Egypt, some 7.5–14 per cent of the population resides in the 1 m affected area, depending on the effects on Alexandria. Thus, roughly between 4 and 7.5 million Egyptians now live in the area that would be affected by a 1 m relative sea-level rise. About 5 per cent of Bangladesh's population currently resides within the area likely to be affected in the 1 m scenario. Thus, some 6.5 million people now reside within the area affected in the 1 m scenario.

Using certain simplifying assumptions about the distribution of economic activities within Bangladesh and Egypt, we estimate that about 14 per cent of the GDP of Egypt currently originates in the area likely to be affected in the 1 m scenario and about 5 per cent of current GDP of Bangladesh originates within the affected area in the same scenario.

Results of a simplified extension of these estimates to potential economic losses suggest that, in the absence of mitigating adjustments, a "worst-case" relative sea-level rise of 1 m by 2050 could impose a cumulative loss in present-value terms of between 1 and 3 per cent of the nation's current GDP. Human adjustments – such as defensive measures, recombinations of productive factors, and institutional and technological adaptation – however, would tend to reduce this loss, which is, therefore, best seen as a bounding value. Even so, the potential scale of economic impact in both countries warrants further careful attention.

Notes

1. The author is indebted to John Milliman, whose original conjecture is reflected throughout. Financial support was provided by The Pew Charitable Trust, the US Environmental Protection Agency, the United Nations University and the Center for Technology, Environment, and Development (CENTED) of Clark University. Thanks, too, to Andrew

Solow, H. Brammer, Roger E. and Jeanne X. Kasperson, Samuel Ratick, Sally Edwards, S. Raper, J. Scheraga, William Nordhaus, M. El Sayed, and participants in the United Nations University conference on Environment and Development at WIDER for constructive ideas. Research assistance in the early stages of this work was provided by Frank Gable and in the later stages by Suzanne Demisch and Sarah Repetto.
2. *Editors' note*: James Broadus died in 1994 and this chapter has consequently not benefited from a final update by the author. Because the key findings have withstood the passage of time, however, the editors have chosen to include the chapter in this volume.

12

Sea-level rise and the North Sea

Timothy O'Riordan

In general, the debate about future sea-level rise in the North Sea is less about whether major losses will occur than about how best to deploy the available money and technology to ensure lasting protection and thereby to minimize anticipatable losses. Unlike the poorer and much more vulnerable regions of the world, the North Sea is surrounded by wealthy countries with the capacity to safeguard their land from a rising and more stormy sea without seriously disrupting their economies. The debate, therefore, is essentially about management or strategy: how much to allow certain areas to flood for conserving nature or achieving recreational benefits; how far to allow natural beach erosion and accretion to take place through manipulative forms of geomorphological surgery; and how far to rely on the durable outer wall of concrete and steel or tidal defence barrier (or barrage) for key estuaries. That debate is essentially about best environmental options where the economics of safeguarding include surrogate values for retaining natural and quasi-natural features as part of a flood-protection strategy, as much as estimates of the benefits for property protection and continuation of daily business and social relationships without tragic disruption.

In this sense, residents of the North Sea states take a concerned but ultimately relaxed view of sea-level rise, at least for the next 150 years or so. Nobody is willing to discuss what may have to be done if sea levels rise higher and faster over the next 100 years, as accumulations of greenhouse-forcing gases cause atmospheric warming effects. Yet 250

years ago these same areas, sheltered behind coastal defences and impressive drainage schemes that pioneered modern civil engineering,
began their mercantile supremacy. The very wealth of the world, which is
currently contributing to its ecological strains and stresses, began in this
pocket of the globe in a past time frame that nobody dares to project into
the future. Beyond 2150 is essentially a void of prediction.

Climate change, global warming, and sea-level rise

Climate change, in the sense of perturbations to synoptic weather patterns, is already well under way (IPCC 1990a, 1996a, b). The causes are
not yet fully proven, though the most likely initiating reason is the natural
"wobble" of the earth's polar axis. This appears to have altered the geo-
dynamics of thermal and kinetic energy throughout the atmosphere, particularly between the equator and temperate latitudes, sufficiently to
create important alterations in the pattern of anticyclones and depressions. Consequently, we have entered an era in which greater extremes of
climate – more intense depressions and tropical storms, long periods of
cold or dry weather, and intense local thunderstorms or tornadoes – are
more likely (IPCC 1997; Mannion 1991: 20–26, 321–322). In general, cli-
matologists now accept that natural variations in the greenhouse effect
(as opposed to human-induced greenhouse forcing) are occurring, notably through fluxes in concentrations of carbon dioxide. These changes in
atmospheric composition have influenced global climates, most especially
in the Quaternary period. These CO_2 fluxes appear to be associated with
variations in organic productivity, probably more on sea than on land,
which in turn are influenced by oscillations in axial angle, changes in the
configuration of the orbit of the earth, and the procession of the equinoxes.
These natural variations provide background "noise" for any analysis of
anthropogenic causes of climate change and illustrate well the complexity of "unravelling" causation: "Our ability to quantify the human influence on global climate is currently limited because the unexpected signal
is still emerging from the noise of natural variability" concludes the Intergovernmental Panel on Climate Change (IPCC 1996a: 5; 2001).

   This is one reason why estimates of sea-level rise are inevitably uncertain and forever will remain so. We cannot know what humanity is doing
to accelerate or decelerate natural climate variability and associated
effects on organic production, the distribution of surface and deep ocean
currents, and the pattern and characteristics of cloudiness and ice-sheet
accretion or melt. Nevertheless, the IPCC (IPCC 1990a, 1996a, b) has
been able to summarize the state of scientific knowledge with remarkable
unanimity and persistence of judgement. Among the key findings of the
second assessment (IPCC 1996a) are the following:

- atmospheric concentrations of greenhouse gases have continued to increase;
- anthropogenic aerosols tend to produce negative radiative forcing;
- climate has changed over the past century and is expected to continue to change in the future;
- the balance of evidence points to a discernible human influence on global climate; and
- many uncertainties persist.

To reduce the uncertainties, the panel (IPCC 1996a: 7) recommends "further work" on several priorities:

- estimation of future emissions and biogeochemical cycling (including sources and sinks) of greenhouse gases, aerosols and aerosol precursors, and projections of future concentrations and radiative properties;
- representation of climatic processes in models – especially feedbacks associated with clouds, oceans, sea ice, and vegetation – in order to improve projections of rates and regional patterns of climate change;
- systematic collection of long-term instrumental and proxy observations of climate-system variables (e.g. solar output, atmospheric energy balance components, hydrological cycles, ocean characteristics, and ecosystem changes) for the purpose of model testing, assessment of temporal and regional variability, and detection and attribution studies.

Studies of climate warming based on best-guess modelling tend to generate estimates of global warming for the period 1765–1990 that are higher (by 0.4°C) than the observed warming of 0.6°C. This may be due to some unknown dampening effect or faulty temperature-trend analysis, or it could be a feature of inadequate modelling. Whatever the reason, Warrick and Wilkinson (1990: 23) regard the 1990 IPCC estimate of 2.5°C for the CO_2 doubling "sensitivity" as too low, suggesting that a figure of 3.0°C would be more accurate, given new assumptions about the thermal behaviour of high-latitude oceans and atmospheres. Meanwhile, however, the more recent IPPC (IPCC 1996a) uses a figure of a 2°C rise for the period 1990–2000 and a range of 1–3.5°C by 2100 (IPCC 1997).

The 1990 best-guess estimate of sea-level rise for the southern North Sea by Warrick and Oerlemans (1990) was 9–29 cm for the period 1990–2030 and a further rise of 4–6 mm per year over the foreseeable future. By 2100, sea-level rise could be in the range 30–110 cm with a best guess of 63 cm, 24–36 cm for 2050, and 16–24 cm by 2030. Using a probabilistic approach, Wilkinson and Warrick (1990: 28) propose a 95 per cent chance that sea levels will be 21–31 cm higher by 2050 (or 12–18 cm by 2030), and a 50 per cent chance of a rise of 23–29 cm by 2050 (14–17 cm by 2030). These figures are broadly in line with Dutch estimates. Table 12.1 summarizes the Wilkinson and Warrick principal predictions.

One can conclude that, in the southern North Sea, sea levels will rise by

Table 12.1 **Sea-level rise predictions for the southern North Sea**

Year	Predicted rise (cm)		
	Low	Median	High
2030	9	14	29
2050	24	28	36
2100	30	63	110

Source: Wilkinson and Warrick (1990).

a significant amount by the middle of the next century, irrespective of measures to combat global warming. With or without global warming, then, adaptive measures are mandatory.

Changes in storm-surge frequencies

Storm surge is the rise in sea surface associated with a deep depression and funnelling winds through a constricted zone of ocean. The surge is partly a function of lower atmospheric pressures and partly a consequence of the pushing of water into a confined area. In the southern North Sea, where a storm surge can add one metre to a tide, timing is critical. Coming at the top of a high tide, the surge can add 30 per cent to a tidal range, causing devastating damage – witness the flood in 1953, which set the basis for the current sea defence in the region.

The normal rule of thumb is to protect agricultural land against a probability (p) of 0.01 of the 1953 flood; for urban areas $p = 0.004$; and for the London Basin $p = 0.001$. However, these calculations were based on return frequencies calculated in the late 1950s and early 1960s, when both climate change and global warming were regarded as not as significant in affecting storm-surge conditions as is the case today.

Given a sea-level rise of around 30 cm in 2030, the frequency of flooding associated with storm surges obviously increases as lesser but more commonly occurring surges begin to press against coastal defences. Figure 12.1 suggests that the likelihood of a surge of the height of the 1953 event could be as frequent as $p = 0.2$, or five times as likely as was suggested in the immediate aftermath of the 1953 flood (Wilkinson 1990: 43).

Climate change, on its own, is likely to be a major reason for increased storminess on the North Sea (Lamb 1971). Some speculation that average wave heights in the north-eastern Atlantic are increasing would suggest that the impact of a given surge could be correspondingly greater. The role of climate warming in all this is almost entirely speculative. Obviously, the rise of sea level is the crucial factor, but it is possible that human-induced

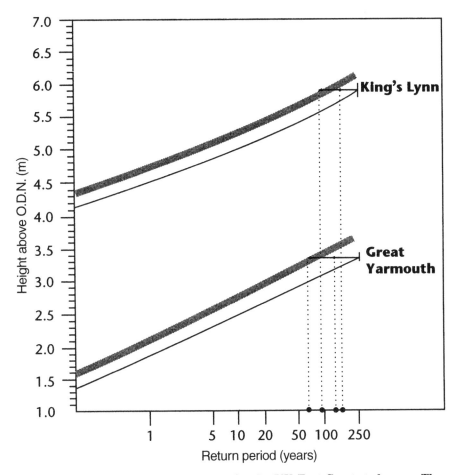

Fig. 12.1 **Frequency distribution curves for the UK East Coast study area. The curves illustrate the shortening of the 1-in-250-year return period, corresponding to the sea-level rise at 2030**

climate change will, in itself, increase storminess. So far that cannot be proved. In one sense, it hardly matters, as the possibility of a damaging flood is now far more likely than it was 20 years ago, irrespective of the storminess linkage. The North Sea has always been mean to its coastal inhabitants; in all probability it will be meaner still to their descendants.

Impacts of sea-level rise in the North Sea

The easiest way to extrapolate the possible effects of sea-level rise on the North Sea coast is to move from the physical to the human, from natural hazardousness to socio-economic vulnerability; in short, to adapt the "sensitivity/vulnerability" approach pursued in the IPCC's second assessment (IPCC 1996b). For the sake of clarity, *hazardousness* in this chapter refers to the propensity of physical systems to inflict damage, whereas *vulnerability* applies to the degree of inability of societies and their governments to adjust to, or cope with, that damage. These definitions, it should be noted, depart from those used in many of the other chapters in this volume (see, for example, chapters 1 and 5–7).

Regarding hazardousness, the key issues for the North Sea States are as follows:

- changing water levels in aquifers, rivers, and coastal lakes, and the likelihood of increased salinity;
- alterations to the configuration of coasts and estuaries, and the impact on coastal geomorphology and ecology; and
- changes in sedimentation rates in coastal areas and offshore zones which, in turn, affect the coastal geomorphology and wave patterns.

All of these are important. All have been extensively studied in the southern North Sea Basin, and most studies point to important changes in the physical and ecological character of the littoral zone.

Coastal hydrogeology

Fresh water in rivers and aquifers holds back salt-water incursion, whereas rising sea levels increase the penetration of brackish and saline water. Figure 12.2 identifies the areas at risk. Climate warming and its associated changes in precipitation and water demand suggest that the availability of fresh water in coastal aquifers is likely to be up to 20 per cent less by 2030 than is the case today (Atkinson and McGuire 1990). This is only partly a consequence of lower precipitation associated with climate change: it will largely be a function of pricing policies that almost certainly will continue to underprice water in terms of its existence values. Existence value, as used here, is the significance of fresh water *in situ* in rivers, lakes, and aquifers in maintaining viable ecosystems, in holding back salty water, and in providing amenity and psychic pleasure for people who appreciate the role of that water. The other factor will be changing consumer behaviour, which tends to increase per capita water use by about one per cent annually (National Rivers Authority 1990). Of course, an aggressive pricing regime could alter consumer behaviour, but prices would have to rise by more than 80 per cent to effect any marked decrease in consumption. Currently, it is unlikely that such increases would be politically acceptable.

The actual change in both the coastal penetration of salt water and the depth of the subsurface saline layer will depend upon the configuration of

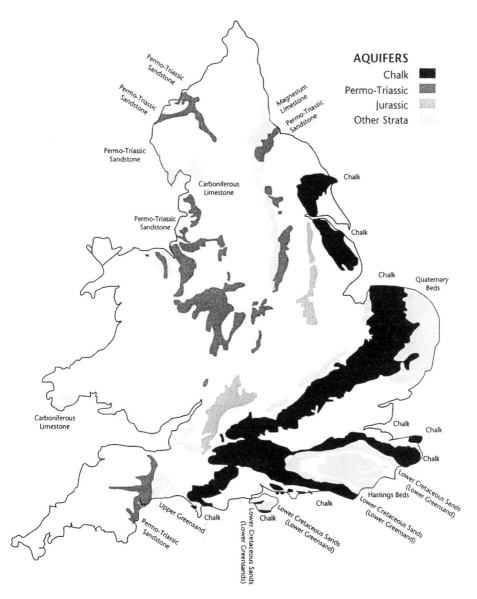

Fig. 12.2 Outcrop areas of major aquifers in England and Wales, showing coastlines that are in hydraulic contact with the sea

the coastline, the nature of the underlying geology, and the probable changes in sea-level rise and freshwater flow reduction. To illustrate, imagine a coastal aquifer of 1 in 1000, with a freshwater flow of 0.1 cubic metres per day, facing a sea-level rise of 50 cm. With sea-level rise alone, the inward progression of salinity would be 500 m, and the reduction in freshwater depth would be some 4 m. Add the 25 per cent reduction in freshwater flow, and the inward progression would be 1 km, and the reduction in freshwater depth would be 11 m – or more than one-third of the original depth of freshwater.

These figures are fairly speculative, but they provide an indication of the magnitude of possible changes to coastal ecosystems that could result from a combination of sea-level rise and alterations in precipitation patterns. The incidence of severe drought could increase by 50 per cent in eastern England, with periods of more than 100 mm of soil-moisture deficit being up to five times more common than is the case today. This would inevitably lead to an increase in irrigation demand, especially for high-value agricultural crops (e.g. carrots, sugar beet, potatoes, and peas/beans) that are grown extensively on the light soils of the East Anglian coastline. The National Rivers Authority has conducted detailed investigations of the probable effects on chalk-fed streams in southern and eastern England to assess what pricing and abstraction limitation measures should be put into effect. It is certain that a combination of the two strategies will be pursued and that agriculture will experience a period of adjustment. This will allow for greater efficiency of water use and lower profit margins for many foodstuffs that are highly dependent on water.

All this has led to renewed interest in water metering as a method of reducing domestic demand, for the possibility of using wastewater flows for non-drinking purposes, and for ambitious programmes of cross-country transfers of water from the surplus regions of the north and west to the relatively wealthy but water-short areas of the south and east. At the very least, it will also focus attention on the value of water as an environmental asset in its own right. Studies have been undertaken to assess the possible environmental consequences to nutrient recycling, freshwater habitats, stream-bank vegetation, and animal life and fish populations of prolonged reductions in river flows, and have provided relevant data to the Intermediate Ministerial Meeting on Agriculture and Environment in Copenhagen in 1993 and at the Intermediate Meeting on the Integration of Fisheries and Environmental Issues in Bergen in 1997. The full social value of *in situ* supplies will be given economic weight as part of a measure for appropriate abstraction pricing.

A statutory liability requires the National River Authority, the official water regulators in England and Wales, to take into account the well-being of ecosystems and scenic amenity when formulating policies for abstraction licensing and minimum available flows. These responsibilities are not absolute in the sense that failure to meet them would result in court action by interested environmental groups; nevertheless, the duties are sufficiently clear that it would be politically difficult for the Authority not to give due emphasis to these environmental considerations in the publication and consultation process that must precede the eventual formulation of water availability.

Change in coastal geomorphology

The coastline of the North Sea has been shaped more by rivers and ice than by the sea (Kay, Clayton, and Vincent 1990). The pattern of hard-rock headland and soft-rock estuaries and salt flats is essentially the result of millions of years of erosion that preceded the final configuration of the coastline in the receding ice age. Figure 12.3 illustrates the areas of England and Wales subject to tidal inundation. The North Sea also has a complicated pattern of channel currents, as shown in figure 12.4. The definition of hazardousness, in the sense of exposure to flood, is not fully accepted by the various responsible authorities in the United Kingdom. Part of the problem lies in the mixture of responsibilities for coastal protection, which is complicated by an awkward combination of local-government, quasi-government, and central-government agencies. This overlapping confusion of responsibilities means that coordination over managing flood risk is cumbersome, time-consuming, and frustrating, even though an element of coordination does occur. There is no doubt that responsibility for flood-hazard management should be more integrated, with a well-established chain of command; this is discussed below.

A study by Bird (1985) suggests that soft-sediment shorelines of the kind commonly found in eastern England are now on the retreat, owing to sea-level rise. For the southern North Sea, this picture is worsened by post-glacial coastal movements that cause the shoreline to sink by about 1.0–1.5 mm per year.

The central government agency primarily concerned with coastal protection is the Ministry of Agriculture, Fisheries and Food (MAFF) and its sister agricultural ministries in Scotland and Wales. At a regional level, the responsible organization is the National Rivers Authority (NRA), which has a central policy directorate in London but which operates on the ground via the regional offices that have a reasonable degree of autonomy over finance and management. Each regional NRA office is advised by a flood-defence committee that is composed of local-authority representatives and members representing land drainage, agriculture, shipping, and conservation.

The NRA offices have mapped coastal hazardousness in the more exposed parts of England and Wales, including a three-stage survey that has involved an evaluation of the following:

Fig. 12.3 **Present flood-risk areas in England and Wales, as defined by Water Authority Section 24 (5) Surveys between 1975 and 1981**

- the type and state of repair of existing coastal defences managed by the NRA;
- all other sea defences;
- all coastal structures, including natural and semi-natural defences that can (or could) play a part in flood alleviation.

All data are, or will be, available on geographic information systems.

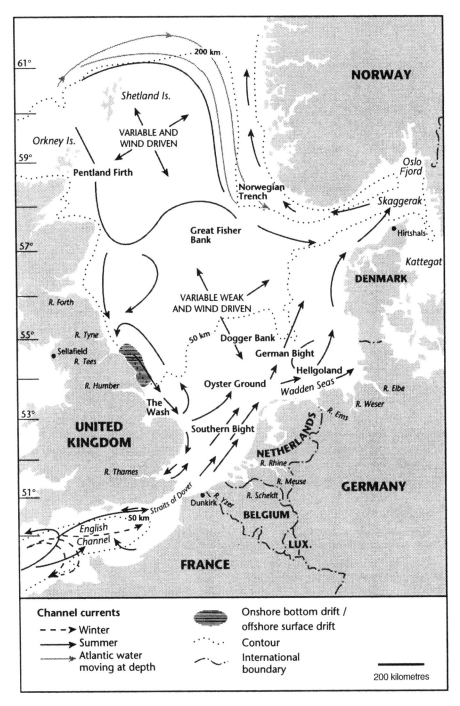

Fig. 12.4 **Channel currents of the North Sea**

Table 12.2 **Examples of erosion scenarios for a cliff currently being eroded at 1 metre/year on the Norfolk coast**[a]

SLR$_S$ (mm/year)	SLR$_P$	E$_P$ (m/year)	Erosion scenarios[b]			
			Low	Best low	Best high	High
1.0	3.0	1.0	65 (1.0)	98 (1.5)	136 (2.1)	201 (3.1)
1.5	2.0	1.0	66 (1.0)	108 (1.7)	147 (2.3)	212 (3.3)
1.0	3.0	1.0	83 (1.3)	146 (2.2)	202 (3.1)	297 (4.6)
1.5	2.0	1.0	100 (1.5)	163 (2.5)	221 (3.4)	318 (4.9)

a. Where E$_P$ = 1 m/year; SLR$_P$ = 2 mm/year or 3 mm/year; SLR$_S$ = 1.0 mm/year or 1.5 mm/year (included within SLR$_P$ above and added to each prediction of future sea-level-rise rate SLR$_F$). SLR, sea-level rise; E, erosion; s, subsidence; P, predicted; F, future.

b. Amount of coastal erosion (in metres) from 1986 to 2050 (E$_{2050}$) and the average rate from 1986 to 2050 (in parentheses).

Kay, Clayton, and Vincent (1990) computed 16 combinations of rates of coastal subsidence and of sea-level rise for a section of the Norfolk coast, an area most at risk from the North Sea. Table 12.2 provides a simplified summary of the conclusions. The range of coastal disappearance runs from 65 to 318 m by 2050 on a coast already eroding at a rate of 1 m per year. The range in estimate not only reflects the uncertainties of sea-level rise but also indicates that coastal erosion patterns are enormously sensitive to the history of sea defence, the degree to which eroding headlands are allowed to nourish downshore beaches with protecting sediment, and the likely mix of management measures. In practice, these calculations are virtually unpredictable because erosion is so much a function of localized relationships among existing defensive measures, physical coastal configurations that are forever changing, and beach nourishment.

Coastal defence and hazard zoning

Britain is proud of its planning legislation, which supposedly makes it impossible to construct any feature on the landscape without obtaining a formal permit from the local planning authority (district council for most matters, county council for strategic issues). In theory, therefore, it should be possible to zone the eroding coastline to establish a *cordon sanitaire* of underdeveloped coastline to allow erosion to continue to feed vital sediment offshore and onshore as part of a natural coastal-defence strategy.

In practice this is far from achieved. A considerable amount of housing and other buildings, much of this dating back to well before the introduction of town and country planning legislation in 1947, already sits on the eroding North Sea coastline. Not much can be done about removing this property without potentially expensive compensation, though some has already been abandoned to the sea. Naturally, most residents and district councillors expect such areas to be protected by expensive sea walls, justified post hoc by the scope for more planning permissions on the defended coast. Again, in theory, the flood-defence authorities are supposed to be notified before any permission is approved but in practice this is not done for eroding coasts (Kay 1990), although the consultation is slightly more rigorous for coasts known to be susceptible to flooding.

In its first assessment, the IPCC (IPCC 1990a) focused attention on the choice of management strategies for endangered coasts, emphasizing the response options that do not rely entirely on expensive civil engineering. The three basic choices are as follows:

1. *Retreat*, where nothing is done and the sea is allowed to enter. This is possible in unpopulated regions and can result in considerable advantages for nature conservation and habitat protection, as well as for recreational and educational values.

2. *Accommodation*, where non-structural measures are put in place – mostly in the form of coastal dunes, offshore beach nourishment, and washland provision – and where both human and non-human life learns to adjust to the increasing frequency of incursion.

3. *Protection*, where hard engineered structures, such as sea walls and embankments, or soft structures, such as sand dunes and vegetation, protect existing development.

In general, planning regimes and insurance schemes are not fully designed to respond to the new thinking required. This is particularly the case in the southern North Sea, where responsibility for action is divided among a number of agencies with very different remits and institutional attitudes to the problem. In any case, the whole issue of sea-level rise is too new to be fed properly into bureaucratic structures that take time to adapt to innovation and social learning.

Britain has witnessed repeated calls for a more coordinated and integrated approach to coastal management (Gubbay 1986; WCC'93 1993). In the mid-1980s, attempts to make the responsibility fall fully on the shoulders of the regional water authorities were shelved, owing to privatization of the water industry in 1989. The process resulted in the creation of the NRA, which has grown in political stature and may well move to adopt more of the central government role (despite its being essentially a regional organisation) with regard to coastal protection (Skjaerseth 1998).

The Royal Town Planning Institute (1990) called for a more integrated approach to planning and coastal management in the face of sea-level rise. This has not yet led to any emergence of new policy, but two county councils (Norfolk in the east and Clwyd in south Wales) have published revisions to their long-range structure plans indicating that a coastal defence zone of up to a kilometre will be set aside for no further development. In the mean time, the Marine Conservation Society has pressed for a fundamental restructuring of coastal protection management and financing. It is likely that all this head of steam has resulted in important changes in the character of monitoring, scientific investigation, forward planning, management responses, and financing of coastal protection.

Impacts on coastal ecology and agriculture

Coastal ecosystems lie parallel to the coast in zones that shift landwards as sea level rises (Wolff 1987). It is, therefore, a concern that coastal habitats will gradually move inland (Boorman, Goss-Custard, and McGorty 1988), as long as the coast is underfunded. Where it is protected, then climate change may alter habitats. Only recently have the nature-conservancy organizations begun a programme of systematic monitoring. Where natural restoration takes place, however, the rate of sea-level rise may be too great for sediment and plant colonization, so the rate of rise and the degree of associated erosion are critical. The coastal habitats most at risk are, in order of importance, salt marshes, sand dunes, intertidal flats, sandy offshore, shingle, soft cliffs, and hard cliffs and rocky shores.

The main issues are the degree of defence, the character of the shoreline, the scope for retreat, and the extent of change. In general, an overall reduction of principal habitat types – notably mudflats, salt marshes, and dunes – will occur and will reduce numbers of invertebrates and increase densities of breedings. The most endangered habitat is the salt marsh, which supports a variety of nationally and internationally important wader species, notably redshank, curlew, and godwit. Salt marsh also helps to stabilize coastal sediment and can absorb more than 70 per cent of the force of breaking waves. The main consequence will be the vertical raising of ecological zones, as well as their lateral compression. It is likely that, with an upper-bound sea-level rise of more than 50 cm, many of Britain's most prized coastal ecosystems will be irreversibly degraded.

New habitats could form, however, especially if there is rear space and the coastal morphology is kind. New brackish lakes and shingle zones should form, as should sand dunes and, eventually, salt marsh. The important task is to identify the most vulnerable sites and to ensure, through planning and appropriate management, that nearby areas exist for landward migration. Currently, this is still a distant possibility, which will especially be necessary for the relocation of endangered sand dunes. It will also be essential to allow for deliberate beach nourishment and coastal shingle and sand accretion. Again, no specific commitment exists as yet to this approach by responsible authorities. Nevertheless, the relevant conservation organizations are actively pursuing research and monitoring strategies to prepare the ground for a more forceful and integrated approach.

Agriculture in coastal Britain is very much dependent on soil type, topography, rainfall, and drainage. In the dry eastern shores bordering the North Sea the farming economy is relatively profitable, especially on the light soils, and farmers are content to grow vegetables and beans, cereals as break crops, and some root crops (e.g. potatoes and sugar beet). The rise in sea level will not substantially affect the agricultural area, but it will affect groundwater salinity and fresh-water availability. It is likely, therefore, that farmers may be constrained by lack of water, sufficient to remove salinity as well as to irrigate crops, and that some migration of agricultural activity away from the coast is likely. Should this be the case, planning permission for non-agricultural land uses, such as golf links or other recreational facilities not involving permanent structures, may become more likely.

Since Britain already overproduces all its temperate foodstuffs, the loss of agricultural production in the North Sea will not be serious from a national economic viewpoint. Farmers may require compensation, however, for reverting some of their more marginal areas to natural habitat. Aiming to control production is a European Community scheme called "set-aside," in which a proportion of land may be removed from production (Countryside Commission 1987). It is possible that this scheme could be amended for the reversion of formerly productive land for nature conservation and washes as part of general-amenity–flood protection. The washes would be areas of marshy land set aside for absorbing high tides on a temporary basis. During the winter, such areas would be attractive for bird life and suitable for reed production. In the summer, wastelands could be flooded for recreational use, notably sailboarding, rowing, and canoeing, which thrive on shallow lagoons.

To justify this, the calculation of cost–benefit analysis would have to be amended and politically accepted. This would involve a combination of weighted values for the protection of non-market attributes of coastal areas, such as wildlife conservation areas, heritage landscapes, nutrient-recycling salt marshes and reed beds, and protective beaches and dunes. The method of valuation will include replacement cost for the same function by human-made investments, contingent valuation studies of willingness to pay for such attributes, and some measure of the intrinsic importance beyond conventional economic measures (Turner 1991). Preliminary findings suggest that some combination of defence, erosion, and natural accretion should result in the highest net benefits. This may appear obvious, but when it is realized that the instinctive funding is to protect shorelines at almost any cost, a combined strategy to protect, abandon, and re-create shorelines is radical indeed.

Towards innovative coastal-zone management

The prospect of sea-level rise from an ever more stormy North Sea has captured some attention. The British Treasury, for example, has accepted a cost–benefit justification for improved coastal defences at Aldeburgh in Suffolk, where the benefits reflect the economic values attributable to nature conservation and open-space recreation (rather than confine such benefits to the safeguarding of agricultural land, as has been the convention in the past). Increasingly, this form of social cost–benefit approach, as laid down by economists a generation ago, is becoming more politically acceptable. Annex A of the UK Government policy statement on the environment (HM Government 1990) contains a full treatment of various approaches to valuing non-market environmental goods. This heralded the current era in which such calculations have become more commonplace.

In eastern England, both MAFF and the NRA have funded research aimed at justifying a multiple-objective approach to coastal protection in the face of sea-level rise (Turner 1991). This approach looks at the combination of hard and soft engineering practices, accommodation, and abandonment as a set of opportunities for which various weights of costs and benefits can be brought to a common net value, based on differing assumptions of the level of protection to be offered, the nature of the land use to be protected, and changing social judgements about the importance of various land uses.

The preliminary results show great sensitivity to the discount rate used to find a common net present value, to the rates of national economic growth, and to the precise configuration of options chosen. These options involve not just engineering but also hydrology, planning, land-use management, and conservation. In general, it is evident that complete erosion control is not the most cost-effective solution. Important social and environmental gains can be realized by changing the character of the coast and the water space. Various flood-protection options, such as washlands to absorb high tides on an intermittent basis, can be used for breeding birds or water recreation. Imaginative comprehensive management can provide a more practicable and cost-effective environmental outcome than sole reliance on expensive hard engineering. Such an approach will require close monitoring of endangered sites, wide-ranging public consultation as to the strategies that will result in the least private and social costs, and imaginative ways of portraying future configurations of coastal defence and abandonment to allow people to become actively involved in designing their own coastal futures.

Sea-level rise in the North Sea provides an opportunity to innovate in the organization, management, and decision-making for determining environmental futures. The techniques of evaluation and prediction are well developed, and communication has become sophisticated. What will be necessary will be the serious analysis of how the structure of governmental organization for the integrated arrangement of the littoral zone should be changed to allow these opportunities to flourish.

Currently, a potentially confusing array of governmental and quasi-governmental agencies, with overlapping responsibilities for coastal management and protection, lacks consistent objectives or budgets. It is likely that a single tier of management authority will eventually emerge, probably based on regional coastal agencies and coupled to a national policy-making and financing agency. This would place the control of coasts at arm's length from both central government departments and local authority "parish pump" politics. The new coastal-protection agencies should have the clout to establish strategic investment plans and be free to act as agents for the planning and conservation aspects of the littoral zone. They would have the resources to undertake their own long-term planning and financing, although they would be expected to operate through informed public consent. Analysts may conjecture, but eventually that decision has to be left to the politicians.

A promising example of proactive adaptation occurred in the early 1990s, partly in response to IPPC estimates regarding sea-level rise. The United Kingdom and the Netherlands both raised their design standards for new sea walls – to 66 cm (100-year planning horizon) and 25 cm (50-year planning horizon), respectively (IPCC 1996b: 314). The overall Dutch experience takes up the next section of this chapter.

The Dutch experience

The Netherlands is peculiarly vulnerable to sea-level rise and increased storminess in the southern North Sea. Almost one-half the nation already lies below mean sea level (fig. 12.5), and the land surface is dropping by 15–20 mm per year owing to post-glacial eustatic effects. More than one-half of the country is protected by coastal walls (dykes), which, depending on the quality of the property being safeguarded, are constructed to protect land from storms of frequencies of 1 in 100 to 1 in 1000.

The consequences of sea-level rise outlined above for England are

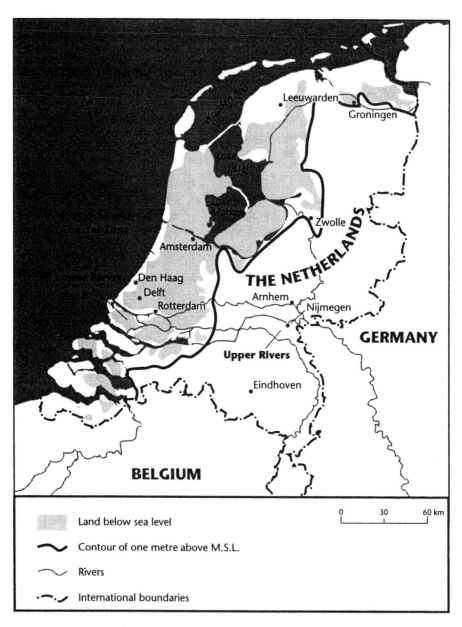

Fig. 12.5 **Low-lying areas of the Netherlands**

even more serious for the Netherlands. Considerable areas of harbour and industrial activity lie seaward of the main dykes, and would require special attention. Large areas of land – at least 17,000 km^2 – still drain naturally and would have to be pump-drained, involving huge expenditures on pumps and power stations. Almost all the rivers and canals would suffer from saline intrusion. The Waddensee, probably the richest nature reserve in northern Europe and a breeding ground for most commercial fish species in the North Sea, would be seriously affected (Waddenvereiniging 1997). Coastal dunes, which protect naturally the remainder of the Dutch coast otherwise not defended by concrete or electricity, would be vulnerable to storm and tidal damage. The Rijkswaterstaat of the Netherlands Ministry of Transport and Public Works (1990) has joined with Delft Hydraulics, a leading environmental engineering consultancy, to propose a set of adaptive responses to the prospect of sea-level rise and increased storminess (Peerbolte et al. 1991). The two agencies set about their task in three ways:

1. They devised scenarios of sea-level rise based on best-guess ranges of uncertainty. These were set on 20 cm, 60 cm, and 100 cm sea-level rise per century, initially to 2100. In addition, the study team produced a worst-case scenario based upon extreme assumptions of various parameters of sea-level rise and storminess.
2. They drew upon a series of models of water management, coastal defence, nutrient states, and nature-conservation management to create specific detail for each of the sea-level scenarios.
3. They applied best-guess cost functions on the basis of construction or avoidance activity (i.e. not planting potentially valuable land with crops or industrial installations) to estimate likely demands on the Dutch economy according to a phased timetable.

Throughout the analysis, the Dutch team tried to avoid committing expenditures either too early or too late by testing the sensitivity of their main strategies to uncertainties in timing and magnitude of sea-level rise. The result was a fascinating example of area-based and time-programmed responses based on geographical information systems (GIS) programming models. Of interest was the decision not to adopt any discount rate to computed costs. This was because any use of the conventional public-sector rate of 8 per cent would be meaningless after 40 years, long before the investments would really prove their worth. Consequently, fixed price figures based on 1990 guilders were adopted.

Sea-level rise

The Dutch models, as noted above, adopted three rates of sea-level rise:
1. 20 cm per century, or the current rate due to natural causes;

2. 60 cm per century, or the most probable "greenhouse enhanced" rate; and
3. 100 cm per century, or the rate unlikely to be exceeded.

For all scenarios, a storminess factor of 5 cm was added to reflect the likelihood of increased cyclonic activity in the North Sea caused by human-induced climate perturbation. Following the work of Oerlemans (1989), the configuration of sea-level rise is skewed to increase more rapidly towards the end of the 100-year period. The research team concluded that a 60 cm rise in sea level over the next century would require spending some US$3.5 billion on raising dykes and other safety infrastructure, US$500 million on preserving dunes, US$900 million on adaptation in flood-prone industrial, residential, and port areas, and US$800 million on water-management facilities. The report also observes that the costs of averting the impacts of an adverse 10 per cent change in the direction of severity of storms may well exceed those of a sea-level rise of 60 cm (Peerbolte et al. 1991).

Dyke raising

The raising of the coastal dykes depends on three parameters: these are the actual estimate of dyke increase estimated to be required; the implementation of construction (conception, planning procedures, design, and construction); and the freeboard available to cope with anticipated storminess and sea-level rise that triggers the timing of construction.

The actual calculation of vulnerability varied with each scenario. For that of a 20 cm rise, the vulnerability factor was 1.3, or an increased likelihood of overtopping by a factor of 1.3. For the most hazardous scenario the vulnerability factor was 2.3. This means, in effect, that a dyke designed to withstand a 1-in-1000 surge today would be able only to withstand a flood of 1-in-434 years' probability in 100 years' time.

Table 12.3 shows the probable costs of dyke raising based on the three-fold scenario and the investment strategy modelled into a programme of action. Timing is critical. No action need be taken for 10 years, by which time clarity should emerge concerning potential sea-level rise. No action needs to be taken for a further 5 years for the scenario of a 60 cm rise. After that, some Dfl300 million (US$180 million) per year for 10 years would be required on critical points. Under a more anticipatory strategy, the work would have to begin within 5 years.

The implementation time is a critical variable. The study envisioned an average period for implementation of 20 years: any delay by 10 years would bring any investment decision forward by 5 years. The report argued that considerable effort should be put into ensuring that the technical issues were properly anticipated, that public opinion was adequately

Table 12.3 **Total cost of raising coastal walls (dykes) over a 100-year period (in US$ billion)**

Sea-level-rise scenario (cm)	Mode[a]			
	Conservative	Anticipating	Progressive	Very progressive
20	1.56	1.74	1.88	2.88
60	3.57	4.56	4.58	5.13
100	5.62	6.32	6.96	7.65
Worst case	8.95	9.62	10.28	10.51

a. In the conservative mode the vulnerability increases by a range of 2.2–7.9; in the anticipating mode the vulnerability increases by 1.4–2.2; in the progressive mode the vulnerability increases only in the worst case; in the very progressive mode the level of protection remains the same as that of today.

prepared, and that planning procedures were put in place to ensure that sufficient land was available for the dyke-improvement work.

The other critical variable is the rise in the sea-level bed as the sea-surface level increases. The models assume that both the sea and the estuaries will fill at rates of 0.5 per cent and 0.8 per cent of the rate of the sea-level increase, respectively. Should this not happen (an unlikely, but possible, outcome), then the onset of construction would have to be advanced by 10 years in all models, and the final cost would rise by 46 per cent, or some US$1.2 billion over the 100-year forecast period.

Some costs can be saved by increasing the frequency of closure of two storm-surge protective barriers, one on the eastern Scheldt and one already ongoing in Rotterdam. These could reduce dyke-raising costs by some $2 billion over the 100-year forecast period, although it is doubtful that the projected increase in closure frequency could be tolerated by both shipping and nature-conservation interests.

In order to maintain the drainage canals by shifting from gravity to pump drainage, the likely costs are $375 million for the 60 cm scenario and $700 million for the 100 cm future. Again, those costs are very sensitive to timing, to geography, and to accepted implementation periods. Much of the Dutch response requires strategic planning, constant surveillance by telemetry and by trained eye, and broad-based public support.

Table 12.4 shows the probable costs of protecting the fresh-water lakes Ijssel and Marken (the remnants of the Zuider Zee) from saline incursion. This would involve dyke raising, sluices, and pumps, programmed to maintain fresh-water levels. According to the rates of sea-level rise, the cost ranges from US$1.06 million to US$8.6 million, and that does not cover the very worst case.

Table 12.4 **Total cost of protecting lakes Ijssel and Marken over 100-year period**

	Total cost (US$ million) of scenario			
Protection	20 cm	60 cm	100 cm	Worst case
Dyke raising	0.81	1.38	2.09	4.10
Extra sluices and pumps	0.07	0.62	1.19	na[a]
Pumps	0.18	1.26	1.87	3.00
Total	1.06	3.26	5.05	≈8.6

a. na, not available.

Table 12.5 **Policy alternatives for the defence of the dune coast**

Retreat	Only where the safety of the land behind the dunes (polders) is threatened will coastal recession be fought; elsewhere along the coast a "controlled retreat" policy applies. This represents the least active realization of the State's legal obligation
Selective preservation	Aside from principles of safety, the protection of special values and interests in the dune areas demands that coastal recession be combated. To a certain extent, current policy may be compared to this alternative
Preservation	Coastal recession is counteracted at all locations
Expansion seaward	Coastal recession is fought all along the coast. At points where the dunes are very narrow or where recession is considerable, the coast will be reinforced by a seaward construction and natural sand accretion will be encouraged

Dune protection

As in the British study, the unprotected coast can be preserved, partially preserved, allowed to retreat, or safeguarded by sea defences. Table 12.5 outlines the basic policy aims, and table 12.6 summarizes the main options. The Dutch government announced in 1990 that it favoured a dynamic preservationist approach. This would mean safeguarding some 3,000 ha of dune areas along 60 km of coastline by a combination of beach nourishment by sand and shingle deposition with strict planning and tourism controls. That decision was taken on the grounds that it must closely meet the principle of sustainable development, even though, in the short term, it was the most costly option. The probable cost will be US$45 million

Table 12.6 **Policy aims and technical realization for the four types of coast**[a]

Type of coast	Aim	Measures of maintenance	Combating erosion
Dune coast	Maintain polder safety A choice, with respect to interests and values in the dune areas	Continue present maintenance	Sand nourishment Local hard measures
Sea dykes	Maintain	Continue present maintenance	Stone revetment and nourishment
Beach flats	Natural development	Limited maintenance	None
"Remaining"	Maintain present maintenance	Continue sand nourishment	Incidental

a. Various defence alternatives are feasible for the 254 km of dune coast. For the 34 km of sea dykes, the 38 km of beach flats, and the 27 km of other hard sea defences, continuation of existing policy is the only realistic option.

annually, or more than $500 million in the 100-year period, depending on the precise rates of sea-level rise (Peerbolte et al. 1991).

Protecting flood-prone areas

In the Netherlands, about 1,000 ha of residential areas and 11,500 ha of industrial zones are not protected by primary coastal-defence works. Raising the actual level of the ground in these areas, as well as embanking their perimeters, would cost US$1.1 billion for the 60 cm scenario, and US$2.1 billion for the 100 cm scenario over 100 years. This investment is estimated to save some US$18 billion worth of damage over the forecast period, a cost–benefit ratio of some 1 to 10. These are, however, very approximate estimates: the variety of protected areas is considerable in terms of both property value and existing levels of protection. A more precise calculation and programming strategy for flood protection is currently being analysed on the basis of area-specific coastal hydraulics.

Nature conservation

A matter of some interest is the decision by the Dutch government to discontinue sand mining in the Waddensea in order to protect the area from further disruption. The additional cost of moving sand mining to other parts of the North Sea coastline is estimated to be US$360 million over the 100-year period – a minimal estimate of worth for this precious natural area.

Regarding the impact of sea-level rise on the marine ecology of the area, the view of the experts is that any increase in sedimentation will be met with increased benthic activity, but that this in turn may be offset by the impact of climate warming. To play safe, therefore, the analysts concluded that the net change of marine species numbers and composition was too uncertain to call either way. Clearly, a careful programme of monitoring and surveillance is needed. Meanwhile, one of the points of the coastal dynamic preservation approach, involving massive movement of sand nourishing, will be to keep most of the existing coastal littoral ecology intact.

Water resources and salinity

Sea-level rise and climate warming are believed to lead to a modest relative cost for the wealthy Dutch agricultural economy. The main problems are salinity caused by incursion of sea water into drainage systems and groundwater tables and increased drought under certain conditions. Even in the worst-case outcomes, however, the US$100 million of expenditure over the 100-year period is small relative to a net wealth of agriculture of some US$18 billion per year. Much of the problem of salinity can be offset by drought management and programmed pumping conducted on the new pumping regimes already likely to be provided for flood-protection purposes. In all scenarios, the benefit–cost analysis is favourable by a factor of at least 4 to 1, based on best estimates of damage saved.

Conclusions

The Dutch approach is much more analytical than its British counterpart. Moreover, it places greater emphasis on technology, investment strategy, and timing with aggressive estimates of cost. Despite fairly large expenditures, totalling some US$25 billion over the 100-year forecast, the Dutch view is that not only is this money necessary but also it is highly cost effective if well planned. The economy should thrive on the jobs and skills created and developed, and no part of the nation's economic activity should be diminished. In that sense, the Dutch approach is an example of a wealthy and well-planned society responding with anticipation and with vigour to a challenge. Of course, some or all of the largely optimistic calculations could prove to be tragically in error.

13

Sea-level rise and the Sea of Japan

Saburo Ikeda and Masaaki Kataoka

The largest island of Japan faces two major sea fronts: one is the Pacific Ocean and the other is the Sea of Japan, a semi-closed marginal sea enclosed by the Asian north-east continent, the Korean Peninsula, and the Japanese islands. The Sea of Japan has two narrow straits that link to the outer oceans – the East China Sea via the Korean Strait, and the Pacific Ocean via the Soya or Tsugaru Strait. Until recently, the Sea of Japan has experienced severe political and military conflicts between both the "East" and the "West," and the "North" and the "South," as a result of the Cold War. The Sea of Japan has been a playing field for "a game of hide and seek" for military vessels and aircraft since the end of the Second World War and has also witnessed several tragic accidents due to commercial maritime traffic.

The introduction of a market economy to Russia and to China and the attendant greater openness has encouraged Japanese local governments to promote the establishment of economic ties with the coastal regions of those countries, as well as with both South and North Korea. Local authorities have been encouraged to seek the formation of an economic bloc across the Sea of Japan or the "East Asian Sea," as it is called by other Asian countries. Incidentally, these regions may benefit from climatic warming owing to reduced severe frost and snow environments for urbanization and industrialization.

Figure 13.1 shows the geographical location of the regions surrounding the Sea of Japan. These are Japan's 14 prefectures (local autonomous

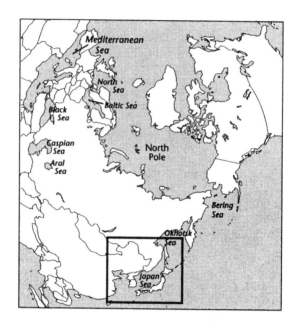

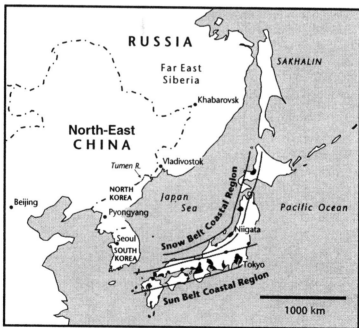

Fig. 13.1 **Geographic location of regions surrounding the Sea of Japan**

units), Russia's 10 Far East Divisions, China's 3 eastern provinces, and South and North Korea. The overall population in these regions numbered approximately 300 million in 1990. These areas are expected to form the "Sea of Japan Coastal Economic Sphere," an important part of the Asian Economic Rim surrounding the Pacific Ocean in future economic development. The share of the major socio-economic indicators for the possible "North-East Asian Economic Bloc Round the Sea of Japan" constitutes about 10 per cent of the total Asian population, 20 per cent of total Asian lands, and approximately US$3 $\times$ 10^{12} of gross national product (GNP) produced in the late 1980s.

The high technology and capital of Japan and South Korea; the energy, minerals, and wood resources of the Far East of Russia; the labour force, energy resources, and agricultural products of the North-East Province of China; and the labour force and mineral resources of North Korea, are expected to be complementary factors promoting cross-regional interactions across the Sea of Japan. Removing the political obstacles to a global economic bloc, with modern technology and investments in the industrial and civil infrastructures, could conquer the severe climatic constraints of frost and heavy snow covering these regions.

In the Japanese case, the climatic character and demographic structure of the region facing the Sea of Japan differ greatly from those of the regions facing the Pacific Ocean. Most of the coastal regions along the Sea of Japan are depopulated hinterlands with relatively poor social infrastructures (sparse networks of highways, railways, harbours, airports, industrial parks, and other civil infrastructures), except for a limited number of locally urbanized areas. The coastal regions facing the Pacific Ocean have made Japan one of the world's economic superpowers, with a dense population and well-stocked social infrastructure. These regions are located mostly along bays, coastal zones, or inland seas (Tokyo Bay, Osaka Bay, Ise Bay, and the Seto Inland Sea) that open to the Pacific Ocean. The critical factor in their concentration in coastal areas is, of course, the great locational advantage for developing modern industries of resource consumption and agglomeration (such as steel, petrochemicals, automobiles, and electrical machinery). In recent years, however, severe climatic conditions – such as heavy snowfall and the shortage of daylight in winter – have become less of a critical constraint for industrialization in these coastal regions owing to the continuing development of transportation throughout Japan.

The potential environmental risks in this region depend not only on the development strategy that both Japanese coastal regions and the outer continental regions adopt but also on the degree of global climatic change that is expected to occur, including the greater variability of ecological systems of the coast, forests, rivers, lakes, and the marginal sea (IPCC

1997, 1998). It is still too early to make quantitative estimates or to adopt comprehensive views concerning future vulnerability of specific regions to global climatic impacts. Even with three completed assessments (IPCC 1990a, 1996a, b, c, 2001) by the Intergovernmental Panel on Climate Change (IPCC), we lack detailed data and information concerning eco-logical sensitivity of marine resources (such as fisheries, minerals, and energy) to global climatic change and associated sea-level rise.

The Sea of Japan has a large environmental capacity for resilience and Japanese soil is said to be more resistant to acid rain than is European soil. None the less, if the coastal areas of the Asian continent adopt a development strategy that neglects the potential environmental hazards of a semi-closed marginal sea, the sea could suffer irreversible environ-mental degradation over the long term. In such a case, international en-vironmental conflict could occur in the area. In Europe, numerous pol-luters and victims are scattered among several countries and regions, each of which has its own cultural, social, and institutional backgrounds. Be it around the North Sea (Argent and O'Riordan 1995; chapter 12, this volume), the Mediterranean (Haas 1990b), or the Baltic (Kohonen 1994; Kristoferson 1996), that complexity has resulted in thorny conflicts of in-terests among the parties involved. Thus far, the Sea of Japan area does not have any international scheme of comprehensive cooperation for its environmental management. The area, in short, urgently needs an in-stitutional arrangement of international risk management for sustainable development.

A systems analytic framework is proposed to analyse a set of scenarios of global environmental impacts based on earlier coastal "economic–ecological" models (Ikeda 1987). The major risk issues addressed in this chapter include ecological fragility, socio-economic vulnerability, and potential societal responses.

Regional vulnerability to climate change

To begin, we define vulnerability and resilience of a regional society to external disturbances based on the concepts presented in chapter 7 and followed by Liverman in chapter 5. Thus, *regional vulnerability* is the degree of inability of a regional society to cope with the occurrence of a hazardous event, and *regional resilience* is defined as the measure of ability of a regional society to absorb and recover from the occurrence of a hazardous event. In the context of current environmental risks in the Sea of Japan, our primary concern focuses on whether we can avert the prospect of the marginal sea becoming a huge graveyard of accumulating dumped wastes. The combination of new economic globalization around

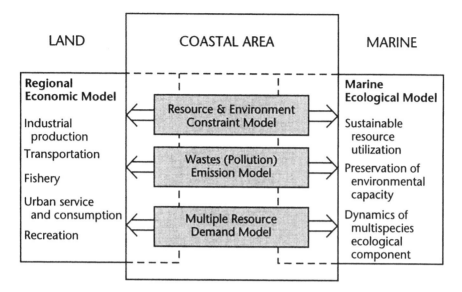

LAND	COASTAL AREA	MARINE

Regional Economic Model

Industrial production

Transportation

Fishery

Urban service and consumption

Recreation

Resource & Environment Constraint Model

Wastes (Pollution) Emission Model

Multiple Resource Demand Model

Marine Ecological Model

Sustainable resource utilization

Preservation of environmental capacity

Dynamics of multispecies ecological component

Fig. 13.2 **Conceptual model of the economic – ecological system in coastal regions**

the sea and the potential rise of sea level could accelerate a risky process before the concrete evidence that has surfaced in many other marginal seas (e.g. the Mediterranean, the Baltic) becomes apparent.

To carry out such a risk analysis of regional vulnerability to global environmental risks in coastal regions, we need a conceptual model of regional society that includes at least the economic system of utilizing coastal resources and the ecosystem to be exploited or preserved. Figure 13.2 presents such a conceptual model, which formulates multifarious human actions and responses between economic and ecological systems as a framework of risk analysis, regional vulnerability, and regional resilience. In this model, we assume that the economic system seeks to maximize the utility and efficiency of the regional society by utilizing marine resources and that the ecological system pursues dynamic resilience in material and energy flows within the system. We analyse major regional responses to global environmental risks through the three submodels, listed below, that work as an interface between economic and ecological models:

1. The Resource and Environment Constraint Model;
2. The Waste (Pollution) Control Model;
3. The Multiple Resource Demand Model.

The first model is concerned with resource sustainability or the environmental capacity of purification in the ecological system as a critical factor in evaluating regional vulnerability and resilience. Regional vulnerability is understood as an interface between the two systems that does not work adequately to respond to the outer disturbances. The second model estimates the amount of wastes to be discharged into the ecological system. The last model calculates various resource-exploitation demands by the multiple sectors in the regional economic system. Water and fishery resources are typical examples of those that will require both quantity and quality of the resources, such as clean water or toxin-free fish.

In building such a regional economic–ecological model, several important management issues of environmental risks must be considered, as shown in table 13.1. The first such issue is how to structure regional vulnerability in the context of the three submodels – the "resource demand," "wastes discharged," and "resource constraint" – that link economic and ecological systems. Here, for simplicity, we focus on only two aspects of vulnerability – the ecological and socio-economic issues that Kasperson, Kasperson, and Dow elaborate in discussing spatial equity problems in chapter 7.

The second issue is that risk information and the detailed context of uncertainty about resource demand play critical roles in the conceptual model but in a different way in the economic and ecological system. In the economic system, market mechanisms do not adequately treat risk information on environmental goods and services, including the limitations of the environment-purifying capabilities of coastal waters. Rather, such information is asymmetrically owned between beneficiaries and victims of environmental degradation, and between the supply side and the demand side of environmental goods and services. Such asymmetry in risk information biases various economic actors (industry, consumers, and regulators) in their decision making to negotiate a social response to regional environmental risks. In the ecological system, the flow of energy and biomass can be described by a simple information unit of entropy or biomass. The distribution of entropy or biomass, however, varies widely in a hierarchical network of ecological space and time in which each component has its own efficiency in utilizing energy or biomass. Therefore, it is not enough to use risk information based on entropy or biomass to regulate the dynamics of heterogeneous ecological processes. Rather, it requires a systemic and holistic view that the regional society is an integrated part of the global economic and ecological world.

The third issue is the accountability of system agents or decision makers in the regional society to respond to the threat of a future total ecological collapse. The role of economic and ecological participants, operating in the complex reality of a multivalued society, departs markedly from a rational economic person who seeks to maximize economic utility, the socially active person who pursues innovation through creative activities,

Table 13.1 **Systems-management issues of environmental risks in constructing economic–ecological models**

System properties		Economic system	Ecological system	Systems approach for integration
System boundaries and constraints	Space	Social structure of production and consumption	Flow of materials and energy	Flow of information and decision-making process
	Time	Medium to very long term	Short to medium term	Short to very long term
	Constraints	Institutions and habits; social infrastructure	Conservation theory; cohabitation; environmental capacity	Decomposition and integration of hierarchical subsystems
System agent		Rational economic person	Passive biological person	Active social person
System information		Free competition; asymmetric circulation of information under externality	Hierarchical order and self-organization under the law of entropy	Cybernetics (control of information and communication through feedback)
Technological environment of the system		Economic growth and technological innovation; alternatives for environment and resource	Vulnerability and resilience; dynamic change in environmental capacity	Technology assessment; risk management

and the biological person who acts passively in natural food linkages. We often find technological optimism that assumes a limitless human ability to solve problems of environmental risk through technological innovation and economic growth. For example, developing countries sometimes adopt economic-development strategies that fail to take environmental risks seriously into account and that assume that economic growth will solve every problem. Others, however, take a pessimistic view of technological progress, arguing that it inevitably brings about irreversible change in the global environment, with catastrophic results. Hence, some institutional mechanisms for disseminating and integrating risk information throughout economic and ecological systems are indispensable.

The systems approach adopted in this chapter takes a cybernetic view of managing environmental risk, one that posits that we should look for responses to maintain a dynamic resilience of the system through feedback of available information even if it is accompanied by a high level of uncertainty. Dynamic resilience of the system can be achieved, not only by increasing research and development to reduce uncertainty but also by expanding resource allocation to the social overhead capital of infrastructure to provide an adequate level of accountability.

The Sea of Japan and its coastal regions

Early history

The Sea of Japan (East Sea) was formed as a result of collision and subduction by the hypothetical Kula–Pacific Ridge from late Cretaceous to Oligocene time. Since then, this marginal sea has experienced at least several important climatic and oceanographic changes in relation to human settlement. The remarkable one happened at the period of "transgression" some 6,000 years ago when the sea level was several metres higher than at present and the early aborigines, called "Jomon people," settled the coastal regions of the Japanese islands. The weather was warmer, owing to the melting of the ice beds and other geological reasons (Kaseno et al. 1991). About 10,000 years ago, however, the sea level was about 40 metres lower than at present. Recent palaeogeological studies provide some understanding of how the oceanographic conditions in the marginal sea and the biological environment in coastal zones changed during the ice age and the "Jomon transgression."

One of the important lessons emerging from studies of these palaeogeological and palaeoclimatic data indicates that the sea-level changes have resulted in the shift of the dominant sea current. This sea has two major currents: one is the prevailing warm Tsushima Current, supplied from the southern Pacific Ocean via the Tsushima Channel between Korea and Kyushu; the other is the cold Oyashio Current, supplied from the northern Pacific Ocean via the Soya Channel between the Sahalin Peninsula and Hokkaido (fig. 13.3). The dominant Tsushima furnishes a variety of marine biological species and accounts for high productivity in the Sea of Japan. When the cold current from the northern channel was dominant around 20,000 years ago, the fish species from the northern Pacific Ocean migrated into (and caused an ecological change in) the sea.

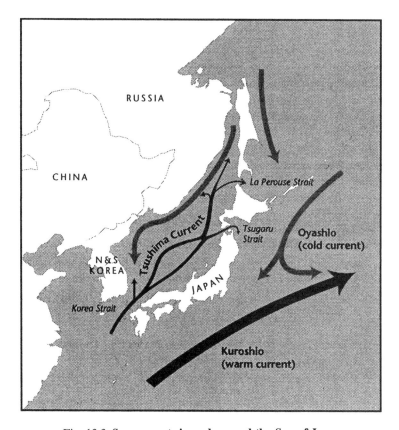

Fig. 13.3 **Sea currents in and around the Sea of Japan**

Hence, even a slight increase in sea level in such a semi-closed marginal sea would involve drastic ecological changes in marine structure and biological productivity.

Meteorology and natural hazards

Asian summer monsoons bring abundant water to the Japanese arc islands and strongly affect the regional climate and associated agricultural activities. In general, the southern and coastal regions facing the Pacific Ocean obtain enough water to cultivate rice during the summer monsoon season. The northern and coastal regions facing the Sea of Japan, however, have a different hydrological character in terms of their annual water balance for agricultural and industrial production. The winter monsoons ("winds of the Westerner") from the northern hemisphere take up a great deal of vapour, generated by the Tsushima warm current, as they pass over the

Sea of Japan, and bring a heavy volume of snow over the northern and coastal regions facing the sea. The snow amassed in winter contributes about 30 per cent of the water that is used for cultivating rice paddies in these regions.

The civil infrastructure for preventing the hazards of heavy snow has gradually emerged in recent years. In particular, means for avoiding disruptions of traffic on major roads and of life-support systems and utilities for citizens have improved remarkably. The packed snow that used to be seen as a hazard may be re-evaluated as a stocked water resource for generating electricity, irrigation, underground water, and a source of thermal heat in summer. Despite the potential use of packed snow, both heavy snows and the lack of daylight are still major obstacles for these coastal regions in competing with the sunbelt region facing the Pacific Ocean.

As for other natural hazards, the coastal regions facing the Sea of Japan have less risk of typhoons, earthquakes, high tides, and tsunamis (earthquake-driven high sea waves) than other coastal regions. If both coastal regions were to confront the same level of natural hazards induced by the potential rise of sea level, then the Sea of Japan would have greater risk because the civil infrastructure of preventing natural hazards has not caught up with the protection levels implemented in the sunbelt regions. This confirms the arguments, put forward in chapters 5–7, regarding vulnerability as a central contributor to risk.

Existing environmental risks

In Japan, modern industrial society has developed along the bays or coastal zones open to the outer oceans, but mostly along the Pacific Ocean, since the end of the Second World War. As local economic centres, the major municipal capitals facing the Sea of Japan (e.g. Kanazawa, Toyama, Niigata, and Akita) are limited in several respects. No typical pattern of industrial and urban pollution has emerged as a salient political issue in these regions, except for several pollution tragedies caused by toxic heavy metals (mercury and cadmium). Figure 13.4 displays a map of the industrial cities that are mandated to establish pollution-control plans for air, water, noise and vibration, soil pollution, and ground subsidence. Among the 39 cities designated as regions of mandated pollution control, however, only three are located along the Sea of Japan.

Beginning in the late 1980s, new types of environmental pollution (primarily in the form of acid rain and oil pollution from both national and international sources) appeared, crossing local, regional, and national boundaries. Figure 13.5 shows the relative acidity of precipitation (CRIEPI 1989). The difference in acidity between the two coastal regions in terms of seasonal variation is apparent. The acidity, measured as grams

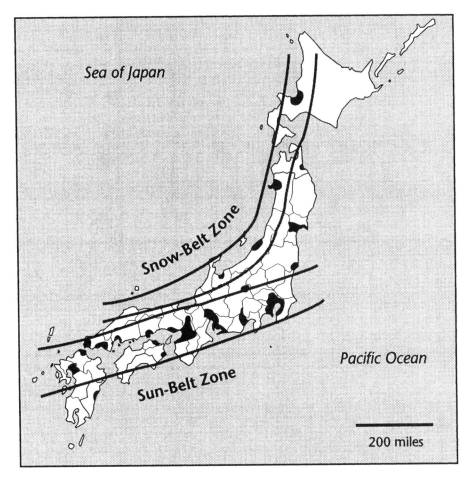

Fig. 13.4 **Industrial cities in Japanese coastal zones mandated to establish pollution-control plans**

of sulphate (SO_4) per square metre, in the coastal region along the Sea of Japan is dominant in winter, owing to the winter monsoons (possibly of continental origin), although the long-range transport of pollutants in East Asian countries needs further study.

Water and soil pollution from the dumping of hazardous industrial wastes is a critical issue from the viewpoint of the interregional transfer of pollution. Hazardous wastes generated in excess of treatment capacity in the sunbelt region are transferred to depopulated regions. The siting of nuclear power plants (such as those in Fukui, Niigata, and Kashiwazaki) has centred mostly on the coastlines facing the Sea of Japan. This could

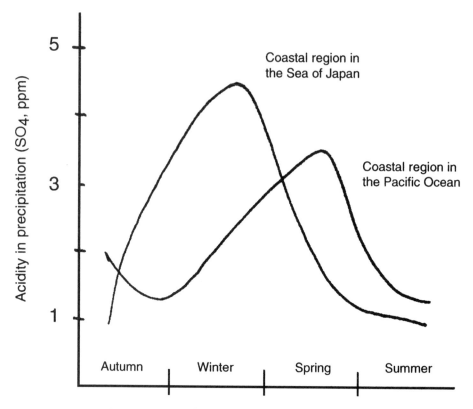

Fig. 13.5 **Acidity in rainfall over the Japanese coast**

add another source of risk, radioactive wastes, to the Sea of Japan. At the same time, as described above, globalization of the economy beyond national borders makes it possible to develop, in the coastal regions facing the marginal sea, industrial activities that formerly were less competitive in the national market. Thus, the risk of sea pollution from dumping industrial wastes, oils, and household wastes is likely to increase dramatically in this marginal sea unless adequate interregional instruments for risk management evolve, such as those for the North Sea described by O'Riordan in the preceding chapter.

Scenario of global climate change in the Sea of Japan

Several models exist by which to estimate global climate changes on the basis of the carbon cycle in the atmosphere and ocean. The important common precondition for these numerical models is that a global equilibrium concentration of CO_2 gas in the atmosphere doubles from the present equilibrium state by 2100. So far, most models simulating global warming provide a wide range of difference with respect to the climatic impacts on East Asia as a whole. Nevertheless, a fairly good consensus exists in terms of the annual mean temperature, precipitation, and pattern of atmospheric pressure (Tokioka 1990), as follows:

- rainfall decreases in the subtropical area, increases in the north-eastern continent, but does not change much over the Japanese islands;
- temperatures in the cold outbreak region (the north-east continent) increase over 10°C, and in the mild region (the Japanese islands) increase around 2–3°C;
- the summer monsoons in the subtropical area intensify, the pattern of rainfall becomes tropical, and the monsoons become unstable in scale and time;
- the winter monsoons from the north-east continent weaken; accordingly, less snowfall is expected in the coastal area of the Sea of Japan.

The grid scale of these model calculations, however, is too crude (400–500 km grids are used in the model developed by the Meteorological Research Institute of Japan) to interpolate their calculated results into the regional scale (100 km) of climate effects surrounding the Sea of Japan. Therefore, it is extremely difficult to predict the future climate on the regional scale by applying global-warming models. One possible means by which we could make reasonable estimates for the regional climate changes is to design an adequate set of impact scenarios that take account of the available scientific knowledge and data. What follows is such a typical scenario of regional climate impacts on the Japanese islands.

Climate impacts on coastal regions

In summer, the Asian monsoon intensifies and rainfall is expected to increase and become tropical, sometimes torrential. Even when there is a downpour, however, the water will not be stored, owing to the steep terrain in the mountains in Japanese river basins. This will bring a decrease not only in water resources but also in soil moisture, particularly in the southern part of the Japan islands.

In winter, the atmospheric pressure pattern of the cold high anticyclone over the northern continent becomes weak, and accordingly snowfall decreases in the coastal zones facing the Sea of Japan and their northern part of the Japanese islands; however, the rainfall in the mountain area increases by about 30 per cent or even more.

Figure 13.6 displays an example of such estimates in terms of the number of days during which the mean temperature exceeds 30°C (summer days)

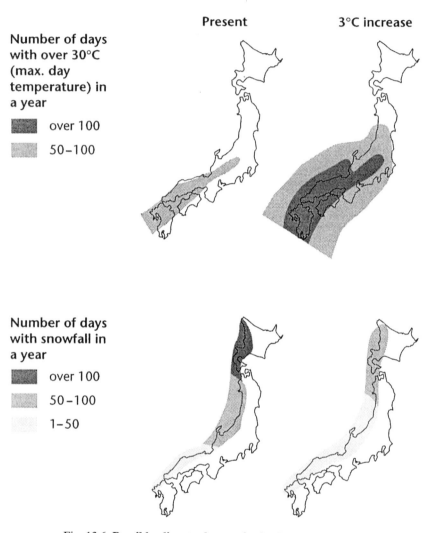

Fig. 13.6 **Possible climate changes in the Japanese islands**

and the number of snow days per year, as projected by the Central Research Institute of the Electric Power Industry (CRIEPI 1991). Obviously, a beneficial trend seems to be apparent in the coastal regions facing the Sea of Japan. However, as climate fluctuations are expected to become greater (e.g. torrential rainfall or drought), natural hazards may increase.

Sea-level rise

No detailed study exists by which to forecast possible sea-level rise in the Sea of Japan, but relevant data do exist in Mimura (1996). In a global sense, the sea level is considered to have risen by 10–20 cm over the past century owing to the thermal expansion of sea water and (possibly) to melting ice. According to monitoring data at the Japanese coast, however, the sea level at the coastlines along the Pacific Ocean peaked in 1950 and has been falling since then (Sugimura 1989). Studies of sea-level variations from 1950 to 1991 in the East Asia region have concluded that these variations were mainly due to plate tectonics (Mimura 1996: 11). Possible future sea-level rise depends not only on the effects of global warming on the sea water at the time of doubling of CO_2 concentration, but also on the long-term coastal land movement in the Japanese islands (Japanese Environment Agency 1989a).

By and large, it seems reasonable to assume a possible rise in sea level of 80–120 cm by the year 2100 along the Sea of Japan (the IPCC working group consensus). If we take a scenario of 100 cm rise in sea level, a considerable portion of Japan's major metropolitan areas, such as Tokyo, Osaka, and Nagoya, might be under water. For example, in Tokyo, approximately 30 per cent of the total urban districts with more than two million residents would be under the zero level, without any defensive measures. This submergence would occur slowly over more than 100 years, however, and would interact with other effects of high waves or high tides associated with monsoons, typhoons, or tsunamis. If sea-level rise were to be 1 metre, Matsui (1993) estimates that more than 15 million people would be endangered. The IPCC (1997: 15) meanwhile, estimates that a 1 m rise would threaten the Japanese coastal zone, on which 50 per cent of its industrial production is located, and about 90 per cent of the remaining sandy beaches in Japan would be in danger of disappearing.

Countermeasures to prevent such naturally occurring high waves or high tides have been implemented for more than 50 years, mostly in the major urban regions along Japanese coastlines (Mimura, Isobe, and Hosokawa 1993). Nevertheless, the hazard potential of damage by high waves or high tides will surely increase in minor urban areas along the Sea of Japan.

Vulnerability analysis

Following the concept of the regional vulnerability noted above, we examine two types of vulnerability: one involves the impacts on the ecological system of the Sea of Japan and the coastal region along the marginal sea; the other involves impacts on the socio-economic system of the regional society. In our conceptual framework, three models are major constituents that merit consideration in the analysis of both ecological and socio-economic vulnerability:

1. Environmental limits of the marginal sea;
2. Waste discharged into the sea and the atmosphere;
3. Resource demand (water, fisheries, minerals, and amenity resources).

The distributional aspects of the environmental risks (potential damages and benefits in their space and time dimensions) are also important factors in gaining a better understanding of scenarios of future regional sustainability around the Sea of Japan.

Ecological vulnerability

On the basis of palaeogeological and oceanographical studies (Nishimura 1990) with respect to the environmental limits of the sea, we can envision the following hypothetical scenario.

The rise in global atmospheric temperature causes a temperature rise in surface water and a corresponding thermal expansion of the body of water, resulting in the increased height of water in the marginal sea. This warming surface water decreases the mixing and upwelling capacity between the surface water (8–17°C) and the cold bottom water (about 0.5°C), which formerly was confined to the deep bottom layer (about 2,000–3,000 m in depth) but now may undergo even less exchange owing to the poor supply from the upper layer. In the long run, the process may eventually induce a change in the flow pattern of the prevailing warm current in the surface layer that provides various species of pelagic fishes (pollack, cuttlefish, sardine, and saurel) in this marginal sea.

In this context, the marginal sea is extremely vulnerable to a possible change in the flow of warm current associated with sea-level rise, and also (possibly) the increased presence of industrial wastes on the sea bed due to the reduced exchange of sea water. Figure 13.7 illustrates this scenario. In the Sea of Japan, the warm water flows in and out through the narrow and shallow channels of the Tsushima and Tsugaru (70 m and 180 m in depth, respectively). It flows only in the upper layer, whereas the cold water in the deep bottom layer is confined for a longer period with decreased upwelling force. Hence, the degree of sensitivity to pollution is one of the greatest among various seas.

Another issue that merits consideration in assessing ecological vulnerability is the environmental nature of the coastal regions. One important potential risk is the dispersion of "acid rain" crossing the Sea of Japan. Figure 13.8 illustrates some examples of long-range transport routes in

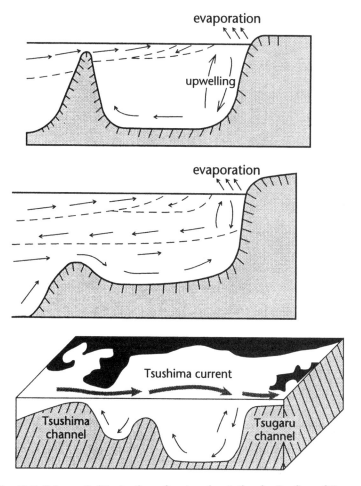

Fig. 13.7 **Schematic illustration of water circulation in the Sea of Japan**

the winter season when the north-west winds of winter monsoons carry suspended particles over several hundred kilometres per day. A study by CRIEPI (1991) has estimated the amount of sulphur dioxide generated by the North-East Asian countries. Table 13.2 shows CRIEPI's estimates of annual emissions by Japan, China, Taiwan, South Korea, and North Korea. For sulphur, the total amount is about 11 million tonnes per year, which equals the amount generated by the North European countries or the north-eastern part of the United States. Among these contributions, China's is the largest – 85 per cent of the total amount generated. Despite clear signs of increasing acidity in the rainfall and soil, so far actual damage to forests and fresh water is not clearly apparent in the Japanese

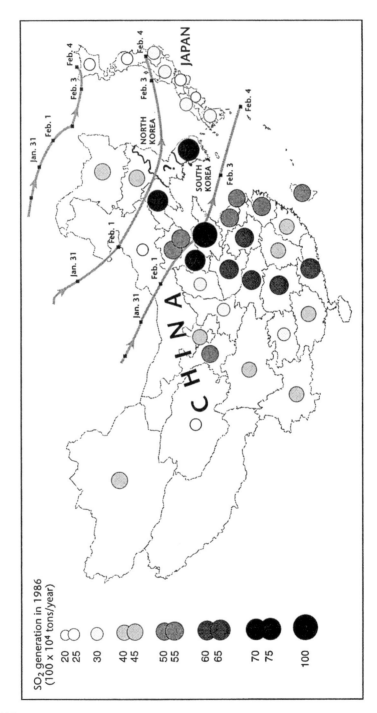

Fig. 13.8 **Long-range transport of SO₂ in water, with estimated sources of SO₂ emission**

Table 13.2 **Estimated anthropogenic sulphur dioxide emissions in East Asia**

Regions	Population ($\times 10^6$)	Area ($\times 10^9 \text{m}^2$)	Primary energy consumption ($\times 10^6$ t/y)	Annual emissions total ($\times 10^{12}$ gS/y)	Year
Japan	120	372	314	0.55 (0.85)[a]	1983
China	1,052	9,597	504	9.85 (0.97)	1986
Taiwan	19	36		0.17 (0.87)	1986
South Korea	41	98	47	0.50 (0.92)	1985
North Korea	20	121	38	0.44?	1986
Total	1,244	10,224		11.54 (0.95)	

Source: Fujita et al. (1991).
a. Estimated ratio of emissions from stationary sources.

coastal regions, because the soils there are more resilient to sulphur acidity in their neutralizing capacity than are continental soils. It is, however, only a matter of time before the trees and soil ecosystems experience significant damage.

Socio-economic vulnerability

Coastal peoples have begun to be aware of their vulnerability to climatic change and sea-level rise, at least of the direct impacts (e.g. damage to urban river systems, harbours, coastal lowlands, and scenic views; and alteration of coastlines by the encroachment caused by erosion and the force of waves). These problems can endanger urban life-support systems as well as transportation, industrial, and agricultural infrastructures.

As for the direct impacts on regional economic systems, there is a varying degree of effects on industry (primarily on agriculture and the climate-dependent industries of brewing and fermentation, recreation and tourism, construction, and transportation). However, in a global economic sense, the following indirect socio-economic vulnerabilities to climatic change are relevant:

- urban life-support systems: vulnerability to increased hazard potentials;
- industrial structure: vulnerability to low energy consumption with high recycle rates; and
- trade structure: vulnerability to the export and import of pollution loads that will be transferred through the atmosphere, river, sea, and other ecological components.

One of the major impacts associated with urban infrastructures is the risk of water pollution from urban non-point sources (garbage and consumer products from households; hazardous wastes from commercial, transportation, and other urban activities). Given an expected decrease in rainfall and a increase in freshwater temperature, the quality of river water and groundwater will generally deteriorate. In addition, sea-level rise will accelerate salt-water intrusion, not only into the river and the groundwater system but also into water-supply and sewage systems. This means that, if the infrastructure of urban life-support systems is not well prepared for a rise in sea level, urban water systems that were recycled from upper to lower stretches of stream will be more vulnerable to contamination by hazardous and toxic wastes from non-point sources.

The second issue is to promote low energy consumption and a resource-sustainable society by adopting an ecologically sound lifestyle and technology. Fortunately, the Japanese part of the coastal region facing the Sea of Japan has retained appropriate technologies that can provide low energy consumption and self-preventive systems in an environment of frost and heavy snow. The technology of utilizing the low entropy of packed snow is now being developed as one of the clean energy sources.

The final risk issue associated with trade structure is critical in the Japanese economy in relation to a new programme of economic globalization around the Sea of Japan. Japan is self-sufficient with regard to only 20 per cent of its energy and 30 per cent of its food needs. The high dependency on the world ecosystem for energy and food supply means that climatic change is certainly a serious issue through its impacts on trade structure (Nishioka 1990). In this sense, socio-economic vulnerability in the East Asian regions around the Sea of Japan is also a critical problem for the Japanese people. Although this new trend is ambitious and open to a number of uncertainties, we may recognize the following scenarios (Niigata Prefecture 1991).

Scenario one

Mutually beneficial economic cooperation will proceed on the basis of the complementary positions among the north-east Asian countries. For example, a fairly simple complementary relationship exists among the coastal countries for economic development:

• technology capitals: Japan and South Korea;
• energy resources: China, Far-East Siberia;
• mineral resources: China, Far-East Siberia, North Korea;
• fishery resources: Far-East Siberia, North Korea, South Korea, and Japan;
• labour: China and North Korea.

Figure 13.9 illustrates such complementary relationships for industrial development, together with recent developments in transportation and communication lines. Several major projects, such as the natural gas in

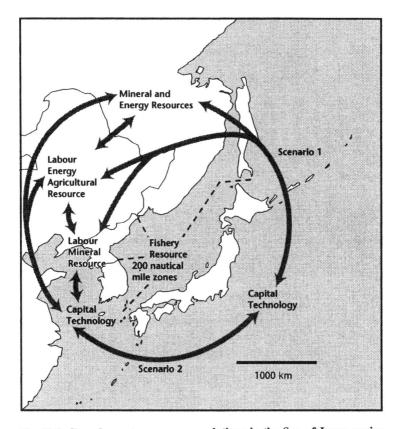

Fig. 13.9 **Complementary resource relations in the Sea of Japan region**

Yakutsku, oil and gas in Saharin (named the Bostokh Plan), and the International Port Development of Yenchi River (Zumoi-Ko), address the development of resources and infrastructure.

Scenario two

A specialization of industrial activities in a shared market base will progress after the major infrastructure of transportation and production facilities is implemented in the North-East Asian countries. Industrialized Japan and South Korea are already moving into a relationship that promotes investment and relocation of their firms in order to form a common market.

It is highly uncertain whether scenario two will have the benefit of a genuine chance in the current political world of North-East Asia. In any case, it is scenario one that is actually proceeding and forms the base for a regional risk-management scheme of the marginal sea associated with the global climatic change and sea-level rise.

To date, no estimated figures of the amount of industrial wastes discharged into the Sea of Japan and the extent of ecological damage, primarily to fishery resources, exist for the marginal sea. These issues are difficult to address. The 1990 Annual Report by the Chinese Government, reported in *Nihon Keizai Shimbun* (the *Nikkei Newspaper*) (1991), states that industrial wastes-to-water systems amount to about 24.9×10^{12} tonnes, except for localized industry. Among this some 2,189 tonnes of heavy metals were included (*Nihon Keizai Shimbun* 1991). The Japanese government has reported that about 0.32×10^{12} tonnes of industrial wastes were generated for waste treatment in 1985 (Japanese Environment Agency 1989b). There are no standardized indices by which to compare environmental risks in this region.

Undoubtedly, first of all, an institutional arrangement is required for monitoring the state of the Sea of Japan, irrespective of the political conditions that have existed for past decades. We could then start to negotiate the future of risk management, as suggested in chapter 1 and by O'Riordan and Timmerman in chapter 14, in this valuable marginal sea as a source of regional sustainability. Socio-economic vulnerability to global climatic change depends not only on future economic development in each coastal country but also on information and monitoring data concerning the ecological conditions of the marginal sea.

Toward interregional risk management

What environmental risks accompany the increased movement toward a new economic globalization among the North-East Asian regions around the Sea of Japan? Could the sea-level rise associated with global warming become a warning signal that the coastal waters will deteriorate, as has occurred for most other industrialized nations in the course of the industrialization of their coastal zones?

In terms of trade structure in this coastal region, Japan should be responsible for its own wastes and for transferring appropriate waste-treatment technologies from other coastal regions. So far, the volume of trade between Japan and Russia has been extremely low, reflecting a tense political relationship. If the new economic globalization follows the course of scenario one, however, Japanese dependency, for energy, mineral, and wood resources will increase, particularly on those from Far East Siberia, the North-East Division of China, and (possibly) North Korea.

Resource-development activities in the major river basins in Siberia and North-East China could lead to a huge output of industrial waste into the Sea of Japan. In addition to the Japanese coastal zone facing the Pacific Ocean, coastal areas along South Korea have already experienced frequent outbreaks of "red tide" (toxic plankton blooms).

Indeed, no definite solution exists to this problem, although the Japanese experience of coastal environmental management, albeit remedial and not anticipatory, may suggest institutional mechanisms. A Japanese regulation programme entitled "Areawide Total Pollution Load Control," for example, was enforced on the basis of the Water Pollution Control Law amended in 1978. The target areas are the major industrialized coastal zones – Tokyo Bay, Ise Bay, and Seto Inland Sea (including Osaka Bay) – where the sea water is so eutrophic and polluted by various inflows of hazardous materials that it brings frequent red tide or "blue tide" (low-oxygen water) that is toxic to fish and shellfish and damages not only the fish-farming industry but also coastal amenities and recreation. The basic principles of the regulation scheme are as follows: (1) setting the total pollution load to be reduced (in terms of biological oxygen demand, or BOD) in order to attain a water-quality standard, depending on the water use; (2) allocating a reduction load for each local government involved in the target area on the basis of its industrial activities; and (3) empowering local government to take steps necessary to reduce the respective amount of pollution load by implementing the pollution-control plan supported by financial aid from the central government. The control measures include the construction of sewage networks, the tightening of effluent standards for industrial waste water, monitoring, and education.

The possibility of introducing such "Areawide Total Pollution Load Control" for the Sea of Japan region would have been blocked by exactly the same kinds of questions as the "CO_2 Emission Quota Policy" might have had in a more regional sense. A shared concern exists, however, that the marginal sea may create an interregional institutional mechanism, possibly linked with fishery regulation. In fact, several Japanese local governments facing the Sea of Japan have proposed an "Interregional Committee or Forum for Marine Resource Research Around the Sea of Japan" which is based on the 200 sea-mile regulation zone.

This is a first initiative to take into consideration the resource and environmental problems around the Sea of Japan. It is certain that various types of regulatory instruments and ventures, whether governmental or non-governmental, will be needed to negotiate future management of the risks of a deteriorating marginal sea.

Conclusions

Ecological vulnerability

Historically, pollution from industrial and public sectors, involving the transfer of pollution disutilities to other industrial sectors or to public sectors without any compensation, has been a major environmental problem. A typical problem is the "public nuisance" from industrial sources (*Kou-gai* in Japanese indicates a variety of health risks and environmental disruptions). The other important "externality" is the spatial or transfrontier pollution crossing regional and national boundaries. In this particular coastal region, several international conflicts, such as the issues of acid rain, hazardous wastes, oil pollution in the sea, and overfishing, have already emerged. The most critical among them will be ecological risks in the Sea of Japan that may be accelerated both by temperature rise and by increases in the disposal of industrial and household wastes.

Socio-economic vulnerability

Because of the high degree of uncertainty involved in the spatial resolution of global climate change, it is extremely difficult to make any concrete estimates concerning potential socio-economic effects on regional societies. The positive or negative impacts on important resources or social structures differ from region to region, but include the following:
1. *Water resources for agricultural irrigation (rice production):* a precipitation pattern may become unstable owing to a change in global wind circulation, with less total snowfall or rainfall but torrential in nature.
2. *Safety measures for natural hazards:* the safety standards for infrastructures at estuaries, bays, and rivers in urban areas may need upgrading.
3. *Fishery resources:* a possible change in ocean flow patterns may disrupt the marine ecosystem through alterations in food-chain operations when the sea level rises. Increased dumping of hazardous materials and various nutrient inflows may trigger large-scale toxic red tides in coastal waters.
4. *Marine amenity resources:* coastal flooding and erosion may change the coastal landscape and induce changes in vegetation.

Depending on the existing and future vulnerability of the coastal resources, various risk profiles in coastal-resource management under differing economic globalization scenarios for the Sea of Japan will need to be developed.

Responses to the potential risks

The regional strategy of forming an economic group of East Asian countries around the Sea of Japan may bring about the critical condition of "being a huge dumping ground," owing to the development of mineral and natural resources in this region. It is apparent that careful and sound environmental management, operating under regional and interregional collaboration, will be needed. In fact, severe industrial pollution and environmentally destructive events have already occurred along the major coastal areas, including that at Minamata (where mercury poisoning developed in people who ate seafood contaminated by industrial discharges into Minamata Bay), red-tide episodes in the Seto Inland Sea, and the intrusion of blue tide into the river fronts of Tokyo Bay.

Social responses for pollution control led to the Areawide Total Pollution Load Control Scheme for Japanese coastal zones (Japanese Environment Agency 1989c). This regulatory instrument has been implemented for nearly two decades to mitigate the eutrophication of sea water resulting from the increased disposal of industrial and household wastes. The institutional setting for allocating both industrial and domestic wastes among the coastal areas involved in the Sea of Japan will be a critical issue in the future risk management of the marginal sea.

Part Four

Global environmental futures

Editors' introduction

Throughout this volume, we and our collaborators have argued that management of global environmental risks cannot be separated from questions about the future we desire and the processes by which we will conceive and decide among visions of alternative risks. Indeed, implicit in the very concept of risk – which we have defined as threats to people and the things they value – is the notion that our actions and technologies should be shaped to realize certain ends or futures and to avoid others. It is instructive that when the Society for Risk Analysis some years ago attempted to arrive at a consensual definition of risk, it quickly discovered that the exercise was laden with value issues and was highly contentious. With the inability to produce consensus came the realization that negotiating risk issues and objectives is inescapably political. A perceptive contribution of the social theorists of risk has been that risk controversies are as much about competing social visions, cultural biases, and world views as about narrower questions of the level of safety or security sought.

So risk matters are inescapably about different futures. Whereas experts use the risk concept as a helpful abstract, component, or metric for assessment, publics undoubtedly decide not on the basis of abstractions but on preferences about desirable or undesirable activities, technologies, facilities, or institutions. Furthermore, it seems clear that the intertwining of risk with broader social choices and alternative futures is certain to increase – indeed, to deepen. It is apparent that the often-quoted paradox, that the peoples of advanced industrial societies are becoming more concerned about risk as they are simultaneously becoming safer and living longer, speaks for a fundamental change in human values. Indeed, the 1992 survey of environmental attitudes in 24 nations (Dunlap, Gallup, and Gallup 1993) found that strong environmental concerns are not restricted to the developed world; indeed, that survey revealed that such concerns were widely distributed throughout the developing world. Ulrich Beck (1992) has argued that the twentieth century marks a period in which the distribution of wealth and capital have dominated public policies and social debate, but that the current century will be the "risk society." Public preoccupation in the twenty-first century may well centre upon the riskiness of life, as part of a growing concern with human security and well-being, so that the allocation of "bads" or dangers will lie at the heart of value concerns, social conflict, and institutional missions. Whether or not the risk society envisioned by Beck actually appears on the scene, it seems likely that we are entering a new period in which risk deliberations and decisions will permeate public agenda and political competition.

With these changes in risk will come an increasing contentiousness around uncertainty. Many of the newer risks – genetic engineering, cybernetic hazards, global change – will be accompanied by high levels of uncertainty. Such uncertainties, and the growing need for anticipation and forecasting of risk, will place greater weight on the assessment of uncertainty. This would appear to call for greater applications of science and formal analysis at a time when society, in its democratic impulses, appears more suspicious and sceptical of both. The current debate over the precautionary principle, which has gained prominence in the 1990s in Europe, suggests what is likely to be a continuing source of tension – whether science and expertise will chart our course to new futures or whether moral principles and the desire to limit the influence of expertise will gain ascendancy. Ironically, the two approaches to risk decisions have greater need for each other than is generally recognized: risk assessment requires normative principles to set goals for decisions, yet normative directives still depend upon analysis and diagnosis for their intelligent application and for a parsimonious use of society's limited resources.

In addition, this grappling with risk futures will become more explicitly international and global in scope. In a world in which regions and places are more tightly integrated in a global economy and social communication system, the panoply of risk confronting nations, regions, and localities will be profoundly more transboundary in nature. The sources of risk will increasingly lie beyond the scope and control of those charged by society with assuring public health, safety, and well-being. Some of these risks will be those associated with systemic global environmental change (see chapter 1); others will reflect the growing reach of technologies, information systems, and a global economy. But the implications of a restructuring of risk is that economic processes, and the response to the risks they generate, will shape futures across the political mosaic of the globe and for people in the future as well as those living now. Thus, notions of "our common future" are not simply wishful rhetoric. Underlying global risk is the issue of who will determine the future of others. What risks should be addressed, how will they be assessed, who will decide, and according to what agenda for the future?

Of course, the social context of risk is changing as well: this dynamism takes the form of changing orientations to economic growth, technological innovation, and the imposition of risk. At the close of the twentieth century, a single economic system – capitalism – had (temporarily, at least) embarked on a new period of global spread. Capitalism is enjoying a new heyday while risk aversion is growing – a situation sure to engender conflict. Furthermore, a world democratic revolution is fuelling increased expectations for openness and transparency, consultation, and involvement in risk decisions. In addition, as Inglehart, Basáñez, and Moreno (1998) have argued, with the trend to greater affluence a shift to post-materialist values emerges – with the propensity toward consumptive lifestyle and growing risk aversion. These trends suggest heightened risk debate and greater centrality of risk issues in major society decisions in the early decades of the coming century.

The three chapters in this section dissect this changing reality from several perspectives. Timmerman and O'Riordan in chapter 14 redefine vulnerability in terms of the future: vulnerability is the loss of effective power in the creation of one's future. So imagining alternative futures is an empowering venture, and certainly part of a more democratic risk analysis. These authors see two particular assets to processes aimed at deliberating alternative futures – such activity can significantly reduce social distrust and it can also reduce uncertainties about managing future global environmental risks. With interesting examples, they demonstrate that such imagining can occur in many forms beside the written text. They also provide further substance to the meaning of emancipatory risk assessment – in expression, in roles and power, and in participation. But whatever the mode of imagining and debate, the core issue surrounding risk adjudication and allocation in their view is that of who will have first and best access to the future, and who will control the shaping of the future.

Robinson, in chapter 15, takes up many of these same issues but reports particularly on an imaginative exercise for debating what a more sustainable environmental future for Canada might look like. The Sustainable Society Project departed from the procedures typically used in forecasting and futures analysis in its emphasis on "backcasting" that ensures that the values implicit in a desired future are debated and that plausible pathways or courses to this future are constructed. Thus, unlike most risk analyses or integrated assessments, the probing of alternative futures is explicitly normative. An interesting outcome of this project is the list of principles of sustainability, provided in the chapter, treating ecology, ethical imperatives, and sociopolitical processes.

Finally, Benjamin and colleagues in chapter 16 consider past approaches to the future to include the need to create diverse new visions of the future. Reviewing the first generation of global models, they argue that most global assessments have had a paucity of creative thinking about the social futures we desire. These models have almost exclusively centred on projections of economic, technological, and environmental variables, whereas the social goals – which, presumably, should be the purpose of economic and environmental change – have somehow been submerged or totally lost. Accordingly, the authors see wide-ranging democratic discussion of diverse social visions of the future as a pressing need for guiding risk deliberations and decisions. With Timmerman and O'Riordan, they call for a broadening of the avenues for getting there, as by use of stories and science fiction as well as more conventional routes for assessment. Assessment, to be successful, will need to break out of the strictures of the IPAT (impact, population, affluence, technology) mode of thinking reviewed in chapter 8 and to embark on a new, enlarged, and explicit consideration of alternative futures. The process for the creation of these richer futures is emblematic of the larger need for a more democratic risk praxis.

14

Risk and imagining alternative futures

Timothy O'Riordan and Peter Timmerman

In considering innovative future research with regard to environmental risk, the authors of this chapter have been drawn, in the first instance, towards presenting examples of research which – as our title indicates – are "imagined," "alternative," and also "future"-oriented in a special way. As will become apparent, these examples have certain elements in common; however, what they share above all is a turning away from models of incremental future risk assessment towards more broadly based evaluations of alternative futures. The differences between these methods are stark, and depend upon a recognition that "risk" and "risk assessment" as concepts and processes already contain (or reflect) implicit assumptions or expectations about what the future is, and what will be necessary to manage it. Reconsidering these, as the introduction to this volume argues, raises basic questions about trust, equity, and the working out of what has been called "our common future" (Nagpal and Foltz 1995; WCED 1987).

The roots of risk

How do the implicit assumptions of risk assessment – which is an artefact of modern technological civilization – affect our consideration of the future, a quasi-artefact constantly in the making?

The continuing escalation of concern about risk, including the polarization of debates about risk and the paralysis of many decision-making bodies when faced with special contentiousness, has had at least one good effect: it has forced a serious reconsideration of the concept of "risk," which has been taken more or less for granted as a technical tool, and which has optimistically been expected to serve as a neutral and objective resolver of conflict. This reconsideration has taken many forms and has drawn upon many disciplines, including philosophy, history, anthropology, and psychology. It has opened up an array of complex issues, many of which are inseparable from fundamental questions involving the nature of modern society and the interpretation of human experience (for a review, see Kasperson et al. 1990).

In particular, it has become clear that "risk" as a concept is bound up with the special needs and obsessions of our own technological culture (Kotze 1999). It is now possible to trace the rise of a "risk culture" from the mercantilism of the Renaissance, through the invention of probability theory in the seventeenth century, the rise of the insurance industry in the eighteenth and nineteenth centuries, all the way to the creation of technical risk analysis towards the middle of the twentieth century. This risk culture has woven together over time some of the following forces:

- the replacement of a theocentric world order in which success was, in principle, based on moral and spiritual considerations, with an anthropocentric world order in which success is based on individual initiative;
- the simultaneous reinterpretation of the relationship of the individual to temporal uncertainty, especially highlighting the roles of fortune, luck, and chance in human life;
- the movement into a capitalist culture, including the rise of longer-term investment in capital stocks and the removal of previous restrictions on charging interest;
- a growing need for long-range administrative and bureaucratic planning in an increasingly complex industrial culture;
- the need to interpret the increasing power of scientific and technical innovation to decision makers;
- the rapid enlargement in the capacities of science and technology to create and detect environmental effects beyond the capacities of unassisted human senses; and
- a growing separation between expert or élite understanding and that of the ordinary citizen in modern society.

These forces (and others) have helped to create the modern risk culture, which can be described roughly as painting a world of probabilistic and atomistic events that surround, bombard, or filter insidiously into the unwary, semi-helpless citizen. The morning coffee, the value of the money in one's wallet, and the most private acts (like the begetting of children) are all laden with risks about which, one suspects, one knows less than one is being told, if anything.

One social critic, Susan Sontag, refers to this alienated paranoia as the expression of an "apocalyptic" culture – a culture in which we fear that certain catastrophes have already occurred, but the news has not yet reached us, and in which contemporary life is tinged by the dark shadow cast back upon us from fearfully projected futures. She notes, in *AIDS and Its Metaphors:*

Future-mindedness is as much the distinctive mental habit, and intellectual corruption, of this century as the history-mindedness that, as Nietzsche pointed out, transformed thinking in the nineteenth century. Being able to estimate how matters will evolve into the future is an inevitable byproduct of a more sophisticated (quantifiable, testable) understanding of process, social as well as scientific. The ability to project events with some accuracy into the future enlarged what power consisted of, because it was a vast new source of instructions about how to deal with the present. (Sontag 1989: 177)

The alternating fear that our grip on the future is faltering or that it is tightening (what Steve Rayner refers to as the question of whether we can "manage the planet" or just "manage to get by" [personal communication 1990]) is part of a continuing debate in modern societies.

The thread of life

Marshall McLuhan once made the famous remark (adapting one of Walter Benjamin's theses on modernity) that we live by driving down the road looking into the rear-view mirror. It is an inescapable fact of life that the future is uncertain, and that our only guide is our past experience. This "existential" fact stretches from our personal lives to the overall trajectory of history on the grand scale, and also encompasses the more narrowly based technique known as "risk assessment." Except for those who have a religious faith, or a faith in the rules of history (such as Marxism), which amounts to a religion, modern peoples rely on a mixed range of predictive techniques or interpretative schemata – some existential, some from the contemporary stock of media-generated myth, some formally institutionalized – to make sense of a life lived forward, but looking backward.

Among the ancient Greeks it was said "Count no man happy until he is dead," partly out of respect for the fickle nature of the gods, but partly out of a sense that there was a shape to a man's life which conformed to (or fell away from) a pattern of the "good life" (MacIntyre 1981). Accordingly, one tried to live a life according to "virtue" or a "good Christian life," and the criteria for that life were openly known and widely shared. Modern societies, on the other hand, are characterized by competing or even non-existent models of what constitutes the appropriately shaped life. By and large, today we are expected to construct meaningful lives of our own. Yet some need to integrate our lives – to find ways of interpreting our lives as if they made overall sense – still remains. On the one hand, this places an intense burden on the few ordering rituals that remain in our culture. On the other, the construction of our life stories by ourselves constantly confronts us with the famous existential burden of choice in a continuing process, which ends only with death.

In the course of a discussion of "activities of risk" (including sport and gambling), Scheibe (1986) notes:

Human identities are considered to be evolving constructions; they emerge out of continual social interactions in the course of life. Self narratives are developed stories that must be told in specific historical terms, using a particular language, reference to a particular stock of working historical conventions and a particular pattern of dominant beliefs and values. (Scheibe 1986: 131)

These self-narratives are inherently fragile in a society in which the stock of conventions and values are themselves changing.

The continuing personal interpretative, integrating process with regard to uncertainty has, as already indicated, institutionalized formal and more tightly focused counterparts, one of which we label conventionally as "risk assessment." Conflicts over the range of applicability of risk assessment and the transfer of its techniques to other areas of life are partly rooted in the still vaguely understood relationship between the broader and narrower process, although the 1996 report *Understanding Risk* by the US National Research Council points to important directions for joining the two (Stern and Fineberg 1996). These conflicts can be set in their larger sociopolitical or sociocultural context through the application of recent work in political theory.

Risk, uncertainty, and politics

With the publication of *The Imperative of Responsibility* (Jonas 1984), the question of how to relate ethics and politics to future global uncertainty received its first serious treatment. The book – which argued for a recasting of our standard ethical principles from a short-term to a long-term focus, and from a categorical to an ontological imperative grounded in prudential behaviour – is part of the burgeoning literature on the nexus between future uncertainty and the creation of political "trust" and "prudence" (e.g. Dunn 1990; Earle and Cvetkovich 1995; Luhmann 1979, 1989).

Most political theory and practice in Anglo-Saxon countries (and increasingly elsewhere) is underpinned by the liberalism of John Locke, whose late seventeenth century writings were designed to cope with the emergence of a "society of strangers," whose relations to each other were characterized by constant uncertainty. In previous societies, the status of people and their roles could be rapidly identified and assessed; in emergent modernism, the creation and sustaining of trust (and predictability) in the roles of others and in political entities generally becomes a central problem (already foreshadowed in Machiavelli's cunning early modernist prince). In Locke, power is entrusted to governments that act as trustees; these governments are temporarily "negatively trusted" – they are open to public scrutiny, subject to internal checks and balances, and their mandate is revocable at some point by the people.

At the root of this entrusting of power in Locke is the need for governments to act beyond their strictly given mandates when faced with uncertainty. A complex modern government cannot function if it needs to return to the citizenry every few hours for new instructions; therefore, the citizenry are compelled to entrust themselves for a period of time to a government, on the understanding that the government's practice will be reviewed retrospectively at the next election.

The Lockean synthesis is part of what is called the "social contract" tradition, which has recently undergone substantial scrutiny and refurbishment, dating perhaps from the publication in 1971 of John Rawls's *A Theory of Justice*. In an attempt to reassess the ethical and philosophical underpinnings of that tradition, Rawls invents a thought experiment, referred to as "The Original Position," which involves choosing a random set of people and allowing them to negotiate the ground rules for future society. The catch is that, by a mysterious process, the negotiators are operating behind a veil of ignorance whereby they have been stripped of all knowledge of their own status in the society, whether they themselves or the others around the table are White, Black, male, female, rich, or poor. Rawls argues that, under such a veil, the negotiators will derive a set of just and equitable principles, among other reasons because they do not know where they will be when the veil of ignorance is lifted, and they will have to live with the consequences of their negotiated social contract. Rawls further argues that the outcome of the negotiation will be a "maximin" principle that optimizes the position of the least-advantaged person in the new society.

Much of the discussion of Rawls's immensely influential book has centred around the participants' attitudes towards risk, and, by extension, their attitudes towards the future generations for whom they are laying down societal foundations. Rawls argues that, although the social founders are under the burden of having to make choices for future generations, the rational choice (of maximin) does not ultimately depend upon attitudes towards risk (whether people are risk prone or risk averse); indeed, he puts these attitudes behind the veil of ignorance as well. Critics have counter-argued that Rawls' argument belongs to that part of liberal tradition that covertly assumes that we are "all alike under the skin", and that in such a tradition the stripping away of the diversities and actual living experiences that make us who we are can be a plausible foundation for a political theory. This liberal rationality is seen as, in some sense, separable from life experience. A rational political system can thereby be devised that is rooted in an Enlightenment or Kantian ideal of the quintessentially autonomous human, self-sufficient and abstracted from lived space and time, but who can be persuaded to sign on to a social system under construction. Under such a system, risk can be subject to neutral, rational calculation, in part because the future within which risk is to operate is itself being constructed by rational projection.

Many of Rawls's critics (the best-known advocate being Alaisdair MacIntyre [1981]) belong to a resurgent political tradition known as the "virtuous" or Aristotelian tradition, which deliberately invokes the premodern (or non-modern) interpretative schema of life described above. This tradition is based on a deliberate rooting of social and political theory in the practices and narratives of ordinary social and political life: it depends not upon an idealized social construct but upon learned, vernacular, and historical knowledge of actual societies in operation. It is noteworthy that much of this tradition focuses on human vulnerability to processes and events beyond one's personal control.

According to variants of this tradition, human beings have long-recognized models or shapes of lives lived happily or virtuously which they can follow; nevertheless, however virtuous they may be, or however closely they follow the patterns of virtue, they are subject to the vagaries of fate and the possibility of tragedy. Unlike the more optimistic liberal model, this tradition – often religious – starts usually from the assumption that our lives are intrinsically constrained (by what is sometimes referred to as "natural law," "the Dao," or "the Dharma"), or shrouded in impenetrable mysteries (like undeserved evil), and that prudential behaviour is therefore our best guide. In *The Practice of the Wild*, Gary Snyder (Snyder 1990: 92) notes: "Preliterate hunting and gathering cultures were highly trained and lived well by virtue of keen observation and good manners ... greed exposes the foolish person or the foolish chicken alike to the ever-watchful hawk of the food web and to early impermanence."

We are also seen to be more fragile, much more preformed, prejudiced, and very much less rational (even in principle) and in control of everything than we have recently been led to believe. Moreover, rather than setting out to construct our future lives, there is often a "best" existing narrative to which we ought to strive to conform, it being the only way in which we can hope to have even limited influence over the future.

By contrast, although Rawls's theory is only one of a number of contemporary approaches to political theory, it does capture very clearly and sympathetically certain general characteristics of modernity. These elements include the powerful idea of beginning society from scratch; an underlying norm of instrumental rationality; and a process of projection into the future that is curiously neutered politically and socially, except in its unspoken devotion to progress. Against this, as just indicated, is now ranged an anti-modernist (or post-modernist) political theory, arguing that the essentially liberal underpinnings of modernism are deracinated (or are based on an implausible model of the self-sufficient individual). Through the attack on Rawls, of course, this argument is directed against the Lockean synthesis that (as has been suggested) underpins at least the Anglo-American version of representative democracy. Clearly, part of the debate now pivots on questions of risk and prudence.

To restate an earlier point, part of the legitimacy of representative democracy is that governments are entrusted with the management of future uncertainty. Niklas Luhmann (1979, 1989) and Earle and Cvetkovich (1995) suggest that trust is one way in which we try to cope with uncertainty over time. This is obviously true at the personal level, where by vows, promises, and manners (the ethic of civility) we try to secure each other's behaviour, but we can also see that our social and political systems operate very largely on the basis of trust (and its shadow, suspicion).

Trust and equity

The issue of trust is now central to many governmental problems revolving about "consent," particularly in the fields of technology assessment and environmental management (Earle and Cvetkovich 1995; Kasperson, Golding, and Tuler 1992). Because science and technology are so complex, citizens are dependent (fearfully) on expertise and promises of happy outcomes. Global environmental issues are, as Funtowicz and Ravetz make clear in chapter 4, very uncertain in their implications, partly because of the long lag times between action and result. Given recent past history, our lack of trust in technology, management, and government may be a reaction against our previous overconfidence. But this legitimation crisis does raise the issue of how to muster a legitimate response.

Whom are we to trust about the warnings and potential responses to global change? Can we trust anyone already part of the status quo to "manage the planet," given their track record over the long term?

We now often find that the citizenry, environmentalists, and the variously marginalized are arrayed on the side of "prudential" behaviour with regard to proposed future risks, whereas governments, administrative experts, and the corporate élite are essentially "risk prone." The trust of experts in their own capacities is undisturbed. It is everyone else who finds themself forced to inject pessimism and uncertainty back into the process, to mandate the consideration of worst cases, and to question whether the system itself may be the largest generator of risk. This can lead us to question, by way of questioning the assurance of the current trajectory of modernism, the theoretical underpinnings of our political system in a new way. Part of the questioning concerns which narrative about the future we are to believe, if any – expertly constructed narratives of "business as usual," or older experience-based "life narratives" generating intuitive senses that something is amiss?

This all feeds into long-standing debates in the risk community. How does technical risk analysis relate to people's experience of risk? Can we delegate our lived understanding of risk to experts whose technical analyses we are forced to trust? What are the social and political implications of attempting to pursue homogeneity in comparing different types of risk?

A more recent and related question, taken up earlier in the Introduction (chapter 1) and elaborated in chapter 7, is that of equity. Equity has three meanings for the purposes of this analysis. First, there is equity as essential justice – the right to participate in and to be respected in the conduct of social affairs. Second, there is equity that takes proper account of the interests of those who are our equals in the order of things – what some would call "equal consideration, if not equal treatment." Finally, there is equity in the sense of giving adequate weighting to the interests of other generations, to those whose interests are, in effect, an extension of our own, and for whom we are custodians and stewards. The phrase "other generations" (rather than future generations) reminds us that we are also the current custodians of the dreams of the past for a better world.

Each one of these interpretations relates inevitably to power, the ability to exert control by right or by threat. Vulnerability can be interpreted as loss of effective power in the creation of one's own future: one is effectively harnessed to be a part of someone else's proposed future. Such a proposal is, at best, usually proferred as the "agenda" from and to which one is supposed to react. At worst, the proposed future is proposed by one voice only, to which all the rest must listen.

If we see social science research on global environmental change as part of an emerging global decision-making mechanism or management structure (in however inchoate a form), then a strategic question is the relationship between expert science and the citizen (as Funtowicz and Ravetz argue in chapter 4). Most of the research to date has been what could be called "monologic" – usually either pure research that interests a particular scholar, or directed research on behalf of a government or industry that has a particular problem or set of problems that it assumes need to be answered (see Robinson's discussion of "mandated" science to follow in chapter 15). These have, by and large, been seen in the past as contributions to a generally paternalistic decision-making process and, in that sense, as contributions to public discussion. But one would be hard-pressed to say that much of this research has been in the business of empowering the broader citizenry to be part of that process. A crucial management issue in global environmental change is how to move from a monologic expert-driven process to one that is dialogic and even multilogic.

It is obvious in many areas of environmental management that the creation of a dialogic process among various stakeholders is absolutely critical. The environmental movement or the environmental situation, and larger cultural shifts, have made necessary the involvement of larger numbers of citizens in the evaluation and decision-making. But learning how to do this is intensely painful and time-consuming. How are we to cope with an influx of new voices, sceptical of science, and by no means expert in areas in which citizens now feel called upon to exercise their franchise? How can we find new ways to make the best use of both experts and the concerned citizenry so that they can mutually enhance each other's contributions?

Here we can borrow a notion from the German philosopher Jurgen Habermas (1970, 1971), who speaks about the current erosion of the public sphere – the public sphere being the space in which discourse about the relations between individuals and communities can be conducted. He points to a number of forces that are carrying out that erosion, including the extension into that space of bureaucracies and scientific experts, followed by (or accompanied by) the abandonment of that space by the citizen who is simultaneously being given models of personal well-being that turn their back on public commitment. That is, in order to protect themselves from being instrumentalized and rationalized, people are retreating into the "last bastions" of personal meaning – the region, the family, and the individual self.

To re-establish the public sphere is essentially to re-establish some basic trust in the process (Kasperson, Golding, and Kasperson 1999), and to re-establish that trust at least two different tasks need to be undertaken. First, questions of equity need to be made central to the debate; this means that, as noted in chapters 1 and 7, if there is to be an equitable outcome to a process, the process itself must be seen to be equitable. Second, we need to re-establish some rules of discourse that encourage participation and engender trust among participants (Renn, Webler, and Wiedemann 1995a, b).

Currently, our models of what constitute equitable process are legalistic in form or content (e.g. conflictual) or have developed haphazardly out of various pieces of legislation. But what is coming to be seen as basic to an equitable process depends upon a much broader understanding of what is entailed in empowerment towards equity. This is true not just for the bureaucrat and the expert but also for the citizen who has lost touch with his or her own participatory function in a democracy.

To touch briefly on one ethical model underpinning this kind of understanding, we can invoke another of Habermas's ideas concerned with the refurbishing of the public sphere, that of the ideal speech situation (Habermas 1990: 43–115). Roughly speaking, his ideal speech situation is an ideal construct from which one can identify those aspects of the rest of our speech situations that fall below that ideal (Webler 1995). The ideal speech situation is one in which all the parties to the discussion are fully present, fully capable of contributing, and are able to make their contributions without obstacle or coercion. The attractive part of the concept is that it derives from an analysis of common speech situations, in which our conversations (when they are being engaged in by equals) have certain characteristics: these include (1) communicability or comprehensibility, (2) sincerity, (3) appropriateness or legitimate representativeness, and (4) truthfulness (adapted from Forester 1987: 213).

These pragmatic considerations still derive from a fundamentally Kantian perspective that all persons are to be presumed worthy of humane treatment, understanding, and respect as individuals. But in addition to this, we are also faced with what Habermas calls "discourse ethics" (Habermas 1990: 66) – that is, we are engaged in discourse with one another. Our ethics are no longer purely individual, they are interpersonal: we are attempting to create a situation in which we can speak as members of a community grounded in mutual respect and trust.

In other words, from an ethical point of view, we begin from a position based on our intuitive and learned understandings of how people are treated humanely by each other in situations in which we are engaged in dialogue, and then gradually work our way out (by reference perhaps to more developed ethical, philosophical, and social arguments) towards the rehabilitation of the public sphere. This means that we begin from already existing norms of human interaction, rather than from formal norms of rationality (although they, too, have a function to play).

One of the interesting challenges to the rebuilding of the public sphere is to apply criteria derived from these characteristics to the evaluation of decision-making processes. On the one hand, we are likely to discover that some of the reasons why we are unable to approach the conditions of an "ideal" or even "reasonable" speech situation is that powers are unevenly distributed among the participants. Here we come up against the expert/lay public dilemma. On the other hand – and requiring, perhaps, less-radical action – we would be likely to discover that many of our assessment processes already embody some of the attributes of "discourse ethics," but that they have been distorted and obscured over time because of the misapplication of criteria derived from essentially bureaucratic rationalities.

One of the unexplored aspects of discourse ethics is the re-establishment of trust in the public sphere through mutual discourse about the future (although see Earle and Cvetkovich 1995). What makes risk assessment (that is, in its institutionalized form) such a flashpoint is that many risk debates are disguised assertions and fears about the future trajectory of the society, as implied by some proposed project and by the nature of the current decision-making process. More broadly, because of the powerful demand in our society for short-term future projections by experts, citizens are at a disadvantage compared with experts, most of whose training is devoted to this task and who fit quite comfortably into adversarial proceedings.

Imagining alternative futures

It is at this point that the idea of "imagining alternative futures" begins to come into its own, as a method of reducing mistrust and uncertainty about the management of future issues, through greater involvement of those potentially affected in the process. The most familiar formal method of alternative imagining is called "backcasting" – a methodology developed by energy analysts in order to overcome the inherent biases of conventional forecasting techniques (see chapter 15). As the term suggests, the technique involves leaping forward some distance into the future and then backcasting to the present. The essential idea is to work backwards from some creatively imagined future (e.g. in 2050) through the range of options and decisions that must be made in order to achieve that future; in this way, as one approaches the present, certain timetables, policies, and decisions become foreclosed or mandated. Among the appeals of this technique is that the built-in assumptions of expert forecasters can be circumvented (or at least minimized) in favour of a wider range of creative alternatives, derived in part from the creativity of non-expert participants. In other words, by going far enough into the future, one can restore the balance of power away from the expert and back towards the citizen. The expert then becomes a tool in the hands of the citizen, and not vice versa. Whereas the expert can make his or her future more detailed, there is nothing intrinsically more meritorious about that future than that provided by an eight-year-old.

Imagining futures is not merely a matter of wishful thinking. It is not a sideshow, for private entertainment, while the crisis of life has to be lived here and now. Seriously imagining futures requires a capacity to have effective control over the means of getting these futures to work in our collective favour. It is a process that relates to power; to the scope for self-reliance; and to a trust that political institutions will provide the mechanisms, respond to the results, and deliver effective outcomes. Imagining futures, as Benjamin and colleagues make clear in chapter 16, cannot be separated from the fundamentals of social justice, civil rights, effective power, and collective self-reliance; it is bound up with the very essence of what will be necessary to make sustainable development succeed.

Imaging

As noted, "backcasting" is a formal approach to imagining alternative futures. To make this imagining even more open, creative, and democratizing, researchers are now turning to board games and other forms of "imaging" – that is, non-written forms of considering alternative futures.

To date, few studies have linked the process of creatively imaging futures to the means of achieving such futures. Early anthropological work was based on storytelling and other oral forms of narrative that allow societies to cope with change and retain social cohesion during times of conflict or stress. It should also be recalled that, although many anthropologists have been able to show profound transformations in narratives due to cultural stress, the general rule in traditional cultures is that these new narrative elements are not accepted as completely new data; rather, they are hidden or put forward as something that just was not said before. In his study of the Dunne-za (the Beaver people of British Columbia), Robin Ridington notes:

The entire fund of Dunne-za adaptive information is reflected and transmitted in their oral tradition ... in traditional Dunne-za culture, every possible human experience or technique is precisely indexed in oral tradition and made accessible through dream and vision; in modern historical culture it is assumed that in the course of a person's life he or she will experience dramatic transformations of past cultural experience. (Ridington 1990: 138–139)

Empowerment in older oral cultures is based on learning the old patterns as the basis for evaluating what is to be expected: that is, how does the future fit into the past? In developed countries, the empowering issue is how the past will change into the future, and who will decide. The research response has been educational (how the past will change) rather than political (who will decide).

Nevertheless, the educational role of imaging futures can be profound. In this narrower frame of reference, it is still worthy of careful scrutiny. Its advantages include the following:

- It compresses enormous amounts of information into visual images that have impact by virtue of providing coded signals that are immediately translatable into a wish to participate;
- It is novel in the sense that the written word is still the dominant formal mode of communication. The use of the creative arts – paintings, narrative, poetry, dance, drama – all have profound means of eliciting deeper understanding of emotional and ethical content that can be cumbersome if confined to the written word;
- It can (and should) be fun, entertaining, memorable, and long-lasting.

One should never underestimate the power of learning through enjoyable engagement, especially in the presence of friends and relatives.

Imaging futures is not, of course, a solution in itself. But it is a device that taps a latent feature of our citizens – namely, a wish to see around the next bend, and to recognize more safety than danger in the prospect. It is also a technique that is both educational and political. In the right hands, it can nurture a sense of security resting on a resilient ability to cope, rather than on a rigid protectionism. In all the high-risk regions examined in this volume (chapters 8–13), this is linked to equity or distributional issues, in part through considerations of vulnerabilities in space, time, and culture (and thereby coincides with the themes of chapter 1 and the elaborations in chapters 5–7). Vulnerability may arise from past or present conditions, but it is always expectantly oriented towards the future.

Imagining alternative futures: Three cases

Two studies that have already sought to make use of such techniques are the energy futures study undertaken in Sweden and the landscapes futures study recently completed in northern England. A third study – on the future of the Great Lakes Basin – proposed in the early 1990s similarly addresses how to make the best use of expert and citizen by combining both a backcasting approach and some elementary interactive television.

Swedish energy scenarios study

We begin with the Swedish and English cases. Both studies relied on paintings of landscapes to portray how they might look on the basis of foreseeable policy and land-use decisions that could be executed within a generation. The point was to create in the respondent's mind an image of change that would apply to familiar landscapes in their own lifetime. The more that the landscape depicted is based on a familiar scene in order for the change to be identified in shocking contrast, the more effective the technique.

The Swedish study examined three energy scenarios – one all nuclear, one related to biomass production, and one all renewables (including biomass conversion) – set in the landscapes of three distinctive parts of Sweden. The painter looked at how these landscapes had changed from 1780 to the present, and how they might look from the present to 2000. There was no continuity in the choice of imaging futures, nor were all of the different energy futures depicted on one landscape. The exercise was more a matter of stimulating the imagination than a formalizing technique.

Nevertheless, the results were impressive: all the main interests in the Swedish energy debate were activated; the schools and the media took the concept to their hearts; the book sold 10,000 copies and helped to reinforce the demand for a non-nuclear future with a high reliance on imported fossil fuels and renewables. Above all, the exercise showed how it can be possible to use pictorial images (no matter how artistic), as opposed to rigorously scientific results, to stimulate the imagination and to create understanding in a population that would not have responded as effectively to the written word.

Yorkshire Dales study

The Yorkshire Dales study took this one stage further. The Yorkshire Dales is an area of the northern Pennines lying some 100 kilometres north of Leeds in England. The dales are farmed for sheep and cattle, the fields being enclosed by stone walls. In the past, each field contained its stone-built barn for housing and feeding livestock and for milking the cattle. The traditional landscape is that of the post-enclosure period, around 1850, of small fields, stone walls and barns, broad-leaved copses (small woods), heather-covered fells, and meadows full of flowering plants that provide the winter feed for the livestock.

All this is changing. Landscapes have never been static, but it is a peculiarity of the human mind to imagine that they are. Landscapes are a product of topography, economy, history, and culture. They are living entities that carry their own social meaning. The aim of the Yorkshire Dales study was to paint seven possible future dales based on realistic changes to the farming economies of northern Britain and the changing pattern of the domestic economy and lifestyles. The landscapes portray neglect, intensification, conscious planning, a leisure society with the capability of living far from centres of work, and deliberately abandoned landscape (reminiscent of the American Adirondacks) where the farmsteads of today become the undergrowth of tomorrow's second-generation forest.

The point here is not to dwell on the landscapes depicted but to emphasize the importance of the method in helping people to imagine futures. The exhibit of seven paintings (plus one of an existing dale) was coupled to a poster display, a view especially made for the occasion, and a board game played on the floor, which allowed people to take an imaginary journey during which they were stopped and asked, in a game, to choose the elements of the landscape they preferred.

A subsequent valuation study undertaken of almost 400 of the 5000 visitors revealed four powerful observations about an imaging-futures experience (Council for the Protection of Rural England 1991). First, the visitors absorbed an enormous amount of complicated detail, even though most of them had relatively little knowledge of landscape change, let alone that in a national park that is supposed to protect stability. Second, those who were already familiar with the changes taking place in the Dales, notably local residents, were still noticeably moved by the visual images to the point of generating fresh concern. Third, those who were relatively unfamiliar with the Dales issues and who came mostly to enjoy a holiday, were even more disturbed by the images created. In that sense, a kind of convergence of emotion and reaction occurred between those who were already activated and those not so initially involved. Finally, this concern, based on a better understanding of the forces that were shaping change, was of sufficient intensity to cause respondents to want to become more involved in the shaping of future national-park policies and land-use change by active involvement in the park-planning process.

Naturally, there is a decay effect. It is unlikely that the powerful images that generated such concern in the aftermath of the visit would remain today. Nevertheless, the exhibit, which was originally experimental, is now to be made part of the permanent display in the park, and the views of participants will be fed into the process of creating the next five-year plan for the park. In addition, the technique of landscape imaging will be converted to a teaching park for schools, polytechnics, and local authorities to use for other future land-use topics, such as the planting and felling of trees, the location of recreating facilities in environmentally sensitive areas, and the choice of traffic options in a smaller town.

Several points must be stressed about these studies: they were designed to be primarily educational, not political; there was no provision for any meaningful translation of views into a process of land-use policy change; they were both experimental, although only the Yorkshire Dales study contained any formal evaluation of its effectiveness. Nevertheless, they showed that the relative formalization of futures images could be translated into a powerful experience.

Where would one go from here? One possibility is to look at ways in which the computer and science communities can be connected to create more meaningful images of future states on a responsive basis. It is possible to produce an image on a computer screen of, say, a coastline subject to sea-level rise. It should then be possible to show how that coastline will change according to various assumptions of global warming and sea-level rise. This could be programmed on software already available. Then it would be possible for participants to assume various policy choices – for example, a reduction in atmosphere-warming gases by such means as pricing, command-and-control, or international protocol – all of which could be "costed." This would then be identified in terms of differences in the configuration of the coast, with images of property saved, new nature reserves created, and the investments necessary for coastal defence. A "game" could be created, of playing futures and identifying how various "costs" in term of possible courses of action could be equated against the "benefits" of more safeguarded habitats and coastlines. Such a method should be tried on the basis of experiment, with careful monitoring of the educational and awareness-raising potential of the techniques. In any case, scientific uncertainty still exists about the disputed outcomes, so it is vital that the early trials be seen as genuinely experimental so that the variety of imponderables can be depicted with sensitivity.

In an age in which the computer game and computer-based instruction are so much part of a contemporary education, there is no intrinsic reason why such an approach should not be welcomed and pursued. However, it is again vital to stress the educational aspects of this approach. It would be wrong, at the outset at least, to try to convert this process into anything more political. Nevertheless, the inevitability of some element of politicization of imaging futures cannot be ignored. It is a powerful technique that carries important messages about how people can relate to the world in which they live.

Passive and active interpretation

These studies point to a changing role for interpretation. Interpretation consists of three phases – informing, explaining, and activating. Each should relate to the other. Information on its own carries no meaning or context, so it cannot excite or stimulate. Explanation is without value if it lacks a fundamental information base. Activating should create informed concern and a sense of solidarity with the landscape being depicted, the history being re-created or re-enacted, or the future circumstances being envisaged.

In its early days, interpretation was essentially a form of guidance, a way of having experts convey history or nature conservation to a passive public. The purpose was largely to create a sense of appreciation as to why a landscape evolved, how it worked as an ecological entity, and hence why it should be preserved. The classical manner of presentation was the illustrated text, long on illustration and short on words. It clearly emphasized the first two phases of interpretation.

This view was epitomized by the American naturalist and planner Freedman Tilden in his widely cited text *Interpreting Our Heritage* (Tilden 1957). Tilden made no bones: conservation through publicly supported preservation was the explicit goal of interpretation. The aim in the Tilden tradition was to retain the past, to bring it to life in some way or other, and to establish a feeling of identification with bygone days or ecosystems that functioned with, or without, the hand of humans. Rarely was there interest in landscapes, buildings, or environmental circumstances as they might evolve. The interpretative act applied to the passive or absorptive recipient, rarely to an engaged and politically activated respondent.

This distinction is important. In the traditional mode, interpretation is facilitative, explanatory, and advisory. The original Tilden approach was to create understanding through information, to generate awareness through explanation, and to generate concern for protection through activation of a sense of identity with the objects under consideration. Actual management of the features concerned took place quite independently, whether involving policy decisions (what to preserve, re-establish, reconstruct), planning (the design, shape and "feel" of the features), or investment (how much to spend, when, and in what sequence).

A participatory, active interpretative strategy operates quite differently. Interpretation is a companion of both policy and management. This is apparent in the application of Tilden's model (fig. 14.1). An extended approach to interpretation can be achieved using the model (fig. 14.2). To begin, one must explain more fully the factors at work that lead to threat, to change, or to possible alternative future outcomes. These have to be set in the context of institutions, scientific prognoses, economic valuation of outcomes, and uncertainties. A need exists to generate images of possible outcomes based on various means of applying scientific prediction and by examining how economic valuation can help to identify the advantages and disadvantages of various courses of action and outcomes.

INTERPRETATION

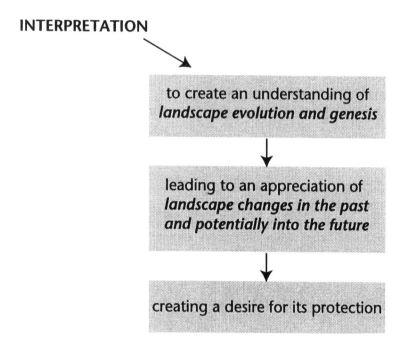

Fig. 14.1 **Tilden's model applied to a research project**

The third component of this "new look" interpretation is essentially to find means of eliciting active responses from participants. These responses might be in the form of *explicit choices* of possible imagined futures. It might be in the form of *negotiated consent* of the most probable outcomes, given intelligent guesses at the costs and benefits of taking action. Or it might come through *group debate* through which villages or wider communities identify their interests in possible futures and work to make it happen.

The Great Lakes Basin study

The third variation seeks to broaden imaging and interactive interpretation through the mixing of expert and citizen in workshops and televised "town meetings" (Robinson 1990a; Robinson and Timmerman 1990). The Great Lakes Basin of North America was the focus of a proposal in the early 1990s that built upon the tradition of work associated with adaptive environmental management (Canadian Climate Programme 1990; Lee 1993; Smit 1993). Specifically, it drew upon the idea of "policy exercises," developed over the years by the International Institute for

INTERPRETATION

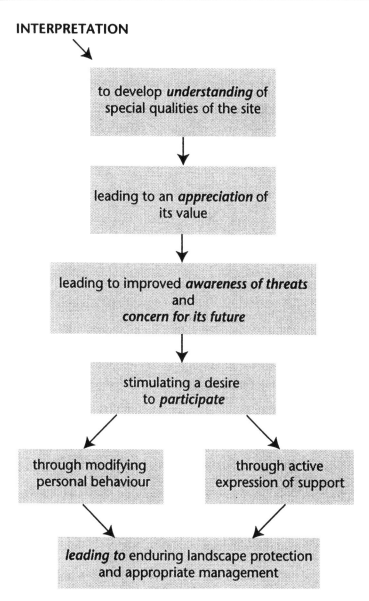

Fig. 14.2 **Tilden's model extended as the basis of the current interpretive experiment**

Applied Systems Analysis (IIASA) in Austria, that enables senior policy makers in certain sectors to explore the future consequences of their decisions through the use of computer-modelling techniques (Toth 1988).

In this proposal, policy exercises were combined with "backcasting" (Mulder and Biesiot 1998), to open up the process to very broad public participation in considering alternative risks and possibilities for the Great Lakes. The specific subject area of study was alternative futures in the Great Lakes Basin, given the possibility of global climate change. The proposal envisioned a project that would operate on two converging tracks: the first would involve the general public and the second would undertake a policy "gaming" exercise with a smaller subset of expert and public participants. Eventually, a detailed process of public participation and citizen advisory committees was mounted, although differing in important respects from the proposed project (Landre and Knuth 1993; Mortsch and Quinn 1996). Here the interest is primarily the proposed study design and the strategy that it set forth.

The initial public-involvement phase sought to include public-information programmes on television, through a sequence of four programmes – general background to the problem, alternative futures imaged through scenarios, the details of global climate change, and possible policy responses. These would be accompanied by publicly distributed questionnaires (including detailed options for preferred alternatives and appropriate policy responses). These would be fed into the policy "gaming" process, which could itself be televised.

The gaming process, which would next begin, would have five formal steps, described below.

Step one: Enlist stakeholder groups

The goal here was to identify and to enlist the involvement of various stakeholder groups with a strong interest in the Great Lakes and in the issue of climate change, including those who would have been identified through the television programming. Such groups would represent the major interests involved in discussions or debates about the future of that sector in the Great Lakes Basin – including, for example, environmental groups, other non-governmental organizations (NGOs), private-sector interests, unions, local or regional politicians, and senior government representatives. The identified groups and individuals would be asked to participate in a series of exchanges and meetings for the purpose of developing what they would consider to be their preferred future scenarios, but participants would not be required to reach a consensus on appropriate future policies on reaching the targets of these scenarios.

Step two: Backcast different preferred futures

In a workshop, these same participants would be asked to backcast a range of different desired alternative futures for the Great Lakes Basin in the year 2050 (or another date sufficiently far off to allow for creative flexibility).

Step three: Further development of scenarios

Through the use of expert analysis and computers, these scenarios would be built up in more detail and undergo some initial feasibility testing. The initial policy implications of each scenario for the sector would be articulated and presented for further discussion to the participants.

Step four: Introducing global climate change

The preferred future scenarios – and their policy implications – would next be confronted with the extra prospect of global climate change. In this step, participants would be presented with a set of scientific findings regarding the likelihood, nature, extent, and timing of global climate change. The degree of scientific uncertainty, and the existence of different and perhaps contradictory views within the scientific community, would be made clear. The presentations would be made by a research team, which would have previously assessed the implications of climate change for various sectors and would be available to respond to questions from participants.

The results of this step would be the development of detailed understanding on the part of participants of the challenges presented by the current assumptions of the prospects of global climate change for their already articulated preferred futures.

Step five: Determining policy responses

The final step would involve asking the participants to devise proposed sets of policies that respond to the challenges identified in Step Four. The goal here would be twofold: (a) to explore ways in which the preferred futures identified in earlier steps *would be* resilient to the potential impacts of climate change, including the uncertainties concerning its prospective existence and characteristics, and (b) to explore ways in which the preferred futures *would be made more* resilient to respond to the potential impacts of climate change.

Whereas participants would be asked to focus on the potential resilience of their own preferred future, a comparison of the relative resilience of the different futures would also permit evaluation of the comparative merits of alternative scenarios with respect to climate change. This would allow for evaluation of the trade-offs both within and between different future scenarios.

In keeping with the overall orientation of the process proposed here, no attempt would be made at the end to reach consensus on a *single* best or preferred scenario. Instead, participants would be encouraged to take away with them whatever they would have learned about their own and other preferences and to use that information in the larger policy arena in which they normally act. In other words, this would not be a predictive process, but a policy-learning process, to assist the participants (and the wider public) in their own exploration of the various trade-offs and implications of certain choices.

As with the earlier examples, this study was to be based on the premise that a major management task of our time was to make better use of expert knowledge in the service of greater public participation. The argument was that these approaches would begin with what the non-expert was best placed to contribute – that is, his or her general concerns or hopes, which could be focused or crystallized through the use of alternative futures as a methodological tool.

Conclusions

The central theme of this chapter is that any citizenry is capable of becoming more emancipated if allowed to imagine or shape its own future. It is thus in line with the more emancipated risk analysis sought for global environmental change in this volume. Emancipation takes on three meanings – emancipation by loosening conceptual and institutional restrictions on freedom of expression, emancipation by permitting a sideways look at roles and power relationships, and emancipation by genuine participation in political discourse. The equity issue then comes, perhaps, down to the question of who will have first and best access to the future.

In the past, risk analysis often embodied one answer to this question as it went about its tasks, often with the best of intentions and within a set of global assumptions that were not themselves often assessed, being, as they were, the securers of the assessors' own world view. Yet global environmental change carries with it potentially the greatest of all risks for humanity – the undermining of basic life-support systems and the destruction of civilized behaviour and values in a struggle to survive. It is this fear that lies rooted in the emerging concern to reassess the risky world we are creating for our descendants, and the equally powerful emerging desire to be directly part of the process through which that future is determined.

15

Exploring a sustainable future for Canada

John B. Robinson

As Canadians navigate the twenty-first century, they share with citizens of other countries an essential priority to develop sustainable and environmentally benign patterns of resource utilization and socio-economic development. This chapter describes one attempt to investigate the feasibility and impacts of a future for Canada. Based upon principles of environmental and sociopolitical sustainability, the Sustainable Society Project (SSP) provides a specific context for exploring the issues addressed by other chapters in this section. For more than six years, from September 1988 to December 1994, the SSP explored the prospects for the development of a future Canadian society that is sustainable in environmental economic, and social terms. The project enlisted scenario analysis to trace the path of Canada from 1990 into a sustainable future around the year 2030. The SSP assessed the relationships among values, lifestyles, and technological and economic development in the scenario, as well as the feasibility and sociopolitical implications of the scenario itself.

The SSP emerged out of a particular Canadian context but reflects a more general interest in normative futures and analysis intended to explore desirable rather than likely futures. This is in keeping with the normative issues entailed in managing global environmental risk and envisioning desirable futures (see chapter 1). After setting out the backdrop for the emergence of the SSP, this chapter moves on to theoretical and methodological issues and then presents some of the initial findings. A more detailed account appears in *Life in 2030: Exploring a Sustainable Future for Canada* (Robinson et al. 1996a).

Background to the project

Concern with environmentally benign futures

The conceptual roots of the SSP originate in the conserver-society concept and research tradition in Canada. The term "conserver society" dates back to a 1973 report on resource policy by the Science Council of Canada (1973). The study urged individual Canadians and their governments, institutions, and industries to "begin the transition from a consumer society preoccupied with resource exploitation to a conserver society engaged in more constructive endeavours" (Science Council of Canada 1973: 39).

The concept of a conserver society first underwent detailed examination in a series of reports (Valaskakis, Sindell, and Smith 1975, 1976; Valaskakis et al. 1979) by the GAMMA (short for *Groupe associé Montreal-McGill pour l'étude de l'avenir*) Institute, a future-studies and strategic-planning organization. The think tank identified and explored three successively more radical conserver-society scenarios – expansion with efficiency ("doing more with less"), a stable industrial state ("doing the same with less"), and a "Buddhist" scenario ("doing less with less"). Since then, the "Gamma approach" has left its imprint on numerous Canadian projects (Valaskakis 1988).

The Science Council of Canada (1977) also took up the concept and published a major report entitled *Canada as a Conserver Society: Resource Uncertainties and the Need for New Technologies*. This report outlined five general policy thrusts: these were concern for the future; economy of design; diversity, flexibility, and responsibility; recognition of total costs; and respect for the regenerative capacity of the biosphere. It also contained a detailed discussion of the specific techniques and technologies that might be adopted in the transition to a conserver society. Generally, these differed little from the changes described in GAMMA's expansion-with-efficiency scenario. The Science Council chose, however, to avoid direct discussion of the sociopolitical dimensions of the conserver-society argument:

We have tried to stick to practical matters, with an incremental approach, to identify some of the technological paths that lead in the right direction, toward sustained relationships with material resources and the biosphere. Whether those paths, about which in our view we do not have much choice, imply other changes can be decided only by Canadians through democratic discussion. (Science Council of Canada 1977: 14)

The importance of the Science Council's emphasis upon technological matters was that it allowed the Council to argue that significant improvements in emissions-reduction, land-use, and resource-development practices, environmental protection, and efficient use of resources and materials were all possible through improved technological development, without significant reductions in material standards of living. This argument came to typify an entire school of analysis in the energy and environmental fields and challenged prevailing notions that environmentalist approaches implied reduced living standards and entailed resisting, rather than embracing, technological progress. All the while, such an approach precluded addressing other versions of the conserver-society argument, such as GAMMA's second and third scenarios, which represented more significant changes in behaviour and development patterns (Robinson and Slocombe 1996: 5–6).

The conserver-society concept was not the only vehicle for environmental arguments in Canada at this time. A parallel exploration of the concept of "ecodevelopment" was proceeding in a series of meetings sponsored by Environment Canada and the Canadian International Development Agency (CIDA) from 1975 to 1979. The term ecodevelopment, coined during the 1972 United Nations Conference on the Human Environment held in Stockholm, referred to the notion of environmentally sustainable development and launched the "environment and development" debate that has been fostered since then by the United Nations Environment Programme (UNEP). Ecodevelopment emphasized the need for (a) meeting the basic human needs of people, (b) enhancing self-reliance, mainly at the community level, and (c) maintaining cultural and biological diversity through what would now be called sustainability practices (Francis 1976; Sachs 1977). Implicitly, ecodevelopment was viewed as something for the third world, whereas the conserver society was for Canada and the industrialized world.

Although the Science Council's report and related activities generated considerable interest both in Canada and abroad, this interest did not translate into any official political or policy response. Instead, the concept itself became part of the ongoing political debate about environmental issues and futures that occurred during the 1970s. Perhaps for this reason, the Science Council downgraded its efforts in this area and, despite publishing a discussion paper on the conserver society (Schrecker 1983), conducted no further research and simply let the concept die a natural death.

The dearth of official interest or follow-up notwithstanding, the conserver society concept significantly influenced the development of environmentally based arguments in areas such as health (Hancock 1980) and agriculture (Hill 1985b) and also provided a favourable context within which to cast environmental arguments in general. Perhaps the most important area of application was in the energy field, which, since 1973, had been the subject of intense political debate and interest. In particular, the "soft-energy path" concept, first articulated by Amory Lovins (1976a, 1977), served as a vehicle for the most significant application of conserver-society principles in Canada.

In short, the soft-path argument suggests that a combination of increased energy efficiency and renewable sources of energy that are diverse, flexible, and matched in geographic distribution and thermodynamic quality to end-use needs can best satisfy demands for the services that energy provides – illumination, warmth, cooling, motive drive, and mobility. This stance entails a massive reorientation of existing energy systems (and thinking) away from primary reliance on large-scale, centralized, and non-renewable energy sources and towards systems characterized by high levels of energy efficiency and the use of small-scale renewables.

The concept prompted various countries to explore the technical and economic feasibility of soft-energy futures. Analyses in Canada culminated in a large, province-by-province, soft-energy-path study by Friends of the Earth Canada (1983–1984) with funding from the federal departments of energy and environment. The research enlisted detailed scenario analysis, to demonstrate the feasibility and cost effectiveness, by 2025, of managing Canada's growing economy using significantly less energy than in the base year of 1978 and by supplying that energy largely through renewable sources of energy. By the time that the study was released in 1983–1984, however, the world oil glut and attendant price drops had the effect of shifting Canadian energy policies back to a traditional concern with oil and gas extraction and revenue issues. Although an updated version of the soft-path study was prepared and presented to the government (Torrie and Brooks 1988), the soft-energy path concept in Canada, and elsewhere, appears to have suffered much the same fate in official circles as the conserver-society concept (Robinson and Slocombe 1996: 7).

The exception to this general rule is that the energy-efficiency arguments first articulated by soft-path analysts have become influential in energy policy formulation across the industrial world (Canadian Electrical Association 1991; International Energy Agency 1987). Increased energy efficiency is now generally perceived to offer the best hope of ameliorating a whole range of environmental problems, including global warming (IPCC 1996a, 2001).

More generally, with the re-emergence of environmental issues typical of the mid-1980s and the publication of the World Conservation Strategy (IUCN, WWF, and UNEP 1980) and the Brundtland report (WCED 1987), many conserver-society ideas have re-emerged under the aegis of the sustainable development concept (Environment Canada 1985). In particular, concern over global issues such as ozone depletion and global climate warming are stimulating renewed interest in exploring environmentally benign patterns of resource use and economic activity. The International Geosphere/Biosphere Programme and the Human Dimensions of Global Change Programme are tangible indications of a growing interest in global environmental problems. On the political level, the 1987 Montreal Protocol on ozone layer depletion (Benedick 1998; Mackey 1988) and subsequent more stringent measures, and the 20 per cent CO_2 reduction target proposed at the Toronto Conference on the Changing Atmosphere (Environment Canada 1988; WMO/OMM and UNEP 1988), and since then adopted by several countries, suggest a reorientation of the political agenda regarding environmental issues.

Methodological and theoretical issues

Perhaps as important as the growing interest in the substance of environmental problems has been the emergence of new approaches, including the emerging developments in global environmental risk analysis that are the subject of this volume. During the 1970s, it became increasingly apparent that traditional approaches to futures analysis and forecasting had a strong bias to the status quo and were essentially incapable of revealing or analysing futures that represented significant departures from business-as-usual trends (Ascher 1978). Again stimulated by the early work of Amory Lovins (1976a, b), energy analysts began undertaking "backcasting" analyses that postulated a future that differed from (and apparently was more desirable than) that revealed by conventional forecasts, and they assessed the feasibility and implications of such a policy path (Robinson 1982b). Others began to develop and use non-predictive scenario-analysis methods intended to alter the manner in which energy industry executives viewed the future (Coates 1997; Wack 1985a, b).

In a parallel development, researchers began to build energy "end-use" models based upon the depiction of physical processes rather than economic relationships. These models had the dual advantage of allowing non-predictive futures analysis of the backcasting or scenario-analysis type and also of permitting a much more detailed look at the potential for increased energy efficiency (Robinson 1982a). More recently, the backcasting and end-use modelling approaches have been generalized beyond energy issues to encompass questions related to global environmental change and sustainable development (Mulder and Biesiot 1998; Robinson 1988, 1990a, 1991).

The growth in popularity of energy end-use models and non-predictive scenario analysis has been rapid. In the electric utility industry in North America, for example, end-use models have increasingly replaced the more traditional econometric models formerly used by load forecasting,

and now dominate the field. More recently, the emergence of environmental target-setting (e.g. for CFCs or for CO_2) at the political level has
created a demand for methods of analysis that are intended to assess the
feasibility and implications of reaching those targets. Jäger et al. (1991)
and Mulder and Biesiot (1998), for example, have reviewed the use of
innovative backcasting methods of analysis.

Backcasting and end-use modelling approaches emerged within the
context of a growing distrust of "business-as-usual" futures and a corresponding emphasis upon openly normative preferred futures (Robinson
and Slocombe 1996: 9). The distrust is part of a more general critique of
traditional positivist social science. That critique argues for more "con-
structivist" approaches to the practice and evaluation of science and its
use in the policy-making process (Robinson 1988, 1990a, 1992). Mindful
of the inescapably value-laden nature of all scientific analysis, such approaches seek to make values explicit and to provide a means for testing
and evaluating them and for discerning policy choices. This is particularly
important with respect to what Salter (1988) has called "mandated
science" – that is, science undertaken in direct support of policy making –
a problem that has been particularly apparent in the past in risk analysis
and which new approaches, such as those in this volume, are now confronting (see also Krimsky and Golding 1992).

Two major examples of mandated science are forecasting and risk
assessment. Thus, it is not surprising that the theoretical and methodological arguments made here with respect to futures studies have their
parallel in the risk field and, indeed, are based in large part upon the
same substantive environmental concerns (e.g. Goodman 1995; Holdgate,
Kassas, and White 1982; Kasperson et al. 1988; O'Riordan and Rayner
1991; Otway 1987; Rayner and Cantor 1987; Wynne 1987). In both cases,
a major emphasis is upon debunking the claims of traditional approaches
to provide definitive technical answers to political questions (What is the
most likely future? What is the objective risk?) and upon the identification of questions of power, credibility, and choice (Who chooses? Who is
expert? How is the problem defined?).

These parallels between the risk and forecasting fields are instructive,
for they illustrate the degree to which the arguments made in these fields
are part of a larger dynamic related to the use of science for policy purposes (see Robinson 1990b, c). In these and other cases of mandated
science, one effect of the growing number of public policy debates about
energy futures or toxic waste disposal has been to call into question
the very science used to support the various claims and counter-claims. In
the energy-forecasting field, for example, a growing mistrust exists on the
part of decision makers about the utility (or even meaning) of large-scale,
energy-modelling efforts. In the risk field, the use of quantitative risk-

assessment tools provoked widespread critique within the risk community about their ability to resolve public policy disputes about human health and environmental hazards.

The "classical" response to this problem has been to focus attention on the scientific adequacy of individual risk assessment or forecasting studies: in so far as problems exist, they have to do with the quality of those studies rather than with the approach itself. However, if problems exist with the way in which energy forecasting and risk analysis are conducted in principle, such individual critiques of specific studies will merely deflect attention from the underlying problems.

A more sophisticated response, as pursued in this volume, is to develop new approaches to risk assessment and forecasting that take into account the critiques of the approaches themselves. "Neoclassical" approaches, however, may introduce only significant methodological improvements, and typically do not alter in any fundamental way the definition of the problem or, therefore, the overall goal of the analysis itself. For example, the emergence of interest in "risk communication," or the growth of more probabilistic treatment of forecast uncertainty, both reflect awareness of the problems with early views of risk and of forecasting. Such reforms, however, do not challenge the view that science in principle tells us true things about the real world, that it should be value neutral and objective, that the goal of science should be prediction, and that the basic problems at issue in public-policy disputes are scientific ones that can be resolved through good scientific analysis – the issues, in short, taken up in chapter 1 of this volume.

An "alternative" approach to mandated science suggests that the problem lies not so much with individual cases of bad analysis, or with improved methods of communicating science to policy makers or the public, as with a misconceived notion of what scientific analysis is and what it can provide to its users. This approach suggests that all mandated science is necessarily value laden in the sense that its findings are influenced strongly by methodological and theoretical values embedded in its methods and approaches. From this perspective, the goal of mandated science is not so much to find out true things as it is to make clear the points of difference, including value differences and competing views of what the problem is. Put another way, the goal is to separate what is uncontroversial from what is not and to make clear the underlying causes of the differences that exist.

The foregoing suggests the existence of at least three views of the role and status of mandated science: a traditional view, in which the mandate and nature of the analysis is unquestioned and clear; a reformist or neoclassical view, which involves changes in approaches and methods in response to both the obvious failure of classical methods and the challenges from critics of mandated science; and a radical or alternative view, which challenges the very role and status of mandated science. The latter, it is argued elsewhere in this volume, offers potentially powerful contributions to understanding global environmental risk.

The SSP, like global environmental risk more generally, unfolded within the context of these theoretical and methodological developments. By design, it reflects what is called here the "alternative" view of the role and meaning of mandated science, in that its focus is upon exploring the feasibility and impacts of explicitly normative, desired futures, within the framework of the perceived undesirability, or "riskiness," of conventional "business as usual" futures. The project went forth amid both a resurgence of interest in environmental issues and a surfacing of new approaches to the study of desirable futures.

Purpose and objectives

A sustainable society has sociopolitical as well as environmental and technological implications. The challenge is to analyse the implications of a sustainable society in an integrated way that accounts for both dimensions. Moreover, the human and technological dimensions of a socio-economic system are dynamically interrelated; sociopolitical structures are reflected in technological and economic development, and vice versa.

The overall purposes of the SSP were to develop a scenario of the future development of Canadian society that is based upon principles of sustainability; to assess this scenario in terms of its feasibility, implications, and implementation requirements; and to contribute to the development of a network of groups and individuals interested in sustainable futures for Canada.

The SSP went beyond previous research in Canada in several interconnected ways. First, the project provided rigorous scenario analysis of all sectors of the economy. Second, the SSP did not confine itself to "technical fix" solutions but included the analysis of lifestyle and social change. Finally, the project produced a detailed analysis of the links between sustainability values and socio-economic and technological developments (fig. 15.1).

Whereas the scenario analysis performed in the SSP involved multiple iterations to elicit a balanced scenario, the design served to test the feasibility and impacts of a single sustainable-society scenario. Analysis of several alternative versions of a sustainable society and comparison with at least one "business as usual" future would have been desirable, but time and resource constraints precluded such an approach (Robinson, Van Bers, and Biggs 1996: 16).

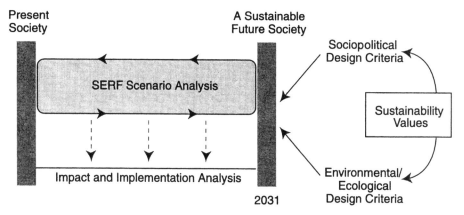

Fig. 15.1 **Sustainable Society Project: Conceptual framework (SERF, Socio-Economic Resource Framework)**

The overall approach taken to integrate the various dimensions of sustainability was to define sustainability as a normative ethical principle having both sociopolitical and environmental/ecological dimensions and then to develop, in each of these two areas, scenario-design criteria that could serve to drive the scenario analysis. An iterative process, taking as many attempts to develop such scenarios as necessary to produce a reasonably consistent picture that conformed to the initial design criteria, facilitated the subsequent assessment of the feasibility, implications, and implementation requirements of a scenario.

The project had specific objectives:

- To develop sociopolitical and environmental/ecological design criteria for the scenario analysis that are based upon sustainability conceived as a normative ethical principle;
- To create quantitative subscenarios of technological and economic development in the modelling system based upon these design criteria and, through an iterative process, to integrate them into one physically consistent scenario of a sustainable future;
- To determine and then evaluate the general social, economic, environmental, and political implications of the final scenario; to assess its overall feasibility; and to analyse the sociopolitical implementation measures required for it to occur;
- to contribute to the development of a network of interested groups and individuals in the field through involvement in the project; and
- to produce and disseminate a final report that summarizes the results of the analysis (Robinson, Van Bers, and Biggs 1996: 16–17.)

Methods

Scope

The SSP addresses sustainable futures at a national level. Thus, its spatial scope is all of Canada. International trade is addressed explicitly through the use of trade and balance-of-trade calculations.

Temporally, the scenario spanned 40 years, from 1990 to approximately 2030, allowing sufficient time for the complete turnover of capital stocks, the generation of new structural relationships in the economy, and the development of new institutional relationships in the political system. Comprehensive descriptions of the physical and technological state of the system are available for each year of the scenario evolution.

Sectorally, the scenario analysis focused on the various consumption sectors (dwellings, consumer goods, health, education, transportation, retail trade, office buildings), the manufacturing sector, and the primary-resource sector (energy, forestry, mining, agriculture, fisheries, and water). Subscenarios of the sectoral subject areas were developed, based upon sustainability principles and scenario-design criteria, and then were integrated into an aggregate scenario.

The impact and implementation analysis involved checking the consistency of the sustainable-society scenario with project-design criteria and assessing the changing institutional relationships and political structures that one might expect to result from (and help to cause) the technological and economic developments described in the quantitative scenarios. This involved an assessment and analysis of the compatibility of these evolving sociopolitical forms with each other, with principles of environmental sustainability, and with value principles basic to Canadian society.

The number and breadth of the subject areas incorporated in the project required a team approach. The project engaged a core research team of five analysts, a contingent of graduate students, a group of other researchers with partial involvement, an advisory committee, and a network of groups and individuals with an interest in the idea of sustainable futures.

Theoretical approach

Environmental problems present important questions about scientific knowledge and technological change, as well as about social and political organization. As SSP researchers viewed science and technology as deeply value laden, it was necessary to examine closely the relationship between technology and social organization and between science and its use in decision-making. It was also necessary to develop technologies and systems of expertise overtly grounded in the values that technology and expertise are intended to serve (Winner 1986).

The result was an openly normative approach to the assessment of technology and social organization. In the SSP, that approach was reflected in the linkages among technological development, social change, and sustainability values, and by the development of a scenario-analysis method that could be enlisted to express and evaluate different values and views of preferred futures. Thus the analysts' role is not to determine the right course of action but to propose both their preferred alternative to current policy (which is explicitly grounded in a set of values) and a means by which others could propose and assess their own views and preferences.

The values forming the normative basis for this project are a version of those associated with the terms "sustainable society" or "sustainable development" (Brown 1981; Clark 1986, 1987; Clark and Munn 1986; WCED 1987). They imply a critical, or even radical, approach to technological and social development (Robinson et al. 1990). To assess the implications of these values and implement the normative approach to science, technology, and analysis described above, it was necessary to adopt an analytical perspective that departs from the predictive and ostensibly value-free orientation common to most futures studies (Robinson 1988). The backcasting approach – as O'Riordan and Timmerman indicate in chapter 14 – involves defining goals, articulating them in terms of preferred future states of the system being analysed, and then attempting to construct a path of technological and social development between the present and the desired end-point (Robinson 1990a). The goal is to assess the feasibility and impacts of normatively defined futures, thereby providing a kind of consistency and feasibility check on the values in terms of which the normatively defined futures are constructed.

A unique feature of the SSP is the use of a computer simulation system – the Socio-Economic Resource Framework (SERF) – for scenario generation and analysis. SERF is an implementation of the design approach to socio-economic modelling (Gault et al. 1987). Because design-approach models are designed to examine the physical feasibility of alternative policy goals over the long term (35–70 years), they are not useful for forecasting; they can, however, be used for backcasting analyses.

David Biggs (1996) has recently described SERF in detail. The framework models numerous aspects of Canadia's socio-economic system and stores the data in 43 separate submodels and over 1,700 multidimensional variables that are based on the extensive Statistics Canada database (fig. 15.2). These models organize energy, labour, and material flows into four major components: demography (population, household formation, and labour force); consumption (housing, consumer goods, health care,

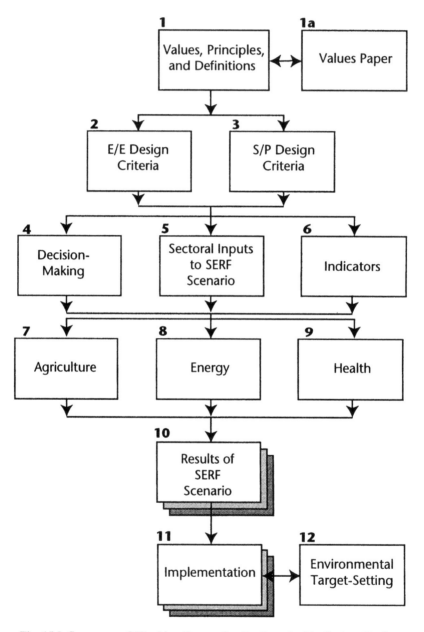

Fig. 15.2 **Sequence of Working Papers for the Sustainable Society Project**

education, transportation, offices, communications, and retail trade); production (detailed input–output and capacity models); and resource extraction (agriculture, forestry, primary energy, minerals, and wildlife harvesting). The disaggregated and comprehensive nature of SERF allows the user to undertake detailed analyses of changing efficiencies, technological substitutions, and labour productivity. The user supplies input scenarios in the form of assumed future values for the approximately 1,700 multidimensional SERF variables, and SERF combines these time-series inputs into integrated scenarios and assesses the physical consistency, over time, of the resultant overall scenario of the evolution of Canadian society.

Whereas the existence of SERF allowed the SSP to undertake detailed quantitative scenario analysis of a kind not hitherto possible, SERF does not encompass all dimensions of Canadian society relevant to the project enquiry. In particular, SERF does not describe the decision-making processes and changing institutional arrangements that are associated with the technological/physical scenarios that it generates. Therefore, as discussed below, analysis of sociopolitical transformation has occurred outside SERF, albeit connected to the scenarios generated in SERF.

Organization of the Sustainable Society Project

The general analytical approach of the project was to articulate sustainability values, derive environmental/ecological and socio-economic scenario design criteria based on these values, develop a qualitative picture of Canada in 2030 consistent with these criteria, construct a quantitative scenario in SERF intended to describe a path between 1981 and 2030 that leads us to that future, and analyse the implications, feasibility, and implementation requirements of that scenario.

The first step was to articulate and elaborate the values and principles underlying the sustainability scenario (Robinson et al. 1990; Robinson, Van Bers, and Biggs 1996). From defining sustainability as a normative ethical principle with both environmental/ecological and sociopolitical dimensions, the next step was to articulate several basic value principles and a set of more specific ecological and social principles, as outlined in table 15.1. These principles were used to generate the environmental/ ecological and sociopolitical scenario design criteria outlined (Lerner 1991; Slocombe and Van Bers 1991), and the description of economic, legal, political, and individual decision-making in Canada in 2030 (Robinson, Francis, and Lerner 1992). Detailed qualitative and quantitative descriptions for each sector of society over the period 1981–2031 have been used to develop inputs to the SERF modelling system

Table 15.1 **Principles of sustainability**

Basic value principles
- The continued existence of the natural world is inherently good. The natural world and its component life forms, and its ability to regenerate itself through its own natural evolution, have intrinsic value.
- Cultural sustainability depends on the ability of a society to claim the loyalty of its adherents through the propagation of a set of values that are acceptable to the populace and the provision of those sociopolitical institutions that make the realization of those values possible.

Definition of sustainability
- Sustainability is the persistence over an apparently indefinite future of certain necessary and desired characteristics of the sociopolitical system and its natural environment.

Key characteristics of sustainability
- Sustainability is a normative ethical principle. It has both necessary and desirable characteristics. There therefore exists no single version of a sustainable system.
- Both environmental/ecological and sociopolitical sustainability are required for a sustainable society.
- We cannot, and do not want to, guarantee persistence of any particular system in perpetuity. We want to preserve the capacity for the system to change. Thus, sustainability is never achieved once and for all, but only approached. It is a process, not a state. It will often be easier to identify unsustainability than sustainability.

Principles of environmental/ecological sustainability
- Life-support systems must be protected. This requires decontamination of air, water, and soil and reduction in waste flows.
- Biotic diversity must be protected and enhanced.
- We must maintain or enhance the integrity of ecosystems through careful management of soils and nutrient cycles, and we must develop and implement rehabilitative measures for badly degraded ecosystems.
- Preventive and adaptive strategies for responding to the threat of global change are needed.

Principles of sociopolitical sustainability
1. Derived from environmental/ecological constraints
- The physical scale of human activity must be kept below the total carrying capacity of the planetary biosphere.
- We must recognize the environmental costs of human activities and develop methods to minimize energy and material use per unit of economic activity, reduce noxious emissions, and permit the decontamination and rehabilitation of degraded ecosystems.
- Sociopolitical and economic equity must be ensured in the transition to a more sustainable society.
- Environmental concerns need to be incorporated more directly and extensively into the political decision-making process, through mechanisms such as improved environmental assessment and an Environmental Bill of Rights.
- There is a need for increased public involvement in the development, interpretation, and implementation of concepts of sustainability.

Table 15.1 (cont.)

- Political activity must be linked more directly to actual environmental experience through allocation of political power to more environmentally meaningful jurisdictions, and the promotion of greater local and regional self-reliance.

2. Derived from sociopolitical criteria

- A sustainable society requires an open and accessible political process that puts effective decision-making power at the level of government closest to the situation and lives of the people affected by a decision.
- All persons should have freedom from extreme want and vulnerability to economic coercion as well as the positive ability to participate creatively and self-directedly in the political and economic system.
- A minimum level of equality and social justice should exist, including equality of opportunity to realize one's full human potential; recourse to an open and just legal system; freedom from political repression; access to high-quality education; effective access to information; and freedom of religion, speech, and assembly.

Source: Robinson et al. (1996b: 46–48).

(McFarlane et al. 1992). The inputs to the SERF scenario take the form of quantitative time series over the period 1981–2031 for the more than 1,700 multidimensional SERF variables, ranging over birth rates, labour force participation rates, housing and consumer-durable consumption patterns, teacher–student ratios, doctor visits per capita, food consumption by food type, communications infrastructure, passenger kilometres by vehicle type, materials use in manufacturing, and primary resource-extraction capacity and operation.

A complete sustainable-society scenario was created and run on SERF in January 1992. The computer simulation of alternative futures detailed energy, labour, and material flows into each of the major sectors and then generated subscenarios. These subscenarios as well as the overall scenario "were rerun as many times as required to resolve physical inconsistencies and to produce a balanced scenario consistent with the project's initial design criteria" (Robinson, Van Bers, and Biggs 1996: 21).

The next step was to consider the longer-term policy options and changes in individual and organizational behaviour that would be associated with successful implementation of the project scenario. In May 1992, project members presented preliminary results to the SSP Advisory Committee. A series of policy workshops have introduced the sustainability scenario and described its implementation requirements with civil servants and other interested and affected parties. Additional work at the University of British Columbia during 1993–1995 involved repeated testing and refinement of the scenario to ensure consistency and compatibility with sustainability values.

Conclusions

The Sustainable Society Project was an ambitious attempt to tie together some of the diverse threads of the environmental arguments that have developed from the conserver-society tradition in Canada. As such, it represented an application of a set of methodological approaches and principles that have emerged out of the same tradition. Indeed, it was the perceived need to develop methods of analysis oriented toward the examination of unconventional futures that led environmentalists and soft-energy analysts to develop backcasting approaches, and Statistics Canada research to develop SERF. The SSP combined these methodological and substantive approaches in an integrated analysis.

This project can provide only a preliminary assessment of the feasibility and impacts of a sustainable future for Canada. In so doing, however, it contributes to the ongoing debate about such futures. Of course, no futures analysis of the type proposed here can be definitive or exhaustive; at best, it can reveal the rough outlines of a desirable future and suggest that such a future seems feasible and, therefore, worth striving for.

This is not to say that the analysis and analysts have no role in the political debate about desirable futures. Futures research and global environmental risk analysis more generally have a major role in informing the policy process, but they should not substitute for decision-making. The proper role of these studies is to explore various versions of possible futures and to indicate their apparent feasibility and implications. The rest belongs, as it should, to the political process.

16

Social visions of future sustainable societies

Patricia Benjamin, Jeanne X. Kasperson,
Roger E. Kasperson, Jacque L. Emel, and
Dianne E. Rocheleau

All manner of social issues receive short shrift in the debates on "environment and development" and "managing the global commons." Family life, community, stratification, and individual dignity and fulfilment all suffer from varying degrees of neglect, while international security, nature/society relations, economic alternatives, and processes of social change are badly in need of new theories. To the extent that the discussion is couched in terms of sustainability, the emphasis has been on reconciling ecological sustainability (planetary life support) with economic sustainability (continued economic growth), while social sustainability (creation of conditions for community and individual well-being) is generally ignored – or equated with economics, which is almost as bad (but note the exception of Robinson in the preceding chapter). Social sustainability, we should note at the outset, does not mean the continuation of existing social structures but, rather, creation and maintenance of the conditions for creativity, empowerment, self-determination, and self-actualization.

Research, both past and present, reflects this imbalance. For example, in a 1989 World Future Society publication, most of the items absent from a list of issues deemed important by a panel of 17 prominent American futurists are social in nature (Coates and Jarratt 1989: 24–25). Current research shows little change: for example, global environmental change is widely construed as a serious global problem with human causes (ICSU 1987; IFIAS 1987; IGES 1999; ISSC 1989; Jacobson and Price 1991), yet the International Human Dimensions Programme has a tiny budget compared with the science-based International Biosphere–Geosphere Programme, a dichotomy largely apparent in the national global change research programmes. In general, little funding is available for collaborative, international, long-term social science research, and the funding allocated to policy work is usually for specific policy problems rather than for understanding more broadly human-induced environmental change. The assessments by the American Council for the United Nations University (ACUNS) in the Millennium Project (Glenn and Gordon 1999; *State of the Future* 1997, 1998, 1999) show important gaps to be filled.

This hesitation to engage social issues is perhaps understandable – ecology and economy alone are complicated enough (Daly 1999); the social and political aspects, as emphasized throughout this volume, are often regarded as intractable and politically touchy; however, the price of ignoring them is high. Social issues are central to problems of environment and development, and are, in any case, worthy of attention in their own right.

Social concerns are critical to environment and development issues in several ways. First, a number of authors, working on an array of environmental-degradation problems, have argued that these issues are fundamentally social in nature. For example, a "downward spiral" linking poverty and environment is widely recognized (e.g. During 1989; Kates 2000; Kates and Haarmann 1992; Leonard 1989; Mellor 1988); Blaikie (1985), Blaikie and Brookfield (1987), and Blaikie et al. (1994) argue that land degradation is a social problem, related to a "simple reproduction squeeze" (e.g. Bernstein 1979). Hunger can be viewed as a distribution rather than a production problem (e.g. Lappé and Collins 1978; Lappé et al. 1998), as an entitlement problem (Sen 1981), or as an outcome of political conflicts (Chen 1990); a few decades ago, it was viewed as a population-control problem (e.g. Ehrlich 1968). To the extent that any of these observations are accurate, social as well as economic and environmental factors will determine whether clashes of environment and economy can be resolved.

Secondly, leaving aside momentarily the thorny problem of implementation, if policy makers do attempt directly to reconcile environment and economy, they still face the task of anticipating or identifying institutions, political structures, value systems, and lifestyles that are consistent with the desired or necessary conditions. What kinds of social structures, belief systems, families, activity patterns, geographies, and aspirations will be compatible with the environment/economy system they recommend?

It is clear that merely making and attempting to implement policy does not necessarily result in actual social change, particularly given the tendency of the policy makers to take existing social structures and institutions for granted. Any changes in resource use, in the structure of the economy, or in nature/society relations must all be implemented through the existing social system and will require major changes in that system. Relatively little is really known, however, about the sources and dynamics of social change, and even less under such conditions as global scale of impacts, rapid change, large populations, complex technologies, diverse and interacting cultures, and responses to subtle and systemic problems with long lag times. It is clear that societal initiatives or responses that work in some cultures do not work in others.

These three concerns – global change as a social phenomenon, identification of social systems compatible with ecological and economic sustainability, and the dynamics of social change – all point to the need for an integral social science engagement in any discussions of sustainable futures. Furthermore, it is clear that the talents of the entire spectrum of human knowledge and creativity – including all the social sciences and the humanities – are needed to address these issues. Discussion and reflection on the social implications at all scales of current trajectories and possible alternatives are sorely needed. The two overriding issues of where to go and how to get there, defined in social terms, need concerted and determined attention.

To return to the role of planning in the context of ecological and economic reconciliation, although there is an argument that responsible public policy must try to manage key environmental dimensions, planning alone is not enough. Furthermore, it cannot be done in a social vacuum: neither rational environmental policy nor planned economic growth necessarily guarantees fundamental human rights such as livelihood security, freedom, dignity, or opportunity; nor do they result in the maintenance of human values, cultures, or communities; nor do they enhance human welfare, particularly for marginal or exploited groups. Securing such values must be a principal motive behind whatever planning does occur, and therefore cannot be the peerogative of current planners. The full plurality and diversity of values that exist in society must be mobilized to inform and guide policy-making.

This leads to a second (and much more critical) point – that social goals and considerations are, a priori, pre-eminent. After all, fulfilment of basic needs propels human material interactions with the rest of the planet: the entire economic system is supposed to be motivated by the enhancement of human welfare. And then there are the broader dimensions of individual and community experience to which people attach great importance – values and aspirations; our relations to each other and to nature; the need for meaning and fulfilment; the institutions through which we interrelate; avenues for creative expression; meanings, symbols, belief systems; in short, family, culture, religion, work, play, artefacts.

Even if global economic planners and managers do devise an efficient, sustainable way to reconcile economic growth and ecological limits, what would that world be like and would anyone want to live in it? In addressing the future, it is equally possible to make a desirable, livable world the starting point, and to strive for resource use and economic systems that foster the attainment of that future. Do we eschew such an approach because of an assumption that everyone wants to live like upper-middle-class North Americans or Europeans, but cannot, owing to resource limits and economic problems? If so, this is surely an assumption worth examining. Gandhi's famous statement, reported by Tolba (1987), may be relevant here: when asked if independent India should be like Britain, he asked, "If it took Britain half the resources of the world to be what it is today, how many worlds would India need?"

It is a question of priorities. Social systems should not be forced to evolve from determinations of optimal capital, material, and energy flows; rather, economic and material systems should be designed to support social ideals. Stating the same point differently, one can argue that it is no more "unrealistic" to expect the economic system to adjust to human needs and ecological limits than it is to expect people to change in order to serve economic imperatives or to expect nature to continue to conform to economic demands.

Not that those are the only options, or even that the global capacity to design and implement ways for reaching solutions is well developed. There is little reason to suppose that change really occurs in this way – by top-down management – no matter what the priorities are. At the very least, as argued in chapters 1, 14, and 15, this exploration needs to be interactive, so that thinking about the future environment and economy is continually tested against thinking about societal and spiritual aspirations. If the direction and weight of current discussion fail to match real priorities, this should be acknowledged and corrected.

Visions and a history of the future

If the preceding line of reasoning has merit, the envisioning of alternative futures is an urgent priority. Although "the concept of alternative futures has become the linchpin of current future thinking" (Coates 1989: 16), much of modern futurist thought has exhibited a very limited conception of alternative futures.

Modern, formal futurism has a relatively short history. Modernists tend to assume that, until very recently, basic survival was an all-consuming endeavour, leaving little time for idle speculations, and that past societies were relatively unchanging and people expected the future to resemble the present and the past. A sense of human influence over one's own destiny grew from the late Middle Ages onward (Polak 1961: 1, 33); however, until the positivists, religious and secular repression kept most Western minds fixed firmly on the past and the present (Adelson 1989). The rapid technological advances of the industrial revolution, along with social changes brought by capitalism and by the revolutions of the late eighteenth century, raised obvious questions about the future of society (McHale 1969), while Darwin and the geologists were expanding the magnitude of conceivable time-scales (Alkon 1987). The idea of constant social and material change has been slowly penetrating popular and official culture since the 1840s. It has influenced American government planning since the First World War, as it influenced the Soviet Plans of the 1920s and 1930s (McHale 1969). Forecasting came of age in the Second World War in the United States, leading to the development of a formal "science" of futurism.

In his *The Image of the Future*, Polak (1961) warned depressed, post-war Europe of the danger of living with no mental horizons beyond the present and of the need to envision alternate futures. Despite this broad view, modern futurism concentrated on narrow post-war military and corporate concerns such as strategic planning, R&D, and marketing (Coates 1989; Coates and Jarratt 1989). Governments and corporations, whose ability to pay and high stakes made them easy clients, became the main constituencies (Adelson 1989). In response to this narrow base of clients, formal futurism has largely abandoned imagery and social imagination and has embraced the technologies of social forecasting and cross-impact analysis. As Elise Boulding (1983) puts it: "Polak asked for visions. Futurists give blueprints." We return to some of these issues below.

Despite its substantial shortcomings, formal futurism points to the importance of envisioning alternative futures. "Visions" may be defined as stories of possible alternative futures (Anderberg 1989; Nagpal and Foltz 1995; Stokke et al. 1991). The process of forming them may be viewed as a purposeful strategic choice, as a tactic for moral suasion, or as an innate propensity of the human psyche.

The first posture is exemplified by Gordon (1989: 26), who states that "without thinking about the future, we abandon our destinies to chance and the decisions of others." Simmonds (1989) shares this view, arguing that, in order for institutions to get where they want to go, they must have a sense of where they are going and an ability to institutionalize problem solving. This is, perhaps, the inspiration behind the global modelling efforts of the 1970s (see below) and the various scenario-development and integrated-assessment efforts of the 1980s and 1990s (Morgan and Dowlatabadi 1996; Rotmans and Dowlatabadi 1998; Timmerman and Munn 1997; Toth 1995). Giddens takes a broader approach, denying that humans are condemned to be swept along by inevitable social forces and calling for the exercise of sociological imagination – being "conscious of the alternative futures potentially open to us" (Giddens 1987: 22).

The second position generally appears in the context of empowering a specific social movement or project. Caldwell (1985, 1990), for example, adopts it in reference to environmental politics, arguing that scientific knowledge alone is ineffective in changing behaviour, because it provide[s] no comprehensive view of the future – or a route-map to reach it – that most people find plausible and persuasive. Rather, scientific knowledge must be converted into "a vision of the possible," to be realized through a programme of action that enlists belief, a popular, political movement with "a vision of the possible that possesses a quasi-religious quality" (Caldwell 1985: 195–196). Similarly, Malone (1988: 289) calls for "unleash[ing] the creative power of human reason" to develop alternatives to nuclear annihilation and planetary environmental collapse; this may be "an unrealistic dream ... but surely at least a vision – and from Proverbs we know 'Where there is no vision, the people perish'."

The third position is an essentialist one, held by many futurists. The claim is that each person carries around a view of the world, built from everyday observation and interpretation, that is projected into the future (e.g. Stokke et al. 1991). According to Adelson (1989: 28), implicit "images of futures within which ... intentions take shape and make sense" are an inherent part of the human psyche which help form our view of the present and are necessary to purposive action. Therefore, neither individual nor group behaviour, culture nor politics can be entirely understood without understanding those images. A better understanding of human behaviour, and particularly its purposive aspects, entails dealing with the key role that images of futures play in shaping it (Adelson 1989: 33). These images also have a pragmatic role, since they are essential to the ability of individuals, organizations, or communities to compose actions, strategies, and policies; to fit them into their context; and to give them meaning (Adelson 1989: 32).

Similarly, Elise Boulding (1983) argues that, from the individual to the societal level, images about "the not-yet" are constantly generated. This imagery inspires our intentions as we move toward that which we consider desirable, as O'Riordan and Timmerman underscore in chapter 14. Innate human creativity allows the imaging of alternatives to the present. Despite repression of these images through social stress and socialization processes (such as education), "all folk experiment with new ways of seeing and doing. The mental imagery precedes the overt act. If this is normal human activity, then utopia-building, in the broadest sense of the word, is a normal human activity" (Boulding 1983).

Taken together, these conceptions of social visions – empowering, inspirational/manipulative, or universalistic – argue for the necessity and power of visions. Polak, taking a mixed idealist/materialist view of history, claims that such images play an active historical role. He argues that forming images of the future requires an awareness of the future, which makes possible a conscious, voluntary, and responsible choice among alternatives. These images are constantly reformulated in a dialectical process between the image itself and the actuality of an unfolding future (Polak 1961, 1: 41–42).

At minimum, it is clear that some people see a need for overall social visions to guide action – whether that vision is articulated as an artistic creation, a corporate plan, or an ideal to guide policy deliberations. Such an enterprise has often been viewed as socially ambiguous. Modern futurism, although "more respectable than it used to be" (Gordon 1989: 21), is tainted by the shady aspects of other futuristic enterprises. Futurism gets a bad name from "mediums and hallucinogens and yuppie stockbrokers ... turn[ing] over tarot cards and pork bellies" (Adelson 1989: 28).

Serious rejection of the value of envisaging societal futures may be found in other quarters. There are some who, except for relatively narrow endeavours such as military or industrial strategic planning and technology assessment (Coates 1989), see no need for visions, finding in technological optimism and progress a belief that things are already fine or getting better. Among the scientifically orthodox, any endeavour so explicitly dependent on imagination is automatically suspect. Among others, a distrust of unscientific, non-material, and (possibly) élitist idealism has lingered since Marx's day. An emphasis on historical specificity and dynamic process rather than static end-point, and a claim that the future is unimaginable because it is impossible for those conditioned by present social conditions to conceive of that which will be produced by human beings not yet born acting under new conditions, have engendered disinterest or suspicion about whether the project of envisioning futures merits human struggle.

Despite these views, it is essential to ponder goals and directions. Creating multiple visions – confronting a diversity of goals, values, and structures – allows for ongoing clarification, communication, contestation, and critique of alternate futures (Nagpal and Foltz 1995). It also provides a forum for examining possible ramifications of current decisions, envisioned actions, and proposed new pathways.

Broadening the discourse

Incorporating social visions into the discourse about the future would serve the purpose of explicitly introducing a number of ignored or neglected elements into the "managing the planet" discusssion. In the process, it would also open up that discussion to critique and to much wider participation.

Futurists and their clients have been accused of being motivated by goals of control and manipulation (e.g. Dublin 1991); however, a number of more benign interests – such as concern for children and grand-children, seeking information on future impacts in order to guide current actions, simple nosiness or curiosity – may also motivate the desire to know the future. Some of us want to know what life will be like after us, and how our time here affects what follows. For others, a deep spiritual connection with the earth and its creatures, religious or ethical notions of stewardship, or a strong sense of responsibility for the earth and its future, inspire concern. Still others harbour a deep respect for what has come before – a sense of history. All of these motives reflect a common desire to know what it would be like to live in various possible futures.

Whereas such issues as population, energy, food, pollution, and the like do tell us something about the constraints under which society is or-ganized, they shed little light on the nature of lived human experience. Just as today's statistics on coal mining, methane emissions, or capital flows fail to capture the experience of most peoples' lives, future scen-arios limited to such factors constrain human inquiry, when we should be seeking to liberate and expand it. Whereas some specialists enter the futures discussion in such topics as material flows, atmospheric gases, capital investment, species richness, energy-conversion technologies, agricultural production, or demographic profiles, and their results are a needed part of knowledge, such analyses are intrinsically unable to explore the nature of society and the human condition. How people will live, and how all these phenomena will be translated through societal institutions into factors affecting daily life and human dignity, remains opaque and unexplored. Understanding what some future circumstance might feel like not only provides empathy or human interest but also makes possible a holistic view, a demonstration of interconnections.

Thus, "dispassionate" social science is not the only, or even the most, valid element in the current discourse. In academic terms, a "passionate" social science and a stronger voice for the humanities are needed. Finding diverse, rich, and textured ways of expressing social visions – through literary or artistic means, for instance – can make the issues accessible to more people and can contribute to the emancipatory assessment, humanistic orientation, and democratic decision processes advocated in this volume (see particularly chapters 1, 7, and 14). The expansion of discourse into the areas of everyday life, tapping our concerns about our grandchildren and about the impacts of our present lives – and piquing our curiosity – can open the possibility of capturing and mobilizing popular attention to important issues, perhaps even enabling the Habermasian dialogue advocated by O'Riordan and Timmerman in chapter 14.

Confronting modellers with issues derived from inquiry outside their bounds of analysis, propelling social choices and social possibilities onto the public agenda, and stimulating explicit thinking about alternative futures can all generate a more powerful knowledge as well as greater awareness of the consequences of present decisions and policies (or of passive acquiescence to them). Public dialogue, conducted through alternate social visions, may also help to clarify the fact that no particular technical, economic, or environmental scenario is inevitable; that a particular set of variables or trajectories does not necessarily translate into a specific social scenario; and that futures (and risks) must be negotiated (see chapter 1). Choice, not determinism, prevails. Viewed in this way, thinking about the future can have an empowering rather than a discouraging effect, and can form an important part of the emancipatory global risk analysis that this volume seeks.

We have asserted that social concerns have been inadequately addressed in past research attempts at envisioning the future. The outstanding efforts of this type were the global models generated in the 1970s. To test this assertion, we turn our attention to an examination of the social content of this first generation of global models.

Global models, futurism, and social visions

Looking at global models

A number of comprehensive efforts to predict the planetary future were mounted in the 1970s. Several factors spurred this work. Mounting concern over population, pollution, and food supplies was expressed in an awakening environmental movement and in such books as Rachel Carson's *Silent Spring* (Carson 1962), Paul Ehrlich's *The Population Bomb* (Ehrlich 1968), Barry Commoner's *The Closing Circle* (Commoner 1971), and E. F. Schumacher's *Small is Beautiful* (Schumacher 1973). The use of the image "spaceship Earth" by Kenneth Boulding (Boulding 1973) and Buckminster Fuller (Fuller 1969) acquired new power, owing to photographs of Earth taken from space and the moon. Developing computer technologies made possible the manipulation of large quantities of data about the present in the service of constructing plausible futures.

Multiple global models appeared. Their distinguishing features included a global scale; a methodology based on extrapolation from the present;

and a focus on the interactions among population, environment, and economy. These models and their constituent features have largely defined the terms of the mainstream global environmental debate up to the current time. The emphasis has been on demography, resources, pollution, and capital. Social commentary proliferated in the 1950s and 1960s (e.g. Bell 1973; Fromm 1955; Galbraith 1958; Heilbroner 1972; Marcuse 1964; Riesman 1950; Whyte 1956), including reports on global futures from the likes of the Commission on the Year 2000 (AAAS 1965–1967; Bell 1969) and the International Peace Research Institute (Jungk and Galtung 1969), and reports addressing the specific issues taken up by modellers [e.g. MIT's Study of Critical Environmental Problems (SCEP 1971) and the Ward and Dubos (1972) report for the 1972 UN Conference on the Human Environment in Stockholm] – yet the models, not the rich variety of research, were adopted as the defining voice of global environment and development issues. The discussion that follows centres on the models, partly because a new generation of global models is now being created [e.g. the various efforts at integrated assessment; the Polestar project of the Stockholm Environment Institute, reported in Raskin et al. (1998); and the Hammond (1998) analysis of the twenty-first-century scenarios].

In the early 1990s, Gordon Goodman, Director of the Stockholm Environment Institute, discussed with members of a Clark University group his plans for Polestar, an analytic process for systematically exploring alternative futures to inform international efforts for coping with global environmental change. The centre-piece of this process as it has developed is an accounting device – the Polestar computer-based tool, building upon the first generations of global models – that allows an ongoing exploration of global environmental change. The Clark group argued that a process such as Polestar should include a component missing from the earlier modelling efforts – an explicit exploration of the social nature of future sustainable societies. Such an inquiry would begin with the desired human conditions and social structures, with processes aimed at fulfilling human development and dignity, and would proceed to elicit diverse social visions from a heterogeneous sample of creative thinkers. The outcomes would have tremendous value in their own right, but the process would also complement modelling of economic and physical phenomena, interacting with them by setting up a dialectic (missing from earlier models) in which diverse social arrangements would drive, be explicitly incorporated in, or be compared with, the models. The process would test not only model feasibility but also the modellers' assumptions and implicit world visions (at least, those aspects of the visions capable of being modelled would be tested, thereby enriching the modelling process).

The Clark group undertook an initial experiment, involving a review of the social content of the first generation of global models. As an aid in comparing the social content of the models, we constructed a set of attributes that (arguably) might be used to characterize a sustainable society. We sought to create a frame of analysis by which very different social visions could be compared (see below). This frame included some components typically treated (e.g. population) as well as some less-obvious attributes (e.g. equity, views of nature). A striking lesson of this process was the difficulty of creating an analytical approach that does not lose the richness of texture and interconnections among attributes or components.

Previous listings of important qualities of social visions include everything from Maslow's hierarchy of needs (Maslow 1959) to transportation systems, from political change to individual spiritual fulfilment. Such a wide range of factors might help to begin to address questions such as those posed above – what it would be like to live in such a society, and whether anyone would want to do so. For the sake of manageability, this wide sweep of characteristics was pared down to a limited set of variables that could be used to construct a comparative matrix. In the interests of "getting a handle" on what we viewed as the multidimensional nature of social visions, however, we narrowed these wide-ranging themes to a set of somewhat conventional attributes (cf. Jessen 1981: 114–116), as well as several that are less conventional. Finding a balance – between having criteria that structure the analysis without overly structuring it – proved difficult. For the limited purpose of reviewing the global models, however, they proved useful in dealing with complexity.

The models

We examined the reports of six global modelling efforts and one related environmentalist tract in an attempt to extract the social visions, if any, contained therein. The models are the two Club of Rome efforts, *Limits to Growth* (Meadows et al. 1972) and the update, *Mankind at the Turning Point* (Mesarovic and Pestel 1974); Kahn and colleagues' Hudson Institute rebuttal from the right, *The Next 200 Years* (Kahn, Brown, and Martel 1976); the Latin American rebuttal from the left, the Latin American World Model or LAWM (Bruckmann 1976; Herrera et al. 1976), also known as the Bariloche model (Gallopin 2001); Leontief's economic model for the UN (Leontief, Carter, and Petri 1977), and the Carter administration's compilation of US government models in the *Global 2000* report (USCEQ 1980). The environmentalist report is *Blueprint for Survival* by Edward Goldsmith and the other editors of the *Ecologist* (1972).

None of these models is explicitly social; they examine neither social structure nor the lived experience of people. All focus primarily on the economic and environmental implications of global development (Leontief, Carter and Petri 1977), on the compatibility of economics and natural resources, and on demography as an input governing economics. The variables used are primarily those for which numerical data were relatively readily available. They all use basically the same variables – population growth, production, income, materials/energy consumption, and sometimes pollution. Each model or report was created for a specific and limited purpose. Appendix 1 provides a brief summary of the purpose and conclusions of each model.

In no case was the primary purpose of a model the elaboration of a social vision. The Latin American and UN (Leontief, Carter, and Petri 1977) reports make no claim to be anything other than economic models; the input–output structure of the Leontief model treats society as a "black box." Other model reports specifically state that they do not address social issues, claiming, for example, to be explicitly "concerned with biophysical matters, as opposed to social, political, and economic developments" (USCEQ 1980, 2: 275), and noting that "no formal model of social conditions" is offered (Meadows et al. 1972: 174).

Even *Blueprint for Survival*, which claims to offer a vision for a radically different and stable society, offers few specifics beyond a call for decentralization; its main emphasis is on resource conservation and a steady-state economy. Its section on ecological disruption is primarily devoted to denouncing technology, while the discussion of social disruption is limited to northern social pathologies such as crime, drug abuse, and suicide. Similarly, the Second Club of Rome model includes "group" and "individual" strata in each regional model but the model results are concerned with the usual issues – population, food, energy, and income. [Nor does this change in the more optimistic second-generation treatment; see Meadows, Meadows, and Randers (1992)].

A preliminary examination forces acknowledgement of the non-social nature of these models. The reports are dominated by a concern with material and readily quantifiable issues. Although the models are, in some sense, views of the future, their anchors in current data, extrapolations therefrom, and moral or political judgement about the result leave the reports with the distinct flavour of the present. In so far as they have social concerns, these seem to centre around an exploration of limits or material and economic breaking points, rather than an exploration of the nature and qualities of human institutions, values, or experiences. They address a set of questions that differ from those posed in chapter 1 of this volume. They ask what limits of stress this society can endure and what will make this society collapse; they do not ask about intrinsic cultural and individual qualities, about human experiences, or about that which is valued. As such, they provide little or no guidance to the central directions by which people can create dignity, liberation, or fulfilment or how these may be possible in different nature–society settings.

Evaluating the models

As mentioned earlier, we formulated a set of criteria for social visions to characterize the essential features of a social vision and to provide a basis for comparing and evaluating the models. We agreed that the overall social vision created by an author or research group could be characterized in terms of scope, bias, scale, tone, and vantage point, among other qualities. Within the vision, major spheres of concern – human/nature relationships, material concerns, social organization, and value systems – could then be analysed. The specific variables we selected include the following: view of nature, view of human nature, view of time, valid modes of understanding, view of society and social goals, agent of change, population, equity, relative rights and responsibilities of individual and society, political organization and authority, institutions, economic system and exchange, technology, lifestyle, spatial linkages, and security. These categories may be loosely grouped as philosophical assumptions, social organization, and material arrangements (see table 16.1).

Table 16.1 **Attributes of social visions**

Philosophical assumptions
- view of nature
- view of human nature
- view of time
- valid modes of understanding

Social organization
- view of society and social goals
- agent of change
- population
- equity
- relative rights and responsibilities of individuals and society
- political organization and authority
- institutions
- lifestyle
- security

Material arrangements
- economic system and exchange
- technology
- spatial linkages

The social content of the models

The results of analysing the models using this framework are summarized in table 16.2. Several points should be noted. For many of the categories, "not addressed" would be the most appropriate notation. If a diligent search of the text revealed even some small mention of the topic, however, it was considered an indicator of the authors' attitude and incorporated into the matrix. As a result, we have in some cases characterized a model on the basis of a brief passage. Mesarovic and Pestel, for example, devote a brief space to their view of nature, waxing eloquent about human survival depending on living in harmony with the web of life. We have repeated these phrases in our characterization, for they express the explicit views of the authors; however, the overall tenor of their report and the model itself treats nature solely as a collection of resources. In other cases, where we found no specific mention in the text but an obvious characterization could be extrapolated from the overall tenor of the text, we have done so. None of the six models, for example, overtly expresses a conception of time, yet all implicitly treat it as linear.

Organizing the attributes into the aforementioned loose categories – philosophical assumptions, social organization, and material arrangements (table 16.1) – permits some general observations. In the sphere of material arrangements, the models are fairly forthcoming, offering moderately detailed visions of economic and technological conditions and at least a general sense of spatial linkages. Philosophical assumptions are, for the most part, not explicitly addressed at all, but implied stances may in many cases be gleaned from the overal tenor of the text, with "view of human nature" being the most difficult to discern. The treatment of social organization is mixed. Equity, and especially population, are central issues in virtually all of the reports and are addressed in some detail. None of the reports, however, offers a comprehensive or coherent picture of social goals, processes, and relations: the attributes listed in table 16.2 are a collection of brief and disparate mentions culled from the texts. The same scatter-shot approach involved in the models applies to politics and institutions, with the difference that these are at least recognized by the modellers as important topics – not that this recognition results in any comprehensive or systematic treatment. Finally, lifestyle and security issues generally receive brief mentions, but are not elaborated.

Further generalizations may be gleaned from a more detailed look at each of the attributes. All of the models treat nature primarily as a bundle of natural resources. None of the reports includes an overt discussion of conceptions of time or notions of legitimate modes of understanding, but implicit in all of them is a linear view of time and a reliance on rational, analytical scientific thought and quantitative data.

The models are fairly evenly split on whether individuals or society as a whole, values, or economic structure drive social change, but none of them explores this issue beyond the level of assertion. The reports also split on the issue of whether population is a problem. All models except those by Kahn, Brown, and Martel (1976) and Goldsmith and colleagues (1972) state explicitly that equity is a major social goal, and the latter implies it. The notion of equity adopted is itself limited, focusing on the admittedly pressing issue of intergenerational equity and particularly apparent distributional inequalities among nation-states. On the other hand, the complexity of equity deliberations such as the valued goods to be distributed, the relevant populations used to structure analysis, and the philosophical principles by which to define social justice receive scant attention. Those studies that propose means of achieving equity advocate economic growth (Kahn, Brown, and Martel 1976; Bruckmann 1976; Leontief, Carter, and Petri 1977; Mesarovic and Pestel 1974) or redistribution (Goldsmith et al. 1972, implied by Bruckmann 1976; Herrera et al. 1976; Leontief, Carter, and Petri 1977). Most of the models advocate economic growth, although Meadows et al. (1972) and Goldsmith et al. (1972) oppose it, and Mesarovic and Pestel (1974) call for a different kind. With the exception of Kahn, Brown, and Martel (1976) and Bruckmann (1976), all agree that technology is not the solution to environmental and economic problems, with Leontief taking a middle position. It is interesting to note that, except on the issue of equity, the positions taken by Kahn, Brown, and Martel (1976) and the advocates of the LAWM model are the most alike: they agree that population growth is not a problem and that resources are not limiting; they advocate economic growth, believe in technological solutions, and argue that environmental concerns are secondary to economic ones.

As mentioned above, for those attributes (e.g. philosophical assumptions, social organization) central to a deeper understanding of society, the models take a haphazard approach, implicitly assigning peripheral importance to them. On the question of "human nature," for example, all but Kahn, Brown, and Martel (1976) – who emphasize struggle and self-interest – imply a belief in the basic goodness of people. Little attention is given to the nature and goals of society or to the structure of the relationship between the individual and political society. Disparate details related to political organization and calls for international cooperation appear in the various reports, but issues of power, authority, conflict, and security are generally not addressed at all, or are addressed only very briefly. The same is true for institutions – only Kahn, Brown, and Martel (1976) really treat them at all, and their analysis is sketchy. The models make more mention of lifestyle, with many general calls for less materialism and more creative leisure activities, but this is for the most part left

Table 16.2 **Comparison of social attributes of first-generation global models**

	Meadows et al. (1972)	Mesarovic and Pestel (1974)	Kahn, Brown, and Martel (1976)	Goldsmith et al. (1972)	LAWM[a] Herrera et al. (1976)	Leontief, Carter, and Petri (1977)	USCEQ[b] (1980)	Clark and Munn (1986)	Brown (1981)
Nature	Resources; waste sink; scarcity	Harmony; web of life; human survival; scarcity	Warehouse; conquer; monitor; manage	System; complexity; stable; self-regulatory; predictable	Resources	Economic factor; re-sources; waste sink; not limiting	Life support; resources	Garden; Gaia; resources	Harmony; resources
Human nature	Culture; creativity; incentive = improve-ment; basic good	Basic goods; can change; do right thing	Self-interest; incentive = struggle; satisfac-tion; test/stimulate; structure/purpose; engineered	Part of nature; adaptation; natural law; community; belonging	Not addressed; solidarity; progress	Not addressed; implied – *Homo econ-omus*	Not addressed; implied – part of nature	Goodness; utilitarian	Rational; do right thing
Time	Mechanical; interval; open future; 70 years	Linear w/ cycles; nar-rowing op-tions; 50 years	Linear: sequential developing progress; slowing change	Not clear; critical point; industrialism as historical aberration	Linear; history open-ended and depends on human will; process as original	Linear	Linear; future as extrapo-lation of present; narrowing options	Linear; holistic	Linear
Valid modes of under-standing	Rational science; analyse, then man-age; science; engaged in morals	Holism; science; objective + subjective; science en-gaged; par-ticipation	Rational; en-gineering; objective; science; morality	Culturally determined; may be many; science serves ecosphere	Analytical; rational; economic models; ideology	Analytical; rational; economic models	Analytical models	Science; inter-disciplinary; surprises; linear argu-ment for non-linearity	Rational

482

Society and goals	Homogeneous system to be managed; open future; participation; can set limits; leisure; freedom from struggle	Sustainable material and spiritual; cooperation; organism; can adapt	Stages: high morale; dynamism; consensus; smooth function	Stability not expansion; analogue of nature; system; hierarchical; functional	Goals = equity; produce for need not profit; social (not state) control	Development; wealth; employment; NIEO; justice; rationality	Not addressed (explicitly); highly evolved system	Not addressed	Not addressed
Agent of change	Moral resources; values; gradual	Individual or society? Values; internal crisis; "public figures"; "political leaders"	Positive images; individual; influence not choice	Economic decentralization; conscious restructuring; education in new values	Not addressed. Locus = property and production – structural changes	Not addressed; economic restructuring as implied driver	Information; planning; analysis plus political will	Managers; also technological, institutions, research; having vision	Individuals; having vision
Population	Out of control; major threat by 2050	Serious problem; 35 years to fix	Rate high but declining; not a problem; economic development will slow growth; colonize space; 7–30 billion	Major threat; wealth will not necessarily slow; ZPG[4]; 3.5 billion maximum; halt immigration	Need to slow; economic development; focus on basic needs; urbanization	Not a big problem; wealth slows growth; economy can keep up	Major problem; reducing carrying capacity; 10 billion under "intensive management"	A problem but not too serious; still time 8–11 billion	Must stabilize at 8 billion
Equity	Social goals: justice = basic needs	Narrow income gap; we're all connected	Gap to persist; inequity normal, healthy, moral	Not addressed; implied goal	Main goal	Main goal	Increased	Not addressed	Need for redistribution, not growth

Table 16.2 (cont.)

	Meadows et al. (1972)	Mesarovic and Pestel (1974)	Kahn, Brown, and Martel (1976)	Goldsmith et al. (1972)	LAWM[a] Herrera et al. (1976)	Leontief, Carter, and Petri (1977)	USCEQ[b] (1980)	Clark and Munn (1986)	Brown (1981)
Rights and responsibilities	Basic needs; produce food; moral resources; behavioural restraint; trade-off of freedoms	Not addressed; share global wealth and resources; change values	All share in market levers of control; individual struggle for meaning; individuals maintain morale	Maintain social economic stability; restrain consumption, reproduction; provide goal structure – prestige	Basic needs; priority = needs; universal rights; work collectively	Basic needs; benefits of development	Not addressed	Not addressed	Individuals responsible to shape society; emphasis on voluntarism
Political organization and authority	Not addressed; nation-state; define values then choose; participation	Nation-state; global co-operation; reduced polarization	Nation-state; no world government; reduced competition; manipulative power; élite rule; public lacks expertise; democracy, if "republican virtues" re-established	Decentralized; local, small scale; participation; peer pressure; only unstable society needs strong authority	International socialism; nation-state; participation	Nation-state; internally restructured LDCs[c]; North surrender power to South	Not addressed; nation-state	More of the same	Not addressed

Institutions	Institutions to constrain growth; humans will develop unknown institutions	Not addressed; institutions put information before public for action; institutions for global crisis prevention	Specialized; technocratic; impersonal; decline of family and community; innovation institutionalized; control of serendipity/synergy; privatization; family, economic exchange superseded	Not addressed; "Movement for Survival" Coalition	Nation-state; cosmopolitan world organization	Not addressed; institutions to close North–South gap	Institutions for long term; global environmental analysis	Better use of existing ones; formal and informal	Not addressed; emphasis on planning
Economic system and exchange	No-growth; services; human needs; planning implied; resource efficiency; balance of consumption/population; surplus for arts and leisure	Organic model; co-operation; planning implied; restrained growth in North	Growth for South to catch up; North model as universal; technology, institutional, morale as key; eventual slowing of demand	Steady state; stock not flow re-source efficiency; minimal disruption; local control; small scale; correspondence of economic/real costs/values	Scarcity = technopolity; human needs; socialist redistribution; social determination of needs; planning; eventual slowing of growth; growth not answer	Develop under-used resources; NIEO[c] to close North–South gap; self-reliance; pollution abatement	Decelerating growth	More of the same	Steady state; intra-generational welfare; durability; high tech, high design; resource efficiency; conservation; labour-intensive; generalization

485

Table 16.2 (cont.)

	Meadows et al. (1972)	Mesarovic and Pestel (1974)	Kahn, Brown, and Martel (1976)	Goldsmith et al. (1972)	LAWM[a] Herrera et al. (1976)	Leontief, Carter, and Petri (1977)	USCEQ[b] (1980)	Clark and Munn (1986)	Brown (1981)
Technology	Not cause or solution; useful technofaith as diversion; shape by social need/priority	Not cause or solution; useful; restrain by social concerns	Good to control nature; main human resource; Faustian bargain	Incompatible with stability; anti-ecological – simplifies complexity; hubris; increases vulnerability; should serve decentralization	Avert scarcity; prevent pollution; social control; progress essential	Narrow North–South gap: expand resource exploitation; pollution abatement	Solution available but politically constrained	Brings environmental changes, good and bad	Decentralized; soft-path renewables
Lifestyle	Art and leisure: leisure time essential for higher pursuits	Standard of living, not just material; less materialism; conservation ethic; social, moral, organizational, scientific growth	Useless work; hedonism; gaming, art, education, ritual, social interaction; secular with romantic, mystic counter-reaction	Personalized; intimate; self-reliance; pleasure of community; non-materialistic; reintegrated work/home	Not based on consumerism; rich to reduce consumption	Lower Northern consumption; higher Southern consumption	Not addressed	Not addressed	Non-materialistic; simple, frugal; personal/social development; integrate home/work; telecommuting and bicycles; rich/poor convergence

Spatial linkages	Not discussed	Diverse regions; interregional cooperation; diversity key to survival, and to moral strength	Homogeneity – "they" become like "us"; modernism; links impersonal, business-like; urban/suburban globe; huge economic scale	Decentralized but linked; small scale; diverse rural/urban mix; material self-sufficiency	Autarchy; North–South inter-connected; economic complementarity	15 political/economic regions; regional interaction; international trade	Food, energy imports; use distant resources; unaware of environmental impacts	Region-to-globe links; global noosphere	Reversed global interdependencies; local self-reliance; population rural and dispersed; decentralized
Security	Not addressed	Disarmament; without reform, conflict will bring nuclear holocaust; confrontation = dysfunctional	Increased violence due to boredom; weakened Northern defences due to false sense of safety	Internal disintegration; chaos of industrial society; chaos leads to dictators, war; deviant behaviour absent; stable society	Not addressed	Not addressed	Tension; vulnerable people; production system; vulnerability to climate change	Not addressed	No war; conflict and fear = environmental not military; UN peace-keeping

a. LAWM, Latin American World Model.
b. USCEQ, US Council on Environmental Quality.
c. NIEO, New International Economic Order.
d. ZPG, Zero population growth.
e. LDCs, Less-developed countries.

to individual decision-making and not linked to social and economic structure or to broad social policies. Finally, spatial linkages, if treated at all, are discussed in general terms, such as degree of centralization and homogeneity, or in terms of North–South economic relations, despite the growing evidence of the importance of regional linkages to the global economy and social movements (see chapter 8).

Model visions

From these observations (and the details in table 16.2), it is clear that – even using relatively conservative and conventional categories and giving the modellers the benefit of the doubt by extracting minor details and extrapolating from the tone of the text – the models fail badly to confront social or philosophical issues, much less to probe them in any depth. The modelling approach is based on quantitative analysis using largely linear extrapolations from the present. Its materialistic (as opposed to relational or experiential) bias results in an avowedly utilitarian approach and a largely managerial orientation. The resulting emphasis is on just a few variables – population, resources, pollution levels, technology, and economy – with some speculations about their social meaning, in terms of wars, crises, international "tensions," "overshoot," or "collapse."

Even within the mode of extrapolation from the present, little or no attention is given to current social trends, with the exception of growing global divisions between rich and poor. The major non-demographic social factors for which data exist – such as health, education, and settlement pattern – receive minimal attention. Structural considerations related to politics, social norms, the organization of political economy, and institutions are only peripherally and sporadically treated; issues such as human rights, cultural diversity, psychological fulfilment, personal relations, and freedom are not addressed at all, and neither are environmental or psychological issues, such as the impacts of an increasingly artificial and homogeneous environment (Mumford 1970: 37 ff) and the social choice of reliability over resilience (Blaikie et al. 1994; Timmerman 1981).

The first-generation models performed the invaluable services of initiating discussion about possible economic and environmental futures; of taking a global instead of a national perspective; of treating the planet as a single, integrated whole; and of taking a long-term view. As our review demonstrates, however, the approach of these models was inherently limited by the specific rationality and methodology employed. Global problems were approached as essentially economic and resource-based, resulting in limited parameters of concern focused largely on material and energy flows. Population, and society in general, were seen as relevant only in so far as they generated (or were affected by) these flows.

Models: The second generation

Efforts to look at the global environment/development picture have become more sophisticated in the 1980s and 1990s. The authors analysed several studies from researchers associated with the International Institute for Applied Systems Analysis (IIASA) and the Worldwatch Institute, applying the same list of attributes used to analyse the models. Table 16.2, above, includes the social visions contained in pieces written in the 1980s by Lester Brown (Brown 1981; Brown, Flavin, and Postel 1990) and by Clark and Munn (1986) on the prospects for a sustainable global future.

The pieces examined demonstrate a more complex view than the models but share a similar focus. Like the modellers, Brown and his Worldwatch colleagues concentrate on population, resources, and material/energy flows. Explicit social concerns are limited to an emphasis on lifestyle (presumably in industrialized countries) and the argument that individual behaviour can reduce demand on global resources. Broader social, organizational, and structural issues are not addressed. The Clark and Munn piece differs from the first-generation models in its non-linear approach; greater concern with institutions; and the treatment of uncertainty, surprise, and discontinuities in the generation of knowledge and its uses. The major concerns, however, remain population and resources, and no distinct social vision is generated.

Although sharing a similar focus with the early models, these more recent efforts attempt to improve on earlier studies by acknowledging the role of uncertainty and surprise in projections and by experimenting with other techniques, such as backcasting and constructing future histories (see also Robinson's discussion in chapter 15). The importance of surprise was emphasized by the failed projections of some of the early models, and has received significant recent attention (Glantz et al. 1998; Kates and Clark 1996; Schneider, Turner, and Morehouse-Garriga 1998; Toth, Hizsnyik, and Clark 1989). Global systems theorists have come to recognize that, for systems near thresholds of change, deterministic analysis no longer works as the system becomes unpredictable and stochastic elements predominate, and inherent indeterminism sets in (see, for example, Gallopin and Raskin 1998). Research concerned with systems undergoing change must assume the prevalence of discontinuous changes and surprise and be more concerned with uncertainty than predictability. Perrings (1987: 11) distinguishes between "the probabilistic uncertainty that assumes away our inability to foresee the effects of our actions" and "uncertainty before ignorance, novelty, and surprise ... [arising] from the system's existence in real, historical, irreversible time." The time in question is, in Georgescu-Roegen's terms, not *time* (the mechanical measurement of an interval) but rather *Time* (the continous succession of moments) (Perrings 1987: 111). Whereas statistical prediction may have some relevance in *time*, it has none in *Time*. The probabilistic approach assumes the future to be a "stationary stochastic process" in which a pre-image of the range of possible outcomes exists and various possible futures become rival hypotheses to be tested (Perrings 1987: 113). But true uncertainty assumes an incomplete set of images of the future and denies that the future is knowable from the past.

The Swedish work on "surprising futures" (Svedin and Aniansson 1987) and the IIASA project to develop "surprise-rich scenarios" (Toth 1995) are both attempts to address these concerns and to face head-on the major shortcomings of projections. Schneider and Turner (1995) hosted an Aspen seminar on surprises in global environmental change in 1994, aimed at identifying a rich panoply of types of surprise, an endeavour to which Kates and Clark (1996) have also contributed. Despite these helpful advances, the primary concern is still with population, economic growth, technology, energy use, and agricultural production. Although much more detailed and sophisticated than earlier modelling efforts, and vigorously developed during the 1990s, the current second generation of models and their scenarios do not include social, cultural, and institutional developments and they lack the requisite variety needed to capture possible worlds reflecting the dynamic interplay and different social and cultural groups (Thompson and Rayner 1998). The models and scenarios still rely heavily on conventional socio-economic indicators, such as population, energy production, SO_2 and CO_2 emissions, labour force, production capacity, industrial emissions, agricultural production, forest production, water demand – plus a long list of environmental indicators, although more social attributes are being worked into the impacts. Future histories, as in *Beyond Hunger in Africa* (Achebe et al. 1990), are richer and explore a wider range of concerns, including cultural ones. Meanwhile, the *process* for developing integrated assessments also remains disputed; none of the second generation of models is being developed in close conjunction with the decision makers and others who will use them. As a result, they remain politically simplistic and naïve and lack the flexibility needed to represent the interests of various interests and social groups over space and time (Rotmans and Dowlatabadi 1998: 341).

Several promising developments in the current global modelling efforts deserve mention. The 2050 Project undertaken by the World Resources Institute during the 1990s intentionally sought to broaden the social visions involved in alternative future worlds. So, in 1993, project members solicited nominations from a range of colleagues and the World Council of Indigenous Peoples and other NGO leaders to prepare essays envisioning positive futures for their regions or locales. Some 52 "envisionaries" from 34 countries eventually submitted essays in five languages, with developing countries substantially over-represented. The results provide a highly suggestive tapestry of the potential richness of global visions (Nagpal and Foltz 1995). Hammond (1998) also subsequently examined three scenarios, or world views, of the future which he entitled the *market world* – a world of rapid economic growth and technological innovation; a *fortress world*, in which future market failures create a future in which enclaves of wealth and prosperity coexist with widening misery, desperation, and conflict; and a *transformed world*, where greater sharing of power and fundamental social change transforms governments and institutions. In a separate effort, Costanza (1999) identifies four "future histories" – "Star Trek," a vision of technological optimism, free competition, and unlimited resources; "Mad Max," the technological scepticist nightmare come true, when technology and consumption go bad; "Big Government," in which protective government policies override the free market; and "Ecotopia," the low-consumption, sustainability vision.

The Polestar Project of the Stockholm Environment Institute, referred to above, has also sought a broader framework that includes the conventional IPAT drivers of change (see chapter 8) but also a broader array of social variables (international equity, national equity, welfare, conflict, poverty, and political instability), designed to capture the "interlocking crises" of concern in the Brundtland report (WCED 1987). Polestar identifies and analyses three archetypal scenarios of the future – *conventional worlds*, a class of scenarios that assumes that current global trends play out without major discontinuity or surprise in the evolution of institutions, environmental systems, and human values; *barbarization*, scenarios in which fundamental social change occurs but brings great human misery and collapse of civilized norms; and *great transitions*, in which fundamental social transformation occurs but a new and higher stage of human civilization is achieved (Gallopin et al. 1997). Events with the capability to redirect beliefs, behaviours, and institutions away from some visions of the future and toward others are posited, as are *sideswipes*, major surprises (e.g. world wars, miracle technologies, pandemics) that greatly alter trends toward particular outcomes (Gallopin and Raskin 1998; Raskin et al. 1998).

Although these and other efforts of the 1980s and 1990s have come a long way in improving on the earlier models, mostly they are still bound by their focus on a limited set of traditional (if important) variables. It seems fair to conclude that the models have unduly set the terms of social and political debate, defining a sustainable global future largely in terms of population, resource and economic relationships. In other words, the debate and the ongoing work in integrated assessment, with several notable exceptions, remain unduly limited in social content and visions. In this regard, the explicit attention by Robinson in chapter 15 to identify social principles of sustainability is a step in the right direction, as are the use of differing cultural perspectives by the RIVM group in the Netherlands to enrich the underlying assumptions of integrated assessment (Asselt and Rotmans 1996) and the efforts of the US Interagency Working Group on Sustainable Development Indicators (1998). In envisioning alternative futures, in short, the net needs to be cast to include a larger array of ideologies, religions, and cultures but also historians, anthropologists, and humanists; of course, some of these will generate dark views of the future (Kaplan 1994).

The ideology of futurism

Although global models have largely set the stage and defined the terms for current debates on global futures, the formal "science" of futurism represents the primary institutionalized means for peering into the future. Most futurists have concentrated on the specific interests of their clientele rather than on broad social alternatives or long-term global futures. This professionalized and institutionalized version of the futurological endeavour has attracted substantial criticism of the narrowness of vision.

One critic equates futurism with the institutionalization of prophecy (i.e. forecasting) in an "attempt by self-appointed experts to rationalize the future" and to impose their own narrow, mechanistic world view. The prophetic endeavour – and its tendency to reflect the needs of corporate, military, and state institutions – trivializes and depersonalizes the future, ignores diversity as too difficult to capture and control, and produces restricted visions that attempt to monopolize the future in the hope that it will resemble the present status quo (Dublin 1991: 248). Critics early on noted a restriction of the range of alternative futures discussed; an ethnocentric preoccupation with a single society (an idealized, postindustrial North America); as well as tendencies towards technological determinism, mystification, and overquantification; a readiness to apply technical fixes to social problems; and an uncritical acceptance of the status quo (Miles 1978). The net effect is to empower further the already powerful, to perpetuate an interpretation of the world that serves particular interests, and to frame problems so that only certain options are considered (Miles 1978).

The critics demand a close scrutiny of futurist ideologies – an explicit identification and evaluation of "what is being sought, by whom, and for what purposes" (Hoos 1983: 61). Although some futurists have called for extension of the "futuristic mission farther into the social domain and toward a much larger constituency of stakeholders" (Adelson 1989: 31), such a "populist futurism" may also be problematic. Although more participation might result in people having increased control over their own destiny and might produce futures more responsive to wider interests, the means to accomplish this "may be actually employed as manipulative tools for legitimating the status quo through pseudoparticipation," exchanging fatalism for people's participation in the management of their own exploitation (Miles 1978: 81).

Forecasting tends to minimize the intrinsic uncertainty of the future by attempting to reduce it to probability. It typically founders on the shoals of institutions and value systems. Although some futurists (e.g. Adelson 1989; Simmonds 1989) argue that current organizations no longer seem to work, most futurists work for existing institutions. Adelson (1989: 35) argues that futurists, instead of designing policies for existing organizations, should concentrate on a transition to new institutions, and that efforts to look to the future should be less bureaucratic and institutional and more "artistic, entrepreneurial, expressive, pragmatic, case-by-case, constructive ... visionary." Adelson (1989: 33) further advocates thinking from the design disciplines, not the sciences – a sentiment echoed by utopian theorists Manuel and Manuel (1979: 813) in their observation that architecture, not science, provides the most imaginative and authentically utopian creations of the age. Another futurist points out that, in the chaotic context of changing value systems, forecasting is not very useful: "All we possess are remnants of previously demolished value systems ... inconsistent collections of irreconcilable bits and pieces from the past" (Michael 1989: 82). Norgaard (1988: 613) describes this as the gradual demise of *Progress* – "a great carpet under which old beliefs and new contradictions were swept for centuries" – and its likely replacement with the meta-belief, *Sustainability* – the "clarion of a new age." Faced with multiple, confusing and uncertain signals, individuals gravitate toward interpretations compatible with their personal world-view – a process that forecasts do not capture.

Some futurists have themselves claimed that the prophetic endeavour, or forecasting, is inherently untenable when applied to society. Forecasting is meant to enhance coherence, but this is impossible in a modern society that is internally fragmented and increasingly incoherent – epistemologically, socially, and psychologically (Michael 1989: 79). For other futurists, a prescriptive or normative futurism offers an alternative to technocratic extrapolations of the status quo. Coates and Jarratt (1989), for example, advocate a prescriptive, ideological, and ethical focus, calling for more emphasis on outcomes, and less on driving forces. Others dismiss normative futurology as a *"rapture of the future"* which, in the case of social prescription, "founders on the axiomatic 'we'" (Kern 1987: 216).

Alternative futures

Underlying these criticisms of formal, mechanistic futurism and its traditional constituency is an association of the prophetic, predictive endeavour with goals of manipulation and control (e.g. Dublin 1991). Such observations should serve not as reasons to disengage from all exploration of future possibilities but rather as warnings. We distinguish between a controlling futurism and the process of creating social visions. We need to eschew blueprints by which some groups control others; rather, we should create means for thinking through the consequences of current actions and for stimulating a sense of open possibilities, of wonder, and of power to act.

Although a desire for control serves as a powerful motive for peering into the future, it is not the only alternative: a sense of wonder or of responsibility may also serve as motives. A sense of wonder permits exploration of alternate possibilities while retaining a strong sense of humility toward the future – and the conviction that it will inevitably be more amazing than we can imagine. A sense of responsibility may guide attempts to add the consideration of long-range impacts to present actions, and it may inspire present activities to be conducted in a manner that leads to morally defensible future directions. Envisioning the future and emancipatory risk assessment should be conducted in such a way as to enrich empowerment and a sense of ability to create one's own life.

Alternative motivations are compatible with a diversity of social visions. Multiple visions encourage continuing clarification and communication, and contestation and critique of alternate futures; they also provide a forum for examining the possible ramifications of present proposals, decisions, and actions. For multiple visions of alternative futures to flourish, it is necessary to bring diverse perspectives to the table. People operating from different basic assumptions will tend to articulate a multiplicity of social visions; those with alternative views of the present will tend to envision alternative futures. Only by such a multiplicity can there be real choice, or even meaningful debate.

Envisioning alternative futures

We have asserted the desirability of an active and broad-based exploration of the widest possible range of imaginable futures. In the context of global environment and development concerns, the envisioning of alternative futures – and exploration of their implications for peoples' lived experience – is an essential component of the hope for a sustainable world. Current processes of envisioning, the content of visions, and the political frame for both imaging and images are incapable of creating a rich range of possible futures. Despite attempts at "backcasting," as reviewed by Robinson in chapter 15 and at greater length in Robinson and Slocombe (1996) and Mulder and Biesiot (1998), the modern approach to the future is symbolized by a technocratic and economic style and a narrow focus to global models and formal futurism. Social institutions, processes of social change, and human values and aspirations are still rarely at centre stage (but see Robinson et al. 1996a).

The reliance on a single approach to examining possible global futures – modelling – has, in our view, promoted a narrowness and sterility in the imaging process. Complementary or conflicting images – derived from anthropology, novels, or paintings, for example – have not yet become relevant to debates on global futures, despite the calls of O'Riordan and Timmerman in chapter 14. Despite the goal of some modellers to initiate public debate and stimulate action (Rotmans and Dowlatabadi 1998), current approaches have too often depicted securing the planetary future as a management challenge properly addressed by experts.

An urgent need exists to enrich this debate by expanding both the content and the form of discussions about global futures. The process of envisioning and debating the future of the planet needs to be democratized and diversified, just as we need the emancipatory risk analysis called for in chapter 1. There is a need to expand the debate into everyday life and to put everyday concerns on the agenda, but also for the discussion itself to become a part of everyday life rather than the exclusive domain of specialists.

In the discussion to follow, we take the modest step of arguing for a broader content in social visions, an expansion that necessitates much wider participation. We argue for the legitimacy of multiple forms of communication, and use the example of science fiction to illustrate the relevance of one non-academic, popular-culture form of vision creation.

Broadening content and form

Enriching the content of future visions and broadening participation may be addressed on several levels. Although the ideal is a wide-ranging democratic discussion – taking place in the world's cities and villages, fields and market-places, bars and watering-holes, schools and workplaces, meetings and media outlets – even the confines of policy circles and academia provide ample scope for expansion.

Content: Social science

Although a few scholars in the social sciences have come forward to participate in discussions about environment and development or human dimensions of global change, many of the pertinent findings of the social sciences other than economics (e.g. sociology, anthropology, political science, human geography, psychology) are still not incorporated. It is not clear how existing research on cultural identity, social change, political cultures, nature–society relations, international politics, or individual development can enrich our understanding of possible futures – a lack which is itself an indictment.

The gap arising from the paucity of attention to prominent social science theory is compounded by the lack of attention to, or visibility of, less-known innovative voices outside the social science mainstream. Those whose views might expand the conventional range of debate, or who question hallowed assumptions, are usually not heard. People doing creative work on a number of theoretical fronts, who may not have published on the subject of planetary futures but who deal with relevant topics, could greatly expand notions of what is possible. For example, critical theorists and oppositional political economists have not chosen, for the most part, to apply their efforts to the economy, institutional structures, politics, or social arrangements of alternative futures, nor have they always been effective when choosing to do so. An encouraging counter-example is the flourishing of a vigorous school of ecological economists, who have brought a new, valuable perspective to economic questions. In general, however, theorists outside the confines of conventional disciplines – such as bio-regionalists and others concerned with regional and cultural diversity, contemplators of the meaning of place, and technology theorists – conduct a largely unacknowledged parallel debate on environment and development.

Beyond the realm of theory, reservoirs of experience have been generated in praxis across a wide spectrum ranging from applied social sciences (such as development, design, architecture, planning, appropriate technology) to social and political activism. A truly enriched debate, one that includes "passionate" as well as "dispassionate" social science, would draw from (for example) proponents of bottom-up development, decentralization, local self-reliance; from advocates for the disadvantaged and people confronting poverty; from activists for grass-roots environmental action, for community empowerment, for indigenous rights – and any number of other sites of activism. The engagement of activists and the participation of those often left out of the discussion (and especially those vulnerable groups who may disproportionately bear global environmental risks) will broaden the set of questions to be asked. Visions generated from the bottom are sure to challenge the view from the top.

Content: Humanities and popular culture

Incorporating social science will bring neglected issues into focus, but the expansion of visions should not stop there. A broader conception should also include the humanities and even popular culture. Expanding the content of social visions to include a wide range of community and individual structures, relations, experiences, and consciousnesses should draw on the richness found in literary, artistic, philosophical, and spiritual expression. And just as visions drawing on the social sciences may gain from mainstream knowledge as well as the more innovative fringes and from the world of experience or praxis, both formal and informal expressions may contribute.

Can the critiques of post-modernist philosophers and literary deconstructionists be directed toward the positive task of envisioning alternative social futures? Can work by ethicists, philosophers, and religious leaders on questions of individual and societal values and aspirations, absent from much of the global-futures discussion, make a contribution? As with social science theory and praxis, there is value in the work of practitioners as well as that of critics in philosophy and the arts; in artists' works and experiences as well as the theories of critics; in the experience of mystics as well as the concepts of theologians. Artists of all kinds can offer a diversity of communicative forms as well as creative new content.

Not only may insights into individual and social meanings be derived from academic work in philosophy, history, literature, and the arts; from the work of writers and artists, performers and composers; but it may also come from popular culture, traditional knowledge, folk wisdom, myths, and prophecies. Work in rural development, in women's studies, and elsewhere has shown the value of folk and women's knowledge (e.g. Richards 1985; Rocheleau, Thomas-Slayter, and Wangari 1996; Shiva 1994). In the West, popular culture has been recognized as an important source for the expression of contemporary social values (e.g. Browne 1984).

Broadening the pool of communicative forms

In Western discussions about the planetary future, discourse has generally occurred in a single communicative form – scientific written texts. To the extent that an enriched and expanded envisioning process is desired, multiple forms of expression – including the qualitative, philosophical, and artistic – must be encouraged. If the goal of creating holistic as well as reductionist, empathetic as well as objectified visions, is a valid one, then personal as well as impersonal styles of communication and visual, oral, or kinetic forms of expression are also legitimate vehicles for conveying these visions. In fact, different forms of expression will be inherent in some of the efforts, suggested above, to broaden the content of visions by including art, popular culture, or traditional knowledge. Future visions abound, for example, in American popular and traditional cultures, such as science-fiction novels, short stories, comics, and films; Hopi prophecies; post-apocalyptic rock videos; and television shows.

Furthermore, as O'Riordan and Timmerman (chapter 14) show in their discussion of the paintings of Swedish energy futures, multiple forms of expression may also be used to gainful (and democratic) effect in the social sciences. Local non-governmental organizations (NGOs) and villagers in rural India have been using maps, models, and time lines (made from local materials) to theorize the present and to imagine possible local futures (Mascarenhas et al. 1991). The Peace 2010 contest of the *Christian Science Monitor* received 1300 essays, written from the perspective of the year 2010, explaining "how peace came to the world" (Foell and Nenneman 1986). Policy exercises like those reported by Achebe et al. (1990) use the literary device of alternative histories to construct alternate futures. The essays from diverse "envisionaries" throughout the world as part of Project 2050 suggest the range of planetary futures that remain to be discovered and articulated (Nagpal and Foltz 1995). Commentators from the South (e.g. Banuri and Marglin 1993) see the roots of global environmental destruction in the power and violence inherent in the Western tradition.

Science, imagination, and story

Within science, the notion of "objective" rationality, untainted by society, personality, or values, tends to be a caricature. Indeed, many scientists have long noted the artificiality of divisions between science and imagination. Popularizers of physics have, for example, pointed out the dissolution of orthodox dualities (matter versus energy, observer versus observed) in the "new" physics – a physics that explicitly acknowledges that a complete understanding of reality lies beyond the capabilities of rational thought. Statements by Einstein and others on the role played in science by serendipity, inspiration, creativity, and a child-like sense of wonder speak to the same point.

Although many scientists embrace qualities such as intuition and imagination, the mechanistic paradigm also survives, zealously protected by orthodox proponents with a tenacity likened by Lynn Margulis (1991: 213) to that of medieval monastic scholars. They reflect the relegation of intuition and imagination to the status of "other" in much of Western society.

Imagination is an "absolutely essential human faculty ... If you truly eradicated it in a child, he would grow up to be an eggplant" (LeGuin 1979: 41–42). This is because of "the mixture of realism and fantasy that lies at the psychic core of all humans" and the key role played by fantasy "in the enlivenment and transformation of culture" (Tuan 1990: 444). In fact, a free but disciplined imagination may "be the essential method or technique of both art and science" (LeGuin 1979: 41).

Storytelling is perhaps the most common imaginative expression. According to novelist Ursula LeGuin (1979: 31), stories are integral to human society:

... [A] person who had never listened to nor read a tale or myth or parable or story, would remain ignorant of his own emotional and spiritual heights and depths, would not know quite fully what is it to be human. For the story ... is one of the basic tools invented by the mind of man, for the purpose of gaining understanding. There have been great societies that did not use the wheel, but there have been no societies that did not tell stories.

The writer's exploration of her own imagination taps into Jungian archetypes – the collective unconscious of deep, shared, experience that makes possible aesthetic, intuitive, and emotional (as well as rational) communication (LeGuin 1979: 78). This type of communication – accessing multiple dimensions of experience – is precisely what we argue is missing from discussions of the planetary future. Story provides one avenue (among many) for achieving it.

Stories of the future: Science fiction

The accessibility and multidimensionality of the story medium serve both to promote a more democratic public discussion and to compensate for the failure of imagination reflected in the sterility of much existing debate about the future. Applied to the future, story adds participant observation to the available methodological repertoire.

A fiction of the future, argues literary scholar Robert Scholes (1975), responds to Sartre's call for literature to be a force for improving the human situation. Such a fiction combines entertainment and intellectual value – wedding idea and story, satisfying both cognitive (content) and speculative (narrative/escapist) needs, and providing suspense with intellectual consequences (Scholes 1975: 41). Writers of such fiction produce imaginative models of the future, alternative projections that can give us some sense of the consequences of present actions and do so "with a power which no other form of discourse can hope to equal" (Scholes 1975: 17, 74). Scholes asserts that what we need in all areas of life is more sensitive and vigorous feedback. A futuristic imagination "will inform mankind of the consequences of actions not yet taken. But it must not merely inform, it must make us feel the consequences of those actions, feel them in our hearts and our viscera. [This] imagination must help us to live in the future so that we can indeed continue to live in the future" (Scholes 1975: 16).

In practice, science fiction is the primary literary genre seeking to accomplish these goals (although Scholes and other critics would deny that all science fiction succeeds). In this visionary role, science fiction has served for decades as a non-academic form of communication in envisioning a wide array of alternate futures. Although there is much to criticize in the genre – including occasional lapses into forecasting – its continued vitality hints at the richness of future visions already existing in society. It provides an excellent example of an existing cultural resource that should be welcomed into policy debates on alternative futures.

Science fiction addresses themes relevant to debates on global futures, such as technology, environment, and sociopolitical alternatives. Within the world of popular culture, science fiction has involved large numbers of people in a decades-long public dialogue (between authors and readers/ fans) about the shape of possible future societies. This public conversation has utilized the power of the story form – a kind of power that the technical narrative of science lacks. By embracing fantasy, these stories achieve an imaginative range lacking in policy debates: they coax a suspension of disbelief, resulting in a creative expansion of the realm of the possible, and they demand emotional involvement. They are, therefore, able to access the multiple dimensions of personal and social experience and to provide an empathetic and holistic assessment of the possible future being envisioned.

Science fiction is, of course, only one of many potential contributors to an enriched global futures debate, within the realm of story, in a Western popular-culture context.

Conclusions

We have emphasized the social content of future visions and the processes of their creation, calling for the ongoing creation of multiple visions by a broad and diverse array of participants. We have argued in favour of multiple ways of knowing and of diverse expressive forms. We have not, however, addressed the politics of vision creation – how to ensure a diversity of views (diverse in both content and form) and how to achieve wide participation – or the politics of turning vision into action. These are, obviously, enormously problematic. To acknowledge the constraints on the creation of multiple, diverse visions (let alone their implementation) is, nevertheless, not to deny the validity of the visioning enterprise.

Currently, a newly globalized society is undergoing rapid political, economic, and cultural shifts. Far from diminishing the validity of, or need for, social visions, chaotic times and conflicting values make them all the more necessary. By stimulating continuous broad-based discourse on alternative futures, the process of creating social visions offers choice and the chance to democratize thinking. The limited visions of alternative futures currently on the table need to be broadened in content to include any and all aspects of human life that various groups of people view as important. Employing multiple forms of expression or modes of discourse will bring not only richer visions but also wider participation and the recognition that envisioning is a dynamic process. In this endeavour, no vision is ever truly completed. There is no possibility of "One Ultimate Vision." This is in line with the emancipated and democratic risk analysis which this volume seeks: we believe that it is worthwhile to ponder goals and directions. Visions of the future offer a means to think through the consequences of current actions: there are stimuli for creativity and the sensing of possibilities, and tools for broadening the range and sharpening the terms of debate on possible global futures.

Appendix: A summary of
first-generation global models

Blueprint for Survival (Goldsmith et al. 1972) was written by the editors of the *Ecologist*; it was designed to complement the modelling work of the Club of Rome. The report criticizes modern society and offers a vision of a radically different and stable society. Industrial society, with its ethos of expansion, is inherently destructive of the environment and society (p. 3) and will eventually self-destruct. In a world of finite resources, unlimited population growth and economic expansion cannot continue (p. 6). "Our task is to create a society that is sustainable and will give the fullest possible satisfaction to its members" (p. 21). Such a society will be based on stability, not expansion. It will be decentralized and governed by four principles: (1) minimum disruption of ecological processes; (2) maximum conservation and an economy of stock rather than flow; (3) zero population growth; (4) a social system in which the individual can enjoy, rather than feel restricted by, the first three conditions.

The *Limits to Growth* (Meadows et al. 1972) model was based on five key factors – population, food production, natural resources, industrial production, and pollution – believed to determine, and ultimately to limit, growth. The model demonstrated that exponential growth of population and capital is incompatible with a world of finite physical resources. If 1972 trends continued, the Earth would reach its limits within the next 100 years (p. 23). Because of the lag time inherent in societal responses, population and industrial production would overshoot those limits, resulting in a sudden and uncontrollable collapse. However, economic and environmental sustainability, meeting the material needs of all and allowing each person an equal opportunity to realize his or her individual potential, are possible; the sooner humanity commits itself to attaining global equilibrium, the greater the chance for success. Policy recommendations (p. 163), centred on limiting population growth and conversion to a steady-state economy, were based on running iterations of the model until overshoot was prevented.

Mankind at the Turning Point (Mesarovic and Pestel 1974) is a refinement of *Limits to Growth*. Instead of looking at a single homogeneous world, it divides the planet into ten regions and concludes that the problem is undifferentiated growth, not growth *per se*, and that organic growth is acceptable. To achieve organic growth, the income gap between North and South, and the gap between "man and nature" needs to be closed. Poor countries should expand their economies, whereas rich ones should cut consumption. Restructuring is required to achieve cooperation and economic coordination among nations and regions and to bring about drastic changes in human values, such as developing a global consciousness, creating a new ethic of conservation, replacing the "conquest of nature" mentality with one of harmony, and developing an identification with future generations. The report concludes that these crises are permanent, and that solutions are possible only at the global level but will not be achieved by traditional means. Cooperation, not confrontation, can provide solutions.

In *The Next 200 Years* (Kahn, Brown, and Martel 1976), Herbert Kahn and his colleagues present issues such as population, economic development, energy, materials, food, and environment as "crucial," but a focus on the long-term, historical perspective reveals these crises to be short- and medium-term "transitional phenomena" (p. vii). Whereas humans were few, poor, and at the mercy of nature in 1776, humans in 2176 will be many, rich, and in control of nature. This may be achieved by either an Earth-based or an outer-space-based scenario. The glass is half full, not half empty: most of our current problems are the result of past and present successes, not failures. Immense progress has been made in the last 200 years to improve living standards and extend affluence to more people. The whole world will become more and more like the United States, which will be "postindustrial." Without growth, the poor are doomed to perpetual poverty. With the exception of long-term environmental constraints, the real concerns are not technological or physical limits; rather, they are psychological, cultural, and social limits, and the real danger is poor decision-making. Energy, food and pollution problems are tractable. More serious, requiring "concern and prudence" (p. 180), are long-term environmental issues such as radiation, climate change, ozone depletion, and toxins. We need to create institutions to serve as an early-warning system for these problems, yet it is also important to look to space in case there is need of a "lifeboat for spaceship earth" (p. 224).

The Latin American World Model (Herrera et al. 1976) tests whether it is materially possible for the world to "be liberated from misery and underdevelopment" (p. 7) and concludes that barriers to achieving this ideal are not physical, but sociopolitical. Modelling resources, energy and pollution – for developed and developing countries – demonstrates that an egalitarian global society is physically viable. The model also estimates the time-scale and conditions required to achieve adequate satisfaction of basic human needs such as nutrition, housing, education, capital goods, and other services. Viability was defined as the ability of all countries to move from the present situation to basic-needs fulfilment within a reasonable time period (p. 8). The report draws the following conclusions: (1) there are not physical limits to growth; (2) population will decrease when basic needs are satisfied and development occurs; and (3) radical social and international organizational changes are required to achieve the ideal society. Specific recommendations include a reduction in nonessential consumption, increased investment, rationalization of urban and agricultural land use, egalitarian distribution of goods and services, and the elimination of trade deficits (see also USCEQ 1980 (2): 637–655).

Leontief's study of the world economy (Leontief, Carter, and Petri 1977) was commissioned by the United Nations to test whether global resource availability and pollution levels/abatement costs would permit attainment of UN development targets making "the world, in its economic and social aspects, a more just and rational home for mankind" (p. 1), or whether that goal needed revision in the face of environmental constraints. The study was later broadened to include other economic aspects of development. There are no physical limits to achieving the UN goal of reducing the North–South income gap from 12:1 in 1970 to 7:1 by 2000 and 1:1 by 2050. However, the model of 45 economic sectors in 15 regions demonstrates that the United Nations' projected growth rates will not close the gap. To do so would require higher growth in the South, lower growth in the North, major social change in the South, and the establishment of the New International Economic Order (NIEO). Problems of resource availability and pollution abatement can be overcome with technology and political will (p. 10). Present limits to sustained economic growth and income redistribution are political, social, and institutional (pp. 10–11).

The *Global 2000 Report to the President* (USCEQ 1980) was intended by the Carter administration to serve as the basis for long-term US government planning. The report combines multiple US government agency models to project "probable changes in the world's population, natural resources, and environment to the end of the century" (p. 453). This study reports on trends, not goals. The famous opening paragraph states: "If present trends continue, the world in 2000 will be more crowded, more polluted, less stable ecologically, and more vulnerable to disruption than the world we live in now. Serious stresses involving population, resources, and environment are clearly visible ahead. Despite greater material output, the world's people will be poorer in many ways than they are today. For hundreds of millions of the desperately poor, the outlook for food and other necessities of life will be no better. For many it will be worse" (p. 1).

References

AAAS (American Academy of Arts and Sciences). 1965–1967. *Working papers of the Commission on the Year 2000 of American Academy of Arts and Sciences.* 5 vols in 6. Boston: Commission on the Year 2000, AAAS.

Achebe, Chinua, Göran Hyden, Christopher Magadza, and Achola Pala Okeyo, eds. 1990. *Beyond hunger in Africa: Conventional wisdom and an African vision.* Nairobi: Heinemann Kenya.

Adams, Robert M., Cynthia Rosenzweig, Robert M. Peart, Joe T. Ritchie, Bruce A. McCarl, J. David Glyer, R. Bruce Curry, James W. Jones, Kenneth J. Boote, and L. Hartwell Allen, Jr. 1990. "Global climate change and US agriculture." *Nature* 345: 219–224.

Adelson, M. 1989. "Reflections on the past and future of the future." *Technological Forecasting and Social Change* 36: 27–37.

Agarwal, Anil. 1995. "Dismantling the divide between indigenous and scientific knowledge." *Development and Change* 26, no. 3: 413–439.

———, and Sunita Narain. 1991. *Global warming in an unequal world.* New Delhi: Centre for Science and Development.

———, and ———. 1999. "Kyoto Protocol in an unequal world: The imperative of equity in climate negotiations." In *Towards equity and sustainability in the Kyoto Protocol*, ed. Karin Hultcrantz. Stockholm: Stockholm Environment Institute, pp. 17–30.

Aguayo, Sergio. 1987. *Chiapas: Las amenazas a la seguridad nacional.* Mexico: Centro Latinoamericano de Estudios Estrategeticos.

Aguilar, Adrián Guillermo, Exequiel Ezcurra, Teresa García, Marisa Mazari Hiriart, and Irene Pisanty. 1995. "The Basin of Mexico." In *Regions at risk: Comparisons of threatened environments*, ed. Jeanne X. Kasperson, Roger

E. Kasperson, and B. L. Turner II. Tokyo: United Nations University Press, pp. 304–366.

Ahmad, Nafis. 1976. *A new economic geography of Bangladesh*. New Delhi: Vikas Publishing House.

Akong'a, J., and Thomas E. Downing. 1988. "Smallholder vulnerability and response to drought." In *The impact of climate variations on world agriculture, Vol. 2, Assessments in semi-arid regions*, ed. Martin L. Parry, Timothy R. Carter, and Nicolaas T. Konijn. Dordrecht, Netherlands: Kluwer Academic Publishers, pp. 221–247.

Alam, M. 1996. "Subsidence of the Ganges–Brahmaputra Delta of Bangladesh and associated drainage." In *Sea-level rise and coastal subsidence: Causes, consequences, and strategies*, ed. John D. Milliman and Bilal U. Haq. Dordrecht, Netherlands: Kluwer, pp. 169–192.

Alcamo, J., G. J. van den Born, A. F. Bouwman, B. J. den Haan, K. Klein Goldewijk, O. Klepper, J. Krabec, R. Leemans, J. G. J. Olivier, A. M. C. Toet, H. J. M. de Vries, and H. J. van der Woerd. 1994. "Modeling the global society–biosphere–climate system: Computed scenarios." In *Integrative assessment of mitigation, impacts, and adaptation to climate change: Proceedings of a Workshop held on 13–15 October 1993 at IIASA, Laxenburg, Austria*, ed. Nebojsa Nakićenović, William D. Nordhaus, Richard Richels, and Ferenc L. Toth. CP 94-9. Laxenburg, Austria: International Institute for Applied Systems Analysis (IIASA), pp. 241–288.

Alexandratos, Nikos. 1995. *World agriculture: Towards 2000*, an FAO study. Rome: Food and Agriculture Organization of the United Nations; Chichester, UK: Wiley.

Ali, Mehtabunisa. 1984. "Women in famine." In *Famine as a geographical phenomenon*, ed. Bruce Currey and Graeme Hugo. Dordrecht, Netherlands: D. Reidel, pp. 113–134.

Alkon, P. 1987. "Origins of futuristic fiction: Felix Bodin's *Novel of the future*." In *Storm warnings: Science fiction confronts the future*, ed. George E. Slusser, Colin Greenland, and Eric S. Rabkin. Carbondale, IL: Southern Illinois University Press, pp. 21–33.

Allan, Nigel J. R. 1986. "Accessibility and altitudinal zonation models of mountains." *Mountain Research and Development* 6, no. 3: 185–194.

———. 1995. "Human aspects of mountain environmental change, 1889–1992." In *Mountains at risk: Current issues in environmental studies*, ed. Nigel J. R. Allan. New Delhi: Manohar, pp. 3–26.

Alperovitz, Gar, Ted Howard, Adria Scharf, and Thad Williamson. 1995. *Index of environmental trends: An assessment of twenty-one key environmental indicators in nine industrialized countries over the past two decades*. Washington: National Center for Economic Alternatives.

Anderberg, S. 1989. "Surprise-rich scenarios for global population, energy and agriculture 1975–2075." In *Scenarios of socioeconomic development for studies of global environmental change: A critical review*, ed. Ferenc L. Toth, Eva Hizsnyik, and William C. Clark. Research Report RR 89-4. Vienna: International Institute for Applied Systems Analysis (IIASA), pp. 230–279.

Appendini, Kirsten, and Diana Liverman. 1996. "Agricultural policy and climate change in Mexico." In *Climate change and food security*, ed. Thomas E. Downing. Berlin: Springer-Verlag, pp. 525–547.

Argent, Julie, and Timothy O'Riordan. 1995. "The North Sea." In *Regions at risk: Comparisons of threatened environments*, ed. Jeanne X. Kasperson, Roger E. Kasperson, and B. L. Turner, II. Tokyo: United Nations University Press, pp. 367–419.

Ascher, William. 1978. *Forecasting: An appraisal for policy-makers and planners.* Baltimore, MD: Johns Hopkins University Press.

Asselt, Marjolein B. A. van, and Jan Rotmans. 1996. "Uncertainty in perspective." *Global Environmental Change* 6, no. 2 (June): 121–157.

Atkinson, Tim, and Frances McGuire. 1990. "Impacts of sea level rise on coastal hydrology." In *The effects of sea level rise on the UK coast*, ed. L. E. J. Roberts and R. C. Kay. Research Report No. 7. Norwich, UK: Environmental Risk Assessment Unit, University of East Anglia, pp. 47–75.

Ayres, Robert U., and Udo E. Simonis, eds. 1994. *Industrial metabolism: Restructuring for sustainable development.* Tokyo: United Nations University Press.

Bailar, John C. 1988. *Scientific inferences and environmental problems: The uses of statistical thinking.* Chapel Hill, NC: Institute for Environmental Studies, University of North Carolina.

Bailey, Norman A., and Richard Cohen. 1987. *The Mexican time bomb.* New York: Priority Press.

Baird, Alec, Phil O'Keefe, K. Westgate, and Ben Wisner. 1975. *Towards an explanation of disaster proneness.* Occasional Paper No. 11. Bradford, UK: Disaster Research Unit, University of Bradford.

Ball, M. C., and G. D. Farquhar. 1984a. "Photosynthetic and stomatal responses of two mangrove species, *Aegiras corniculatus* and *Avicennia marina*, to long-term salinity and humidity conditions." *Plant Physiology* 74: 1–6.

———, and ———. 1984b. "Photosynthetic and stomatal responses of the gray mangrove, *Avicennia marina*, to transient salinity conditions." *Plant Physiology* 74: 7–11.

Bandyopadhyay, Jayanta. 1989. *Natural resource management in the mountain environment: Experiences from the Doon Valley, India.* ICIMOD Occasional Paper No. 14. Kathmandu: ICIMOD (International Centre for Integrated Mountain Development).

Banskota, Mahesh. 1989. *Hill agriculture and the wider market economy: Transformation processes and experience of Bagmati zone in Nepal.* Kathmandu: ICIMOD (International Centre for Integrated Mountain Development).

———, and Narpat S. Jodha. 1992a. "Mountain agricultural development strategies: Comparative perspectives from the countries of the Hindu Kush–Himalayan region." In *Sustainable mountain agriculture, vol. 1: Perspective and issues*, ed. Narpat S. Jodha, Mahesh Banskota, and Tej Partap. New Delhi: Oxford and IBH Publishing Co, pp. 83–114.

———, and ———. 1992b. "Investment, subsidies, and resource transfer dynamics: Issues for sustainable mountain agriculture." In *Sustainable mountain agriculture, vol. 1: Perspective and Issues*, ed. Narpat S. Jodha, Mahesh Banskota, and Tej Partap. New Delhi: Oxford and IBH Publishing Co, pp. 185–203.

————, P. Sharma, S. Sharma, B. Bhatta, K. Banskota, and T. Tenzing. 1990. *Economic policies for sustainable development in Nepal*. Kathmandu: ICIMOD (International Centre for Integrated Mountain Development).

Banuri, Tariq, and Frédérique Apfell Marglin, eds. 1993. *Who will save the forests? Knowledge, power, and environmental destruction*. London: Zed.

Barbault, Robert, S. Sastrapradia, K. Hindar, Y. Michalakis, B. Schaal, W. J. Leverich, J. Hanski, J. Clobert, W. Reid, M. Loreau, H. Kawanabe, M. Higashi, E. Alvarez-Buylla, and F. Renaud. 1995. "Generation, maintenance and loss of biodiversity." In *Global biodiversity assessment*, ed. V. H. Heywood. Cambridge: Cambridge University Press for the United Nations Environment Programme (UNEP), pp. 196–274.

Barbier, E. B., G. Brown, S. Dalmazzone, C. Folke, M. Gadgil, N. Hanley, C. S. Holling, W. H. Lesser, K.-G. Mäler, T. Panayotou, C. Perrings, R. K. Turner, and M. Wells. 1995. "The economic value of biodiversity." In *Global biodiversity assessment*, ed. Cambridge: Cambridge University Press for the United Nations Environment Programme (UNEP), pp. 823–914.

Barbour, Alan G. 1996. *Lyme disease: The cause, the cure, the controversy*. Baltimore, MD: Johns Hopkins University Press.

Barkin, David, and Blanca Suarez. 1983. *El fin de principio: Las semillas y la seguridad alimentaria*. Mexico: Ediciones Oceano.

Barnett, T. P. 1984. "The estimation of 'global' sea level change: A problem of uniqueness." *Journal of Geophysical Research* 89, no. C5: 7980–7988.

Barney, Gerald O. 1999. *Threshold 2000: Critical issues and spiritual values for a global age*. Arlington, VA: Millennium Institute.

Barth, Michael C., and James G. Titus, eds. 1984. *Greenhouse effect and sea level rise: A challenge for this generation*. New York: Van Nostrand Reinhold Company.

Baudrillard, Jean. 1988. "Simulacra and simulations." In *Jean Baudrillard: Selected Writings*, ed. Mark Poster. Stanford, CA: Stanford University Press, pp. 166–184.

Baumann, H., and T. Rydberg. 1994. "Life-cycle assessment: A comparison of three methods for impact analysis and evaluation." *Journal of Cleaner Production* 2, no. 1: 13–20.

BCAS (Bangladesh Centre for Advanced Studies). 1991a. "Bangladesh: Threatened by sea level rise." *Clime Asia* 1, no. 2. CANSA (Climate Action Network-South Asia) Newsletter.

————. 1991b. "Environmental alterations deplete fisheries." *Bangladesh Environmental Newsletter* 2, no. 1.

Beanlands, Gordon E., and Peter N. Duinker. 1983. *An ecological framework for environmental impact assessment*. Quebec, Canada: Institute for Resource and Environmental Studies, Dalhousie University, in Cooperation with the Federal Environmental Assessment Review Office.

Beard, J. S. 1946. "Lox clímax de vegetación en la América tropical." *Revista Facultad Nacional de Agronomía, Medellío* 6: 225–293.

Beck, Ulrich. 1992. *The risk society: Towards a new modernity.* London: Sage.

Behm, H., and J. Vallin. 1980. "Mortality differentials among human groups." In *Biological and social aspects of mortality and the length of life,* ed. Samuel H. Preston. Liège, Belgium: Ordina Editions, pp. 11–37.

Bell, Daniel, ed. 1969. *Toward the year 2000: Work in progress.* Boston: Beacon Press.

———. 1973. *The coming of post-industrial society.* New York: Basic Books.

Benedick, Richard E. 1998. *Ozone diplomacy: New directions in safeguarding the planet.* Enl. Ed. Cambridge, MA: Harvard University Press.

Bentkover, Judith D., Vincent T. Covello, and Jeryl Mumpower, eds. 1986. *Benefits assessment: The state of the art.* Dordrecht, Netherlands: D. Reidel.

Bernstein, H. 1979. "African peasantries: A theoretical framework." *Journal of Peasant Studies* 6: 420–444.

Berry, Brian J. L. 1991. *Long-wave rhythms in economic development and political behavior.* Baltimore, MD: Johns Hopkins University Press.

Bhatia, B. M. 1963. *Famines in India: A study in some aspects of the economic history of India 1860–1965.* Bombay: Asia Publishing House.

Biggs, David. 1996. "Appendix B: The socio-economic resource framework." In *Life in 2030: Exploring a sustainable future for Canada,* by John B. Robinson, David Biggs, George Francis, Russel Legge, Sally Lerner, D. Scott Slocombe, and Caroline Van Bers. Vancouver: UBC Press, pp. 152–157.

Bird, Eric C. F. 1985. *Coastline changes.* Chichester, UK: Wiley Interscience.

———. 1986. "Potential effects of sea level rise on the coasts of Australia, Africa, and Asia." In *Effects of changes in stratospheric ozone and global climate, Vol. 4: Sea level rise,* ed. James G. Titus. Washington: US Environmental Protection Agency, pp. 83–98.

———, and Maurice L. Schwartz, eds. 1985. *The world's coastline.* New York: Van Nostrand Reinhold.

Blaikie, Piers M. 1985. *The political economy of soil erosion in developing countries.* London: Longman.

———, and Harold C. Brookfield, eds. 1987. *Land degradation and society.* London: Methuen.

———, and Daniel Coppard. 1998. "Environmental change and livelihood diversification in Nepal: Where is the problem?" *Himalayan Research Bulletin* 18, no. 2: 28–39.

———, Terry Cannon, Ian Davis, and Ben Wisner. 1994. *At risk: Natural hazards, people's vulnerability, and disasters.* London: Routledge.

Bohle, Hans Georg. 1993. "The geography of vulnerable food systems." In *Coping with vulnerability and criticality: Case studies on food-insecure people and places,* ed. Hans-Georg Bohle, Thomas E. Downing, John O. Field, and Fouad N. Ibrahim. Freiburger Studien zur Geographischen Entwicklungsforschung, vol. 1. Saarbrücken, Germany: Verlag Breitenbach, pp. 15–29.

———, Thomas E. Downing, and Michael J. Watts. 1994. "Climate change and social vulnerability: Toward a sociology and geography of food insecurity." *Global Environmental Change* 4, no. 1: 37–48.

Bolin, Bert, and Richard Warrick. 1986. *The greenhouse effect and ecosystems.* New York: John Wiley and Sons.

Boorman, L. A., J. Goss-Custard, and S. McGorty. 1988. *Ecological effects of climate change*. Furzebrook, Dorset: Institute of Terrestrial Ecology.

Boulding, Elise. 1983. "Shaping a viable future with our imaginations." *UNU Newsletter* 7, no. 3: 8.

Boulding, Kenneth E. 1973 (© 1966). "The economics of the coming spaceship earth." In *Toward a steady-state economy*, ed. Herman E. Daly. San Francisco: Freeman, pp. 121–132.

Bowden, Martyn J., Robert W. Kates, Paul A. Kay, William E. Riebsame, Richard A. Warrick, Douglas L. Johnson, Harvey A. Gould, and Daniel Weiner. 1981. "The effect of climatic fluctuations on human populations: Two hypotheses." In *Climate and History*, ed. T. M. L. Wigley, M. Ingram, and G. Farmer. Cambridge: Cambridge University Press, pp. 479–513.

Bowonder, B. 1981. "Environmental risk assessment issues in the Third World." *Technological Forecasting and Social Change* 19, no. 1: 99–127.

———, and Jeanne X. Kasperson. 1988. "Hazards in developing countries: Cause for global concern." *Risk Abstracts* 5, no. 4: 103.

Brammer, H. 1993. Geographical complexities of detailed impact assessment for the Ganges–Brahmaputra–Meghna Delta of Bangladesh. In *Climate and sea level change: Observations, projections, and implications*, ed. R. A. Warrick, E. M. Barrow, and T. M. L. Wigley. Cambridge: Cambridge University Press, pp. 246–262.

Bravo, Raul E., Luis Cañadas-Cruz, Washington Estrada, Tom Hodges, Gregory Knapp, Andres C. Ravelo, Ana M. Planchuelo-Ravelo, Oscar Rovere, Tarquino Salcedo Solis, and Trajano P. Yugcha. 1988. The effects of climatic variations on agriculture in the central Sierra of Ecuador. In *The impact of climatic variations on agriculture, Vol. 2: Assessments in semi-arid regions*, ed. Martin L. Parry, Timothy K. Carter, and Nicolaas T. Konijn. Dordrecht, Netherlands: Kluwer Academic Publishers, pp. 381–493.

Bread for the World Institute. 1999. *The changing politics of hunger: Hunger 1999*. Silver Springs, MD: Bread for the World Institute.

Brennan, L. 1984. "The development of the Indian Famine Code." In *Famine as a geographical phenomenon*, ed. Bruce Currey and Graeme Hugo. Dordrecht, Netherlands: D. Reidel, pp. 91–112.

Brenner, M. Harvey. 1980. "Industrialization and economic growth: Estimates of their effects on the health of populations." In *Assessing the contributions of the social sciences to health*, ed. M. Harvey Brenner, Anne Mooney, and Thomas J. Nagy, Boulder, CO: Westview Press, pp. 65–115.

———. 1984. *Estimating the effects of economic change on national health and social well-being: A study prepared for the use of the Subcommittee on Economic Goals and Intergovernmental Policy of the Joint Economic Committee Congress of the United States*, June 15. Senate Report 98–198. Washington: USGPO.

Broadus, James M., John Milliman, Steven Edwards, David Aubrey, and Frank Gable. 1986. "Rising sea level and damming of rivers: Possible effects in Egypt and Bangladesh." In *Effects of changes in stratospheric ozone and global climate, Vol. 4: Sea level rise*, ed. James G. Titus. Washington: U.S. Environmental Protection Agency, pp. 165–189.

Brookfield, Harold, Lesley Potter, and Yvonne Byron. 1995. *In place of the forest: Environmental and socioeconomic transformation in Borneo and the eastern Malay Peninsula.* Tokyo: United Nations University Press.

Brooks, Elizabeth, and Jacque L. Emel. 1995. "The Llano Estacado of the American Southern High Plains." In *Regions at risk: Comparisons of threatened environments*, ed. Jeanne X. Kasperson, Roger E. Kasperson, and B. L. Turner II. Tokyo: United Nations University Press, pp. 255–303.

———, and ———, with Brad Jokisch and Paul Robbins. 2000. *The Llano Estacado of the US Southern High Plains: Environmental transformation and the prospect for sustainability.* Tokyo: United Nations University Press.

Browder, John O., ed. 1989. *Fragile lands of Latin America: Strategies for sustainable development.* Boulder, CO: Westview Press.

Brown, James H., and Arthur C. Gibson. 1983. *Biogeography.* St Louis: C. V. Mosby.

Brown, Lester C. 1981. *Building a sustainable society.* New York: Norton and Co.

———, Christopher Flavin, and Sandra Postel. 1990. "Earth Day 2030." *Worldwatch* 3, no. 2: 12–21.

Brown, P. 1987. "Popular epidemiology: Community response to toxic waste-induced disease in Woburn, Massachusetts." *Science, Technology, and Human Values* 12: 78–85.

Browne, R. B. 1984. "Popular culture as the new humanities." *Journal of Popular Culture* 17, no. 4: 1–8.

Bruckmann, Gerhart, ed. 1976. *Latin American world model: Proceedings of the second IIASA Symposium on Global Modelling, Baden, Austria, October 7–10, 1974.* Laxenburg, Austria: International Institute for Applied Systems Analysis (IIASA).

Brush, S. B. 1988. "Traditional agricultural strategies in the hill lands of tropical America." In *Human impacts on mountains*, ed. Nigel J. R. Allan, Gregory W. Knapp, and Christoph Stadel. New Jersey: Rowman and Littlefield, pp. 116–126.

Bruun, Per. 1986. "Worldwide impact of sea level rise on shorelines." In *Effects of changes in stratospheric ozone and global climate, Vol. 4: Sea level rise,* ed. James G. Titus. Washington: US Environmental Protection Agency, pp. 99–128.

Bullard, Robert D. 1990. *Dumping in Dixie: Race, class, and environmental quality.* Boulder, CO: Westview Press.

Burgess, T. L., and A. Shmida. 1988. "Succulent growth forms in arid environments." In *Arid lands. Today and tomorrow,* Proceedings of an International Research and Development Conference, Tucson, Arizona, USA, October 20–25, 1985, ed. E. E. Whitehead, C. F. Hutchinson, B. N. Timmermann, and R. G. Varady. Boulder, CO: Westview Press, pp. 383–395.

Burton, Ian. 1988. "Dimensions of a new vulnerability: The significance of large-scale urbanization in developing countries." In *The earth's fragile systems: Perspectives on global change,* ed. Thorkil Kristensen and Johan Peter Paludan. Boulder, CO: Westview Press, pp. 30–50.

———, Robert W. Kates, and Gilbert F. White. 1978. *The environment as hazard.* New York: Oxford University Press.

——, ——, and ——. 1993. *The environment as hazard*, 2nd edn. New York: Guilford.

Byers, Douglas S., ed. 1967. *The prehistory of the Tehuacán Valley. vol. 1: Environment and subsistence*. Austin, TX. University of Texas Press for the Robert S. Peabody Foundation: Andover, MA.

Caballero, J. 1990. "El uso de la diversidad vegetal en México: tendencias y perspectivas." In *Medio ambiente y desarrollo en México, vol. 1*, ed. E. Leff. Mexico: M. A. Porrúa Editores, pp. 257–298.

Caldwell, John C. 1986. "Routes to low mortality in poor countries." *Population and Development Review* 12, no. 2: 171–220.

Caldwell, Lynton K. 1985. "Science will not save the biosphere but politics might." *Environmental Conservation* 12, no. 3: 195–197.

——. 1990. *Between two worlds: Science, the environmental movement, and policy choice*. Cambridge: Cambridge University Press.

Camilieri, J. C., and G. Ribi. 1986. "Leaching of dissolved organic carbon (DOC) from dead leaves, formation of flakes from DOC, and feeding on flakes by crustaceans in mangroves." *Marine Biology* 91: 337–344.

Camino, A., J. Recharte, and P. Bidegaray. 1985. "Flexibilidad calendárica en la agricultura tradicional de las vertientes orientales de los Andes." In *La tecnología en el mundo andino, Vol. 1*, ed. Heather Lechtman and Ana M. Soldi. Mexico, D. F.: Universidad Nacional Autónoma de México, pp. 169–194.

Canadian Climate Programme. 1990. *Climate change adaptation strategies workshop*. Downsview, Ontario: Canadian Climate Centre, Environment Canada.

Canadian Electrical Association. 1991. *Demand site management in Canada – 1991*. Department of Energy, Mines and Resources and the Canadian Electrical Association, Ottawa.

Cannon, Terry. 1994. "Vulnerability analysis and the explanation of 'natural' disasters." In *Disasters, development and environment*, ed. Ann Varley. Chichester, UK: John Wiley and Sons, pp. 13–30.

Carpenter, R. A., K. R. Smith, L. Habbeger, and C. P. B. Claudio. 1990. *Environmental risk assessment*. Manila: Asian Development Bank.

Carson, Rachel. 1962. *Silent spring*. Boston: Houghton Mifflin.

Castaneda, J. G. 1986. "Mexico's coming challenges." *Foreign Policy* 64: 120–139.

Chambers, Robert. 1984. "Participatory rural appraisal (PRA): Challenges, potentials, and paradigms." *World Development* 22, no. 10: 1437–1454.

——. 1987. *Sustainable rural livelihood: A strategy for people, environment and development*. IDS Discussion Paper 240. Sussex, UK: Institute of Development Studies.

——. 1989. "Editorial introduction: Vulnerability, coping and policy." *IDS Bulletin* 21: 1–7.

Changnon, Stanley A., ed. 1981. *METROMEX, a review and summary*. Boston, MA: American Meteorological Association.

Chen, Robert S., ed. 1990. *The hunger report, 1990*. Providence, RI: Alan Shawn Feinstein World Hunger Program, Brown University.

——, and Robert W. Kates. 1994. "World food security: Prospects and trends." *Food Policy* 19, no. 2: 192–208.

———, Elise Boulding, and Steven H. Schneider, eds. 1983. *Social science research and climate change: An interdisciplinary appraisal.* Dordrecht, Netherlands: Reidel.

CIMMYT (Centro Internacional de Mejoriamento de Maíz y Trigo). 1974. *The Puebla Project: Seven years of experience, 1966–1973.* El Batan, Mexico: CIMMYT.

Clark, William C. 1980. *Witches, floods, and wonder drugs: Historical perspectives on risk management.* R-22. Vancouver: Institute of Resource Ecology, University of British Columbia.

———. 1985. *On the practical implications of the carbon dioxide question.* WP-85-43. Laxenburg, Austria: International Institute for Applied Systems Analysis (IIASA).

———. 1986. "Sustainable development of the biosphere: Themes for a research program." In *Sustainable development of the biosphere,* ed. William C. Clark and Ralph E. Munn. Cambridge: Cambridge University Press, 5–48.

———. 1987. *Sustainable development of the biosphere: Interaction between the world economy and the global environment.* Wallace W. Atwood Lecture Series, No. 2. Worcester, MA: Clark University.

———. 1990. "Towards useful assessments of global environmental risks." In *Understanding global environmental change: The contributions of risk analysis and management,* ed. Roger E. Kasperson, Kirstin Dow, Dominic Golding, and Jeanne X. Kasperson. Worcester, MA: Clark University, The Earth Transformed Program, pp. 5–22.

———. 1991. "Energy and environment: Strategic perspectives on policy design." In *Energy and the environment in the 21st century,* ed. Jefferson W. Tester, David O. Wood, and Nancy A. Ferrari. Cambridge, MA: MIT Press, pp. 63–78.

———, and Ralph E. Munn, eds. 1986. *Sustainable development of the biosphere.* Cambridge: Cambridge University Press.

Claussen, Eileen, and Lisa McNeilly. 1998. *Equity and global climate change.* Washington: Pew Center on Global Climate Change.

Coates, Joseph F. 1989. "Forecasting and planning today plus or minus twenty years." *Technological Forecasting and Social Change* 36: 15–20.

———. 1997. *2025: Scenarios of global society reshaped by science and technology.* Greensboro, NC: Oakhill Press.

———, and J. Jarratt. 1989. *What futurists believe.* A World Future Society Book. Mt Airy, MD: Lomond.

Cody, M. L. 1984. "Branching patterns in columnar cacti." In *Being alive on land,* no. 13, ed. N. Margaris, M. Arianoutsou-Faraggitaki, and W. Oechel. The Hague, Netherlands: W. Junk, pp. 201–236.

Cohen, Stewart, David Demeritt, John Robinson, and Dale Rothman. 1998. "Climate change and sustainable development: Towards dialogue." *Global Environmental Change* 8, no. 4: 341–371.

Coll de Hurtado, Atlántida. 1982. *Es México un país agrícola?: Un análisis geográfico.* Mexico: Siglo Veintiuno.

Collins, J. J., R. T. Lundy, and D. Grahn. 1983. "A demographic model for performing site-specific health risk projections." *Health Physics* 45, no. 1: 9–20.

Committee on Shipborne Wastes (US National Research Council). 1995. *Clean ships, clean ports, clean oceans: Controlling garbage and plastic wastes at sea.* Washington: National Academy Press.

Commoner, Barry. 1971. *The closing circle: Nature, man and technology.* New York: Knopf.

Connell, J. H., and W. P. Sousa. 1983. "On the evidence needed to judge ecological stability or persistence." *American Naturalist* 121, no. 6: 789–824.

Conway, Gordon. 1985. "Agroecosystem analysis." *Agricultural Administration* 20: 31–55.

Cook, Brian, Jacque Emel, and Roger E. Kasperson. 1990. "Organizing and managing radioactive waste as an experiment." *Journal of Policy Analysis and Management* 9, no. 3 (Summer): 339–366.

Copans, Jean, ed. 1975. *Sécheresses et famines du Sahel.* 2 volumes. Paris: F. Maspero.

Costanza, Robert. 1999. "Four visions of the century ahead: Will it be Star Trek, Ecotopia, Big Government, or Mad Max? (technological optimism and skepticism)." *The Futurist* 33, no. 6: 23–30.

———, Ralph D'Arge, Rudolf de Groot, Stephen Farber, Monica Grasso, Bruce Hannon, Karin Limburg, Shahid Naeem, Robert V. O'Neill, Jose Paruelo, Robert G. Raskin, Paul Sutton, and Marjan van den Belt. 1997. "The value of the world's ecosystem services and natural capital." *Nature* 387, no. 6630 (15 May): 253–260.

Cothern, C. Richard, ed. 1996. *Handbook for environment risk decision making: Values, perceptions, and ethics.* Boca Raton, FL: Lewis.

Council for the Protection of Rural England. 1991. *Water on demand: The case for demand management in water charging policy.* London: CPRE.

Countryside Commission. 1987. *New directions for the countryside.* Cheltenham: Countryside Commission.

Covello, Vincent T. 1991. "Risk comparisons and risk communication: Issues and problems in comparing health and environmental risks." In *Communicating risks to the public: International perspectives,* ed. Roger E. Kasperson and Pieter-Jan M. Stallen. Dordrecht, Netherlands: Kluwer Academic Publishers, pp. 79–124.

———, and R. Scott Frey. 1990. "Technology-based environmental health risks in developing nations." *Technological Forecasting and Social Change* 37, no. 2 (April): 159–179.

———, and Jeryl Mumpower. 1986. "Risk analysis and risk management: A historical perspective." In *Risk evaluation and management,* ed. V. T. Covello, Joshua Menkes, and Jeryl Mumpower. New York: Plenum Press, pp. 519–540.

———, and Kazuhiko Kawamura, with Mark Borush, Saburo Ikeda, Paul F. Lynes, and Michael S. Minor. 1988. "Cooperation versus confrontation: A comparison of approaches to environmental risk management in Japan and the United States." *Risk Analysis* 8, no. 2: 247–260.

CRIEPI (Central Research Institute of the Electric Power Industry). 1989. *The review of acid rain in Japan* (in Japanese). *Institute Review 22.* Tokyo: CRIEPI.

———. 1991. *Technology for protecting the earth* (in Japanese). Tokyo: CRIEPI.

Cronon, William. 1991. *Nature's metropolis: Chicago and the Great West.* New York: W.W. Norton.

Cropper, M., and P. R. Portney. 1990. "Discounting and the evaluation of life-saving programs." *Journal of Risk and Uncertainty* 3, no. 4: 369–379.

Crosby, Alfred W, Jr. 1972. *The Columbian exchange: Biological and cultural consequences of 1492.* Westport, CT: Greenwood.

CSE (Centre for Science and Environment). 1985. *The state of India's environment 1984–1985: The second citizen's report.* New Delhi: CSE.

Currey, Bruce, and Graeme Hugo, eds. 1984. *Famine as a geographical phenomenon.* Dordrecht, Netherlands: D. Reidel.

Cutter, Susan L. 1995. "The forgotten casualties: Women, children, and environmental change." *Global Environmental Change: Human and Policy Dimensions* 5, no. 3: 181–194.

———. 1996. "Vulnerability to environmental hazards." *Progress in Human Geography* 20 no. 4: 529–539.

Dahlberg, Kenneth A. 1987. "Redefining development priorities: Genetic diversity and agro-eco development." *Conservation Biology* 1, no. 4: 311–322.

Dalby, Simon. 1992. "Ecopolitical discourse: Environmental security and political geography." *Progress in Human Geography* 16: 503–522.

Daly, Herman E. 1999. *Ecological economics and the ecology of economics: Essays in criticism.* Cheltenham, UK: Edward Elgar.

Dankelman, Irene, and Joan Davidson. 1988. *Women and environment in the Third World: Alliance for the future.* London: Earthscan.

Davies, J. Clarence, ed. 1996. *Comparing environmental risks: Tools for setting government priorities.* Washington: Resources for the Future.

Davies, Susanna. 1996. *Adaptable livelihoods: Coping with food security in the Malian Sahel.* Basingstoke, UK: Macmillan.

Denevan, William H. 1989. "The geography of fragile lands in Latin America." In *The fragile lands of Latin America: Strategies for sustainable development,* ed. John O. Browder. Boulder, CO: Westview Press, pp. 3–25.

Deoja, B., and B. Thapa. 1991. *Manual of mountain risk engineering.* Kathmandu: International Centre for Integrated Mountain Development (ICIMOD).

DESFIL (Development Strategies for Fragile Lands). 1988. *Development of fragile lands: theory and practice.* Washington: DESFIL.

Deudney, Daniel H., and Richard A. Matthew. 1999. *Contested grounds: Security and conflict in the new environmental politics.* Albany, NY: State University of New York Press.

DeWalt, Billie R. 1979. *Modernization in a Mexican ejido.* Cambridge: Cambridge University Press.

Dinkel, R. H. 1985. "The seeming paradox of increasing mortality in a highly industrialized nation: The example of the Soviet Union." *Population Studies* 39: 87–97.

Done, T., J. Ogden, and W. Wiebe. 1995. "Coral reefs." In *Global biodiversity assessment.* Cambridge: Cambridge University Press for the United Nations Environment Programme (UNEP), ch. 6.1.10, pp. 381–387.

Douglas, Mary. 1985. *Risk acceptability to the social sciences.* New York: Russell Sage Foundation.

Dovring, F. 1970. "Land reform and productivity in Mexico." *Land Economics* 46: 264–274.

Dow, Kirstin. 1992. "Exploring differences in our common future(s): The meaning of vulnerability to global environmental change." *Geoforum* 23: 417–436.

———, and Thomas E. Downing. 1995. "Vulnerability research: Where things stand." *Human Dimensions Quarterly* 1, no. 3 (Summer): 3–5.

Dowlatabadi, Hadi. 1995. "Integrated assessment models of climate: An incomplete overview." *Energy Policy* 23, no. 4: 1–8.

———, and M. Granger Morgan. 1993. "Integrated assessment of climate change." *Science* 259, no. 5103 (26 March): 1813, 1932.

Downing, Thomas E. 1991a. *Assessing socio-economic vulnerability to famine: A report to the US Agency for International Development (AID), Famine Early Warning System (FEWS) project.* Washington: AID/FEWS.

———. 1991b. "Vulnerability to hunger in Africa: A climate change perspective." *Global Environmental Change* 1, no. 5 (December): 365–381.

———. 1992. *Climate change and vulnerable places: Global food security and country studies in Zimbabwe, Kenya, Senegal, and Chile.* Research Report No. 1. Oxford: Environmental Change Unit, University of Oxford.

———, ed. 1996. *Climate change and world food security.* NATO ASI Series 1: Global Environmental Change, vol. 37. Berlin: Springer-Verlag.

———, and Karen Bakker. 2000. "Drought discourse and vulnerability." In *Drought: A global assessment*, ed. Donald A. Wilhite. London: Routledge, vol 2: pp. 213–230.

———, Kangethe W. Gitu, and Crispin M. Kamau, eds. 1989. *Coping with drought in Kenya: National and local strategies.* Boulder, CO: Lynne Rienner.

———, Michael J. Watts, and Hans G. Bohle. 1995. "Climate change and food insecurity: Toward a sociology and geography of vulnerability." In *Climate change and world food security*, ed. Thomas E. Downing. Berlin: Springer-Verlag, pp. 183–226.

———, Megan J. Gawith, Alexander A. Olsthoorn, Richard S. J. Tol and Pier Vellinga. 1999. "Introduction." In *Climate, change and risk*, ed. Thomas E. Downing, Alexander A. Olsthoorn and Richard S. J. Tol, 1–18. London: Routledge.

Downs, R. E., Donna O. Kerner, and Stephen P. Reyna, eds. 1991. *The political economy of African famine.* Food and Nutrition in History and Anthropology, vol. 9. Langhorne, PA: Gordon and Breach.

Drèze, Jean, and Amartya K. Sen. 1990a. *Hunger and public action.* Oxford: Clarendon Press.

———, and ———. 1990b. *The political economy of hunger.* Oxford: Clarendon Press.

Dublin, Max. 1991 (© 1989). *Futurehype: The tyranny of prophecy.* New York: Dutton.

Dunlap, Riley E., George H. Gallup, Jr., and Alec M. Gallup. 1993. "Health of the planet: Results of a 1992 international opinion survey of citizens in 24 countries." Princeton, NJ: The George H. Gallup International Institute.

Dunn, John. 1990. *Interpreting political responsibility: Essays 1981–1989.* Princeton, NJ: Princeton University Press.

Durning, Alan T. 1989. *Poverty and the environment: Reversing the downward spiral.* Worldwatch Paper No. 92. Washington: Worldwatch Institute.

Earle, Timothy C., and George T. Cvetkovich. 1995. *Social trust: Toward a cosmopolitan society.* Westport, CT: Praeger.

Easterling, William E. 1997. "Why regional studies are needed in the development of full-scale integrated assessment modeling of global change processes." *Global Environmental Change* 7, no. 4: 337–356.

Eckholm, E. P. 1975. "The deterioration of mountain environments." *Science* 139: 764–770.

———. 1976. *Losing ground.* New York: Worldwatch Institute.

Economist. 1991. "Bangladesh in troubled water." 11 May: 29–30.

Eddy, John A. 1991. "Global change: Where are we now? Where are we going?" *EarthQuest* 5, no. 1: 1–3.

Edgerton, Sylvia A., Kirk R. Smith, Richard A. Carpenter, Toufig A. Siddiqi, Steven G. Olive, Corazon Pe Benito Claudio, Vincent T. Covello, Donald J. Fingleton, Kwi-Gon Kim, and Bruce A. Wilcox. 1990. "Priority topics in the study of environmental risk in developing countries: Report of a workshop held at the East–West Center, August, 1988." *Risk Analysis* 10, no. 2 (June): 273–283.

Edwards, Ward, and Detlof von Winterfeldt. 1986. "Public disputes about risky technologies: Stakeholders and arenas." In *Risk evaluation and management,* ed. V. T. Covello, J. Menkes, and J. Mumpower. New York: Plenum Press, pp. 69–92.

Ehrenfeld, John R. 1997. "The importance of LCAs: Warts and all." *Journal of Industrial Ecology* 1, no. 2 (Spring): 41–49.

Ehrlich, Paul R. 1968. *The population bomb.* New York: Ballantine.

El Sayed, Mahmoud Kh. 1991. "Implications of relative sea level rise on Alexandria." In *Impact of sea level rise on cities and regions, Proceedings of the First International Meeting "Cities on Water," Venice, December 11–13, 1989,* ed. Roberto Frassetto. Venice: Marsilio Editoria, pp. 183–189,

———. 1996. "Rising sea-level and subsidence of the northern Nile Delta: A case study." In *Sea-level rise and coastal subsidence: Causes, consequences, and strategies,* ed. John D. Milliman, and Bilal U. Haq. Dordrecht, Netherlands: Kluwer, pp. 215–233.

Eleri, Ewah Otu. 1997. "Africa and climate change." In *International politics of climate change: Key issues and critical actors,* ed. Gunnar Fermann. Oslo: Scandinavian University Press, pp. 265–284.

Eleuterius, L. N., and C. K. Eleuterius. 1979. "Tide levels and salt marsh zonation." *Bulletin of Marine Sciences* 29: 394–400.

Enge, Kjell I., and Scott Whiteford. 1989. *The keepers of water and earth: Mexican rural social organization and irrigation.* Austin: University of Texas Press.

Engi, Dennis. 1995. *Historical and projected costs of natural disasters.* SAND 94-2686/UC-900. Albuquerque, NM and Livermore, CA: Sandia National Laboratories.

Englander, Tibor, Klara Farago, Paul Slovic, and Baruch Fischhoff. 1986. "A comparative analysis of risk perception in Hungary and the United States." *Social Behaviour* 1, no. 1: 55–66.

Enterprise for the Environment. 1998. *The environmental protection system in transition: Toward a more sustainable future.* Washington: CSIS Press.

Environment Canada. 1985. *Sustainable development.* A Submission to the Royal Commission on the Economic Union and Development Prospects for Canada. Ottawa: Environment Canada.

———. 1988. *Conference Statement – The Changing Atmosphere: Implications for Global Security.* Conference sponsored by the Government of Canada, Toronto, 27–30 June 1988. Toronto: Environment Canada.

Environmental Almanac. Annual. Boston: Houghton-Mifflin.

Environmental Data Report: United Nations Environment Programme. Biennial. Oxford: Basil Blackwell.

Escalante Pliego, Patricia, and Jorge Llorente. 1985. "Riqueza y endemismo de aves y mariposas como un criterio para determinar áreas de reserva. Datos del estado de Nayarit, México." In *Primer Simposio Internacional de Fauna Silvestre.* Mexico: The Wildlife Society of Mexico, pp. 335–363.

———, Adolf G. Navarro Sigüenza, and A. Townsend Peterson. 1993. "A geographic, ecological, and historical analysis of land bird diversity in Mexico." In *Biological diversity of Mexico: origins and distributions,* ed. T. P. Ramamoorthy, Robert Bye, Antonio Lot, and John Fa. Oxford: Oxford University Press, pp. 281–307.

ESSC (Earth System Science Committee), and NASA (National Aeronautics and Space Administration) Advisory Council. 1988. *Earth system science: A closer view.* Washington: NASA.

Ezcurra, E. 1990. *De las chinampas a la megalópolis: El medio ambiente en la cuenca de México.* Mexico: Fondo de Cultura Económica.

———, and Marisa Mazari Hiriart. 1996. "Are megacities viable?: A cautionary tale from Mexico City." *Environment* 38, no. 1 (Jan/Feb 1996): 6–15, 26–35.

———, R. S. Felger, A. D. Russell, and M. Equihua. 1988. "Freshwater islands in a desert sand sea: The hydrology, flora, and phytogeography of the Gran Desierto oases of Northwestern Mexico." *Desert Plants* 9: 35–63.

———, Marisa Mazari Hiriart, Irene Pisanty, and Adrián Guillermo Aguilar. 1999. *The Basin of Mexico: Critical environmental issues and sustainability.* Tokyo: United Nations University Press.

Ezrahi, Yaron. 1990. *The descent of Icarus: Science and the transformation of contemporary democracy.* Cambridge, MA: Harvard University Press.

Falkenmark, Malin. 1986. "Fresh water: Time for a modified approach." *Ambio* 15, no. 4: 192–200.

———. 1996. "Approaching the ultimate constraint: Water shortage in the Third World." In *Resources and population,* ed. Bernardo Colombo, Paul Demeny, and Max Perutz. Oxford: Clarendon Press, pp. 70–81.

———, and G. Lindh. 1993. "Water and economic development." In *Water in crisis: A guide to the world's fresh water resources,* ed. Peter Gleick. Oxford: Oxford University Press, pp. 80–91.

———, and Jan Lundqvist. 1997. *World freshwater problems: Call for a new realism.* Comprehensive Assessment of the Freshwater Resources of the World, Background Reports, 1. Stockholm: Stockholm Environment Institute.

FAO (Food and Agriculture Organization of the United Nations). 1984. *Land, food and people*. Rome: FAO.

————. 1997. *The state of food and agriculture 1997*. Rome: FAO.

————. 1998. *The state of food and agriculture 1998*. Rome: FAO.

Farhar, Barbara C. 1979. *The influence of the St Louis rain anomaly on human activities*. IBS Report. Boulder. CO: Institute of Behavioral Science, University of Colorado.

Farvar, M. Taghi, and John P. Milton, eds. 1972. *The careless technology: Ecology and international development*. Garden City, NY: The Natural History Press.

Felger, R. S., and G. P. Nabhan. 1976. "Tierras desérticas. Una aridez engañadora." *Ceres* (March–April).

Fermann, Gunnar. 1997. "The requirement of political legitimacy: Burden-sharing criteria and competing conceptions of responsibility." In *International politics of climate change: Key issues and critical actors*, ed. Gunnar Fermann. Oslo: Scandinavian University Press, pp. 179–192.

Feshbach, Murray. 1995. *Environmental and health atlas of Russia/Atlas okruzhaiuschaia sreda i zdorove Russii*. Moscow: Paims Publishing House.

FEWS (Famine Early Warning System). 1999. *FEWS current vulnerability guidelines*. Washington: US Agency for International Development.

Field, John A. 2000. "Drought, the famine process, and the phasing of interventions." In *Drought: A global assessment*, vol. 2, ed. Donald A. Wilhite. London: Routledge, pp. 271–284.

Finkel, Adam M., and Dominic Golding, eds. 1994. *Worst things first? The debate over risk-based national environmental priorities*. Washington: Resources for the Future.

Fischer, Frank. 1993. "The greening of risk assessment. Towards a participatory approach." In *Business and the environment: Implications of the new environmentalism*, ed. Denis Smith. New York: St. Martin's Press, pp. 98–115.

Flavin, Christopher. 1989. *Slowing global warming: A worldwide strategy*. Worldwatch Paper 91. Washington: Worldwatch Institute.

Florescano, Enrique. 1969. *Precios del maíz y crisis agrícolas en México 1708–1810: Ensayo sobre el movimionto de los precios y sus consucuencia económic y sociales*. Mexico: El Colegio de México.

————. 1980. "Una historia olvidada: La sequia en México." *Nexos* 32: 9–18.

Flores-Villela, Oscar. 1991. "Análisis de la distribución de la herpetofauna en México." PhD dissertation. Universidad Nacional Autónoma de México, Mexico.

————, and Patricia Gerez. 1988. *Conservacion en México: Sintesis sobre vertebrados terrestres, vegetacion y uso del suelo*. Xalapa, Ver.: Instituto Nacional de Investigaciones sobre Rescursos Bioticos.

Flynn, James, James Chalmers, Doug Easterling, Roger Kasperson, Howard Kunreuther, C. K. Mertz, Alvin K. Mushkatel, K. David Pijawka, and Paul Slovic. 1995. *One hundred centuries of solitude: Redirecting America's high-level nuclear waste policy*. Boulder, CO: Westview Press.

————, Roger E. Kasperson, Howard Kunreuther, and Paul Slovic. 1997. "Overcoming tunnel vision: Redirecting the U.S. high-level nuclear waste program." *Environment* 39, no. 5 (April): 6–11, 25–30.

Foell, Earl W., and Richard A. Nenneman. 1986. *How peace came to the world.* Cambridge, MA: MIT Press.

Forester, John, ed. 1987. *Critical theory and public life.* Cambridge, MA: MIT Press.

Francis, G. 1976. *Prospective on ecodevelopment, national development and international cooperation policy.* Report of a Workshop. CIDA and Environment Canada, Ottawa.

Franke, Richard W., and Barbara H. Chasin. 1980. *Seeds of famine: Ecological destruction and the development dilemma in the west African Sahel.* Montclair, NJ: Allanheld, Osmun.

Frederick, K. D. 1994. *Balancing water demands with supplies: The risk of management in a world of increasing scarcity.* Technical Paper No. 189, The World Bank. Washington: World Bank.

Freudenburg, William. 1988. "Perceived risk, real risk: Social science and the art of probabilistic risk assessment." *Science* 242: 44–49.

Friends of the Earth Canada. 1983–1984. *2025: Soft energy futures for Canada.* 12 volumes. Ottawa: Energy Mines and Resources Canada.

Fromm, Erich. 1955. *The sane society.* Greenwich, CT: Fawcett.

Fujita, S. I., Y. Ichikawa, R. K. Kawaratani, and Y. Tonooka. 1991. "Preliminary inventory of sulphur oxide emissions in East Asia." *Atmospheric Environment* Part A, 25 A, no. 7: 1409–1411.

Fuller, R. Buckminster. 1969. *Operating manual for spaceship earth.* Carbondale, IL: Southern Illinois University Press.

Funtowicz, Silvio O., and Jerome R. Ravetz. 1985. "Three types of risk assessment: A methodological analysis." In *Risk analysis in the private sector,* ed. Chris Whipple and Vincent T. Covello. New York: Plenum Press, pp. 217–232.

———, and ———. 1990a. *Global environmental issues and the emergence of second order science.* London: Council for Science and Society.

———, and ———. 1990b. *Uncertainty and quality in science for policy.* Dordrecht, Netherlands: Kluwer Academic Press.

———, and ———. 1991a. "A new scientific methodology for global environmental issues." In *Ecological economics,* ed. R. Costanza. New York: Columbia University Press, pp. 137–152.

———, and ———. 1991b. "Three types of risk assessment and the emergence of post-normal science." In *Social theories of risk,* ed. Dominic Golding and Sheldon Krimsky. New York: Greenwood Press, pp. 251–273.

Gadgil, S., A. K. S. Huda, N. S. Jodha, R. P. Singh, and S. M. Virmani. 1988. "The effects of climatic variations on agriculture in dry tropical regions of India." In *The impact of climatic variations on agriculture, Vol. 2: Assessments in semi-arid regions,* ed. Martin L. Parry, Timothy R. Carter, and Nicolaas T. Konijn. Dordrecht, Netherlands: Kluwer, pp. 495–578.

Galbraith, John K. 1958. *The affluent society.* Boston: Houghton Mifflin.

Gallopin, Gilberto. 1994. *Impoverishment and sustainable development: A systems approach.* Winnipeg: International Institute for Sustainable Development.

———. 2001. "The Latin American World Model (a.k.a. the Bariloche Model): Three decades ago." *Futures* 33, no. 1 (January): 77–89.

———, and Paul Raskin. 1998. "Windows on the future: Global scenarios and sustainability." *Environment* 40, no. 3 (April): 6–11, 26–31.

———, P. Gurman, and H. Maletta. 1989. "Global impoverishment, sustainable development and the environment: A conceptual approach." *International Social Science Journal* 121: 375–397.

———, Allen Hammond, Paul Raskin, and Rob Swart. 1997. *Branch points: Global scenarios and human choice.* Polestar Series Report No. 7. Stockholm: Stockholm Environment Institute.

Garcia, Rolando V. 1981. *Nature pleads not guilty, vol. 1: Drought and man.* Oxford: Pergamon Press.

Gates, David M. 1980. *Biophysical ecology.* New York: Springer-Verlag.

Gault, F., K. E. Hamilton, R. B. Hoffman, and B. C. McInnis. 1987. "The design approach to socio-economic forecasting." *Futures* 19, no. 1: 3–25.

Gentry, A. H. 1982. "Patterns of neotropical plant species diversity." *Evolutionary Biology* 15: 1–84.

German Advisory Council on Global Change. 1999. *World in transition: Ways toward sustainable management of fresh water resources.* Berlin: Springer-Verlag.

Ghan, S. J., W. T. Pennell, K. L. Petersen, E. Rykiel, M. J. Scott, and L. W. Vail, eds. 1993. *Regional impacts of global climate change: Assessing change and response at the scales that matter.* Columbus, Ohio: Battelle Press.

Gibbs, Michael J. 1984. "Economic analysis of sea level rise: Methods and results." In *Greenhouse effect and sea level rise,* ed. Michael C. Barth and James G. Titus. New York: Van Nostrand Reinhold, pp. 215–251.

———. 1986. "Planning for sea level rise under uncertainty: A case study of Charleston, South Carolina." In *Effects of changes in stratospheric ozone and global climate, Vol. 4: Sea level rise,* ed. James G. Titus. Washington: US Environmental Protection Agency, pp. 57–72.

Gibson, A. C., and K. E. Horak. 1978. "Systematic anatomy and phylogeny of Mexican columnar cacti." *Annals of the Missouri Botanical Gardens* 65: 999–1057.

Giddens, Anthony. 1987. *Sociology,* 2nd edn. San Diego: Harcourt, Brace, Jovanovich.

GIESCC (Grupo Internacional de Expertos Sobre Cambios Climáticos). 1990. *El cambio climático: Un tema clave de alcance mundial.* WMO-UNEP, Report WGES 251 (Spanish version). Geneva: World Meteorological Organization–United Nations Environmental Programme.

GIEWS (Global Information and Early Warning System). 2000. *Global watch: The global information and early warning system on food and agriculture.* Rome: Food and Agriculture Organization of the United Nations (FAO). Brochure available electronically at http://www.fao.org under Global Watch and Economics.

Ginsburg, Norton S., Bruce Koppel, and T. G. McGee, eds. 1990. *The dispersed metropolis: A phase of the settlement transition in Asia.* Honolulu: University of Hawaii Press.

Gladfelter, E. H., R. K. Monahan, and W. B. Gladfelter. 1978. "Growth rates of five reef-building corals in the north-eastern Caribbean." *Bulletin of Marine Science* 28: 728–734.

Glantz, Michael H., ed. 1988. *Societal responses to regional climatic change: Forecasting by analogy*. Boulder, CO: Westview Press.

——. 1994. *Drought follows the plow: Cultivating marginal areas*. Cambridge: Cambridge University Press.

——, and W. Degefu. 1991. Drought issues for the 1990s. In *Climate change: Science, impacts and policy: Proceedings of the Second World Climate Conference, Geneva, 1990*, ed. Jill Jäger and H. L. Ferguson. Cambridge: Cambridge University Press, pp. 253–263.

——, Martin F. Price, and Maria E. Krenz. 1990. *Report of the workshop on Assessing Winners and Losers in the Context of Global Warming, St Julians, Malta, 18–21 June 1990*. Boulder, CO: National Center for Atmospheric Research.

——, D. G. Streets, T. R. Stewart, N. Bhatti, C. M. Moore, and C. H. Rosa. 1998. *Exploring the concept of climate surprises: A review of the literature on the concept of surprise and how it is related to climate change*. ANL/DIS/TM-46. Argonne, IL: Argonne National Laboratory.

Glazovsky, Nikita F. 1995. "The Aral Sea basin." In *Regions at risk: Comparisons of threatened environments*, ed. Jeanne X. Kasperson, Roger E. Kasperson, and B. L. Turner, II. Tokyo: United Nations University Press, pp. 92–139.

Gleick, Peter H. 1988. "The effects of future climatic changes on international water resources: The Colorado River, the United States, and Mexico." *Policy Sciences* 21: 23–39.

——. 1989. "Climate change and international politics: Problems facing developing countries." *Ambio* 18, no. 6: 333–339.

——. 1992. "Water and conflict." In *Occasional paper series on environmental change and acute conflict*, no. 1 (September): 1–27. Boston, MA and Toronto, Ontario, Canada: International Security Studies Program, American Academy of Arts and Sciences, and Peace and Conflict Studies Program, University College, University of Toronto.

——. 1993a. *Water in crisis: A guide to the world's fresh water resources*, ed. Peter H. Gleick. Oxford: Oxford University Press.

——. 1993b. "Water in the 21st century." In *Water in crisis: A guide to the world's fresh water resources*, ed. Peter H. Gleick. Oxford: Oxford University Press, pp. 105–119.

——. 1998. *The world's water: The biennial report of fresh water resources*. Washington: Island Press.

——. 2000. *The world's water 2000–2001: The biennial report on freshwater resources*. Washington: Island Press.

Glenn, Jerome C., and Theodore J. Gordon, eds. 1999. "Special issue: The Millennium Project." *Technological Forecasting and Social Change* 61, no. 2 (June): 97–208.

Glickman, Theodore S., Dominic Golding, and Emily D. Silverman. 1992. *Acts of God and acts of man: Recent trends in natural disasters and major industrial accidents*. CRM 92-02. Washington, DC: Center for Risk Management, Resources for the Future.

Global Water Partnership. 1996. *Summary report: Technical Advisory Committee, regional meeting, Namibia 4–8 November 1996.* Stockholm: Global Water Partnership.

Glynn, P. W. 1991. "Coral reef bleaching in the 1980s and possible connection with global warming." *Trends in Ecology and Evolution* 6, no. 6: 175–179.

Goenaga, C. 1991. "The state of coral reefs in the wider Caribbean." *Interciencia* 16, no. 1: 12–20.

Goldman, Benjamin A., and Laura Fitton. 1994. *Toxic wastes and race revisited: An update of the 1987 report on the racial and socioeconomic characteristics of communities with hazardous waste sites.* Washington: Center for Policy Alternatives.

Goldsmith, Edward, Robert Allen, Michael Allaby, John Davoll, and Sam Lawrence. 1972. *Blueprint for survival.* Harmondsworth, UK: Penguin.

Golley, Frank B., 1993. *A history of the ecosystems concept in ecology.* New Haven: Yale University Press.

González-Pacheco, Cuauhtémoc. 1983. *Capital extranjero en la selva de Chiapas 1863–1982.* Mexico: Instituto de Investigaciones Economicas.

González-Quintero, L. 1968. *Tipos de vegetación del Valle del Mezquital, Hgo.* Mexico: Instituto Nacional de Antropología e Historia.

Goodman, Gordon. 1980. "Some criteria for the priority ranking and selection of urgent environmental issues." Unpublished mimeo. Stockholm, Sweden: Beijer Institute.

———. 1995. *Confronting the worldwide environment and development crisis: Celebrating the inauguration of Clark University's Environmental School.* Address by honorary degree recipient Gordon T. Goodman, 7 October, 1995. Worcester, MA: Clark University.

Gordon, T. J. 1989. "Futures research: Did it meet its promise? Can it meet its promise?" *Technological Forecasting and Social Change* 36: 21–26.

Goreau, Thomas F., Nora I. Goreau, and Thomas J. Goreau. 1979. "Corals and coral-reefs." *Scientific American* 241, no. 2 (August): 110–120.

Goreau, Thomas J. 1991. "Coral bleaching in Jamaica." *Nature* 343: 417.

Gould, S. J. 1991. "En défense de l'écureuil rouge." *Alliage* 1, no. 7–8: 199–204.

Gow, David D. 1989. "Development of fragile lands: An integrated approach reconsidered." In *Fragile lands of Latin America: Strategies for sustainable development,* ed. John O. Browder. Boulder, CO: Westview Press, pp. 25–43.

Grainger, Alan. 1982. *Desertification: How people make deserts, how people can stop, and why they don't.* London: International Institute for Environment and Development.

———. 1990. *The threatening desert: Controlling desertification.* London: Earthscan.

Greenberg, Michael R. 1986. "Disease competition as a factor in ecological studies of mortality: The case of urban centers." *Social Science and Medicine* 23, no. 10: 929–934.

Grubb, Michael, James Sebenius, Antonio Magalhes, and Susan Subak. 1992. "Sharing the burden." In *Confronting climate change: Risks, implications and responses,* ed. Irving M. Mintzer. Cambridge: Cambridge University Press, pp. 305–322.

Gubbay, S. 1986. "Nature conservation in the coastal zone of Great Britain." *Journal of Shoreline Management* 2: 241–257.

Guillet, D. G. 1983. "Towards a cultural ecology of mountains: The central Andes and the Himalaya compared." *Current Anthropology* 24: 561–574.

Gumley, P., and S. V. Inamdar. 1987. "Employing PSA as a safety tool in developing countries." *Nuclear Engineering International* 32, no. 390: 25–26.

H.M. Government. 1990. *This common inheritance: Britain's environmental strategy*. London: HMSO.

Haas, Peter. 1990a. "Obtaining international environmental protection through epistemic consensus." *Millennium* 19: 347–364.

————. 1990b. *Saving the Mediterranean: The politics of international environmental cooperation*. New York: Columbia University Press.

Habermas, Jurgen. 1970. *Knowledge and human interests*, trans. J. Shapiro. Boston: Beacon Press.

————. 1971. *Toward a rational society*, trans. J. Shapiro. Boston: Beacon Press.

————. 1988. *On the logic of the social sciences*. Cambridge, MA: MIT Press.

————. 1990. *Moral consciousness and communicative action*, trans. Christian Lenhardt, and Sherry Wiber Nicholson. Cambridge, MA: MIT Press.

Haffer, J. 1982. "General aspects of the refuge theory." In *Biological diversification in the tropics*, ed. Gillian T. Prance. New York: Columbia University Press, pp. 6–24.

Hall, Bob, and Mary Lee Kerr. 1991. *1991–1992 Green index: A state-by-state guide to the nation's environmental health*. Washington: Island Press.

Hamilton, D. P. 1990. "Death of the Nile Delta?" *Science* 250: 1084.

Hamilton, Lawrence S. 1995. "The protective role of mountain forests." In *Mountains at risk: Current issues in environmental studies*, ed. Nigel J. R. Allan. New Delhi: Monahar, pp. 49–72,

Hammond, Allen L. 1998. *Which world? Scenarios for the 21st century*. Washington: Island Press.

————, Eric Rodenburg, and William R. Moomaw. 1991. "Calculating national accountability for climate change." *Environment* 33, no. 1 (January/February): 10–15, 33–35.

Hancock, T. 1980. "The Conserver Society: Prospects for a healthier future." *Canadian Family Physician* 26: 320–321.

Harney, T. 1992. "A vital breadbasket undergoes rapid change by nature and man." *Smithsonian Institution Research Reports* 67 (Winter).

Harries, M. 1991. "Environmental pollution in Siberia: Prevent the crisis through international cooperation." *Nihon Keizai Shinbun (Nikkei Newspaper)*, 15 January.

Harris, Richard J. 1975. *A primer of multivariate statistics*. New York: Academic Press.

Hazell, Peter B. R. 1984. "Sources of increased instability in Indian and U. S. cereal production." *American Journal of Agricultural Economics* 66: 302–311.

————. 1991. *The Green Revolution reconsidered: The impact of high-yielding varieties in Southern India*. Baltimore, MD: Johns Hopkins.

Heilbroner, Robert. 1972. *An inquiry into the human prospect*. New York: Norton.

Hekstra, G. P., and Diana M. Liverman. 1986. "Global food futures and desertification." *Climatic Change* 9: 59–66.

Hernández-Xolocotzi, E. 1985. *Biología agrícola*. CECSA, México.

Herrera, Amílcar O., Hugo D. Scolnik, Graciela Chichilnisky, Gilberto C. Gallopin, Jorge E. Hardoy, Diana Mosovich, Enrique Oteiza, Gilda, L. de Romero Brest, Carlos E. Suárez, and Luis Talavera. 1976. *Catastrophe or new society? A Latin American world model.* IDRC-064e. Ottawa, Ontario, Canada: International Development Research Centre.

Hewitt de Alcántara, Cynthia. 1976. *Modernising Mexican agriculture: Socioeconomic implications of technological change, 1940–1970.* UNRISD Studies on the Green Revolution, vol. 11. Geneva: United Nations Research Institute for Social Development (UNRISD).

Hewitt, Kenneth. 1984a. "The idea of calamity in a technocratic age." In *Interpretations of calamity from the viewpoint of human ecology,* ed. Kenneth Hewitt. Boston: Allen and Unwin, pp. 3–32.

———, ed. 1984b. *Interpretations of calamity from the viewpoint of human ecology.* Boston: Allen and Unwin.

———. 1988. "The study of mountain lands and peoples: A critical overview." In *Human impact on mountains,* ed. Nigel J. R. Allan, Gregory W. Knapp, and Christoph Stadel. Totowa, NJ: Rowman and Littlefield, pp. 6–23

———. 1995. "Hazard and disaster in mountain environments: Problems in the geography of risk." In *Mountains at risk: Current issues in environmental studies,* ed. Nigel J. R. Allan. New Delhi: Monahar, pp. 98–128.

———. 1997. *Regions of risk: A geographical introduction to disasters.* Essex, UK: Addison Wesley Longman.

———, and Ian Burton. 1971. *The hazardousness of place: A regional ecology of damaging events.* Research Publication No. 6. Toronto: University of Toronto, Department of Geography.

Higgins, G. M., A. H. Kassam, L. Naiken, G. Fischer, and M. M. Shah. 1983. *Potential population supporting capacities of lands in the developing world.* Technical Report of Project FPA/INT/513, "Land resources for populations of the future," undertaken by the Food and Agriculture Organization of the United Nations, Rome, in collaboration with the International Institute for Applied Systems Analysis, Vienna for the United Nations Fund for Population Activities, New York. Rome: FAO.

Hill, K. 1985a. "The pace of mortality decline since 1950." In *Quantitative studies of mortality decline in the developing world.* World Bank Staff Working Paper No. 683. Washington: World Bank, pp. 53–95.

Hill, S. 1985b. "Redesigning the food system for sustainability." *Alternatives* 12, no. 3–4: 32–36.

Hjorth, Ronnie. 1994. "Baltic Sea environmental cooperation: The role of epistemic communities and the politics of regime change." *Cooperation and Conflict* 29, no. 1: 11–31.

Hoffman, J. S., J. Wells, and J. G. Titus. 1986. "Future global warming and sea level rise." In *Iceland coastal and river symposium, proceedings: Papers presented at Symposium on Coastal Geomorphology, Sedimentary Budgets, Coastal and River Hydraulics, 2–4 September 1985,* ed. C. Sigbjarnarson. Reykjavik, Iceland: Organizing Committee of the Symposium, pp. 245–266.

Hofsten, E., and H. Lundström. 1976. *Swedish population history: Main trends from 1750 to 1970.* Urval No. 8: 179. Stockholm: Statistiska Centralbyrån.

Hohenemser, Christoph, Roger E. Kasperson, and Robert W. Kates. 1985. "Causal structure." In *Perilous progress: Managing the hazards of technology,* ed. Robert W. Kates, Christoph Hohenemser, and Jeanne X. Kasperson. Boulder, CO: Westview Press, pp. 43–66.

———, Robert W. Kates, and Paul Slovic. 1985. "A causal taxonomy." In *Perilous progress: Managing the hazards of technology,* ed. Robert W. Kates, Christoph Hohenemser, and Jeanne X. Kasperson. Boulder, CO: Westview Press, pp. 67–89.

———, Robert Goble, Jeanne X. Kasperson, Roger E. Kasperson, Robert W. Kates, Peggy Collins, Abe Goldman, Paul Slovic, Baruch Fischnoff, Sarah Lichtenstein, and Mark Layman. 1983. *Methods for analyzing and comparing technological hazards: Definitions and factor structures.* CENTED Research Report No. 3. Worcester, MA: Clark University Hazard Assessment Group.

Holdgate, Martin W., Mohammed Kassas, and Gilbert White, eds. 1982. *The world environment 1972–1982: A Report by the United Nations Environment Programme.* Dublin: Tycooly International Publishers.

Holdren, John P. 1993. "A brief history of IPAT (Impact = Population × Affluence × Technology)." September, unpublished paper.

Holling, C. S. 1973. "Resilience and stability of ecological systems." *Annual Review of Ecology and Systematics* 4: 1–23.

———. 1986. "The resilience of terrestrial ecosystems: Local surprise and global change." In *Sustainable development of the biosphere,* ed. W. C. Clark and R. E. Munn. Cambridge: Cambridge University Press, pp. 292–317.

Holmberg, John, Ulrika Lundqvist, Karl-Henrik Robert, and Mathis Wackernagel. 1999. "The ecological footprint from a systems perspective of sustainability." *International Journal of Sustainable Development and World Ecology* 6, no. 1 (March): 17–33.

Homer-Dixon, Thomas F. 1999. *Environment, scarcity, and violence.* Princeton, NJ: Princeton University Press.

Homewood, Katherine. 1996. "Pastoralist production systems and climate change." In *Climatic change and world food security,* ed. Thomas E. Downing. NATO ASI Series 1: Global Environmental Change, vol. 37. Berlin: Springer-Verlag, pp. 505–524.

Hoos, Ida R. 1983. *Systems analysis in public policy: A critique.* Rev. edn. Berkeley: University of California Press.

Houston, M. A. 1985. "Patterns of species diversity on coral reefs." *Annual Review of Ecology and Systematics* 16: 149–177.

Hubbard, J. P. 1974. "Avian evolution of the arid-lands of North America." In *The living bird. Thirteenth Annual of the Ornithological Society of America,* ed. Douglas A. Lancaster. Ithaca, NY: Laboratory of Ornithology, Cornell University, pp. 155–196.

Human Development Report. Annual. New York: United Nations Development Programme (UNDP).

Hunter, M. L., G. L. Jacobson, and T. Webb. 1988. "Paleoecology and the coarse-filter approach to maintaining biological diversity." *Conservation Biology* 2, no. 4: 375–385.

Hurley, Andrew. 1995. *Environmental inequalities: Class, race, and industrial pollution in Gary, Indiana 1945–1980.* Chapel Hill: University of North Carolina Press.

Hutchison, T. W. 1977. *Knowledge and ignorance in economics.* Chicago: University of Chicago Press.

ICIMOD (International Centre for Integrated Mountain Development). 1990a. *Agricultural development experiences in Himachal Pradesh: MFS workshop report No. 1 on agricultural development experiences in the Hindu Kush–Himalayas.* Kathmandu: ICIMOD.

———. 1990b. *Agricultural development experiences in West Sichuan and Xizang, China: MFS workshop report No. 2 on agricultural development experiences in the Hindu Kush–Himalayas.* Kathmandu: ICIMOD.

———. 1990c. *Agricultural development experiences in Nepal: MFS workshop report No. 3 on agricultural development experiences in the Hindu Kush–Himalayas.* Kathmandu: ICIMOD.

———. 1990d. *Agricultural development experiences in Pakistan: MFS workshop report No. 4 on agricultural development experiences in the Hindu Kush–Himalayas.* Kathmandu: ICIMOD.

ICSU (International Council of Scientific Unions). 1987. *The International Geosphere Biosphere Program: A study of global change.* Paris: ICSU.

IDNDR (International Decade for Natural Disaster Reduction). 1999a. *Final report of the Scientific and Technical Committee of the International Decade for Natural Disaster Reduction (IDNDR).* IDNDR Online Documents, available @ http://www.landr.org/docs.

———. 1999b. *International Decade for Natural Disaster Reduction: Addendum,* final report of the Scientific and Technical Committee of the International Decade for Natural Disaster Reduction. Report a/54/132/Add. 1. New York: United Nations.

IFIAS (International Federation of Institutes of Advanced Study). 1989. *The human dimensions of global change: An international programme on human interactions with the Earth. Report of the Tokyo International Symposium on the Human Response to Global Change, Tokyo, 19–22 September, 1988.* Toronto: IFIAS.

IGES (Institute for Global Environmental Strategies). 1999. *Climate Change Research Project: Discussion papers in FY 1988 for the design of effective framework of Kyoto mechanisms.* Hayama, Kanagawa, Japan: IGES.

——— and Environmental Agency of Japan). 1999. *Quality of the environment in Japan 1999: Environmental messages toward sustainable development in the 21st century.* Kamigamaguchi: IGES.

Ikeda, Saburo. 1987. "Economic–ecological models in regional total systems." In *Economic–ecological modeling,* ed. Leon C. Braat and Wal F. J. van Lierop. Amsterdam: Elsevier, pp. 185–203.

Informe Ganadero (Buenos Aires). 1983. "Brasil: La recesión económica y el ciclo ganadero." 2, no. 34: 23–26.

Inglehart, Ronald, Miguel Basáñez, and Alejandro Menéndez Moreno. 1998. *Human values and beliefs: A cross-cultural sourcebook: Political, religious, sexual, and economic norms in 43 societies: Findings from the 1990–1993 world value survey.* Ann Arbor, MI: University of Michigan Press.

Inhaber, Herbert. 1998. *Slaying the NIMBY dragon*. New Brunswick, NJ: Transaction Publishers.

International Energy Agency. 1987. *Energy conservation in IEA countries*. Paris: OECD.

IPCC (Intergovernmental Panel on Climate Change). 1990a. *Climate change: The IPCC scientific assessment*. Cambridge: Cambridge University Press.

———. 1990b. *IPCC first assessment report, Vol. 2*. Geneva: WMO/UNEP.

———. 1990c. *Climate change: The IPCC scientific assessment*. Final Report of Working Group I. Geneva: IPCC.

———. 1990d. *Strategies for adaption to sea level rise*. Report of the Coastal Zone Management Subgroup. Geneva: IPCC.

———. 1991. *The seven steps to the assessment of the vulnerability of coastal areas to sea level rise: A common methodology*. Response Strategies Working Group, Advisory Group on Assessing Vulnerability to Sea Level Rise and Coastal Management. Geneva: IPCC.

———. 1996a. *Climate change 1995: The science of climate change, Contribution of Working Group I to the second assessment report of the Intergovernmental Panel on Climate Change*, ed. J. T. Houghton, L. G. Meira Filho, B. A. Callander, N. Harris, A. Kattenberg, and K. Maskell. Cambridge: Cambridge University Press for the Intergovernmental Panel on Climate Change.

———. 1996b. *Climate change 1995. Impacts, adaptations and mitigation of climate change: Scientific–technical analyses, contribution of Working Group II to the second assessment report of the Intergovernmental Panel on Climate Change*. Cambridge: Cambridge University Press for the Intergovernmental Panel on Climate Change.

———. 1996c. *Climate change 1995: Economic and social dimension of climate change, contribution of Working Group III to the second assessment report of the Intergovernmental Panel on Climate Change*, ed. J. P. Bruce, Hoesung Lee, and E. Haites. Cambridge: Cambridge University Press for the IPCC.

———. 1997. *The regional impacts of climate change: An assessment of vulnerability*, ed. Robert T. Watson, Marufu C. Zinyowera, and Richard H. Moss. A special report of IPCC Working Group II. Geneva: IPCC.

———. 1998. *The regional impacts of climate change: An assessment of vulnerability, A special report of IPCC Working Group II*, ed. Robert T. Watson, Marufu C. Zinyowera, and Richard H. Moss. Cambridge: Cambridge University Press for the IPCC.

———. 2001. *Climate change 2001: Synthesis report, third assessment report of the Intergovernmental Panel on Climate Change*. Cambridge: Cambridge University Press.

ISSC (International Social Science Council). 1989. "Plan of action for research on the human dimensions of global environmental change." Draft 3.1. Paris, 27 October 1989.

IUCN (International Union for the Conservation of Nature), WWF (World Wildlife Fund) and UNEP (United Nations Environment Programme). 1980. *World conservation strategy: Living resource conservation for sustainable development*. Gland, Switzerland: IUCN.

Ives, Jack D. 1989. "Deforestation in the Himalayas: The cause of increased flooding in Bangladesh and northern India?" *Land Use Policy* 6, no. 3 (July): 187–193.

———, and Bruno Messerli. 1989. *The Himalayan dilemma: Reconciling development and conservation.* London: Routledge.

Ives, Jane H., ed. 1985. *The export of hazard.* Boston: Routledge and Kegan Paul.

Izrael, Yu. A., and R. E. Munn. 1986. "Monitoring the environment and renewable resources." In *Sustainable development of the biosphere,* ed. William C. Clark and R. E. Munn. Cambridge: Cambridge University Press, pp. 360–375.

Jacobson, Harold K., and Martin F. Price. 1991. *A framework for research on the human dimensions of global environmental change.* ISSC/UNESCO, 3. Paris: ISSC with the cooperation of UNESCO.

Jacobson, Jodi L. 1989a. *Environmental refugees.* Worldwatch Paper No. 86. Washington: Worldwatch Institute.

———. 1989b. "Swept away." *Worldwatch* 2, no. 1 (January/February): 20–26.

Jäger, Jill, Nicholas Sonntag, David Bernard, and Werner Kurz. 1991. *The challenge of sustainable development in a greenhouse world: Some visions of the future. Report of a Policy Exercise held in Bad Bleiberg, Austria, 2–7 September 1990.* Stockholm: Stockholm Environment Institute.

Jahnke, Marlene. 1999. "Climate change: Plan of action adopted." *Environmental Policy and Law* 29, no. 1: 2–8.

Janzen, D. H. 1967. "Why mountain passes are higher in the tropics." *American Naturalist* 101: 233–249.

Japanese Environment Agency. 1989a. "Interim report of the sub-group on impacts and response strategies" (in Japanese). The Advisory Committee on Climate Change. Tokyo: Japanese Environment Agency.

———. 1989b. *The state of the environment.* Tokyo: Japanese Environment Agency.

———. 1989c. *Toward a regeneration of Tokyo Bay* (in Japanese). Tokyo: Japanese Environment Agency .

Jasanoff, S. 1990. *The fifth branch.* Cambridge, MA: Harvard University Press.

———, and Brian Wynne. 1998. "Science and decisionmaking." In *Human choices and climate change, vol. 1: The societal framework,* ed. Steve Rayner and Elizabeth L. Malone. Columbus, OH: Battelle Press, pp. 1–87.

JEEM (*Journal of Environmental Economics and Management*) 1990. (Entire issue.) 18 (2).

Jeffery, Robert W. 1989. "Risk behaviors and health: Contrasting individual and population perspectives." *American Psychologist* 44, no. 9 (September): 1194–1202.

Jeffery, Susan E. 1981. *Our usual landslide: Ubiquitous hazard and socioeconomic causes of natural disaster in Indonesia.* Natural Hazards Research Working Paper no. 40. Boulder, CO: Institute of Behavioral Science, University of Colorado.

———. 1982. "The creation of vulnerability in natural disaster: Case studies from the Dominican Republic." *Disasters* 6, no. 1: 38–43.

Jessen, P. J. 1981. "The role of energy ideologies in developing environmental policy." In *Environmental policy formulation,* ed. D. E. Mann. Lexington, MA: Lexington Books, pp. 103–123.

Jiang, Hong. 1999. *The Ordos Plateau of China: An endangered environment.* Tokyo: United Nations University Press.

———, Peiyuan Zhang, Du Zheng, and Fenghui Wang. 1995. The Ordos Plateau of China. In *Regions at risk: Comparisons of threatened environments,* ed. Jeanne X. Kasperson, Roger E. Kasperson, and B. L. Turner, II. Tokyo: United Nations University Press, pp. 420–459.

Jochim, M. A. 1981. *Strategies for survival: Cultural behaviour in an ecological context.* New York: Academic Press.

Jodha, N. S. 1989. "Potential strategies for adapting to greenhouse warming: Perspectives from the developing world." In *Greenhouse warming: Abatement or adaptation,* ed. Norman J. Rosenberg, William E. Easterling, Pierre R. Crosson, and Joel Darmstadter. Washington: Resources for the Future, pp. 47–158.

———. 1990a. "Mountain agriculture: The search for sustainability." *Journal of Farming Systems Research Extension* 1: 55–75.

———. 1990b. *Sustainable agriculture in fragile resource zones.* MFS Discussion Paper no. 3. Kathmandu, Nepal: International Centre for Integrated Mountain Development (ICIMOD).

———. 1991. "Sustainable agriculture in fragile resource zones: Technological imperatives." *Economic and Political Weekly (Quarterly Review of Agriculture)* 26, no. 13 (30 March): A15–A30.

———. 1992a. *Common property resources: A missing dimension of development strategies.* World Bank Discussion Papers, 169. Washington: World Bank.

———. 1992b. "Mountain perspective and sustainability: A framework for development strategies." In *Sustainable mountain agriculture, vol. 1: Perspectives and issues,* ed. Narpat S. Jodha, Mahesh Banskota, and Tej Partap. New Delhi: Oxford and IBH Publishers, pp. 41–82.

———. 1992c. "Understanding and responding to global climate change in fragile resource zones." In *The regions and global warming: Impacts and response strategies,* ed. Jurgen Schmandt and Judith Clarkson. HARC Global Change Studies vol. 1. Oxford: Oxford University Press, pp. 203–218.

———. 1995. "The Nepal middle mountains." In *Regions at risk: Comparisons of threatened environments,* ed. Jeanne X. Kasperson, Roger E. Kasperson, and B. L. Turner II. Tokyo: United Nations University Press, pp. 140–185.

———, and W. J. Maunder. 1991. "Climate, climate change and the economy." In *Climate Change: Science, impacts and policy. Proceedings of the Second World Climate Conference, Geneva, 1990,* ed. Jill Jäger and H. L. Fergnan. Cambridge: Cambridge University Press, pp. 409–414.

———, M. Banskota, and T. Partap. 1992. "Strategies for the sustainable development of mountain agriculture: An overview." In *Sustainable mountain agriculture, vol. 1: Perspectives and issues,* ed. Narpat S. Jodha, Mahesh Banskota, and Tej Partap. New Delhi: Oxford and IBH Publishers, pp. 3–40.

Johannes, R. E., and B. C. Hatcher. 1986. "Shallow tropical marine environments." In *Conservation biology: The science of scarcity and diversity,* ed. M. E. Soulé. Sunderland, MA: Sinauer Associates, pp. 371–382.

Johnson, Branden B., and Vincent T. Covello, eds. 1987. *The social and cultural construction of risk.* Dordrecht, Netherlands: D. Reidel.

Johnson, Kirsten J. 1977. "'Do as the land bids': A study of Otomi resource-use on the eve of irrigation." PhD dissertation. Graduate School of Geography, Clark University, Worcester, MA.

———, E. A. Olson, and S. Manandhar. 1982. "Environmental knowledge and response to natural hazards in mountainous Nepal (Kathmandu)." *Mountain Research and Development* 2, no. 2: 175–178.

Jonas, Hans. 1984. *The imperative of responsibility*. Chicago: University of Chicago Press.

Juma, Calestous. 1989. *The gene hunters: Biotechnology and the scramble for seeds*. Princeton, NJ: Princeton University Press.

Jungk, Robert, and Johan Galtung, eds. 1969. *Mankind 2000*. Futures Research Monographs, no. 1. Oslo: Universitetsforlaget.

Kahn, Herman, W. Brown, and L. Martel. 1976. *The next 200 years: A scenario for America and the world*. New York: Morrow.

Kahneman, Daniel, Paul Slovic, and Amos Tversky, eds. 1982. *Judgment under uncertainty: Heuristics and biases*. New York: Cambridge University Press.

Kalhok, Sarah, and Glynn Gomez. 1999. "Box 2: Emerging issues in retrospect, and environmental surprises since 1950." *Ambio* 28, no. 6 (September): 465.

Kandlikar, Milind, and Ambuj Sagar. 1999. "Climate change research and analysis in India: An integrated assessment of a South–North divide." *Global Environmental Change* 9: 119–138.

Kaplan, Robert D. 1994. "The coming anarchy." *Atlantic Monthly* 273(2) (February): 44ff.

Kariel, Herbert G., and Dianne L. Draper. 1995. "Outdoor recreation and tourism in mountain environments." In *Mountains at risk: Current issues in environmental studies*, ed. Nigel J. R. Allan. New Delhi: Mandar, pp. 220–237.

Karim, Z. 1996. "Agricultural vulnerability and poverty alleviation in Bangladesh." In *Climate change and world food security*, ed. Thomas E. Downing. Berlin: Springer-Verlag, pp. 307–346.

Kaseno, Y., T. Takano, T. Ohkuma, A. Yamane, T. Nakagawa, I. Isobe, H. Ishida, and K. Kanda. 1991. "Geotechnical engineering in the shorelines of Japan Sea" (in Japanese). *Tsuchi-To-Kiso* 39, no. 3 (398): 51–86.

Kasperson, Jeanne X., and Roger E. Kasperson. 1987. "Priorities in profile: Managing risks in developing countries." *Risk Abstracts* 4, no. 3: 113–118.

———, ———, and B. L. Turner II, eds. 1995. *Regions at risk: Comparisons of threatened environments*. Tokyo: United Nations University Press.

———, ———, and ———. 1996. "Regions at risk: Exploring environmental criticality." *Environment* 38, no. 10 (December): 4–15, 26–29.

Kasperson, Roger E., ed. 1983. *Equity issues in radioactive waste management*. Cambridge: Oelgeschlager, Gunn, and Hain.

———. 1986. "Six propositions on public participation and their relevance to risk communication." *Risk Analysis* 6, no. 3 (September): 275–282.

———. 1992. "The social amplification of risk: Progress in developing an integrative framework of risk." In *Social theories of risk*, ed. Sheldon Krimsky and Dominic Golding. New York: Praeger, pp. 153–178.

———, and Kirstin Dow. 1991. "Developmental and geographical equity in global environmental change." *Evaluation Review* 15: 149–171.

————, and Jeanne X. Kasperson. 1991. "Hidden hazards." In *Acceptable evidence: Science and values in hazard management*, ed. D. C. Mayo and R. Hollander. Oxford: Oxford University Press, pp. 9–28.

————, and ————. 1996. "The social amplification and attenuation of risk." *Annals of the American Academy of Political and Social Science* 545 (May): 95–105.

————, and Murdo Morrison. 1982. "A proposal for international risk management research." In *Risk in the technological society*, ed. Christoph Hohenemser and Jeanne X. Kasperson. Boulder, CO: Westview Press, pp. 303–331.

————, Dominic Golding, and Jeanne X. Kasperson. 1999. "Trust, risk, and democratic theory." In *Social trust and the management of risk*, ed. George Cvetkovich and Ragnar Löfstedt. London: Earthscan, pp. 22–41.

————, ————, and Seth Tuler. 1992. "Social distrust as a factor in siting hazardous facilities and communicating risks." *Journal of Social Issues* 48, no. 4: 161–187.

————, Jeanne X. Kasperson, and B. L. Turner, II. 1999. "Risk and criticality." *Ambio* 28 (6) (September): 562–568.

————, Kirstin Dow, Dominic Golding, and Jeanne X. Kasperson, eds. 1990. *Understanding global environmental change: The contributions of risk analysis and management*. CENTED, Clark University, Worcester, MA.

————, Ortwin Renn, Paul Slovic, Halina S. Brown, Jacque Emel, Robert Goble, Jeanne X. Kasperson, and Samuel Ratick. 1988. "The social amplification of risk: A conceptual framework." *Risk Analysis* 8, no. 2: 177–187.

Kates, Robert W. 1978. *Risk assessment of environmental hazard*. SCOPE 8. Chichester, UK: John Wiley and Sons.

————. 1986. "Managing technological hazards: Success, strain, and surprise." In *Hazards: Technology and fairness*, ed. National Academy of Engineering. Washington: National Academy Press, pp. 206–220.

————. 1997. "Climate change 1995: Impacts, adaptation, and mitigation." *Environment* 39, no. 9 (November): 29–33.

————. 2000. "Cautionary tales: Adaptation and the rural poor." *Climatic Change* 45, no. 1 (April): 5–17.

————., and William C. Clark. 1996. "Environmental surprise: Expecting the unexpected." *Environment* 38, no. 2 (March): 6–11, 28–34.

————, and Viola Haarmann. 1992. "Where the poor live: Are the assumptions correct?" *Environment* 34, no. 4: 4–11, 25–28.

————, and Jeanne X. Kasperson. 1983. "Comparative risk analysis of technological hazards (a review)." *Proceedings of the National Academy of Sciences of the USA* 80: 7027–7038.

————, Jesse H. Ausubel, and Miriam Berberian, eds. 1985. *Climate impact assessment: Studies of the interaction of climate and society*. SCOPE. New York: John Wiley and Sons.

————, Christoph Hohenemser, and Jeanne X. Kasperson, eds. 1985. *Perilous progress: Managing the hazards of technology*. Boulder, CO: Westview Press.

Kavanagh, Barbara, and Steve Lonergan. 1992. *Environmental degradation, population displacement and global security*. Report IR92-1. Ottawa: Canadian Global Change Program.

Kay, Robert C. 1990. "Development controls on eroding coastlines: Reducing the future impact of greenhouse-induced sea level rise." *Land Use Policy* 7: 172–196.

———, K. M. Clayton, and C. E. Vincent. 1990. "Sea level rise and physical coastal change." In *The effects of sea level rise on the UK coast*, ed. L. E. J. Roberts, and R. C. Kay. Research Report No. 7. Norwich, UK: Environmental Risk Assessment Unit, University of East Anglia, pp. 75–150.

Keeney, Ralph L., and Detlof von Winterfeldt. 1986. "Why indirect health risks of regulations should be examined." *Interfaces* 16, no. 6: 13–27.

Kern, G. 1987. "News vs. fiction: reflections on prognostication." In *Storm warnings: Science fiction confronts the future*, ed. Slusser, Greenland, and Eric Rabkin. Carbondale, IL: Southern Illinois University Press, pp. 211–231.

Keyfitz, Nathan. 1988. "Letters: Support for IIASA." *Science* 242: 496.

Kim, K. G. 1987. *Risk assessment in urban planning and management*. Paris: UNESCO.

Kingdon, John W. 1995. *Agendas, alternatives, and public policies*, 2nd edn. New York: Harper Collins.

Kirkby, Anne V. T. 1973. "The use of land and water resources in the past and present, Valley of Oaxaca, Mexico." *Memoirs of the Museum of Anthropology* No. 5. Ann Arbor: Museum of Anthropology, University of Michigan.

Kjellstrom, T., and L. Rosenstock. 1990. "The role of environmental and occupational hazards in the adult health transition." *World Health Statistics Quarterly* 43, no. 3: 188–196.

Kloppenburg, J., and D. L. Kleinman. 1987. "The plant germplasm controversy." *Bioscience* 37, no. 3: 190–198.

Koch, F. H. 1988. "Land development in the Ganges–Brahmaputra estuary in Bangladesh." In *Sea-level rise*, ed. P. C. Schröder. Delft: Delft Hydraulics, pp. 1–14.

Kohonen, T. 1994. "Regional environmental policies in Europe: Baltic/Nordic cooperation." In *Environmental cooperation in Europe: The political dimension*, ed. Otmas Hall. Boulder, CO: Westview Press, pp. 209–222.

Kondratiev, N. D. 1925. "The major economic cycles." *Voprosy Kon'iunktury* 1: 28–79. Translated in *Review* 4: 519–562 (Spring 1979).

Koshland, Daniel E., ed. 1987. "Risk assessment issue." *Science* 236 (4799): 241, 267–300.

Kotlyakov, Vladimir M. 1991. "The Aral Sea basin: A critical environmental zone." *Environment* 33, no. 1 (January/February): 4–9, 36–38.

Kotze, Astrid von. 1999. "A new concept of risk." In *Risk, sustainable development and disasters*, ed. Ailsa Holloway. Rondebosch, South Africa: Periper Publications, University of Cape Town, pp. 33–40.

Krimsky, Sheldon, and Dominic Golding, eds. 1992. *Social theories of risk*. New York: Praeger.

Krings, Thomas. 1993. "Structural causes of famine in the Republic of Mali." In *Coping with vulnerability and criticality*, eds H.-G. Bohle, T. E. Downing, J. O. Field, and F. N. Ibrahim. Freiburg Studies in Development Geography. Saarbrücken: Verlag breitenbach, pp. 129–144.

Kristoferson, Lars. 1996. *Baltic 21: Creating an agenda for the Baltic Sea region.* 3 vols. Stockholm: Stockholm Environment Institute.

Krutilla, J. V. 1979. "Economics and the environment: A time for taking stock." RFF Reprint 171. Washington: Resources for the Future.

Kuhn, Thomas S. 1962. *The structure of scientific revolutions.* Chicago: University of Chicago Press.

Lamartine Yates, Paul. 1981. *Mexico's agricultural dilemma.* Tucson: University of Arizona Press.

Lamb, Hubert H. 1971. *Climate: Present, past and future.* London: Methuen.

Landre, Betsy K., and Barbara A. Knuth. 1993. "Success of citizen advisory committees in consensus-based water resources planning in the Great Lakes Basin." *Society and Natural Resources* 6, no. 3 (July–September): 229–257.

Lane, Charles R., ed. 1998. *Custodians of the commons: Pastoral land tenure in East and West Africa.* London: Earthscan.

Lappé, Frances M., and J. Collins. 1978. *Food first: Beyond the myth of scarcity,* rev. edn. New York: Ballantine.

——, ——, and Peter Rosset, with Luis Esparza. 1998. *World hunger: 12 myths,* 2nd edn. London: Earthscan.

Latour, Bruno. 1988. *The pasteurization of France.* Cambridge, MA: Harvard University Press.

Lave, Judith, and Lester B. Lave. 1985. "Adapting risk management methods to the developing countries." In *Workshop on Risk Analysis in Developing Countries: An evaluation of the suitability and applicability of health and environmental risk analysis methods and approaches, 27 October–1 November 1985, Hyderabad, India: Conference papers.* Aberdeen, Scotland: Centre for Environmental Management and Planning, University of Aberdeen.

Lave, Lester B., ed. 1982. *Quantitative risk assessment in regulation.* Washington: Brookings Institution.

Lee, Kai. 1993. *Compass and gyroscope: Integrating science and politics for the environment.* Washington: Island Press.

Lee, R. D. 1986. "Malthus and Boserup: A dynamic synthesis." In *The state of population theory: Forward from Malthus,* ed. David A. Coleman and Roger S. Schofield. Oxford: Basil Blackwell, pp. 96–130.

Lees, S. H. 1976. "Hydraulic development and political response in the Valley of Oaxaca, Mexico." *Anthropological Quarterly* 49: 107–210.

LeGuin, Ursula K. 1979. *Language of the night: Essays on fantasy and science fiction.* New York: Putnam.

Leonard, H. J. 1989. *Environment and the poor: Development strategies for a common agenda.* New Brunswick: Transaction Books.

Leontief, Wassily W., Anne P. Carter, and Peter A. Petri. 1977. *The future of the world economy: A United Nations study.* New York: Oxford University Press.

Leopold, A. S. 1950. "Vegetation zones of Mexico." *Ecology* 31: 507–518.

Lerner, Sally. 1991. *Socio-political design criteria for a sustainable Canadian society.* Working Paper no. 3, Sustainable Society Project, Department of Environment and Resource Studies, University of Waterloo, Waterloo (July).

Lesbirel, S. Hayden. 1998. *NIMBY politics in Japan: Energy siting and the management of environmental conflict.* Ithaca, NY: Cornell University Press.

Levy, Marc A. 1995. "Is the environment a national security issue?" *International Security* 20, no. 2 (Fall): 35–62.

Lewis, James. 1999. *Development in disaster-prone places: Studies of vulnerability.* London: Intermediate Technology Publications.

Lind, R. C., K. J. Arrow, G. R. Corey, P. Dasgupta, A. K. Sen, T. Stauffer, J. E. Stiglitz, J. A. Stockfisch, and R. Wilson. 1982. *Discounting for time and risk in energy policy.* Washington: Resources for the Future.

Liverman, Diana M. 1990a. "Seguridad y medio ambiente en Mexico." In *Busca de la seguridad perdida: Aproximaciones a la seguridad nacional mexicana,* ed. Sergio Aguayo Quezada, and Bruce M. Bagley. Mexico: Siglo Veinteuno, pp. 233–263.

———. 1990b. "Vulnerability to drought in Mexico: The cases of Sonora and Puebla in 1970." *Annals of the Association of American Geographers* 80, no. 1: 49–72.

———. 1990c. "Vulnerability to global environmental change." In *Understanding global environmental change: The contributions of risk analysis and management,* ed. Roger E. Kasperson, Kirstin Dow, Dominic Golding, and Jeanne X. Kasperson. Worcester, MA: Clark University, The Earth Transformed Program, pp. 27–44.

———. 1992. "The regional impact of global warming in Mexico: Uncertainty, vulnerability and response." In *The regions and global warming: Impacts and response strategies,* ed. Jurgen Schmandt and Judith Clarkson. Oxford: Oxford University Press, pp. 44–68.

———. 1994. "Environment and security in Mexico." In *Mexico: In search of security,* ed. Bruce M. Bagley and Sergio Aguayo. New Brunswick, NJ: Transaction Publishers, pp. 211–246.

———. 2000. "Adaptation to drought in Mexico." In *Drought: A global assessment,* vol. 2, ed. Donald A. Wilhite. London: Routledge, pp. 35–45.

———, and Karen L. O'Brien. 1991. "Global warming and climate change in Mexico." *Global Environmental Change* 1, no. 5: 351–364.

Lobato, R. 1980. "Social stratification and destruction of the Lacandona Forest in Chiapas." *Ciencia Forestal* 5: 21–38.

Lonergan, Steve. 1997. "Global environmental change and human security." *Changes,* Issue No. 5. Ottawa: Canadian Global Change Program.

———. 1998. *The role of environmental degradation in population displacement.* Global Environmental Change and Human Security Project, Research Report 1. Victoria, BC: Department of Geography, University of Victoria.

———. 1999. *Global Environmental Change and Human Security (GECHS) Science Plan.* IHDP Report No. 11. Bonn: International Human Dimensions Programme.

López-Portillo, J., and Exequiel Ezcurra. 1985. "Litter fall of *Avicennia germinans* L. in a one-year cycle in a mudflat at the Laguna de Mecoacán, Tabasco, Mexico." *Biotropica* 17, no. 3: 186–190.

———, and ———. 1989a. "Response of three mangrove species to salinity in two geoforms." *Functional Ecology* 3: 355–361.

————, and ————. 1989b. "Zonation in mangrove and salt-marsh vegetation in relation to soil characteristics and species interactions at the Laguna de Mecoacán, Tabasco, Mexico." *Biotropica* 21, no. 2: 107–114.

Lovei, Magda. 1998. *Phasing out lead from gasoline: Worldwide experience and policy implications*. World Bank Technical Paper No. 397. Washington: World Bank.

Lovins, Amory. 1976a. "Energy strategy: The road not taken?" *Foreign Affairs* 55 (October): 65–96.

————. 1976b. "Exploring energy-efficiency futures for Canada." *Conserver Society Notes* 1, no. 4: 3–16.

————. 1977. *Soft energy paths*. New York: FOE/Ballinger.

Low, Nicholas, and Brendan Gleeson. 1998. *Justice, society, and nature: An exploration of political ecology*. London: Routledge.

Lowrance, R., B. R. Stinner, and G. J. House, eds. 1984. *Agricultural ecosystems: Unifying concepts*. New York: John Wiley and Sons.

Lubchenko, J., A. M. Olson, L. B. Brubaker, S. R. Carpenter, M. M. Holland, S. P. Hubbell, S. A. Levin, J. A. MacMahon, P. A. Matson, J. M. Melillo, H. A. Mooney, C. H. Peterson, H. R. Pulliam, L. A. Real, P. J. Regal, and P. G. Risser. 1991. "The sustainable biosphere initiative: An ecological research agenda." *Ecology* 72, no. 2: 371–412.

Luhmann, Niklas. 1979. *Trust and power*, ed. T. Burns, and G. Poggi. Chichester, UK: John Wiley and Sons.

————. 1986 *Ökologische Kommunikationen: Kann die moderne Gesellschaft sich auf ökologische Gefährdungen einstellen?* Opladen, Germany: Westdeutscher Verlag.

————. 1989. *Ecological communication*, trans. J. Bednarz, Jr. Chicago: University of Chicago Press.

Lumbers, J. P., and P. W. Jowitt. 1981. "Risk analysis in the planning, design, and operational control of water pollution prevention schemes: A perspective." *Water Science Technology* 13: 27–34.

Luna-Vega, M. I. 1984. "Notas fitogeográficas sobre el bosque mesófilo de montaña en México. Un ejemplo en Teocelo-Cosautlán-Ixhuacán, Veracruz, México." Unpublished BSc thesis. Fac. de Ciencias, UNAM, Mexico.

MacIntyre, Alaisdair. 1981. *After virtue*, 2nd edn. 1983. Notre Dame, IN: University of Notre Dame Press.

Mackey, H. 1988. Historic global agreement signed to reduce CFCs. *CO_2/Climate Report*, Canadian Climate Centre, Downsview, Ontario, 88-1 (Winter), [pp. 2–3 (unnumbered)].

MacLane, S. 1988. "Letters." *Science* 241: 114 and 242: 1623–1624.

MacLean, Douglas. 1986. "Social values and the distribution of risk." In *Values at risk*, ed. Douglas MacLean. Totowa, NJ: Rowman and Allanheld, pp. 75–93.

Maguire, Andrew, and Janet Welsh-Brown. 1986. *Bordering on trouble: Resources and politics in Latin America*. Bethesda, MD: Adler and Adler.

Mahtab, F. U. 1989. *Effect of climate change and sea-level rise on Bangladesh*. Expert Group on Climate Change and Sea-Level Rise. London: Commonwealth Secretariat.

Malone, Thomas F. 1986. "Mission to planet earth: Integrating studies of global change." *Environment* 28 (October): 6–11, 39–42.

———. 1988. "A dance of death or a celebration of life?" *Environmental Conservation* 15, no. 4: 289–290.

Mangelsdorf, P. C., R. S. MacNeish, and W. C. Galinat. 1967. "Prehistoric wild and cultivated maize." In *The prehistory of the Tehuacán Valley.* vol. 1: *Environment and subsistence.* ed. Douglas S. Byers. R. S. Peabody Foundation. Austin, TX: University of Texas Press, pp. 178–200.

Mannion, Antoniette M. 1991. *Global environmental change: A natural and cultural environmental history.* London: Longmans.

Manuel, F. E., and F. P. Manuel. 1979. *Utopian thought in the western world.* Cambridge, MA: Harvard University Press.

Marchetti, C. 1980. "Society as a learning system: Discovery, invention, and innovation cycles revisited." *Technological Forecasting and Social Change* 18: 267–282.

Marcuse, Herbert. 1964. *The one-dimensional man.* Boston: Beacon Press.

Mardia, Kantilar Varichand, John R. Kent, and John M. Bibby. 1979. *Multivariate analysis.* London: Academic Press.

Mares, M. A. 1979. "Small mammals and creosote bush: Patterns of richness." In *Larrea,* ed. E. L. Campos, T. J. Mabry, and S. T. Fernández. Mexico: Consejo Nacional de Ciencia y Tecnología, pp. 57–94.

Margalef, Ramon. 1968. *Perspectives in ecological theory.* Chicago, IL: University of Chicago Press.

Margulis, L. 1991. "Big trouble in biology: Physiological autopoiesis versus mechanistic neo-Darwinism." In *Doing science: The reality club*, ed. J. Brockman. New York: Prentice Hall, pp. 211–235.

Marsh, George Perkins. 1864. *Man and nature; or, physical geography as modified by human action.* New York: Charles Scribner.

Marshall, Michael. 1987. *Long waves of regional development.* New York: St. Martin's Press.

Martin, T. 1991. Personal communication. Team Leader, FAP-19 GIS, Bangladesh Flood Action Plan, with information from the Cyclone Protection Project (FAP-7). 11 November 1991.

Martínez, E., and C. H. Ramos. 1989. "Lacandoniaceae (Triuridales): Una nueva familia de México." *Annals of Missouri Botanical Garden* 76: 128–135.

Mascarenhas, J., P. Shah, S. Joseph, R. Jayakaran, J. Devavaram, V. Ramachandran, A. Fernandez, R. Chambers, and J. Pretty, eds. 1991. *Participatory rural appraisal: Proceedings of the February 1991 Bangalore PRA Trainers Workshop.* RRA Notes, No. 13. London: IIED.

Maslow, Abraham. 1959. *New knowledge in human values.* New York: Harper.

Mather, John R., and Galina V. Sdasyuk, eds. 1991. *Global change: Geographic approaches.* Tucson: University of Arizona Press.

Mathews, Jessica T. 1989. "Redefining security." *Foreign Affairs* 68, no. 2: 162–177.

———. 1993. "Nations and nature: A new view of security." In *Threats without enemies: Facing environmental security,* ed. Gwyn Prins. London: Earthscan, pp. 25–37.

Matsui, Takafumi. 1993. "Evolution of the terrestrial atmosphere" (in Japanese). *Journal of Geography* 102, no. 6: 633–644.

May, R. M. 1989. "How many species are there on Earth?" *Science* 241: 1441–1448.

Mazzucchelli, Sergio, and Viviana Burijson. 1999. "Issues of sustainability in the Kyoto Protocol." In *Towards equity and sustainability in the Kyoto Protocol*, ed. Karin Hultcrantz. Stockholm: Stockholm Environment Institute, pp. 1–11.

McCormick, John. 1991. *Urban air pollution.* UNEP/GEMS Environment Library no. 4. Nairobi: United Nations Environment Programme (UNEP).

McFadden, D. 1984. "Welfare analysis of incomplete adjustment to climatic change." *Advances in Applied Microeconomics* 3: 133–149.

McFarlane, D., L. Kalbfleisch, C. Van Bers, S. Lerner, and J. Robinson. 1992. *Canada in 2030: Sectoral inputs to the sustainability scenario.* Working Paper no. 5, Sustainable Society Project, Department of Environment and Resource Studies, University of Waterloo, Waterloo (April).

McHale, John. 1969. *The future of the future.* New York: G. Braziller.

Meadows, Donella H., Dennis L. Meadows, and Jørgen Randers. 1992. *Beyond the limits: Confronting global collapse, envisioning a sustainable future.* Mills, VT: Chelsea Green.

———, ———, ———, and William W. Behrens, III. 1972. *The limits to growth.* New York: Universe Books.

Mellor, John W. 1988. "The intertwining of environmental problems and poverty." *Environment* 30, no. 9: 8–13, 28–30.

Mesarovic, M., and E. Pestel. 1974. *Mankind at the turning point.* New York: Dutton.

Messerli, Bruno, and Jack D. Ives. 1997. *Mountains of the world: A global priority.* New York: Parthenon.

Metcalfe, S. E. 1987. "Historical data and climate change in Mexico: A review." *Geographical Journal* 153, no. 2: 211–222.

Meyer, William B. 1996. *Human impact on the earth.* Cambridge: Cambridge University Press.

———, and B. L. Turner, II. 1992. Human population growth and global land-use/cover change. *Annual Review of Ecology and Systematics* 23: 39–61.

Michael, Donald N. 1989. "Forecasting and planning in an incoherent context." *Technological Forecasting and Social Change* 36, nos. 1–2 (August): 79–87.

Michaels, P. J. 1979. "The response of the Green Revolution to climatic variability." *Climatic Change* 5: 255–279.

Micklin, Philip P. 1988. "Desiccation of the Aral Sea: A water-management disaster in the Soviet Union." *Science* 241: 1170–1176.

Miles, Ian. 1978. "The ideologies of futurists." In *Handbook of futures research*, ed. Jib Fowles. Westport, CT: Greenwood Press, pp. 67–97.

Mileti, Dennis S. 1999. *Disasters by design: A reassessment of natural hazards in the United States.* Washington: Joseph Henry Press.

Miller, Roberta B., and Harold K. Jacobson. 1992. "Research on the human components of global change: Next steps." *Global Environmental Change* 2, no. 3 (September): 170–182.

Milliman, J. D. 1987. Personal communication. Senior Scientist, Geology and Geophysics Department, Woods Hole Oceanographic Institution.

———, and R. Meade. 1983. "World-wide delivery of river sediment to the oceans." *Journal of Geology* 91, no. 1: 1–21.

———, J. M. Broadus, and F. Gable. 1989. "Environmental and economic impact of rising sea level and subsiding deltas: The Nile and Bengal examples." *Ambio* 18, no. 6: 340–345.

Mimura, Nobuo. 1996. *Data book of sea-level rise.* Tsukuba: Center for Global Environmental Research, National Institute for Environmental Studies, Environment Agency of Japan.

———, M. Isobe, and Y. Hosokawa. 1993. "Impacts of sea level rise on Japanese coastal zones and response strategies." In *Global climate change and the rising challenge of the sea: Proceedings of the third IPPC CZMS workshop, Margarita Island, 9–13 March 1992*, ed. J. O'Callahan. Silver Spring, MD: National Oceanic and Atmospheric Administration, pp. 329–349.

Miranda, F., and A. J. Sharp. 1950. "Characteristics of the vegetation in certain temperate regions of Eastern Mexico." *Ecology* 31: 313–333.

Mitchell, James K. 1988. "Confronting natural disasters: An international decade for natural hazard reduction." *Environment* 30, no. 2: 25–29.

———. 1989. *Risk assessment of global environmental change.* WP-13. Honolulu: Environment and Policy Institute, East–West Center.

———, ed. 1999. *Crucibles of hazard: Mega-cities and disasters in transition.* Tokyo: United Nations University Press.

Mittermeir, R. A. 1988. "Primate diversity and the tropical forest: Case studies from Brazil and Madagascar and the importance of the megadiversity countries." In *Biodiversity*, ed. E. O. Wilson. Washington: National Academy Press, pp. 145–154.

Mix, A. C. 1989. "Influence of productivity variations on long term atmospheric CO_2." *Nature* 337: 541–544.

Moltke, Konrad von, and Atiq Rahman. 1996. "External perspectives on climate change: A view from the United States and the Third World." In *Politics of climate change: A European perspective*, ed. Tim O'Riordan and Jill Jäger. London: Routledge, pp. 330–345.

Monasterio, M., G. Sarmento, and O. T. Solbrig, eds. 1985. "Comparative studies on tropical mountain ecosystems: Planning for research." *Biology International (Special Issue)* 14 (B5).

Moomaw, William, Kilaparti Ramakrishna, Kevin Gallagher, and Tobin Freid. 1999. "The Kyoto Protocol: A blueprint for sustainable development." *Journal of Environment and Development* 8, no. 1 (March): 82–90.

Mooney, H. A., J. Lubchenco, R. Dirzo and O. E. Sala. 1995. "Biodiversity and ecosystem functioning." In *Global biodiversity assessment*, ed. V. H. Heywood. Cambridge: Cambridge University Press for the United Nations Environment Programme (UNEP), pp. 275–325.

Morgan, M. Granger, and Hadi Dowlatabadi. 1996. "Learning from integrated assessment." *Climate Change* 34: 337–368.

———, Baruch Fischhoff, Lester Lave, and Paul Fischbeck. 1996. "A proposal for ranking risk within federal agencies." In *Comparing environmental risks: Tools for setting government priorities*, ed. J. Clarence Davies. Washington: Resources for the Future, pp. 111–148.

———, ———, Keith Florig, Karen Jenni, Kara Morgan, Michael L. Dekay, and Paul Fishbeck. 1997. "A method for ranking risks." Paper presented at the Annual Meeting of the Society for Risk Analysis, Washington, DC, 7–10 December.

Mortsch, Linda D., and Frank H. Quinn. 1996. "Binational Great Lakes–St. Lawrence Basin Project Implementation Plan." *Adaptations* 1, no. 1: 3–5.

Moser, Susanne C. 1998. *Talk globally, walk locally: The cross-scale influence of global change on coastal zone management in Maine and Hawaii.* Harvard University, John F. Kennedy School of Government, Environment and Natural Resources Program, E-98-16. Cambridge, MA: Harvard University.

Mourelle, Christine. 1997. "Biodiversidad de la familia Cactaceae: Un enfoque biogeográfico." Ph.D. dissertation. Centro de Ecología, Universidad Autónoma de México.

Moyers, Bill D. 1990. *Global dumping ground: The international traffic in hazardous waste.* Washington: Seven Locks Press.

Mueller, M. W. 1970. "Changing patterns of agricultural output and productivity in the private and land reform sectors in Mexico 1940–1960." *Economic Development and Cultural Change* 18: 262–266.

Mulder, Henk A. J., and Wouter Biesiot. 1998. *Transition to a sustainable society: A backcasting approach to modelling energy and ecology.* Cheltenham, UK: Edward Elgar.

Mumford, Lewis. 1970. *The pentagon of power.* New York: Harcourt Brace Jovanovich.

Munich Re. 1999. *Topics: Annual review of natural catastrophes 1999.* Munich, Germany: Münchener Rück.

Munn, R. E. 1988. "Environmental prospects for the next century: Implications for long-term policies and research strategies." *Technological Forecasting and Social Change* 33, no. 3: 203–218.

———, (Ted), Anne Whyte, and Peter Timmermann. 1999. "Emerging environmental issues: A global perspective of SCOPE." *Ambio* 28, no. 6 (September): 464–471.

Murdock, Steven H., Richard S. Krannich, and F. Larry Leistritz. 1999. *Hazardous wastes in rural America: Impacts, implications, and options for rural communities.* Lanham, MD: Rowman and Littlefield.

Murray, Charles J. L., and Alan D. Lopez, eds. 1996a. *The global burden of disease: A comprehensive assessment of mortality and disability from diseases, injuries, and risk factors in 1990 and projected to 2020.* Global Burden of Disease and Injury Series, vol. 1. Cambridge, MA: Harvard School of Public Health on behalf of the World Health Organization and the World Bank, distributed by Harvard University Press.

———, and ———, eds. 1996b. *The global burden of disease: A comprehensive assessment of mortality and disability from diseases, injuries, and risk factors in 1990 and projected to 2020, Summary.* Global Burden of Disease and Injury

Series, vol. 1. Cambridge, MA: Harvard School of Public Health on behalf of the World Health Organization and the World Bank, distributed by Harvard University Press.

————, and ————. 1996c. *Global health statistics: A compendium of incidence, prevalence and mortality estimates for over 200 conditions.* Global Burden of Disease and Injury Series, vol. 2. Cambridge, MA: Harvard School of Public Health on behalf of the World Health Organization and the World Bank, distributed by Harvard University Press.

Myers, D. K., N. E. Genter, and M. M. Werner. 1984. "Comparison of the health benefits and health risks of energy development." In *Risks and benefits of energy systems: Proceedings of an International Symposium on the Risks and Benefits of Energy Systems, organized by the International Atomic Energy Agency in co-operation with the United Nations Environment Programme and the World Health Organization, and held in Jülich, 9–13 April 1984.* Vienna: International Atomic Energy Agency, pp. 289–303.

Myers, Norman. 1981. "The hamburger connection: How Central America's forests became North America's hamburgers." *Ambio* 10: 3–8.

————. 1984. *The primary source: Tropical forests and our future.* New York: W. W. Norton.

————. 1986. "Environmental repercussions of deforestation in the Himalayas." *Journal of World Forest Resource Management* 2: 63–72.

————. 1989. "Environment and security." *Foreign Policy* 74: 23–41.

————. 1995. *Environmental exodus: An emergent crisis in the global arena.* Washington: Climate Institute.

Nabhan, G. P. 1985. *Gathering the desert.* Tucson, AZ: University of Arizona Press.

————. 1989. *Enduring seeds.* San Francisco: North Point Press.

Nagpal, Tanvi, and Camilla Foltz, eds. 1995. *Choosing our future: Visions of a sustainable world.* Washington: World Resources Institute.

Nahmad, Salomón, Alvaro Gonzalez, and Martha W. Rees. 1988. *Tecnologías indígenas y medio ambiente: Análisis crítico en anco regiones etnicas.* Mexico: Centro de Ecodesarrollo.

Nanda, Arun, Anatoly Nossikov, Remigins Prokhorskas, and Mirvet H. Abou Shabanah. 1993. "Health in the central and eastern countries of the WHO European Region: An overview." *World Health Statistics Quarterly* 46, no. 3: 158–165.

National Rivers Authority. 1990. *Review of rivers suffering from low flow.* London: National Rivers Authority.

Neev, D., N. Bakler, and K. O. Emery. 1987. *Mediterranean coasts of Israel and Sinai.* New York: Taylor and Francis.

Neilson, R. P., G. A. King, R. L. DeVelice, J. Lenihan, D. Marks, J. Dolph, B. Campbell, and G. Glick. 1989. *Sensitivity of ecological landscapes and regions to global climatic change.* EPA Report 600/3-89/073. Environmental Research Laboratory, Corvallis, Oregon.

Netherlands Ministry of Transport and Public Works. 1990. *A new coastal defence policy for the Netherlands.* Rijkswaterstaat, Netherlands.

Newell, Norman D. 1972. "The evolution of reefs." *Scientific American* 226, no. 6: 54–65.

New Scientist. 1981. "UN acts to save the world's seeds." 92, no. 1284 (17 December): 786.

NGS (National Geographic Society). 1989. *Endangered Earth* [map]. New York: NGS.

Nguyen, D. T. 1979. "The effects of land reform on agricultural production, employment and income distribution: A statistical study of Mexican states 1959–1969." *Economic Journal* 89: 624–635.

Nickerson, Colin. 1991. "Bangladesh tragedy follows a bitter history." *Boston Globe* 13 May, section 1: 5.

Nihon Keizai Shinbun (Nikkei Newspaper). 1991. "Current eye." 22 July.

Niigata Prefecture. 1991. "Summary procedures of the Niigata International Forum of the Japan Sea, Niigata, Japan" (English summary).

Nishimura, S. 1990. *The formation of the Japan Sea: Biogeographical approach.* Tokyo: Tsukuji Publishers (in Japanese).

Nishioka, S. 1990. "The Japanese response to global warming: Background, policy, and research work." In *Responding to Global Warming, Options for the Pacific and Asia, Proceedings of a workshop sponsored by Argonne National Laboratory and Environment and Policy Institute, East–West Center, June 21–27, 1989: Honolulu, Hawaii,* ed. D. G. Streets and T. A. Siddiqi. ANL/EAIS/ TM-17. Argonne, IL: Argonne National Laboratory, pp. 6-11–6-29.

Nobel, P. S. 1988. *Environmental biology of agaves and cacti.* New York: Cambridge University Press.

Nordhaus, William D., and D. Popp. 1997. "What is the value of scientific knowledge? An application to global warming using the PRICE model." *Energy Journal* 18, no. 1: 1–45.

Norgaard, R. B. 1988. "Sustainable development: A co-evolutionary view." *Futures* 20, no. 6: 606–620.

O'Brien, Karen L. 1998. *Sacrificing the forest: Environmental and social struggles in Chiapas.* Boulder, CO: Westview Press.

Oden, B. 1982. "A historical perspective on the risk panorama in a changing society." In *On making use of history: Research and reflections from Lund,* ed. J. Zitomersky. Lund Studies in International History 15. Lund, Sweden: Scandinavian University Books, pp. 153–178.

OECD (Organisation for Economic Co-operation and Development). 1997a. *OECD environmental performance reviews: A practical introduction.* Paris: OECD.

———. 1997b. *Sustainable consumption and production.* Paris: OECD.

———. 1998. *Environmental performance reviews: Mexico.* Paris: OECD.

——— environmental data. "Données OCDE sur l'environnement." Biennial. Paris: Organisation for Economic Co-operation and Development (OECD).

Oerlemans, J. 1989. "A projection of future sea level." *Climate Change* 15: 151–174.

O'Looney, John. 1995. *Economic development and environmental control: Balancing business and community in an age of NIMBYs and LULUs.* Westport, CT: Quorum Books.

Olshansky, S. J., and A. B. Ault. 1986. "The fourth stage of the epidemiologic transition: The age of delayed degenerative diseases." *Milbank Quarterly* 64, no. 3: 355–391.

Omran, Abdel R. 1971. "The epidemiologic transition: A theory of the epidemiology of population change." *Milbank Memorial Fund Quarterly* 49, no. 4: 509–538.

O'Riordan, Timothy, and Steve Rayner. 1990. "Chasing a specter: Risk management for global environmental change." In *Understanding global environmental change: The contributions of risk analysis and risk management, a report on an international workshop, Clark University, October 11–13, 1990*, ed. Roger E. Kasperson, Kirstin Dow, Dominic Golding, and Jeanne X. Kasperson. Worcester, MA: The Earth Transformed Program, Clark University, pp. 45–62.

———, and ———. 1991. "Risk management for global environmental change." *Global Environmental Change* 1, no. 2: 91–108.

———, Chester L. Cooper, Andrew Jordan, Steve Rayner, Kenneth R. Richards, Paul Runci, and Shira Yoffe. 1998. "Institutional frameworks for political action." In *Human choice and climate change. Vol. 1: The societal framework*, ed. Steve Rayner. Columbus, OH: Battelle Press, pp. 345–439.

Ortiz, Monasterio Fernando. 1987. *Tierra profanada: Historia ambiental de México*. Mexico: Instituto Nacional de Antropología e Historia (INAH)/ Secretaría de Desarrollo Urbano y Ecología (SEDUE).

Ostrom, Elinor. 1990. *Governing the commons: The evolution of institutions for collective action*. Cambridge: Cambridge University Press.

———, Joanna Burger, Christopher B. Field, Richard B. Norgaard, and David Policansky. 1999. "Revisiting the commons: Local lessons, global challenges." *Science* 284 (9 April): 278–282.

OTA (Office of Technology Assessment). 1995. *Policy tools: A user's guide*. OTA-ENV-634. Washington: OTA.

Otway, Harry. 1987. "Experts, risk communication and democracy." *Risk Analysis* 7, no. 2: 125–129.

Owens, J. W. 1997. "Life-cycle assessment: Constraints on moving from inventory to impact assessment." *Journal of Industrial Ecology* 1, no. 1 (Winter): 37–50.

PAI (Population Action International). 1998. *Educating girls: Gender gaps and gains*. Washington: PAI.

Palm, Risa. 1990. *Natural hazards: An integrative framework for research and planning*. Baltimore: Johns Hopkins University Press.

Palmlund, Ingar. 1992. "Social drama and risk evaluation." In *Social theories of risk*, ed. Sheldon Krimsky and Dominic Golding. Westport, CT: Praeger, pp. 197–212.

Paranjpye, V. 1988. *Evaluating the Tehri dam: An extended cost benefit appraisal*. New Delhi: Indian National Trust for Art and Cultural Heritage.

Parikh, J. K. 1992. "IPCC strategies unfair to the South." *Nature* 360: 507–508.

———, and J. P. Painuly. 1994. "Population, consumption patterns and climate change: A socioeconomic perspective from the South." *Ambio* 23: 434–437.

Parker, Frank L., Torsten L. Anderson, Roger E. Kasperson, and Stephen A. Parker. 1987. *Technical and socio-political issues in radioactive waste disposal*. Stockholm: The Beijer Institute.

Parker, Jonathan, and Chris Hope. 1992. "The state of the environment: A survey of reports from around the world." *Environment* 34, no. 1: 18–20, 39–44.

Parry, Martin L. 1985. "The impact of climate variations on agricultural margins." In *Climate impact assessment: Studies of the interaction of climate and society*, ed. Robert W. Kates, Jesse H. Ausubel, and Miriam Berberian. Chichester, UK: Wiley, pp. 351–367.

———, Timothy R. Carter, and Nicolaas T. Konijn, eds. 1988. *The impact of climatic variations on agriculture*, 2 volumes. Dordrecht, Netherlands: Kluwer Academic.

Pastor, R. A., and J. G. Castaneda. 1988. *The limits to friendship: The United States and Mexico*. New York: A. Knopf.

Pe Benito Claudio, Corazon. 1988. "Risk analysis in developing countries." *Risk Analysis* 8, no. 4: 475–478.

Pearse, Andres. 1980. *Seeds of plenty, seeds of change*. Oxford: Clarendon Press.

Pearson, Charles S. 1987. *Multinational corporations, environment, and the Third World: Business matters*. Durham, NC: Duke University Press.

Peck, S. C., and T. J. Teisberg. 1992. *Global warming uncertainties and the value of information: An analysis using CETA*. Palo Alto, CA: Electric Power Research Institute.

———, and ———. 1995. "International CO_2 emissions control, and analysis using CETA." *Energy Policy* 23: 297–308.

Peerbolte, E. B., J. G. de Ronde, L. P. M. de Vrees, M. Mann, and G. Barse. 1991. *Impacts of sea-level rise on society: A case study for the Netherlands*. Delft and the Hague: Delft Hydraulics and Ministry of Transport, Public Works, and Waste Management.

Perrings, C. 1987. *Economy and environment: A theoretical essay on the interdependence of economic and environmental systems*. Cambridge: Cambridge University Press.

Perrow, Charles, and Mauro F. Guillén. 1990. *The AIDS disaster: The failure of organizations in New York and the nation*. New Haven, CT: Yale University Press.

Peto, Richard, Alan Lopez, Jillian Boreham, Michael Thun, and Clark Heath, Jr. 1994. *Mortality from smoking in developed countries, 1950–2000: Indirect estimates from national vital statistics*. Oxford: Oxford University Press.

Pezzoli, Keith. 1998. *Human settlements and planning for ecological sustainability: The case of Mexico City*. Cambridge, MA: MIT Press.

Pianka, E. R. 1966. Latitudinal gradients in species diversity: A review of concepts. *American Naturalist* 100: 33–46.

Polak, Frederick L. 1961. *The image of the future: Enlightening the past, orientating the present, forecasting the future* (2 vols.) Leiden: A. W. Sythoff.

Polanyi, M. 1958. *Personal knowledge: Towards a post-critical philosophy*. London: Routledge and Kegan Paul.

Porter, J. W. 1976. "Autotrophy, heterotrophy and resource partitioning in Caribbean reef-building corals." *American Naturalist* 110: 731–742.

Postel, Sandra. 1997. "Water for food production: Will there be enough in 2025?" Paper presented at the 9th Congress of the International Water Resources Association, Montreal, Canada, 3–7 September.

Potter, Lesley, Harold Brookfield, and Yvonne Byron. 1995. "The Eastern Sundaland region of Southeast Asia." In *Regions at risk: Comparisons of threatened environments*, ed. Jeanne X. Kasperson, Roger E. Kasperson, and B. L. Turner, II. Tokyo: United Nations University Press, pp. 460–518.

PRB (Population Reference Bureau). 2000. *World population data sheet 2000*. Washington: PRB.

Preston, Samuel H. 1975. "The changing relation between mortality and level of economic development." *Population Studies* 29, no. 2: 231–248.

———. 1976. *Mortality patterns in national populations: With special reference to recorded causes of death*. New York: Academic Press.

———. 1980. "Causes and consequences of mortality declines in less developed countries during the twentieth century." In *Population and economic change in developing countries*, ed. Richard A. Easterlin. Chicago: University of Chicago Press, pp. 289–360.

———. 1985. "Mortality and development revisited." In *Quantitative studies of mortality decline in the developing world*, ed. Julie Davanzo. World Bank Staff Working Paper no. 683. Washington: World Bank, pp. 97–122.

Price, Larry W. 1981. *Mountains and man: A study of process and environment*. Berkeley: University of California Press.

Price, Martin F. 1990. *The human aspects of global change: Final report*. Environment and Societal Impacts Group. Boulder, CO: Center for Atmospheric Research.

———. 1995. "Patterns of the development of tourism in mountain communities." In *Mountains at risk: Current issues in environmental studies*, ed. Nigel J. R. Allan. New Delhi: Manohar, pp. 199–219.

———, and John R. Haslett. 1995. "Climate change and mountain ecosystems." In *Mountains at risk: Current issues in environmental studies*, ed. Nigel J. R. Allan. New Delhi: Manohar, pp. 73–97.

Puente, Sergio, and Jorge Legorreta, eds. 1988. *Medio ambiente y calidad de vida*. Mexico: Plaza y Valdes.

Rabinowitz, D. 1975. "Planting experiments in mangrove swamps of Panama." In *Proceedings of the International Symposium on Biology and Management of Mangroves, East–West Center, Honolulu, Hawaii, October 8–11, 1974, vol. 1*, ed. Gerald E. Walsh, Samuel C. Snedaker, and Howard J. Teas. Gainesville: Institute of Food and Agricultural Sciences, University of Florida, pp. 385–393.

———. 1978. "Dispersal properties of mangrove propagules." *Biotropica* 10, no. 1: 47–57.

———, S. Cairns, and T. Dillon. 1986. "Seven kinds of rarity." In *Conservation biology*, ed. M. E. Soulé. Sunderland, MA: Sinauer, pp. 182–204.

Ramírez-Pulido, J., and C. Müdespacher. 1987. "Estado actual y perspectivas del conocimiento de los mamíferos de México." *Ciencia* 38: 49–67.

Rangasami, A. 1985. "Failure of exchange entitlements." *Economic and Political Weekly* 20 no. 41: 1747–1752.

Rapoport, E. H. 1975. *Areografía: Estrategias geográficas de las especies*. México: Fondo de Cultura Económica.

Rappaport, Roy A. 1984. *Pigs for the ancestors*, New edn. New Haven: Yale University Press.

———. 1988. Toward postmodern risk analysis. *Risk Analysis* 8 no. 2: 189–191.

Raskin, Paul, Michael Chadwick, Tim Jackson, and Gerald Leach. 1996. *The sustainability transition: Beyond conventional development.* Polestar Series Report No. 1. Stockholm: Stockholm Environment Institute.

———, Peter Gleick, Paul Kirshen, Gil Pontius, and Kenneth Strzepek. 1997. *Water futures: Assessment of long-range patterns and problems.* Comprehensive Assessment of the Fresh Water Resources of the World, Background Report, 3. Stockholm: Stockholm Environment Institute.

———, Gilberto Gallopin, Pablo Gutman, Al Hammond, and Rob Swart. 1998. *Bending the curve: Toward global sustainability: A Report of the Global Scenario Group.* Polestar Series Report No. 8. Stockholm: Stockholm Environment Institute.

Ravetz, Jerome R. 1971. *Scientific knowledge and its social problems.* Oxford: Oxford University Press.

———. 1990. "Usable knowledge, usable ignorance: Incomplete science with policy implications." In *The merger of knowledge with power.* London: Cassell, pp. 260–283.

Rawls, John. 1971. *A theory of justice.* Oxford: Oxford University Press.

Rayner, Steve, and Robin Cantor. 1987. "How fair is safe enough? The cultural approach to society technology choice." *Risk Analysis* 7, no. 1: 3–13.

———, Francis Bretherton, Stanley Buol, Michael Fosberg, Wolf Grossman, Richard Houghton, Rattan Lal, Jeffrey Lee, Stephen Lonergan, Jennifer Olson, Richard Rockwell, Colin Sage, and Evert van Imhoff, 1991. "A wiring diagram for the study of land use/cover change: Report of Working Group A." In *Changes in land use and land cover: A global perspective*, ed. William B. Meyer and B. L. Turner, II. Cambridge: Cambridge University Press, pp. 13–53.

Rees, William E., and Mathis Wackernagel. 1994. "Ecological footprints and appropriated carrying capacity: Measuring the natural capital requirements of the human economy." In *Investing in natural capital: The ecological economics approach to sustainability*, ed. Ann Mari Jansson, Monica Hammer, Carl Folke, and Robert Costanza. Washington: Island Press, pp. 362–390.

Reid, Walter V. 1997. "Strategies for conserving biodiversity." *Environment* 39, no. 7 (September): 16–20, 39–43.

Renn, Ortwin, Thomas Webler, and Peter Wiedemann, eds. 1995a. *Fairness and competence in citizen participation: Evaluating models for environmental discourse.* Dordrecht: Kluwer.

———, ———, and ———. 1995b. "The pursuit of fair and competent citizen participation." In *Fairness and competence in citizen participation: Evaluating models for environmental discourse*, ed. Ortwin Renn, Thomas Webler, and Peter Wiedemann. Dordrecht: Kluwer, pp. 339–368.

Rhoades, Robert E. 1992. Thinking globally, acting locally: Technology for sustainable mountain agriculture. In *Sustainable development of mountain agriculture, vol. 1: Perspectives and issues*, ed. Narpat S. Jodha, Mahesh Banskota, and Tej Partap. New Delhi: Oxford and IBH Publishers, pp. 253–272.

———. 1997. *Pathways toward a sustainable mountain agriculture for the 21st century: The Hindu Kush–Himalayan experience.* Kathmandu: International Centre for Integrated Mountain Development.

Richards, Paul. 1985. *Indigenous agricultural revolution: Ecology and food production in West Africa.* Boulder, CO: Westview Press.

Ridington, Robin. 1990. *Little Bit know something: Stories in a language of anthropology.* Iowa City: University of Iowa Press.

Riesman, David. 1950. *The lonely crowd.* New Haven, CT: Yale University Press.

Rivers, J. P. W. 1982. "Women and children last: An essay on sex discrimination in disasters." *Disasters* 6, no. 4: 256–267.

Robinson, A. H. 1979. "Coral: The living rock." In *Virgin Islands National Park: The story behind the scenery.* Las Vegas: K. E. Publications, pp. 23–31.

Robinson, J. C. 1986. "Philosophical origins of the economic valuation of life." *Milbank Quarterly* 64, no. 1: 133–155.

Robinson, John (B). 1982a. "Bottom-up methods and low-down results: Changes in the estimation of future energy demands." *Energy – The International Journal* 7, no. 7: 627–635.

———. 1982b. "Energy backcasting: A proposed method of policy analysis." *Energy Policy* 10, no. 4: 337–345.

———. 1988. "Unlearning and backcasting: Rethinking some of the questions we ask about the future." *Technological Forecasting and Social Change* 33, no. 4: 325–338.

———. 1990a. "Climate change in the Great Lakes Basin Project." Memorandum to Environment Canada.

———. 1990b. "Futures under glass: A recipe for people who hate to predict." *Futures* 22, no. 9: 820–843.

———. 1990c. "Risks, futures and other optical illusions: Some comparisons between the social construction of risk and of forecasting issues." Paper presented at the Annual Conference of the International Sociological Association, Madrid, 9–13 July.

———. 1991. "Modelling the interactions between human and natural systems." *International Social Science Journal* 130: 629–647.

———. 1992. "Of maps and territories: The use and abuse of socioeconomic modelling in support of decision making." *Technological Forecasting and Social Change* 42, no. 2: 147–164.

———, and D. Scott Slocombe. 1996. Exploring a sustainable future for Canada. In *Life in 2030: Exploring a sustainable future for Canada,* by John B. Robinson, David Biggs, George Francis, Russel Legge, Sally Lerner, D. Scott Slocombe, and Caroline Van Bers. Vancouver, BC: UBC Press, pp. 3–12.

———, and Peter Timmerman. 1990. "Climate change in the Great Lakes Basin Project." Proposal to Environment Canada, May 1990.

———, George Francis, and Sally Lerner. 1992. "Canada as a sustainable society: Environmental and socio-political dimensions." Working Paper no. 4, Sustainable Society Project, Department of Environment and Resource Studies, University of Waterloo, Waterloo (April).

———, Caroline Van Bers, and David Biggs. 1996. "The Sustainable Society Project." In *Life in 2030: Exploring a sustainable future for Canada,* by John B. Robinson, David Biggs, George Francis, Russel Legge, Sally Lerner, D. Scott Slocombe, and Caroline Van Bers. Vancouver: UBC Press, pp. 13–25.

————, George Francis, Russel Legge, and Sally Lerner. 1990. "Defining a sustainable society: Values, principles and definitions." *Alternatives* 17, no. 2: 36–46.

————, David Biggs, George Francis, Russel Legge, Sally Lerner, D. Scott Slocombe, and Caroline Van Bers. 1996a. *Life in 2030: Exploring a sustainable future for Canada*. Vancouver, BC: UBC Press.

————, George Francis, Sally Lerner, and Russel Legge. 1996b. "Defining a sustainable society." In *Life in 2030: Exploring a sustainable future for Canada*, by John B. Robinson, David Biggs, George Francis, Russel Legge, Sally Lerner, D. Scott Slocombe, and Caroline Van Bers. Vancouver, BC: UBC Press, pp. 26–52.

Rocheleau, Dianne (E.), Patricia Benjamin, and Alex Diang'a. 1995. "The Ukambani region of Kenya." In *Regions at risk: Comparisons of threatened environments*, ed. Jeanne X. Kasperson, Roger E. Kasperson, and B. L. Turner, II. Tokyo: United Nations University Press, pp. 186–254.

————, Barbara P. Thomas-Slayter, and Esther Wangari, eds. 1996. *Feminist political ecology: Global issues and local experience*. London: Routledge.

Rodda, Annabel. 1991. *Women and the environment*. Atlantic Highlands, NJ: Zed Press.

Rodgers, G. B. 1979. "Income and inequality as determinants of mortality: An international cross-section analysis." *Population Studies* 33, no. 2: 343–351.

Rogers, Peter. 1990. "Socio-economic development of arid regions: Alternative strategies for the Aral Basin." Paper presented to the International Symposium on The Aral Crisis: Causes, Consequences and Ways of Solution, Nukus, Karakalpak Autonomous Republic, Uzbek SSR, USSR, 1–7 October 1990.

————. 1991. "Commentary: The Aral Sea." *Environment* 33, no. 1: 2–3.

Rosenzweig, Cynthia, and Daniel Hillel. 1998. *Climate change and the global harvest: Potential impacts of the greenhouse effect on agriculture*. Oxford: Oxford University Press.

Rotmans, Jan, and Bert de Vries, eds. 1997. *Perspectives on global change: The TARGETS approach*. Cambridge: Cambridge University Press.

————, and Hadi Dowlatabadi. 1998. "Integrated assessment modeling." In *Human choice and climate change. vol. 3: Tools for policy analysis*, ed. Steve Rayner and Elizabeth L. Malone. Columbus, OH: Battelle Press, pp. 291–378.

————, M. B. A. van Asselt, A. J. de Bruin, M. G. J. den Elzen, J. de Greef, H. Hilderink, A. T. Hoekstra, N. A. Janssen, H. W. Köster, W. J. M. Martens, L. W. Niessen, and H. J. M. de Vries. 1994. *Global change and sustainable development: A modelling perspective for the next decade*. Bilthoven, Netherlands: National Institute for Public Health and Environmental Protection (RIVM).

Royal Town Planning Institute. 1990. *Planning implications of sea level rise*. London: Royal Town Planning Institute.

Ruckelshaus, William D. 1983. "Science, risk and public policy." *Science* 221.

Russell, J. 1990. *Environmental issues in eastern Europe: Setting an agenda*. London: Royal Institute of International Affairs.

Rzedowski, J. 1978. *Vegetación de México*. Mexico: Editorial Limusa.

Sachs, I. 1977. "Environment and development: A new rationale for domestic policy." Formulation and International Cooperation Policies. Joint Project on Environment and Development 2, Ottawa.

Sagan, Leonard A. 1987. *The health of nations*. New York: Basic Books.

Sagar, Ambuj, and Milina Kandlikar. 1997. "Knowledge, rhetoric, and power: The international politics of climate change." *Economic and Political Weekly* 32, no. 49: 3139–3148.

Sagasti, Francisco, R., and Michael E. Colby. 1995. "Eco-development and perspectives on global change from developing countries." In *Global accord: Environmental challenges and international responses*, ed. Nazli Choucri. Cambridge, MA: MIT Press, pp. 175–203.

Sagoff, Mark. 1995. "Carrying capacity and ecological economics." *Bioscience* 45: 610–620.

Salter, L. 1988. *Mandated science: Science and scientists in the making of standards*. Boston and Dordrecht, Netherlands: Kluwer Academic.

Sanderson, Steven E. 1986. *The transformation of Mexican agriculture: International structure and the politics of rural change*. Princeton, NJ: Princeton University Press.

Sanwal, M. 1989. "What we know about mountain development: Common property, investment priorities, and institutional arrangements." *Mountain Research and Development* 9, no. 1: 3–14.

Satterthwaite, David. 1997. "Environmental transformations in cities as they get larger, wealthier and better managed." *Geographical Journal* 163, no. 2 (July): 216–224.

SCEP (Study of Critical Environmental Problems). 1971. *Man's impact on the global environment: Report of the study of critical environmental problems*. Cambridge, MA: MIT Press.

Scheibe, K. E. 1986. "Self-narratives and adventure." In *Narrative psychology: The storied nature of human conduct*, ed. Theodore R. Sarbon. New York: Praeger, pp. 129–151.

Schelling, Thomas C. 1983. "Climatic change: Implications for welfare and policy." In *Changing climate: Report of the Carbon Dioxide Assessment Committee*. Washington: National Academy Press, pp. 449–482.

Schellnhuber, Hans-Joachim, Arthur Block, Martin Cassel-Gintz, Jürgen Kropp, Gerhard Lammel, Wiebke Lass, Roger Lienenkamp, Carsten Loose, Matthias K. B. Lüdeke, Oliver Moldenhauer, Gerhard Petschel-Held, Matthias Plöchl, and Fritz Reusswig. 1997. "Syndromes of global change." *GAIA* 6, no. 1: 19–34.

Schmandt, Jurgen, and Judith Clarkson, eds. 1992. *The regions and global warming*. New York: Oxford University Press.

Schneider, Stephen H. 1983. "CO_2, climate and society: A brief overview." In *Social science research and climate change: An interdisciplinary appraisal*, ed. Robert S. Chen, Elise Boulding, and Stephen H. Schneider. Dordrecht, Netherlands: D. Reidel, pp. 9–15.

———, and B. L. Turner, II. 1995. "Summary: Anticipating global change surprise." In *Elements of change, 1994*, ed. Susan J. Hassol and John Katzenberger. Aspen, CO: Aspen Global Change Institute, pp. 130–190.

———, ———, and Holly Morehouse Garriga. 1998. "Imaginable surprise in global change science." *Journal of Risk Research* 1, no. 2 (April): 165–185.

Scholes, Robert. 1975. *Structural fabulation: An essay on fiction of the future.* Notre Dame, IN: Notre Dame University Press.

Schrecker, T. 1983. "The Conserver Society revisited." A Science Council of Canada Discussion Paper, Ottawa.

Schroeder, Richard A. 1987. *Gender vulnerability to drought: A case study of the Hausa social environment.* Natural Hazard Research Working Paper, 40. Boulder, CO: Institute of Behavioral Science, University of Colorado.

Schumacher, E. F. 1973. *Small is beautiful.* New York: Harper and Row.

Schwab, Margo. 1988. "Differential exposure to carbon monoxide among sociodemographic groups in Washington, DC." Ph.D. thesis, Graduate School of Geography, Clark University, Worcester, MA.

Science Council of Canada. 1973. *Natural resource policy in Canada.* Report #19. Ottawa: Science Council of Canada.

———. 1977. *Canada as a conserver society: Resource uncertainties and the need for new technologies.* Report no. 27. Ottawa: Science Council of Canada.

Scott, James C. 1976. *The moral economy of the peasant.* New Haven: Yale University Press.

Seabrook, Jeremy. 1993. *Victims of development: Resistance and alternatives.* London: Verso.

Sears, P. B., and H. Clisby. 1955. "Pleistocene climate in Mexico." *Bulletin of the Geological Society of America* 66: 521–530.

Sen, Amartya. 1981. *Poverty and famine: An essay on entitlement and deprivation.* Oxford: Oxford University Press.

———. 1990. "Food, economics and entitlements." In *The political economy of hunger,* vol. 1, ed. J. Drèze and A. Sen. Oxford: Clarendon, pp. 34–50.

Shakman, R. A. 1980. *Poison-proof your body: Food, pollution, and your health.* Westport, CT: Arlington House Publishers.

Sharma, P., and M. Banskota. 1992. "Population dynamic and sustainable agricultural development in mountain areas." In *Sustainable mountain agriculture, vol. 1: Perspectives and issues,* ed. Narpat S. Jodha, Mahesh Banskota, and Tej Partap. New Delhi: Oxford and IBH Publishers, pp. 165–184.

Shaw, Daigee, ed. 1996. *Comparative analysis of siting experience in Asia.* Taipei: Institute of Economics, Academia Sinica.

Sheridan, Thomas E. 1988. *Where the dove calls: The political ecology of a peasant corporate community in northwestern Mexico.* Tucson: University of Arizona Press.

Shiva, Vandana. 1994. *Close to home: Women reconnect ecology, health, and development worldwide.* London: Earthscan.

Shlyakhter, Alexander, L. James Valverde, Jr., and Richard Wilson. 1995. "Integrated risk analysis of global climate change." *Chemosphere* 30 no. 8: 1585–1618.

Shrader-Frechette, Kristin S. 1985. *Risk analysis and scientific method: Methodology and ethical problems with evaluating societal hazards.* Dordrecht, Netherlands: D. Reidel.

Shutain, G., and H. Chunru. 1988. *Problems of the environment in Chinese agri-

culture and a strategy for its ecological development: An overview. Beijing: Ministry of Agriculture and Beijing Agricultural University.

Shy, C. 1990. "Lead in petrol: The mistake of the XXth century." *World Health Statistics Quarterly* 43, no. 3: 168–176.

Simmonds, W. H. C. 1989. "Gaining a sense of direction in futures work." *Technological Forecasting and Social Change* 36: 61–67.

Singh, S. C., ed. 1989. *Impact of tourism on the mountain environment.* Meerat, India: Research India.

Skjaerseth, Jon Birger. 1998. "The making and implementation of North Sea commitments: The politics of environmental participation." In *The implementation and effectiveness of international environmental commitments: Theory and practice,* ed. David G. Victor, Kal Raustiala, and Eugene B. Skolnikoff. Cambridge, MA: MIT Press, pp. 327–380.

Skolnikoff, Eugene B. 1999. "The role of science in policy: The climate change debate in the United States." *Environment* 41, no. 5 (June): 17–20, 42–45.

Slocombe, D. Scott, and Caroline Van Bers. 1991. "Ecological design criteria for a sustainable Canadian society." Working Paper no. 2, Sustainable Society Project, Department of Environment and Resource Studies, University of Waterloo, Waterloo (March).

Slovic, Paul. 1987. "Perception of risk." *Science* 236: 280–285.

———, Baruch Fischhoff, and Sarah Lichtenstein. 1985. "Characterizing perceived risk." In *Perilous progress,* ed. Robert W. Kates, Christoph Hohenemser, and Jeanne X. Kasperson. Boulder, CO: Westview Press. pp. 91–125.

Smit, Barry, ed. 1993. *Adaptation to climate variability and change: Report of the Task Force on Climate Adaptation, The Canadian Climate Program.* Occasional Paper no. 19. Guelph, Ontario, Canada: Department of Geography, University of Guelph.

Smith, Kirk R. 1988. "Air pollution: Assessing total exposure in developing countries." *Environment* 30, no. 10 (December): 16–20, 28–35.

———. 1990a. "The risk transition." *International Environmental Affairs* 2, no. 3: 227.

———. 1990b. "The risk transition and global warming." *Journal of Energy Engineering* 116, no. 3: 178.

———. 1995. *The potential of human exposure assessment for air pollution regulation.* Geneva: WHO Collaborating Centre for Studies of Environmental Risk and Development.

———, and Yok-Shin F. Lee. 1993. "Urbanization and the environmental risk transition." In *Third World cities: Problems, policies, prospects,* ed. John D. Kasarda and Allan M. Parnell. Newbury Park, CA: Sage Publishers, pp. 161–179.

———, Richard A. Carpenter, and M. Susanne Faulstich. 1988. *Risk assessment of hazardous chemical systems in developing countries.* Occasional Paper No. 5. Honolulu: EWEAPI.

Smith, Nigel J. H., E. Adilson, S. Serrão, Paulo T. Alvim, and Italo C. Falesi. 1995a. *Amazonia: Resiliency and dynamism of the land and its people.* Tokyo: United Nations University Press.

———, ———, ———, ———, and ———. 1995b. "Amazonia." In *Regions at risk: Comparisons of threatened environments*, ed. Jeanne X. Kasperson, Roger E. Kasperson, and B. L. Turner, II. Tokyo: United Nations University Press, pp. 42–91.

Smith, V. Kerry., ed. 1986. *Advances in applied micro-economics, Vol. 4: Risk, uncertainty and the valuation of benefits and costs*. Stamford, CT: JAI Press.

Snedaker, S. C. 1989. "Overview of ecology of mangroves and information needs for Florida Bay." *Bulletin of Marine Science* 44, no. 1: 341–347.

Snyder, Gary. 1990. *The practice of the wild: Essays*. San Francisco: North Point Press.

Sokona, Youha, Stephen Humphreys, and Jean-Philippe Thomas. 1999. "Sustainable development: A centerpiece of the Kyoto Protocol – An African perspective." In *Towards equity and sustainability in the Kyoto Protocol*, ed. Karin Hultcrantz. Stockholm: Stockholm Environment Institute, pp. 31–44.

Solow, Andrew R. 1987. "The application of eigenanalysis to tide-gauge records of relative sea level." *Continental Shelf Research* 7, no. 6: 629–641.

———, Samuel J. Ratick, and R. Mohan Srivastava. 1994. "Conditional simulation and the value of information." In *Geostatistics for the next century: An international forum in honour of Michel David's contribution to geostatistics, Montreal, 1993*, ed. Roussos Dimitrakopoulos. Quantitative Geology and Geostatistics, 4. Dordrecht, Netherlands: Kluwer, pp. 209–217.

Sontag, Susan. 1989. *Illness as metaphor* and *AIDS and its metaphors*. New York: Anchor Books [combined edition of works first published in 1978 and 1989 (New York: Farrar, Straus and Giroux)].

South Commission. 1990. *Challenge to the South*. Oxford: Oxford University Press.

Stanley, D. J. 1988. "Subsidence in the northeastern Nile Delta: Rapid rates, possible causes, and consequences." *Science* 240: 497–500.

START (Global Change SysTem for Analysis, Research, and Training. 1999. *Start Network News: Special issue: Capacity building*. Issue No. 4. Washington: International START Secretariat.

State of the Future. Annual. Washington: American Council for the United Nations University.

State of the World. Annual. Washington: Worldwatch Institute.

State of the World's Children. Annual. New York: United Nations Children's Fund (UNICEF).

Steinberg, Paul F. 1998. "Defining the global biodiversity mandate: Implications for international policy." *International Environmental Affairs* 10, no. 2 (Spring): 113–130.

Stern, Paul C., and Harvey V. Fineberg, eds. 1996. *Understanding risk: Informing decisions in a democratic society*. Washington: National Academy Press.

———, and William E. Easterling, eds. 1999. *Making climate forecasts matter*. Washington: National Academy Press.

———, Oran Young, and Daniel Druckman, eds. 1992. *Global environmental change: Understanding the human dimensions*. Washington: National Academy Press.

———, Thomas Dietz, Vernon W. Ruttan, Robert H. Socolow, and James L. Sweeney, eds. 1997. *Environmentally significant consumption: Research directions*. Washington: National Academy Press.

Stevens, G. C. 1989. "The latitudinal gradient in geographical range: How so many species coexist in the tropics." *American Naturalist* 133: 240–256.

Stokke, Per R., Thomas A. Boyce, William K. Ralston, and Ian H. Wilson. 1991. "Visioning (and preparing for) the future: The introduction of scenario-based planning into Staf oil." *Technological Forecasting and Social Change* 40: 73–86.

Sugimura, T., ed. 1989. *Interim report of the Sub-Group on Impacts and Response Strategies, Advisory Committee on Climate Change.* Tokyo: Japan Environment Agency.

Sullivan, R. E., and P. Weng. 1987. "Comparison of risk estimates using life-table methods." *Health Physics* 53, no. 2: 123–134.

Susman, Paul, Phil O'Keefe, and Ben Wisner. 1984. "Natural disasters: A radical interpretation." In *Interpretations of calamity from the viewpoint of human ecology*, ed. Kenneth Hewitt. Boston: Allen and Unwin, pp. 263–283.

Svedin, Uno, and Britt Aniansson. 1987. *Surprising futures: Notes from an international workshop on long term world development.* Stockholm, Sweden: Swedish Council for Planning and Coordination of Research.

Szekely, Miguel, and Ivan Restrepo. 1988. *Frontera agrícola y colonizacíon.* Mexico: Centro de Ecodesarrollo.

Tapia, N.-E. 1992. "Mountain agricultural development strategies." In *Sustainable mountain agriculture, vol. 1: Perspectives and issues*, ed. N. S. Jodha, M. Banskota, and Tej Partap. New Delhi: Oxford and IBH Publishing Co, pp. 115–128.

Teitelbaum, M. S. 1975. "Relevance of demographic transition theory for developing countries." *Science* 188: 420–425.

Thompson, Michael. 1983. "Why climb Everest? A critique of risk assessment." *Journal of Mountain Research and Development* 3: 299–302.

———, and Steve Rayner. 1998. "Cultural discourses." In *Human choices and climate change, vol. 1: The societal framework*, ed. Steve Rayner and Elizabeth L. Malone. Columbus, OH: Battelle Press, pp. 265–343.

———, Michael Warburton, and Tom Hatley. 1986. *Uncertainty on a Himalayan scale.* London: Milton Ash Editions.

Tilden, Freeman. 1957. *Interpreting our heritage.* Chapel Hill, NC: University of North Carolina Press.

Timmerman, Peter M. 1981. *Vulnerability, resilience and the collapse of society.* Environmental Monograph No. 1. Toronto: Institute for Environmental Studies, University of Toronto.

———, and R. E. Munn. 1997. "The tiger in the dining room: Designing and evaluating integrated assessments of atmospheric change." *Environmental Monitoring and Assessment* 46: 45–58.

Titus, James G., ed. 1986. *Effects of changes in stratospheric ozone and global climate, vol. 4: Sea level rise.* Washington: US Environmental Protection Agency.

Tobin, Graham A., and Burrell E. Montz. 1997. *Natural hazards: Explanation and integration.* New York: Guilford.

Tokioka, T. 1990. "Impacts on East Asian climate." *Pollution and Countermeasures* 26: 1060–1067 (in Japanese).

Tolba, Mostafa Kamal. 1987. *Sustainable development: Constraints and opportunities.* London: Butterworth Heinemann.

Toledo, Victor M. 1982. "Pleistocene changes of vegetation in Tropical Mexico." In *Biological diversification*, ed. G. T. Prance. New York: Columbia University Press, pp. 93–111.

———. 1989. *Naturaleza, producción y cultura. Ensayos de ecología política*. Mexico: Universidad Veracruzana.

Topping, J. C., A. Qureshi, and S. A. Sherer. 1990. "Implication of climate change for the Asian and Pacific Region." Paper for the Asian Pacific Seminar on Climate Change, 23–26 January, 1991, Nagoya, Japan. Washington: Climate Institute.

Torrie, Ralph., and David B. Brooks. 1988. *2025: Soft energy futures for Canada: 1988 update*. Report prepared for Canadian Environmental Network Energy Caucus, for submission to the Energy Options Policy Review, Energy, Mines and Resources Canada, Ottawa.

Torry, William I. 1979. "Hazards, hazes and holes: A critique of *The Environment as Hazard* and general reflections on disaster research." *Canadian Geographer* 23, no. 4: 368–383.

Toth, Ferenc. 1988. "Policy exercises." *Simulation and Games* 19: 235–276.

———. 1994. "Practice and progress in integrated assessments of climate change: A review." In *Integrative assessment of mitigation, impacts, and adaptation to climate change*, ed. Nebojsa Nakićenović, William D. Nordhaus, Richard Richels, and Ferenc L. Toth. Laxenburg, Austria: International Institute for Applied Systems Analysis, pp. 3–31.

———. 1995. "Practice and progress in integrated assessments of climate change: A workshop report." *Energy Policy* 23, no. 4/5: 253–268.

———, Eva Hizsnyik, and William C. Clark. 1989. *Scenarios of socioeconomic development for studies of global environmental change: A critical review*. Laxenburg, Austria: IIASA.

Troll, Carl. 1988. "Comparative geography of high mountains of the world in the view of landscape ecology: A development of three and a half decades of research and organization." In *Human impact on mountains*, ed. Nigel J.R. Allan, Gregory W. Knapp, and Christoph Stadel. Totowa, NJ: Rowman and Littlefield, pp. 36–56.

Tuan, Yi-fu. 1990. "Realism and fantasy in art, history, and geography." *Annals of the Association of American Geographers* 80, no. 3: 435–446.

Turner, B. L., II, and Patricia Benjamin. 1994. "Fragile lands and their management." In *Agriculture, environment and health: Toward sustainable development in the 21st century*, ed. Vernon W. Ruttan. Minneapolis: University of Minnesota Press, pp. 101–145.

———, William C. Clark, Robert W. Kates, John F. Richards, Jessica T. Mathews, and William B. Meyer, eds. 1990a. *The earth as transformed by human action: Global and regional changes in the biosphere over the past 300 years*. Cambridge: Cambridge University Press with Clark University.

———, Roger E. Kasperson, William B. Meyer, Kirstin M. Dow, Dominic Golding, Jeanne X. Kasperson, Robert C. Mitchell, and Samuel J. Ratick. 1990b. "Two types of global environmental change: Definitional and spatial-scale issues in their human dimension." *Global Environmental Change* 1, no. 1 (December): 14–22.

———, Jeanne X. Kasperson, Roger E. Kasperson, Kirstin Dow, and William B. Meyer. 1995a. "Comparisons and conclusions." In *Regions at risk: Comparisons of threatened environments*, ed. Jeanne X. Kasperson, Roger E. Kasperson, and B. L. Turner, II. Tokyo: United Nations University Press, pp. 519–586.

———, D. L. Skole, Steven Sanderson, G. Fischer, L. Fresco, and R. Leemans, eds. 1995b. *Land-use and land-cover change: Science/research plan.* IGBP Report No. 35, HDP Report No. 7. Stockholm and Geneva: International Geosphere–Biosphere Programme and Human Dimensions Programme.

Turner, R. Kerry. 1991. *Economic appraisal of the consequences of climate induced sea level rise.* Norwich: Environmental Appraisal Groups, University of East Anglia.

Turney, John. 1990. "End of the peer show?" *New Scientist* 127, no. 1735 (22 September): 38–42.

Twigg, John, and Mihir R. Bhatt, eds. 1998. *Understanding vulnerability: South Asian perspectives.* London: Intermediate Technology Publications on behalf of Duryog Nivaran.

Twiley, R. R., A. E. Lugo, and C. Patterson-Zucca. 1986. "Litter production and turnover in basin mangrove forests in Southwest Florida." *Ecology* 67, no. 3: 670–683.

Tyler, M., and K. R. Smith. 1991. *Vulnerability to natural hazards and global warming: Lessons for development.* Honolulu: Environment and Policy Institute, East–West Center.

UCC (United Church of Christ). 1987. *Toxic wastes and race in the United States: A national report on the racial and socio-economic characteristics of communities with hazardous waste sites.* New York: Public Data Access, Inquiries to the Commission.

UK Interdepartmental Liaison Group on Risk Assessment. 1998. *Risk assessment and risk management: Improving policy and practice within government departments.* London: Health and Safety Executive.

UNCTAD (United Nations Conference on Trade and Development). 1987. *The promotion of risk management in developing countries.* New York: Trade and Development Board, United Nations Conference on Trade and Development.

UNDP (United Nations Development Programme). 1996. *Capacity 21 in review.* New York: UNDP.

UNEP (United Nations Environment Programme). 1995. *Global biodiversity assessment.* Cambridge: Cambridge University Press for UNEP.

UNIDO. 1987. *Workshop on hazardous waste management, industrial safety and emergency planning.* Vienna: UNIDO.

United Nations. 1992. *Agenda 21, Rio declaration, forest principles.* New York: United Nations.

———. 1999. "Secretary-General to Decade for Natural Disaster Reduction: Despite dedicated efforts, number and cost of natural disasters continue to rise." Press Release SG/SM/7060, 6 July 1999. New York: United Nations.

Urquhart, John, and Klaus Heilmann. 1985. *Risk watch: The odds of life.* New York: Facts on File Publishers.

USAID (US Agency for International Development). 1987. "Environmental health and safety fundamental to development." In *The Environment.* Washington: USAID, pp. 23–30.

———. 1988. *Energy efficient stoves in east Africa: An assessment of the Kenya ceramic Jiko (Stove) program.* A Report of the Office of Energy, Bureau for Science and Technology and Regional Economic Development Services Office for East and Southern Africa.

USCEQ (US Council on Environmental Quality). 1980. 3 vols. *The global 2000 report to the President.* Washington, DC: USCEQ.

USDHHS (US Department of Health and Human Services). 1986. *Determining risks to health.* Dover, MA: Auburn House Publishing Co.

USDOE (US Department of Energy). 1998. *Accelerating cleanup: Paths to closure.* DOE/EM-0342. Washington: DOE.

USEPA (US Environmental Protection Agency). 1984. *Risk assessment and management: Framework for decision making.* Washington: EPA.

———. 1986. "Risk and exposure assessment guidelines." *Federal Register* 51, no. 185: 33992–34054.

———. 1987. Office of Policy Analysis and Office of Policy, Planning and Evaluation. *Unfinished business: A comparative assessment of environmental problems.* Overview Report and Appendices I–IV. Washington: USEPA.

———. 1988. *Future risk: Research strategies for the 1990s.* SAB-EC-88-040. Washington: EPA.

———. 1990. *Reducing risk: Setting priorities and strategies for environmental protection.* SAB-EC-90-021. Washington: USEPA.

———. 1992. *Environmental equity: Reducing risk for all communities.* EPA230-R-92-008 and 008A. 2 vols. Washington: USEPA.

US Interagency Working Group on Sustainable Development Indicators. 1998. *Sustainable development in the United States: An experimental set of indicators.* Washington: US Council on Environmental Quality.

US National Decade for Natural Disaster Reduction. 1991. *A safer future: Reducing the impacts of natural disasters.* Washington: National Academy Press.

USNRC (US National Research Council). 1983a. *Changing climate: Report of the Carbon Dioxide Assessment Committee.* Washington: National Academy Press, pp. 449–482.

———. 1983b. *Risk assessment in the federal government: Managing the process.* Washington: National Academy Press.

———. 1984. *Science and judgment in risk assessment.* Washington: National Academy Press.

———. 1986. *Ecological knowledge and environmental problem-solving: Concepts and case studies.* Washington: National Academy Press.

———. 1987a. *Confronting natural hazards.* Washington: National Academy Press.

———. 1987b. *Responding to changes in sea level: Engineering implications.* Marine Board. Washington: National Academy Press.

———. 1990. *Sea level change.* Geophysics Study Committee. Washington: National Academy Press.

———. 1993. *Measuring lead exposure in infants, children, and other sensitive populations.* Washington: National Academy Press.

———. 1994. *Science and judgment in risk assessment.* Washington: National Academy Press.

US Nuclear Regulatory Commission. 1984. *Probabilistic risk assessment (PRA): Status report and guidance for regulatory application.* NUREG-1050/Draft. Washington: Nuclear Regulatory Commission.

US President's Council on Sustainable Development. 1996. *Sustainable America: A new consensus for prosperity, opportunity, and a healthy environment for the future.* Washington: US Government Printing Office.

Valaskakis, Kimon. 1988. "At the crossroads of 'futurism' and 'prospective': Towards a Canadian synthesis?" *Technological Forecasting and Social Change* 33, no. 4 (July): 339–353.

———, Peter S. Sindell, and J. Graham Smith. 1975. *Tentative blueprints for a Conserver Society in Canada.* Montreal: GAMMA.

———, ———, and ———. 1976. *Conserver Society project: Phase II report.* 4 volumes. Montreal: GAMMA.

———, I. Martin, Peter S. Sindell, and J. Graham Smith. 1979. *The Conserver Society: A workable alternative for the future.* Toronto: Fitzhenry and Whiteside.

Van Der Hammen, T., and A. M. Cleef. 1986. "Development in the High Andean Páramo flora and vegetation." In *High altitude tropical biogeography*, ed. Francois Vuilleumier, and Maximina Monasterio. Oxford: Oxford University Press, pp. 153–201.

Venezian, E. L., and W. K. Gamble. 1969. *The agricultural development of Mexico: Its structure and growth since 1950.* New York: Praeger.

Victor, David G., K. Raustiala, and Eugene B. Skolnikoff. 1998. *The implementation and effectiveness of international environmental commitments: Theory and practice.* Cambridge, MA: MIT Press.

Vizcáino, Murray Francisco 1975. *La contaminación en México.* Mexico: Fondo de Cultura Economica.

Vogel, Coleen. 1998. "Vulnerability and global environmental change." *LUCC Newsletter* no. 3 (March): 15–18.

Vuilleumier, Francois. 1986. "Origins of the tropical avifaunas of the high Andes." In *High altitude tropical biogeography*, ed. Francois Vuilleumier and Maximina Monasterio. Oxford: Oxford University Press, pp. 586–623.

Wack, P. 1985a. "Scenarios: Uncharted waters ahead." *Harvard Business Review*, no. 5: 72–89.

———. 1985b. "Scenarios: Shooting the rapids." *Harvard Business Review*, no. 6: 139–150.

Waddenvereiniging. 1997. *Memorandum: Toward a trilateral Wadden Sea Convention: Wadden Sea region, one and indivisible.* Harlingen, The Netherlands: The Wadden Society.

Walker, Peter. 1989. *Famine early warning systems: Victims and destitution.* London: Earthscan.

Ward, Barbara, and Rene Dubos. 1972. *Only one earth: The care and maintenance of a small planet.* New York: Norton.

Warrick, Richard A., and J. Oerlemans. 1990. "Sea level rise." In *Climate change: The IPCC scientific assessment*, ed. John Houghton, G. J. Jenkins, and J. J. Ephraums. Cambridge: Cambridge University Press, pp. 61–70.

————, and Angela Wilkinson. 1990. "The enhanced greenhouse effect." In *The effects of sea level rise on the UK coast*, ed. L. E. J. Roberts and R. C. Kay. Research Report No. 7. Norwich: Environmental Risk Assessment Unit, University of East Anglia, pp. 18–27.

Waterstone, Marvin. 1987. "Reducing groundwater pollution by toxic substances: Procedures and policies." *Environmental Management* 11, no. 6: 793–804.

Watts, Michael. 1983. *Silent violence: Food, famine and peasantry in Northern Nigeria*. Berkeley: University of California Press.

————. 1987. "Drought, environment and food security: Some reflections on peasants, pastoralists and commoditization in dryland West Africa." In *Drought and hunger in Africa*, ed. Michael H. Glantz. Cambridge: Cambridge University Press, pp. 171–211.

————. 1991. "Heart of darkness: Reflections on famine and starvation in Africa." In *The political economy of African famine*, ed. R. E. Downs, Donna O. Kerner, and Stephen P. Reyna. Food and Nutrition in History and Anthropology, vol. 9. Langhorne, PA: Gordon and Breach, pp. 23–68.

————, and Hans G. Bohle. 1993. "The space of vulnerability: The causal structure of hunger and famine." *Progress in Human Geography* 17, no. 1: 43–67.

WCC'93. 1994. *Preparing to meet the coastal challenges of the 21st century. Conference report, World Coast Conference, Noordwijk, The Netherlands, 1–5 November 1993*, ed. Luitzen Bijlsma. The Hague: Ministry of Transport, Public Works and Water Management, National Institute for Coastal and Marine Management/RIKZ, Coastal Zone Management Centre.

WCED (World Commission on Environment and Development). 1987. *Our common future*. Oxford: Oxford University Press.

Webler, Thomas. 1995. "'Right' discourse in citizen participation: An evaluative yardstick." In *Fairness and competence in citizen participation: Evaluating models for environmental discourse*, ed. Ortwin Renn, Thomas Webler, and Peter Wiedemann. Dordrecht: Kluwer, pp. 35–86.

Weinberg, Alvin. 1972a. "Science and trans-science." *Minerva* 10: 209–222.

————. 1972b. "Letters." *Science* 180: 1124.

Weiss, Edith Brown. 1989. *In fairness to future generations: International law, common patrimony, and intergenerational equity*. New York: Transnational Publishers and the United Nations University.

————. 1990. "In fairness to future generations." *Environment* 32 no. 3: 7–11, 30–31.

————. 1995. "Intergenerational equity: Toward an international legal framework." In *Global accord: Environmental challenges and international responses*, ed. Nazli Choucri. Cambridge: MIT Press, pp. 333–353.

Weiss, Thomas G., and K. M. Campbell. 1991. "Military humanitarianism." *Survival* 33, no. 5: 451–465.

Wellhausen, E. 1976. "The agriculture of Mexico." *Scientific American* 235, no. 3: 128–150.

Wells, M. S., ed. 1988. *Coral reefs of the world, vol. 1: Atlantic and Eastern Pacific*. Geneva: United Nations Environmental Programme (UNEP) and International Union for the Conservation of Nature and Natural Resources (IUCN).

Whetten, Nathan L. 1948. *Rural Mexico*. Chicago: University of Chicago Press.

Whipple, Chris. 1985. "Redistributing risk." *Regulation* 9, no. 3: 37–44.

White, Gilbert F., ed. 1974. *Natural hazards, local, national global.* Oxford: Oxford University Press.

———. 1980. "Environment." *Science* 209: 183–190.

Whiteman, Peter T. S. 1988. "Mountain agronomy in Ethiopia, Nepal, and Pakistan." In *Human impact on mountains,* ed. Nigel J. R. Allan, Gregory W. Knapp, and Christoph Stadel. Totowa, NJ: Rowman and Littlefield, pp. 57–82.

Whitmore, Thomas, B. L. Turner, II, Douglas L. Johnson, Robert W. Kates, and Thomas R. Gottschang. 1990. "Long-term population change." In *The earth as transformed by human action: Global and regional changes in the biosphere over the past 300 years,* ed. B. L. Turner, II, William C. Clark, Robert W. Kates, John F. Richards, Jessica T. Mathews, and William B. Meyer. Cambridge: Cambridge University Press with Clark University, pp. 25–39.

Whittemore, A. S. 1983. "Facts and values in risk analysis for environmental toxicants." *Risk Analysis* 3, no. 1: 23–33.

WHO (World Health Organization). 1984. *Risk assessment and its use in the decision making process for chemical control.* ICP/RCE 903 (28). Working Group on Health and the Environment. Geneva: WHO.

———. 1995. *Community water supply and sanitation: Needs, challenges and health objectives.* 48th World Health Assembly, A48/INF.DOC./2,28 (April). Geneva: WHO.

———. 1996a. Water and sanitation. *Fact Sheet* no. 112, (Reviewed) November 1996. Accessible on line at *http://www.who.org.*

———. 1996b. *Water supply and sanitation sector monitoring report: Sector status as of 1994.* WHO/EOS/96.15. Geneva: WHO.

———, and UNEP (United Nations Environment Programme). 1987. *Global pollution and health: Results of health-related environmental monitoring.* London: Yale University Press.

———, and ———. 1988. *Assessment of urban air quality worldwide.* Geneva: WHO.

Whyte, Anne V., and Ian Burton, eds. 1980. *Environmental risk assessment.* SCOPE no. 15. Chichester, UK: John Wiley and Sons.

Whyte, William H. 1956. *The organization man.* New York: Simon and Schuster.

Wijkman, Anders, and Lloyd Timberlake. 1984. *Natural disasters: Acts of God or acts of man?* London: Earthscan.

Wilcoxen, P. J. 1986. "Coastal erosion and sea level rise: Implications for Ocean Beach and San Francisco's Westside Transport Project." *Coastal Zone Management Journal* 14, no. 3: 173–191.

Wildavsky, Aaron. 1979a. "No risk is the highest risk of all." *American Scientist* 67, no. 1: 32–37.

———. 1979b. *Speaking truth to power.* Boston: Little Brown.

———. 1988. *Searching for safety.* New Brunswick, NJ: Transaction Books.

Wilde, G. J. S. 1982. "The theory of risk homeostasis: Implications for safety and health." *Risk Analysis* 2, no. 4: 209–225.

Wilken, Gene C. 1987. *The good farmers: Traditional agricultural resource management in Mexico and Central America.* Berkeley: University of California Press.

Wilkinson, Angela. 1990. "The greenhouse effect and storm surge frequencies."

In *The effects of sea level rise on the UK coast*, ed. L. E. J. Roberts, and R. C. Kay. Research Report No. 7. Norwich: Environmental Risk Assessment Unit, University of East Anglia, pp. 38–45.

———, and Richard A. Warrick. 1990. "Sea level rise predictions." In *The effects of sea level rise on the UK coast*, ed. L. E. J. Roberts, and R. C. Kay. Research Report No. 7. Norwich: Environmental Risk Assessment Unit, University of East Anglia, pp. 28–37.

Wilson, Richard. 1979. "The environmental and public health consequences of replacement electricity supply." *Energy* 4: 81–86.

Winner, Langdon. 1986. *The whale and the reactor: A search for limits in an age of high technology*. Chicago: University of Chicago Press.

WMO/OMM (World Meteorological Organization), Government of Canada, Environment Canada, and UNEP (United Nations Environment Programme). 1988. *The changing atmosphere/L'atmosphère en évolution: Implications for global security/Implications pour la sécurité du globe. Proceedings, World Conference, Toronto, Canada, June 27–30, 1998/Actes, Conference Mondiale, Toronto, Canada 27–30 juin 1988*. WMO/OMM no. 710. Geneva: WMO/OMM.

Wolff, W. J. 1987. "Ecological effects of rapid relative increase of sea level." In *Impact of sea level rise and society*, ed. H. G. Wind. Rotterdam, Balkema, pp. 153–155.

Wood, William B. 1990. "Tropical deforestation: Balancing regional development demands and global environmental concerns." *Global Environmental Change* 1 (December): 23–41.

Woodlands. 1991. Conference on the regional impacts of global warming. Woodlands, Texas, 3–6 March, 1991.

World Bank. 1985. *Manual of industrial hazard assessment techniques*. Washington: World Bank.

———. 1995. *Monitoring environmental progress: A report on work in progress*. Washington, DC: World Bank.

World Development Report. Annual. Washington: World Bank.

World Disasters Report. Annual. Geneva: International Federation of Red Cross and Red Crescent Societies (IFRC).

World Health Report. Annual. Geneva: World Health Organization (WHO).

World Resources. Biennial. Oxford: Oxford University Press.

Wynne, Brian. 1987. *Risk management and hazardous waste: Implementation and the dialectics of credibility*. Berlin: Springer-Verlag.

———. 1996. "May the sheep safely graze? A reflexive view of the expert–lay knowledge divide." In *Risk, environment, and modernity*, ed. Scott Lash, Bronslaw Szerszynski, and Brian Wynne. London: Sage, pp. 44–83.

Yaffee, Steven L., Ali F. Phillips, Irene C. Frentz, Paul W. Hardy, Sussanne M. Maleki, and Barbara E. Thorpe. 1996. *Ecosystems management in the United States: An assessment of current experience*. Washington: Island Press.

Yohe, G. 1990. "The cost of not holding back the sea: Toward a national sample of economic vulnerability." *Coastal Management* 18: 403–431.

Young, Oran. 1989. *International cooperation*. Ithaca, NY: Cornell University Press.

————. 1994. *International governance: Protecting the environment in a stateless society.* Ithaca, NY: Cornell University Press.

————, ed. 1997. *Global governance: Drawing insight from the environmental experience.* Cambridge, MA: MIT Press.

————, Arun Agrawal, Leslie A. King, Peter H. Sand, Arild Underdal, and Merrilyn Wasson. 1999. *Institutional dimensions of global environmental change (IDGEC) Science Plan,* IHDP Report No. 9. Bonn: International Human Dimensions Programme.

Zaman, Munir, ed. 1983. River basin development. In *Proceedings of the National Symposium on River Basin development.* Dublin: Tycooly International.

Zimmerer, Karl S. 1995. "Havens on high: The fate of biodiversity in mountain agriculture." In *Mountains at risk: Current issues in environmental studies,* ed. Nigel J. R. Allan. New Delhi: Manohar, pp. 129–151.

Contributors

Bhavik Bakshi
Assistant Professor, Chemical
  Engineering Department, The Ohio
  State University, Columbus, Ohio,
  USA.

Patricia Benjamin
Doctoral candidate in the Graduate
  School of Geography, Clark
  University, Worcester,
  Massachusetts, USA.

Jo Anne Berkenkamp
Management consultant, St Paul,
  Minnesota, USA.

Sherry A. Bishko
After a stint working for the United
  States Environmental Protection
  Agency, she is pursuing a law
  degree.

James M. Broadus[†]
Prior to his death in 1994, was
  Director of the Marine Policy
  Center at the Woods Hole
  Oceanographic Institution, Woods
  Hole, Massachusetts, USA.

William C. Clark
Harvey Brooks Professor of
  International Science,
  Public Policy, and Human
  Development, Harvard University,
  Cambridge, Massachusetts,
  USA.

Kirstin Dow
Assistant Professor of Geography,
  University of South Carolina,
  Columbia, South Carolina, USA.

Jacque L. Emel
Associate Professor of Geography and
  Director of Women's Studies at
  Clark University, Worcester,
  Massachusetts, USA.

Exequiel Ezcurra
Deputy Director of Research and
  Collections, San Diego Natural
  History Museum, San Diego,
  California, USA.

563

Oscar Flores-Villela
Principal Researcher, Museo de
  Zoología, Facultdad des Ciencias,
  Universidad Nacional Autónoma
  de México (UNAM); is currently
  at the Centro de Investigaciones
  Biológicas, Universidad Autónoma
  del Estado de Hidalgo, Mexico.

Silvio O. Funtowicz
Senior Scientist, the Joint Research
  Centre of the European
  Commission, Ispra, Italy.

Saburo Ikeda
Professor at the Institute of Policy and
  Planning Sciences at the University
  of Tsukuba, Japan.

N. S. Jodha
Policy analyst, MEI Division,
  International Centre for Integrated
  Mountain Development (ICIMOD),
  Kathmandu, Nepal.

Jeanne X. Kasperson
Research Associate Professor and
  Research Librarian at the George
  Perkins Marsh Institute, Clark
  University, Worcester,
  Massachusetts, USA; is a visiting
  scholar at the Stockholm
  Environment Institute.

Roger E. Kasperson
After more than 30 years at Clark
  University, Worcester,
  Massachusetts, USA, where he was
  University Professor (Government
  and Geography) and Director of the
  George Perkins Marsh Institute, has
  joined the Stockholm Environment
  Institute as its Executive Director.

Masaaki Kataoka
Associate Professor in the Faculty of
  Policy Management, Keio
  University at Shonan Fujisawa,
  Japan.

Mark D. Koehler
Former Research Assistant to
  Professor Vicki Norberg-Bohm.

Diana M. Liverman
Director, Latin American Studies
  Program, and Professor of
  Geography, University of Arizona.

Jennifer A. Marrs
Captain Marrs is Acquisition Manager
  for the Assistant Secretary of the
  Air Force for Acquisitions, United
  States Air Force, the Pentagon,
  USA.

James K. Mitchell
Professor of Geography and Director
  of the Graduate Program in
  Geography at Rutgers University,
  Piscataway, New Jersey, USA.

Chris P. Nielsen
Executive Director of the University
  Committee on Environment China
  Project, Harvard University,
  Cambridge, Massachusetts, USA.

Vicki Norberg-Bohm
Director of the Energy Technology
  Innovation Project, Belfer Center
  for Science and International
  Affairs, Harvard University,
  Cambridge, Massachusetts, USA.

Timothy O'Riordan
Associate Director, Centre for Social
  and Economic Research on the
  Global Environment (CSERGE)
  and Professor of Environmental
  Sciences at the University of East
  Anglia, Norwich, UK.

Samuel J. Ratick
Associate Professor of Geography
  and Director of the Environmental
  Science and Policy Program at
  Clark University, Worcester,
  Massachusetts, USA.

Jerome R. Ravetz
Director of the Research Methods
Consultancy Ltd, London, UK.

John B. Robinson
Director, Sustainable Development
Research Unit, University of British
Columbia, Vancouver, BC, Canada.

Dianne E. Rocheleau
Associate Professor of Geography,
Clark University, Worcester,
Massachusetts, USA.

Ambuj Sagar
Research Fellow with the Energy
Technology Innovation Project,
Belfer Center for Science and
International Affairs, Harvard
University, Cambridge,
Massachusetts, USA.

Kirk R. Smith
Professor and Chair, Division of
Environmental Health Sciences,
University of California, Berkeley,
California, USA.

Peter Timmerman
Research Associate at the Institute for
Environmental Studies, University
of Toronto, Canada.

B. L. Turner, II
Milton P. and Alice C. Higgins
Professor of Environment and
Society, Graduate School of
Geography and the George Perkins
Marsh Institute, Clark University,
Worcester, Massachusetts, USA.

Alfonso Valiente-Banuet
Principal Researcher, Instituto de
Ecologiá, Universidad Nacional
Autónoma de México (UNAM),
Mexico City, Mexico.

Ella Vázquez-Domínguez
Researcher, Comision Nacional pana
Conocimiento y Uso de la
Biodiversidad (CONABIO),
Mexico.

Index